SYSTEMS ENGINEERING: AN APPROACH TO INFORMATION-BASED DESIGN

Prentice Hall International Series in Industrial and Systems Engineering

W.J. Fabrycky and J.H. Mize, Editors

Amos/Sarchet—Management for Engineers
Amrine/Ritchey/Moodie/Kmec—Manufacturing Organization and Management, 6/E
Asfahl—Industrial Safety and Health Management, 3/E
Babcock—Managing Engineering and Technology, 2/E
Badiru—Comprehensive Project Management
Badiru—Expert Systems Applications in Engineering and Manufacturing
Banks/Carson/Nelson—Discrete Event System Simulation, 2/E
Blanchard—Logistics Engineering and Management, 4/E
Blanchard/Fabrycky—Systems Engineering and Analysis, 2/E
Burton/Moran—The Future Focused Organization
Bussey/Eschenbach—The Economic Analysis of Industrial Projects, 2/E
Buzacott/Shanthikumar—Stochastic Models of Manufacturing Systems
Canada/Sullivan—Economic and Multi-Attribute Evaluation of Advanced Manufacturing Systems
Canada/Sullivan/White—Capital Investment Analysis for Engineering and Management, 2/E
Chang/Wysk—An Introduction to Automated Process Planning Systems
Chang/Wysk/Wang—Computer Aided Manufacturing
Delavigne/Roberston—Deming's Profound Changes
Eager—The Information Payoff: The Manager's Concise Guide to Making PC Communications Work
Eberts—User Interface Design
Eberts/Eberts—Myths of Japanese Quality
Elsayed/Boucher—Analysis and Control of Production Systems, 2/E
Fabrycky/Blanchard—Life-Cycle Cost and Economic Analysis
Fabrycky/Thuesen—Economic Decision Analysis
Fishwick—Simulation Model Design and Execution: Building Digital Worlds
Francis/McGinnis/White—Facility Layout and Location: An Analytical Approach, 2/E
Gibson—Modern Management of the High-Technology Enterprise
Gordon—Systematic Training Program Design
Graedel/Allenby—Industrial Ecology
Hall—Queuing Methods: For Services and Manufacturing
Hansen—Automating Business Process Reengineering
Hammer—Occupational Safety Management and Engineering, 4/E
Hazelrigg—Systems Engineering
Ignizio/Cavalier—Linear Programming
Kroemer/Kroemer/Kroemer-Elbert—Ergonomics: How to Design for Ease and Efficiency
Kusiak—Intelligent Manufacturing Systems
Landers/Brown/Fant/Malstrom/Schmitt—Electronic Manufacturing Processes
Leemis—Reliability: Probabilistic Models and Statistical Methods
Michaels—Technical Risk Management
Mundel/Danner—Motion and Time Study: Improving Productivity, 7/E
Ostwald—Engineering Cost Estimating, 3/E
Pinedo—Scheduling: Theory, Algorithms, and Systems
Prasad—Concurrent Engineering Fundamentals, Vol. I: Integrated Product and Process Organization
Pulat—Fundamentals of Industrial Ergonomics
Shtub/Barb/Globerson—Project Management: Engineering Technology and Implementation
Taha—Simulation Modeling and SIMNET
Thuesen/Fabrycky—Engineering Economy, 8/E
Turner/Mize/Case/Nazemetz—Introduction to Industrial and Systems Engineering, 3/E
Turtle—Implementing Concurrent Project Management
Walesh—Engineering Your Future
Wolff—Stochastic Modeling and the Theory of Queues

SYSTEMS ENGINEERING: AN APPROACH TO INFORMATION-BASED DESIGN

George A. Hazelrigg

National Science Foundation

PRENTICE HALL, Upper Saddle River, NJ 07458

Library of Congress Cataloging-in-Publication Data

Hazelrigg, George A.
Systems engineering: an approach to information-based design/
George A. Hazelrigg
p. cm.--(Prentice Hall international series in industrial
and systems engineering)
Includes bibliographical references and index.
ISBN 0-13-461344-9
1. Systems engineering. 2. Engineering design--Mathematical
models. I. Title. II. Series.
TA168.H39 1996
620'.0042--dc20 96-42370
CIP

Acquisitions editor: **MARCIA HORTON**
Editorial/production supervision
and interior design: **SHARYN VITRANO**
Copy editor: **BARBARA DANZIGER**
Art contributers: **C. PELLETREAU, P. TICOLA, AND P. GUTRIERREZ**
Manufacturing buyer: **DONNA SULLIVAN**
Editorial assistant: **DELORES MARS**

A Pearson Education Company
Upper Saddle River, NJ 07458

Printed in the United States of America

10 9 8 7 6 5 4 3 2 1

0-13-461344-9

Prentice-Hall International (UK) Limited,London
Prentice-Hall of Australia Pty. Limited, Sydney
Prentice-Hall Canada Inc., Toronto
Prentice-Hall Hispanoamericana, S.A., Mexico
Prentice-Hall of India Private Limited, New Delhi
Prentice-Hall of Japan, Inc., Tokyo
Pearson Education Asia Pte. Ltd., Singapore
Editora Prentice-Hall do Brasil, Ltda., Rio de Janeiro

"Would you tell me, please, which way I ought to walk from here?"

"That depends a good deal on where you want to get to," said the Cat.

"I don't care much where____" said Alice.

"Then it doesn't matter which way you walk," said the Cat.

"____ so long as I get *somewhere*," Alice added as an explanation.

"Oh, you're sure to do that," said the Cat, "if you only walk long enough."

Lewis Carroll
Alice In Wonderland

Contents

ACKNOWLEDGMENTS ***xvii***

FOREWORD ***xix***

PREFACE ***xxi***

1 Introduction to Engineering Design and Decision Making ***1***

1.1 The Engineering Process 1

1.2 Historical Perspective 6

1.3 The Decision-Making Process 8

1.4 Beans in a Jar 9

1.5 Summary 13

References 14

Problems 15

2 Design Options ***17***

2.1 The Purpose of this Chapter 17

2.2 The Role of Options in Decision Making 18

2.3 System Objectives 20

2.4 Options in Physical Design 21
Packaging Complexity, 23

2.5 System Manufacture and Deployment 23
Cartoons, 24

2.6 Options in System Operation 25
Normal Operation, 26
Maintenance and Repair, 27

2.7 End-of-Life Options 28

2.8 Paring the Option Space 28
Feasibility Analysis, 29

2.9 Summary 29
Class Project, 30

Reference 30

Problems 30

3 Fundamentals of Probability Theory **32**

3.1 The Purpose of this Chapter 32

3.2 The Concept of Probability 32

3.3 Properties of a Random Variable 39

3.4 The Mathematics of Random Variables 42

3.5 Some Important Distributions 46
Discrete Distributions, 46
Continuous Distributions, 51

3.6 Bayes' Formula 55

3.7 Hypothesis Testing 57

3.8 Summary 59

References 60

Problems 60

4 Monte Carlo Modeling **63**

4.1 The Purpose of this Chapter 63

4.2 The Notion of a Monte Carlo Model 63

4.3 Random Number Generators 65

4.4 Sampling a Distribution 69

4.5 Formulating Monte Carlo Models 71

4.6 Analysis of Monte Carlo Model Results 74

4.7 An Alternative Analysis of Monte Carlo Results 75

4.8 Monte Carlo Models with Embedded Decisions 76

4.9 Summary 76

Reference 77

Problems 77

5 *Optimization* ***78***

5.1 The Purpose of this Chapter 78

5.2 The Need for Optimization in Systems Engineering 78

5.3 Notions of Minimum and Maximum 79

5.4 Maximation of a Function 81

5.5 Search Methods 83
Zeroth-Order Methods, 84
First-Order Methods, 86
Second-Order Methods, 88
Conjugate Gradient Methods, 89

5.6 Constrained Optimization 91

5.7 Linear Programming 94

5.8 Integer Programming 99
The Assignment Problem, 101

5.9 Network Flow Optimization 101

5.10 The Transportation Problem 104

5.11 Calculus of Variations 105
Necessary Conditions, 109
Problem Formulation in the Calculus of Variations, 111
The Open Time Case, 111
The Maximum Principle, 113
Example, 118

5.12 Summary 120
References 121
Problems 121

6 Engineering Microeconomics **124**

6.1 The Purpose of this Chapter 124

6.2 The Concept of Preference 124
Basic Human Needs, 125
The Concept of Value, 126
Indifference Curves, 127
Preferences Under Conditions of Limited Resources, 129
An Engineering Example, 130
Preferences Involving Time, 130

6.3 Equilibrium Economics 131
Demand Curves, 131
Supply Curves, 135
Pricing Theory, 136
The Effect of Changing the Supply Curve, 138
The Effect of Product Research and Development, 139
Regulation, 140
The Effect of Taxes on Prices, 140

6.4 Discounting and Present Value 141
The Social Rate of Discount, 141
Present Value, 142

6.5 Interest and Annuities 143
Annuities, 144

6.6 Inflation and Deflation 145

6.7 The Value of a Forecast 146

6.8 Resource Economics 148

6.9 Shadow Prices 151

6.10 Summary 152
References 153
Problems 153

7 Utility Theory **155**

7.1 The Purpose of this Chapter 155

7.2 Rationality 155

7.3 The Notion of Utility 158
The Law of Diminishing Marginal Utility, 159
The Law of Equal Marginal Utilities, 160
Measuring Utility, 160

7.4 Multiattribute Utility Functions 161
Linearly Additive Utility, 163
Multiplicative Utility, 163
Log-Linear Utility, 164
Lexicographic Preference Ordering, 164

7.5 Arrow's Impossibility Theorem 164

7.6 Pareto Optimality 166

7.7 Decision Making in the Presence of Risk 167
Von Neumann–Morgenstern Lotteries, 167
The Axioms of Von Neumann–Morgenstern Utility, 169

7.8 The Value of Better Information 170

7.9 The Value of Improved Safety 172

7.10 Summary 174

References 175

Problems 176

8 Forecasting **178**

8.1 The Purpose of this Chapter 178

8.2 Types of Forecasts 179

8.3 Regression Analysis 181

8.4 Logistic Curves 190

8.5 Forecasting Uncertainty 192

8.6 Forecasting Uncertainty with Embedded Decisions 197

8.7 Shortcomings of Forecasts 199

8.8 Summary 199

References 200

Problems 200

9 Engineering Systems Modeling **202**

9.1 The Purpose of this Chapter 202

9.2 Engineering Systems and System Models 203

9.3 The Model-Building Process 210

9.4 Sources of Error in Symbolic Models 211
Relative Error, 215
Resolution in the Case of Uncertainty, 216
Nonintuitive Cases, 217

9.5 Design Models 217

9.6 System Life Cycle Modeling 220
Design Validation Costs, 220
Manufacturing Costs, 221
Maintenance Costs, 221
Distribution Costs, 222
Sales Costs, 222
Disposal Costs, 223
Liability Costs, 223

9.7 Objective Functions 224

9.8 Summary 226

Reference 226

Problems 226

10 Analysis of System Reliability **228**

10.1 The Purpose of this Chapter 228

10.2 Notions of System Reliability 228
Mean Time Between Failure, 229
The Impact of a Failure on System Utility, 230

10.3 System Reliability Diagrams 232

10.4 Cartoons and Scenarios 241

10.5 Decision and Event Trees 46

10.6 Two Different Monte Carlo Approaches 249
Time-Driven Monte Carlo Models, 250
Event-Driven Monte Carlo Models, 252
Comparison, 252

10.7 Probabilistic Risk Assessment 252

10.8 An Alternative to Probabilistic Risk Assessment 255

10.9 Summary 256

References 256

Problems 257

11 Cost and Benefit Analysis **260**

11.1 The Purpose of this Chapter 260

11.2 Costs 260

11.3 Modeling System Cost 263
Top-Down Cost Estimating, 264
Bottom-Up Cost Estimating, 266
Cost Learning, 267
Cost Spreading, 268
Life Cycle Cost Estimating, 270
Opportunity Costs, 270

11.4 Probabilistic Cost Analysis 270

11.5 Benefits 273
Monetary Benefits, 274
Benefits of Government Funded Projects, 278
Externalities, 279
Miscellaneous Comments on Benefit Estimation, 279
The Benefits of Research and Development, 280
Timing of Benefits, 281

11.6 Probabilistic Benefit Analysis 281

11.7 Benefit-Cost Analysis 282

11.8 Project Selection 283

11.9 Summary 286

References 286

Problems 287

12 Methods of Decision Analysis **291**

12.1 The Purpose of this Chapter 291

12.2 Decisions and Decision Analysis 291

12.3 Backward Induction 293

12.4 Expected Utility Analysis 298
Determination of Utility, 299
An Engineering Example, 300

12.5 The Use of Decision/ Event Trees with an Infinite Number of Possible Outcomes 302

12.6 Confidence Level Decision Making 306

12.7 Minimax Decision Making 306

12.8 Regret 307

12.9 Bayes Solutions 310

12.10 Summary 311

References 312

Problems 312

13 State Transition Matrix Models **315**

13.1 The Purpose of this Chapter 315

13.2 The State Transition Matrix Model 315

13.3 Example Problems 316

13.4 Including Uncertainty in State Transition Matrix Models 321

13.5 Summary 321

Problems 322

14 Modeling the Research and Development Process **324**

14.1 The Purpose of this Chapter 324

14.2 Attributes of Research and Development Activities 324

14.3 A Simulation Approach to Evaluating Research and Development Activities 325

Modeling the Research Process, 327
Modeling Technological Breakthroughs, 329
Modeling Interrelationships among Subprojects, 332
Modeling Performance of Alternative Approaches, 333
Modeling the Decision Process, 335
Typical Management Results, 337

14.4 Summary of the Simulation Approach 338

Reference 339

Problems 339

15 Information 340

15.1 The Purpose of this Chapter 340

15.2 A Quantitative Measure of Information 340

15.3 Summary 343

Problems 343

16 System Life-Cycle Modeling and Optimization 345

16.1 The Purpose of this Chapter 345

16.2 Objective Functions of a Firm 346

16.3 Objective Functions of Consumers 348

16.4 Phases of a Product Life Cycle 349
Research to Proof of Concept, 349
System Engineering and Design, 351
System Test and Evaluation, 351
System Manufacturing, 352
Distribution and Sales, 352
System Operation, Maintenance, and Repair, 353
System Disposal, 354
New Product Development, 354

16.5 System Optimization and Improvement 355

16.6 Long-Term Planning 356

16.7 Model Validation 357

16.8 The Use of Model Results in the Design Process 358

16.9 Summary 359

Problems 360

17 Game Theory **361**

17.1 The Purpose of this Chapter 361

17.2 The Description of a Game 362

17.3 Types of Games 364

17.4 Example Games 364
A Game of Dominant Strategies, 364
A Game of Iterated Dominance, 367
Getting to the Marketplace, 368

17.5 Information in the Context of Game Theory 369
Information Sets, 371
Categories of Information, 372
Bayesian Games, 372

17.6 Mixed Strategy Games 376

17.7 The Nash Model 379
Nash Equilibrium, 380
Evolutionary Stable Strategies, 380
Nash Existence Theorem, 381
Nash Bargaining Solution, 383

17.8 The Minimax Theorem 385

17.9 Solution of a Two-Player, Zero-Sum Game by Linear Programming 388

17.10 Summary 389

References 390

Problems 391

18 Management of Engineering Systems Design and Operation **393**

18.1 The Purpose of this Chapter 393

18.2 The Management of Engineering Systems Design 394

18.3 The Management of Engineering Systems Operation 398

18.4 Continuing Product Improvement 400

18.5 Summary 405

Problems 405

19 Case Studies **407**

19.1 The Purpose of this Chapter 407

19.2 Solution of the Bean Jar Problem 407

19.3 An Undersea Cable Design 413

19.4 Summary of the Systems Engineering Approach 434

20 Concluding Remarks **436**

Appendix A Vectors and Matrices **439**

Appendix B A Test for the Utility Independence of Attributes **450**

Appendix C Determination of the Weighting Factors in a Linearly Additive Utility Function **452**

NAME INDEX **459**

TOPIC INDEX **463**

ABOUT THE AUTHOR **469**

Acknowledgments

All significant works build upon a base of contributions made by those who precede the work, and this book is no exception. It is certainly not the first book to address issues related to engineering design. Some recent texts include: *Rational, Descriptions, Decisions and Designs,* by Myron Tribus; *Methodology for Large Scale Systems,* by Andrew P. Sage; *Systems Engineering and Analysis*, by B. S. Blanchard and W. J. Fabrycky; *System Planning and Design*, by R. de Neufville and D. H. Marks; *The Principles of Design*, by Nam P. Suh; *Patterns of Problem Solving*, by Moshe F. Rubinstein and Iris R. Firstenberg; and *Optimal Engineering Design*, by James N. Siddall. The topics covered in these references are much the same as the topics covered in this book. The significant difference, however, lies in the fact that this book takes as its underlying premise and starting point the notion that engineering design is a decision-making process. This premise has evolved over some 25 years of engineering practice by the author and interaction with experts from a broad cross-section of both the hard and soft sciences. These people have contributed to this book through their endless hours of discussion and critique. Some are deserving of mention: Joel S. Greenberg, David Elliott, Hyung S. Hahn, Frank L. Huband, Allen R. Stubberud, and Carl W. Hall. Finally, I would like to give special mention to J. Preston Layton (1920–1992), who served as my faculty advisor during my Ph.D. studies at Princeton University. His vision of systems engineering led to the establishment of The Aerospace Systems Laboratory in the Department of Aerospace and Mechanical Sciences, where much of the original inspiration for this book derived.

This effort has been made possible by the support of the Institute for Advanced Engineering, a joint institute of the Daewoo Group and Ajou University, Seoul, Korea, and the National Science Foundation, Arlington, Virginia, U.S.A. Special appreciation is given to Drs. KunMo Chung and Joseph Bordogna. It was written as part of my efforts while on sabbatical developing and teaching courses in systems engineering.

Foreword

This book presents a theory of engineering design that is both comprehensive and consistent. It is comprehensive in that it provides a framework for the consideration and comparison of design options, accounting for all phases in the life cycle of a product or service, and it enables rational choice among design options, including cases in which uncertainty and risk are important. It is consistent in its application of a rigorous mathematical approach to design decision making. However, we do not intend to imply that the theory is complete or fully developed. To be sure, we present only the fundamental framework of a theory of design. Nor is the material presented here particularly new. Much of it has existed in the literature of nonengineering disciplines for decades. What we believe to be new here is the collection of independent theories and their unification into a consistent and comprehensive theory of design. Many extensions and applications of this theory are possible and expected.

The underlying concept behind this theory is the basic notion that engineering design is a decision-making process. Good design requires good decision making, and good decision making requires good design options and good information to enable comparison of the options. We focus on three questions: What information is needed for good design? How do we get that information? How do we use it when we get it? The answers to these questions comprise the text of this book.

Finally, we also focus on the true objective of engineering design, which, more often than not, is simply to make money. Better design is often obtained by paying attention to this goal, and by optimizing both physical and nonphysical design against this objective. Often, there is much value to be gained by doing this.

Who should pay attention to systems engineering, when, and why? Our answer to this question is that anyone who is responsible for design decisions of significance can benefit from the systems engineering approach, and it should be applied whenever the design matters. Application of the systems engineering approach frequently will lead to better design decisions, higher profits, and greater benefit to society.

Preface

I began my engineering education in the late 1950s, at a time when many engineering professors held as their highest degree the Master of Science. Many had been practicing engineers at some point in their careers, and they brought with them the experience of "real" engineering. This was still the case as I began graduate school in the early 1960s. But throughout this period, a transition in engineering education from a base of experience and practice to a base of science and problem solving was taking place. A notion inherent in this transition is that the physical sciences provide analytical models that enable engineers to examine mathematically the consequences of their designs, and these models can be taught to engineering students. Engineering institutions began to place high value on the development of discipline-oriented analytical methods, and the role that research plays in engineering institutions and engineering education gained in importance. Over the same period, the credit given to practical experience gained through actual engineering practice waned, and engineering institutions turned to recent Ph.D. graduates for their new faculty.

By the time I emerged from my formal engineering education, I held the view that all problems or questions have uniquely correct answers, and to perform my job as an engineer well all I must do is find those answers. I felt that I had been taught that these answers can be found through proper application of the scientific method. As I studied to learn new disciplines, I found myself reading in search of correct approaches and answers. I read different texts on a subject, but only to understand better the method in question, not to gain a new or different perspective. After all, if there is only one correct answer or one best approach; once it is found, there is no need to continue the search. The value in reading more than one text on a subject, in this view, is not to gain new information but only to seek the most easily understood explanation of the uniquely correct method. I was taught the axiom, *You do not understand something well unless you can explain it in a way that it can be understood by a 12-year old.* This I still believe.

I was presented with the concept of a continuum ranging from art to science. In this continuum, art is the result of a human endeavor that lacks an explicit logic. Good art requires some amount of creativity, and it cannot be taught. Science, at the opposite extreme, is nature understood and explained. When I think of applying science to engineering design, I think of taking the art and creativity out of the design process, and replacing them with laws of nature. Two thousand years ago, engineering was an art. One learned engineering through an apprenticeship, by doing engineering. In more recent times, however, we learned how to model elements of the physical world and predict their behavior. We have grown to use this predictive capability as the basis for engineering design. This approach to engineering has been referred to, by people such as Dr. Joseph Bordogna, as the reductionist approach. It is a disciplinary approach to engineering.

The strength to engineering of the reductionist approach is that it regularizes many of the activities performed by an engineer. By determining underlying physical principles, it is possible to reduce much of the art of engineering to science. And it is possible to teach this science, whereas many people argue that art cannot be taught. Thus it is that engineering science has become the focus of engineering education. Unfortunately, modern engineering science has focused on physical phenomena, and thus excludes many areas of design that contain significant nonphysical elements as shown in this book.

More recently, engineers have begun to view engineering design as a multidisciplinary activity. This view is holistic and integrative, although to some extent, it emerges through buzzword approaches—just in time, zero defects, total quality control—that tend simply to be band-aid fixes to glaring problems. And some engineers see the multidisciplinary view as opposite to that of the reductionist view, and they are inclined to take one side or the other. The approach taken in this book, however, is that these views are merely different sides of the same coin, and both are necessary for engineering design. Indeed the purpose of this book is to put these views together into a single, comprehensive approach to engineering design.

I was introduced in graduate school to another concept that I find most helpful, this regarding the learning process. It is the notion that learning is an individual process, and that an educator should do three things for the student: define the boundaries of a discipline, provide order for the learning process, and assist the student in self testing. Some people add to this list motivating the student. The boundaries of a discipline can be defined as the list of topics that comprise it. When the student has mastered this set of topics, he or she then has mastery of the discipline. The learning process can be easier if the topics comprising the discipline are learned in a particular order. The educator can provide this order, but recognize that it is not unique, and that a discipline generally can be learned taking topics in other orders. Finally, the student needs to know when a topic has been mastered. This is the purpose of testing. The educator can impose tests on the student, but this should be only the last phase in the testing process. The student also needs to know how to test himself or herself.

This book is designed to assist the student in the three ways noted above. The first chapter is intended to define *systems engineering* and to give an idea of the topics that comprise the discipline. The next several chapters strengthen knowledge of related disciplines, and the final chapters treat the topics that comprise the discipline of systems engineering.

Finally, I need to emphasize that this book is itself an experiment. To be sure, the discipline *systems engineering* is yet to be defined, and this book offers but a single point of view. To become a discipline, the boundaries of the discipline must be set by consensus of a broad community. This has not been done. Many books have been written on the subject of systems engineering. But, by and large, their treatment is a qualitative one. This book examines systems engineering from a quantitative point of view. It assembles knowledge from related fields, and gives a rigorous mathematical treatment to the engineering decision-making process. My best hope is that it will serve as a guide for books to come.

George A. Hazelrigg

1

Introduction to Engineering Design and Decision Making

1.1 THE ENGINEERING PROCESS

As engineers, we frequently think of ourselves as problem solvers. Indeed, we are taught problem-solving skills as the major element of our education and, throughout our lives, we seek to hone and improve these skills. To be sure, problem-solving capabilities are important to engineering. Yet, an underlying concept behind this book is that problem solving is not the principal activity of engineering; rather it is decision making. If we accept this premise, a door to a vast new array of engineering skills opens to us. These skills can be taught, and the purpose of this book is to present them.

The decision-making approach to engineering may seem foreign to many experienced engineers, and not everyone will agree with the approach taken, but it leads to fruitful quantitative methods. As benefit is obtained from these methods, the decision-making view of engineering is likely to gain acceptance.

We begin with a set of definitions. This is necessary in order to discuss the subject matter clearly. Much can be gained through careful and consistent definition of terms alone. Perhaps you will not agree with these definitions. This is part of the path-breaking nature of this book. It is all right if you do not agree. Take the time to create your own set of definitions. But be sure that they are precise and self consistent, and see if they lead to productive concepts. The definitions presented here, to the best of my ability, meet these criteria.

We begin by drawing a distinction between science and engineering, and we will define science first. Our interest here is in the physical sciences.

> ***Science (physical science)*** is the process of rationally and methodically seeking to understand nature, with the principal objective of developing a predictive or problem-solving capability.

Science deals with problem solving, and we need to define very precisely what we mean by the word *problem.*

> *Problem*—In the physical sciences or engineering, a problem is a question posed in the following form: given a state of nature at one time, determine or predict the state of nature at another time, or given elements of the state of nature, determine others. For example, given the loads placed on a structural element in a spacecraft, determine the stresses in the element. Then determine whether the element will support the loads for ten years.

This is a very restrictive use of the word problem, but we shall adhere to it throughout this book. To solve a problem, you must know (1) the relevant state of nature, which engineers sometimes refer to as the initial condition or the boundary condition; (2) the laws of nature that apply; and (3) have the necessary computational capabilities to apply the laws of nature.

Key to our problem-solving capability is an understanding of the laws of nature. It is this that scientists seek. The process of conducting science is referred to as research.

> *Research* is the activity by which a scientist (that is, one who practices science) seeks to understand nature and thus improves our predictive capability. It is a methodical process that builds on the existing knowledge base, mainly by formulating testable hypotheses and testing these hypotheses by means of repeatable experiments. Research includes measurement and modeling aimed at the development of accurate predictive models.

Equally important, we should understand what science is not. Science builds on an existing knowledge base. Science is not accepting theories without proof. Therefore, it does not jump thither and yon randomly. Sometimes established theories are disputed, and they may fall. But they must be replaced methodically with new theories that provide advantages over the previous theories. For example, relativistic mechanics replaced Newtonian mechanics early in this century, and it provided an accurate explanation of the motion of the planet Mercury, in addition to accurate explanations of the motion of the other planets and other systems. It not only contains all of Newtonian mechanics, but also is a broader theory that applies to systems for which Newtonian mechanics does not apply.

Science is not dishonest. Science does not prejudge nor jump to unwarranted conclusions. Science must take into account all sources of information, not merely those that the scientist might want to accept. Science is not for personal gain. To be a good scientist, one must seek total honesty with nature. Man has created many laws according to which society functions. Yet, man may on occasion violate man-made laws. Nature, too, has a set of laws according to which everything physical behaves. We may understand some of these laws, and many remain to be understood. Nonetheless, no man may ever violate a law of nature. Nature is the ultimate truth, which the scientist seeks. This is why it is important that new scientific theories be accepted through a consensual process that shuns all forms of dishonesty.

Frequently, scientific theories are expressed in a mathematical form. Sometimes this takes the form of an equation. But the key point is that the expression should be precise. We refer particularly to mathematical expressions of theories as models.

Engineers, unlike scientists, do not have as their major activity the search for knowledge regarding nature. Rather, they make use of the extant knowledge.

> ***Engineering*** involves the manipulation of nature to create systems for the benefit of at least some segment of mankind.

This definition of engineering embeds several notions. First, nature is recognized essentially as a set of laws and a boundary condition. Man cannot change nature; it is allowed only that man can manipulate nature. Second, the definition implies that through the engineering process something physical is created. And, more than this, that whatever is created has some value, at least to someone. The process of creating something physical requires allocation of nature's resources; therefore, engineering design is, essentially, the effective allocation of resources. The allocation of resources is, by definition, decision making.

> *Decision*—An irrevocable allocation of resources. The selection of design parameters for an engineering system such as a computer or an automobile constitutes an allocation of resources. Design is a decision-making process, and the selections of design parameters represent decisions.

Taking the view that engineering is decision making is a major step. Decision making is very different from problem solving as defined earlier. Decision making is an act that demands responsibility. Poor decision making can pose a threat to society. A medical doctor, as decision maker, prescribes treatment to a patient. A lawyer, as decision maker, represents a client. An engineer, as decision maker, creates systems. In any of these cases, people can suffer from poor decision making. It is largely for this reason that society generally requires that people who are in positions of decision-making authority be licensed (with the notable exception of politicians). Problem solving, on the other hand, is not in itself a threat to society, and scientists are not generally licensed.

To make a decision one must have options or choices from which to choose; one must have for each option an expectation of possible outcomes, that is what will happen; and one must place values on the outcomes to determine which outcome is most desired. The first element of decision making is self evident. For without choices there can be no decision. It is less clear that the other two elements are a necessary part of every decision. I contend that they are, even though their consideration might not be explicit. Certainly, each time you cross the street, you do not do a careful analysis of outcomes and values. Yet, when you set foot off the curb, you observe traffic and quickly assess the feasibility of safely crossing the street. Furthermore, you must sense some value in getting to the other side of the street. Why else would you cross? Again, this is not totally obvious. Suppose you are merely out for a walk, and the street is just in your way. Perhaps you really do not want to cross the street, but to get on with your walk you must. Crossing the street provides value by making it possible to continue with your walk.

As in the case of crossing the street, virtually all decisions are made under conditions of uncertainty, that is, under the condition that the outcomes cannot, in advance of the decision, be determined precisely. Whenever there is uncertainty in the outcomes, the decision maker is exposed to the possibility of making a decision that will result in an outcome that is not the most desired outcome. The same concepts apply in engineering, and they lead us to a concise definition of technology.

> ***Technology*** is the ability to make good engineering decisions and to carry them through to their desired conclusions.

This definition carefully distinguishes between the ability to do something and the act of doing it by defining technology as an ability, not as an act. The technology to do something is the ability to create a workable design, to carry out that design into a workable product, device, or service, and to place it successfully into service or action. Each of these things requires that appropriate decisions be made at all levels of design, manufacture, and use. A technology, therefore, is an ability to make good engineering decisions with respect to a particular engineering system.

The distinction between problem solving and decision making leads to a distinction between knowledge and information.

> *Knowledge*—An agreed upon set of facts. Knowledge is essentially the contents of a database, with elements such as $F = ma$. For purposes of this discussion, knowledge can be defined as an understanding of the laws of nature and the ability to apply those laws to predict the behavior of physical systems.

Knowledge gives us the ability to create choices. For example, the knowledge that $F = ma$ enables one to conceive of reaction propulsion: a rocket.

> *Information*—Information relates to a specific decision. Quantitatively, it can be measured as the probability that the preferred choice in a decision will lead to the most desired outcome.

Information enables us to make effective choices in engineering design. It is virtually never the case that we would make a decision without some expectations on the outcome, that is, without some information. Furthermore, with perfect information, we would never make a bad decision. But, even with perfect information, a better decision might be made if better choices became available.

Now that we have distinguished between knowledge and information, it is possible to distinguish between different types of research. It is widely held that there are two fundamentally different types of research: basic and applied. But no clear distinction between these has existed. Here I choose to define three types of research and provide a clear distinction among them.

> *Basic Research*—Research that seeks to improve knowledge, that is, man's understanding of nature, and man's ability to predict the behavior of physical systems.

Basic research seeks to provide a fundamental understanding of nature that leads to better predictive capabilities and, hence, to an ability to create and use new options for engineering design.

> *Applied Research*—Research that seeks to improve information, that is, to improve the ability to make choices in decision making.

Applied research enables the engineer to create designs that better achieve the goals set for the system. For example, better information reduces the need for overdesign that has weight, cost, and performance penalties.

The third type of activity that could be classified as research does not fit under these definitions of either basic or applied research. It may be called "look-see research."

> *Look-See Research*—Research that is performed as a process of simply trying things, not to better understand nature nor to improve information for decision making. Look-see research consists of trying things, more or less at random, but perhaps with some insights, to accomplish a goal. Look-see research frequently occurs immediately following the invention of a new measurement device, as scientists are eager to take a new look at nature afforded by the device. Quite frequently, the view of nature that new instruments give leads to the discovery of new phenomena. A good example is the use of new microscopes. Invention of the optical microscope quickly led to the discovery of bacteria, and subsequently to the theory that bacteria are responsible for disease. As new phenomena are found or new results obtained from look-see research, scientists turn quickly to basic research in order to understand them.

Look-see research is driven by technology, both by technological needs and by technological advances. Engineering achievements frequently drive look-see research. It is the mode of research that is performed when either no other opportunity presents itself or when special new opportunities for look-see research appear, for example, as in the case of the invention of a new instrument. Look-see research leads to new areas for basic and applied research, with the possibility of creating new options. However, it does not provide information for decision making. It must be followed by some amount of applied research to enable application.

As defined previously, science is not a process of decision making; engineering is. Thus, by these definitions, research ends where decision making, that is, engineering, takes over. The engineering process begins by developing an understanding of the objectives of a project. The engineer must develop an understanding of the benefit that the project will provide, and how this benefit will depend on the engineering design itself. In essence, the engineer must create a set of values by which design options will be judged.

Next, the engineer seeks conceptual designs to accomplish the relevant objectives. Conceptual designs comprise a general description of the system, including both physical and nonphysical elements, but generally do not include specifics such as dimensions. In the case of an airplane, for example, the conceptual design variations might include high wing versus low wing, jet versus turboprop, aluminum versus composite structure, and so on. In the case of a space vehicle, the conceptual design might include the nature of the power source, propulsion, and structure. Some people refer to the creation of conceptual designs as engineering synthesis.

Each conceptual design contains a number of design parameters. Again, for an airplane, these could include the wingspan, engine thrust, gross weight, and so on. The choice of these parameters is a decision-making process that engineers often refer to as optimization. The goal for each conceptual design is to pick those values of the parameters that lead to a design that maximizes the net benefit of the system. But uncertainties cloud both the choice of the values for each parameter and the computation of the benefit that these choices produce. One goal of research is to reduce these uncertainties, but some forms of uncertainty cannot be resolved entirely by research. For example, the fatigue life of an airplane depends on the weather through which it flies and the turbulence it encounters, the extent to which it is loaded each time it flies, and the care with which the pilots handle it. These parameters can be known only within limits, and no amount of research will resolve them any better.

On the other hand, research can resolve many uncertainties. For example, the finest details of the design of a load-bearing member of a structure can affect stress concentrations that could lead to failure. Improved theories and methods of stress computation may improve prediction of such conditions and enable better design. In the absence of such theories, the engineer typically chooses to be conservative, that is, the engineer overdesigns the member to allow for uncertainty. I know of no comprehensive theory of overdesign to compensate for uncertainty, but engineers make occasional reference to concepts such as statistical design, factors-of-safety, and fudge factors that implicitly allow for uncertainty.

Once the engineer has obtained an adequate state of information for decision making, both the system conceptual design and its design parameters are chosen. Frequently, this is an iterative process. It is generally expensive to obtain accurate information for an engineering design, so many conceptual designs may be eliminated at an early stage in the process, and only those remaining will be compared with a high degree of accuracy. Then, once a specific design concept is chosen, it may be subject to much more in-depth analysis.

The engineering process described here relies heavily on the results of scientific research to produce design options that emerge as part of conceptual designs and to select values of design parameters and predict the performance of the resulting system. But it includes also synthesis and decision making, which are not processes that can be addressed by the physical sciences. We now can add one final definition.

> ***Systems engineering*** is the treatment of engineering design as a decision-making process.

It is this definition that defines the subject of this book. From this definition alone, we will build the discipline of systems engineering.

Science rarely produces results that, in themselves, hold value to society. Results that themselves are of value to society relate to advances or discoveries that are called *cosmological*. The Copernican theory of the solar system, that the earth revolves around the sun and therefore is not the center of the universe, is an example of a cosmological advance. All other scientific advances lend value to society only through their implementation, typically in engineering systems. *Engineering is a process that makes science of value to society.*

1.2 HISTORICAL PERSPECTIVE

The preceding concepts and definitions deviate rather substantially from those taught to and perceived by most engineers. It is interesting to consider from where the present notions derived.

Engineering has been practiced for millennia, but until rather recently it was taught through a process of apprenticeship. The current, formalized approach to engineering education may be attributed largely to the French. Military engineering was taught in France at least as early as 1689 and although the emphasis was on military needs, French engineering included civil needs such as bridges, roads, and waterways. In the mid-1700s, King Louis XV appointed Jean Rodolphe Perronet chief engineer of bridges and highways and gave him authority to establish a school of engineering. This he did in 1747 within the Corps des Ponts et Chaussées. By 1775, the curriculum had developed into a three-year course, and the school was renamed École des Ponts et Chaussées. This is widely considered to be the first formal school of engineering. Later, in 1794, Napoleon created École des Travaux Publics to train engineers for both military and civilian service. This institute later became École Polytechnique. At the time there existed a well-developed university system, based on the Greek system, and into which it would have seemed logical to include engineering education. However, the university system did not deem it appropriate to incorporate the new, applied curriculum.

The word "university" derives from the Greek "universe," meaning all or everything, and the Greeks thought of the university as the place where one would learn about everything. But, oddly, apprenticed skills such as engineering were not thought to be included in the topics appropriate for a university. The word "polytechnic" derives from the Greek words "polys," meaning many, and "technik," meaning art or craft: hence, many arts. Engineering was taught in the French polytechnic, but, interestingly, not agricultural engineering.

Toward the end of the Napoleonic wars, the *military engineers* who had been educated in the polytechnic found themselves in the public disfavor. They felt that their education made them useful to society, yet there was a stigma attached to them. So they found a new name to call themselves, a name that drew as great a distinction between them and the military as they could manage. They called themselves *civil engineers*.

The French polytechnic system was copied in the United States as the land-grant college system, formalized by the Morrill Act of 1862, which included curricula in both military engineering and agriculture: hence, the A&M colleges. Rensselaer Polytechnic Institute became the first U.S. institution to award a degree in engineering in 1824. From that time through World War II, there was little change in the teaching of engineering. Engineering was taught largely by engineers educated to the Master's level.

But World War II changed forever the perception of engineering. The war was widely recognized as a technological battle that was won by engineers and scientists working on mega-projects such as radar and the atomic bomb. These projects advanced technology so rapidly that the Allies gained an awesome advantage in military power, and the possibilities afforded through concentrations of scientific and engineering research and development became entirely apparent. In 1950, the National Science Foundation was created, and together with other Federal agencies, provided a continuing stream of funds for advancing technology.

Around 1950, Karl Compton, a famous Nobel physicist, became the president of the Massachusetts Institute of Technology. He observed that technology had begun to advance too fast to pass down from generation to generation by a process that still had a substantial element of apprenticeship. In 1952, Gordon Brown, chairman of the Electrical Engineering Department at MIT, implemented Compton's idea and began the turn to basics—mathematics, physics, and chemistry. Other major schools followed immediately, and the rush to a science base for engineering was on. This rush intensified through the 1960s and 1970s, and during this time engineering faculties turned to the hiring almost exclusively of Ph.D.s. The new Ph.D. engineers were specialists and researchers, and they were, in kind, more scientists than engineers. Some engineering schools even eliminated the word "engineering" from their names and became schools of applied science. Today the influence of science on engineering is profound. Most engineering institutions teach applied science as their engineering curriculum. This education ends with the notion that, upon graduation, an engineer must obtain experience (that is, become an apprentice) to round out his or her education.

This book seeks to dispel the myth that engineering is applied science or problem solving, and provides a basis for formally teaching much of engineering that is yet thought of as an art.

1.3 THE DECISION-MAKING PROCESS

The three key elements of decision-making processes are identified earlier: identification of options or choices, development of expectations on the outcomes of each choice, and formulation of a system of values for ordinally ranking the outcomes and thereby obtaining the preferred choice. Two things complicate almost all engineering decision-making processes: first, uncertainty on both outcomes and values and, second, knowing what to include in the universe of phenomena that needs to be considered in determining the outcomes of the options. If indeed systems engineering is the treatment of engineering design as a decision-making process, then it will be seen that systems engineering is an activity that knows no disciplinary bounds. This is in part why some engineers claim that systems engineering is not a discipline in itself, but rather a collection of disciplines, and it is one of the things that makes engineering decision making so difficult to teach.

On October 4, 1960, a turboprop airliner was on takeoff from Logan Airport in Boston, Massachusetts. During the takeoff, the engines ingested some 30,000 starlings, which resulted in a loss of power in three of the four engines. One minute after takeoff, the airliner crashed into Winthrop Bay, killing 62 of the 72 people on board. In the subsequent investigation, it was easily determined that loss of power was the probable cause of the accident, and that the loss of power was the result of starling ingestion. But why did 30,000 starlings fly into the engines of the airliner? It was eventually determined that a small, power takeoff shaft on the engines, used to drive hydraulics, electrical, and other systems, developed a vibration when the engine was accelerated to takeoff power. This vibration, although inconsequential structurally, produced a noise very similar to that of the mating call of the male cricket, a favorite food of the starling. Redesign of the shaft to eliminate this vibration eliminated bird ingestion problems. Here is an example that shows how far afield from the conventional engineering disciplines an engineer might need to go to fully analyze a design. The lesson is that engineering design should always be thought of as a multidisciplinary activity. Indeed, it must include consideration of all disciplines: it is *omnidisciplinary*.

1.4 BEANS IN A JAR

We will close this chapter with an example that illustrates the nature of engineering design and decision making. The example will introduce several concepts and quantitative approaches, and it will tie together the concepts outlined earlier. We will return throughout this book to the concepts and methods presented here, and explain them in detail. For now, the important notion is that engineering is decision making, and decision making is amenable to rational, quantitative analysis. If you do not understand all of the approaches discussed here, do not be concerned. They will be more fully developed later.

Consider a jar that is filled with beans and you are asked the question: how many beans are in the jar? As in the case of most engineering decisions, there are many methods by which you could arrive at an answer. First, and most obvious perhaps, is to pour the beans out of the jar and count them. It might appear that this method could give an exact solution, but this is not necessarily the case. In counting several hundred or thousand beans, it would be easy to make a mistake. Also, beans break in half. Should each half be counted separately as a full bean, as half a bean, or not at all? And bags of beans frequently contain small, beanlike stones. Should they be included? Then, upon pouring the beans back into the jar, they might not fit. Even by counting, it may not be possible to obtain a unique and correct solution. The same is true of many engineering systems.

Often, in real engineering systems, it is also the case that we are not permitted to count the beans. Merely to determine the stresses in the structural members of a proposed bridge design, we are not permitted to build the bridge and test it. We have to find other approaches to the estimation of such stresses. And, even if we have very accurate methods of prediction given the loads placed on the bridge, we cannot know precisely what will be the loads. Wind, water, and traffic loads might someday be quite different than we thought. There will always be uncertainty in cases such as this.

At the other end of the spectrum, it is possible simply to guess. Although this method is less likely to be as accurate as counting the beans, it is simple, easy, and takes little time. Here we see another important difference between problem solving and decision making. To solve a problem, you need to have some idea of what you are doing. This is not the case with decision making. Everyone can make decisions. It's just that people who do not know what they are doing are likely to make rather poor decisions. Again, we see that it is important that the public be protected from people who might make poor engineering decisions.

In between the extremes of counting and guessing lies a wide range of methods that could be broadly referred to as modeling. Before beginning to examine models, however, it is useful to introduce the concept of an option space.

The option space is the set of possible system configurations and decisions. It is the full range of choices available to the decision maker. In the case of the bean-jar example, the option space is easy to define: it is the set of real, positive integers. Knowing this greatly clarifies the decision-making assignment. You are asked simply to pick a single number from this set. It is much more complex a task to define the option space for most engineering systems, but it can be broken into two subtasks: design synthesis and parameterization. Design synthesis considers the broad configurational possibilities. Again, for an airplane, these could include the general configuration of the airframe—high wing, low wing, T-tail or straight tail, number of engines, wing- or fuselage-mounted engines, types of materials, and so on. The second part is parameterization, which means identifying the parametric alternatives. Some parameters are very limited. For example, if an existing engine is to be used, we are limited to available engines. Other parameters, such as the wingspan, may take on a range of parameter values.

> *In engineering, defining the option space requires two steps: design synthesis or the definition of system configurations, and parameterization or the identification of the parameters and their permissible ranges that define the specific implementation of a system configuration.*

Regardless of the ease or difficulty in establishing the option space, it is an important step in a decision-making process. Later, it will be necessary to evaluate all options and to make a choice. This is a much easier process if the number of options to be evaluated is fixed at the outset. Indeed, the level of analysis necessary may well be determined by the nature of the option space.

Next is the problem of determining for each option the outcome space. Although determination of the outcome space is what most of engineering education focuses on, it is rarely presented as such, and thus the results are lacking. The outcome space is more than an engineering design or, in our example, an estimate of the number of beans in the jar. It is the set of possible outcomes for all system options and their *associated probabilities of occurrence*. In our example, the system options consist of the positive integers. The outcome space consists of our final win/loss position with regard to each possible guess. Thus, to determine the outcome space, we need to ask about the payoff associated with a guess.

We can look at a trivial case that provides an interesting insight. Suppose you have to pay \$10 to make a guess, but you win \$250 if you guess correctly. Of course, a correct guess can be defined in many different ways. It could mean an exact guess. It could mean a guess within 10 beans of the exact number. It could mean a guess that is lower than the actual number of beans. Let us take the last case. If we win simply by guessing low, then a guess of one bean wins the prize. We need go no further, and we need do no analysis on how many beans are actually in the jar. A guess of one bean is clearly a good guess. In engineering terms, it is a good design. We could refer to it even as a best design in the sense that no other guess provides a higher probability of winning and, therefore, no other guess is better. We can see from this example that there are good designs and there are bad designs, and what differentiates them is the outcome. Furthermore, the outcome can be relatively independent of the actual physical nature of the system. In our trivial example, the actual number of beans in the jar is not relevant, so long as there is at least one bean in the jar.

Our example is a bit less trivial if we take a payoff function that is not so simple. Let us take the nonsymmetrical payoff function of Fig. 1.1. This function provides a maximum prize of \$250 for an exact guess. But the prize diminishes linearly to zero for a guess that is low by 200 beans, and it is zero if the guess is high by even a single bean. Non-symmetrical payoff functions are common in engineering. The design of a beam is an example. The thicker or stronger the beam, the more it costs and the less its net value.

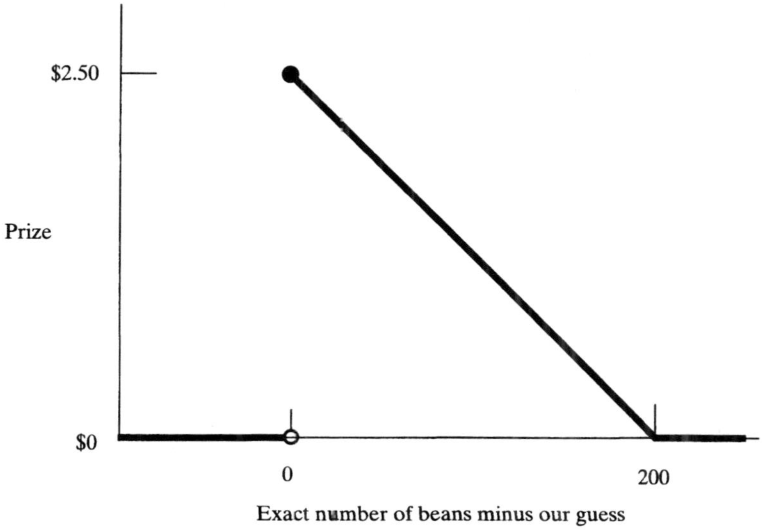

Figure 1.1 A nonsymmetrical payoff function.

So up to a point, the thinner and weaker we design the beam, the higher will be its net value. But, if we design it the slightest bit too weak it will fail, and we will sustain a large economic loss. We can write the payoff function corresponding to Fig. 1.1 as follows:

$$J\{u\} = \begin{cases} -\$10 & \text{if } u > N \\ \$250\left(1 - \dfrac{N-u}{200}\right) - \$10 & \text{if } N - 200 < u \leq N \\ -\$10 & \text{if } u \leq N - 200 \end{cases}$$

where the payoff, $J\{u\}$, is a function of the number of beans actually in jar, N, and the guess, u.

Obviously, we would have no problem choosing a value for u if we knew N. The problem is that we do not know N; we can only obtain an estimate, n. If we simply guess at how many beans are in the jar, say 1,214 beans, we are in essence saying that we are sure this is the correct number. On the other hand, if we say that this guess is correct with probability 0.01 (that is, we think there is a one-in-a-hundred chance that this is the correct number, and this is actually a pretty high probability given the example), then we have to recognize that this guess is wrong with probability 0.99. This begs the question: what is the right answer? What we need is the set of possible values of n together with their associated probabilities of being correct. If 1,214 is our best guess, that is, if it is the guess with the highest probability of being correct, then all other guesses will have probabilities that are at best no higher. So, for example, n could include the pairs (1,212; 0.0098), (1,213; 0.0099), (1,214; 0.01), (1,215; 0.00995), (1,216; 0.0099), and many more. The first number in each pair is the numerical estimate of the number of beans, and the second number is our assessment of the probability that this number is correct. The sum of these probabilities for all pairs should equal one. All pairs that have non-zero probability comprise the full assessment of n, and it is this full set that represents the information we have on how many beans are in the jar. From this estimate, for each possible guess, u, we can derive the outcome space, which will consist of the set of possible outcomes with their associated probabilities. That is to say, we will determine $J\{u\}$ as a set of values, each with an associated probability. We will use the preceding equations for J, taking for N each possible value of n and its associated probability, which is the probability of the resulting value of J.

Now we see that the choice of u is not so simple. The uncertainty in our ability to estimate how many beans are in the jar leads to uncertainty in the prize, and comparison of alternative values of u must take this uncertainty into account. There is a well-developed theory of decision making under conditions of uncertainty, and we shall discuss this theory later in this book. First, however, it is clear that the better we can estimate the number of beans in the jar, the easier it is to choose a good value for u. Getting better estimates of physical parameters is what much of engineering analysis seeks to do, and the tool we use is called modeling.

A simple model for the number of beans in a jar is the following:

$$n = \frac{V_j}{V_b} f \qquad V_b = \frac{4}{3} \pi r_b^3$$

V_j is the volume of the jar, V_b is the average volume of a bean, and f is a packing factor or the fraction of the jar that is filled with beans as opposed to air. Assuming the beans to be spherical, the average volume of a bean is given in terms of the average radius, r_b. Of course, this is just one model. Many other models might be constructed to estimate the number of beans in the jar, some much more complex than this. Take a moment now to see how you might construct a different model. How many different approaches can you think of in five minutes? For example, in a simple extension of the model given above, the volume of a bean could be computed assuming that beans are oblate spheroids, and models could be developed for the computation of f. We could even account for edge effects, so that f is dependent on the size and shape of the jar.

In the above model, V_j, V_b, and f can be estimated with reasonable accuracy. What the model has done is to disaggregate the variable n into three variables (in this case r_b is used in place of V_b), each of which can be estimated with an accuracy such that their aggregation via the model into n leads to an estimate that contains less uncertainty than a direct estimate of n itself. *The purpose of engineering models is to reduce uncertainty for decision making.* But we also see that this model provides only a single number for n, when what we need is a set of numbers and their associated probabilities. This is a problem of virtually all engineering models; they do not provide information. Indeed, a numerical estimate of any sort provides information only when it has an associated probability of correctness. Extending ordinary engineering models into models that provide information for decision making is a topic to which much of this book is devoted.

Although much analysis remains to be done, we cannot complete our study of how many beans are in the jar and make a choice, u, until we develop some fundamental concepts. So we will leave this problem for now. Suffice it to say that, in the end, the methods we will develop will not give a unique best choice anyway. Indeed, there is generally no unique best choice. But there are good choices and there are bad choices. Our goal will be to find the good choices and to understand what makes one choice better or worse than another.

1.5 SUMMARY

We have begun our exploration of engineering as a decision-making process, and uncovered many flaws or deficiencies in conventional forms of engineering analysis. First, we found that engineering decision making is inherently an omnidisciplinary activity; it cannot be reduced to a single discipline. Thus, it cannot be reduced to a disciplinary model. But this does not mean that we cannot create a discipline of systems engineering. By defining systems engineering as the treatment of engineering design as a decision-making process, we can draw on methods and procedures that have been well developed for other applications. Second, we found that the design process encompasses three distinct activities: the identification of options, the development of expectations on outcomes for each option, and the use of values to select the option that has the range of outcomes and associated probabilities that are most desired. Recognizing that uncertainty is an inherent aspect of virtually all decision making, we found the conventional engineering modeling process lacking. In particular, it tends not to address the uncertainty inherent in most aspects of engineering. We saw that, in order to provide information for decision making, a model must produce both a prediction, such as a numerical quantity, and an associated probability regarding the accuracy of the prediction. Finally, we saw that good engineering design depends on the intended purpose of the system as well as the physical characteristics of the elements that comprise the system.

We now see engineering design as a process of synthesis and integration. We see it as an activity that knows no disciplinary bounds. We see systems engineering as a process of dealing with information, knowing what information is needed for the design of a specific system, knowing how to get that information, and knowing how to use it. These three elements comprise the discipline of systems engineering. The remainder of this book draws from well-established disciplines such as mathematics, economics, operations research, physics, chemistry, and the conventional engineering disciplines to develop a theory and methodology of systems engineering.

As you read this book, you will see emerging a difference between the conventional approach to engineering and the systems engineering approach. In the conventional approach, the design of a system is accomplished by dividing the overall system into manageable parts. Each part tends to have a disciplinary focus, and it can be designed by a single engineer or, at most, a few engineers. At the highest level of division, we call the parts subsystems, and we refer to them as components or devices at finer levels of division. The overall system is first partitioned into subsystems by the imposition of constraints, such as budgets—a weight budget, a power budget, a thermal budget, and so on. As the design progresses, the constraints that divide the system become more and more detailed, and eventually the design emerges as an artifact of the constraints. Program management and communication focuses on the constraints, and the optimality of the design is restricted by the constraints. This is a reductionistic or disciplinary approach to design.

In the systems engineering approach, the system is divided into disciplinary models, and it is the job of the systems engineer to integrate these models into a model of the whole system. Project management focuses on the modeling process, which also is the basis for communication between engineers working on the project. And the design emerges in increasing detail as the models are refined and extended. The systems engineering approach is an approach of synthesis, using models as its basis. Being less restricted by constraints than the conventional approach, the systems engineering approach has the potential to lead to more optimal designs, better response to the marketplace, lower risk to the system producer, and greater benefit to society.

Buried in this approach is a new philosophy toward engineering design. Instead of viewing design as a process of providing functionality at minimum cost while satisfying constraints, systems engineering views engineering design as a process of availing ourselves of an opportunity. Often it is the opportunity to serve society or simply to make money. We view good design as doing this well.

In the simplest sense, we can say that engineering design requires only two steps: enumeration of possible design alternatives and selection from among these of the best alternative. We have already seen that neither of these steps is trivial. Indeed, we have much yet to learn about how to accomplish them effectively and efficiently.

This book begins with several chapters that are intended to provide the basic skills needed to discuss the principles of systems engineering. The substance of these chapters is drawn from existing disciplines outside of engineering. In each case, however, we pay special attention to the requirements of system engineering, and we find that there are nuggets to be found by mining these fields. Although it may appear that the initial chapters of this book could be arranged in almost any order, the student may find it easier to study them in the order presented. The later chapters develop specific skills of systems engineering.

REFERENCES

1. Blanchard, B. S. and W. J. Fabrycky, *Systems Engineering and Analysis*, 2nd ed., Prentice Hall "International Series in Industrial and Systems Engineering," Englewood Cliffs, NJ, 1990.
2. Bush, Vannevar, *Science—The Endless Frontier*, July 1945, reprinted by the National Science Foundation, NSF 90-8, Washington, D.C., 1990.
3. *Engineering Education and Practice in the United States, Engineering in Society*, National Academy Press, Washington, D.C., 1985.
4. *Engineering Education and Practice in the United States, Engineering Infrastructure Diagramming and Modeling*, National Academy Press, Washington, D.C., 1985.
5. Fishburn, P. C., "Utility Theory," *Management Science*, vol. 14, no. 5, 1968, pp. 335–378.

6. Ferguson, E. S., "How Engineers Lose Touch," *American Heritage of Invention and Technology*, vol. 8, no. 3, Winter 1993, pp. 16–24.

7. Gero, D., *Aviation Disasters, The World's Major Civil Airliner Crashes Since 1950*, Patrick Stephens Limited, Sparkford, Yeovil, Somerset, Great Britain, 1993.

8. Grayson, Lawrence P., *The Making of an Engineer: An Illustrated History of Engineering Education in the United States and Canada*, John Wiley, Inc., New York, 1993.

9. Hazelrigg, G. A., 1988, "The Engineering Decisionmaking Process," Proceedings of the Cambridge Symposium on Optical and Optoelectronic Engineering, sponsored by The International Society for Optical Engineering, Cambridge, MA, Nov. 6–11.

10. Koen, Billy Vaughn, *Definition of the Engineering Method*, American Society for Engineering Education, Washington, D.C., 1985.

11. Lewis, W. P. and A. E. Samuel, *Fundamentals of Engineering Design*, Prentice Hall, Sydney, Australia, 1989.

12. Lomask, Milton, *A Minor Miracle: An Informal History of the National Science Foundation*, U.S. Government Printing Office, NSF 76-18, Washington, D.C., 1976.

13. Roederer, Juan G., "Tearing Down Disciplinary Barriers," *Astrophysics and Space Science*, 144, 1988, pp. 659–667.

14. Rubinstein, Moshe F. and Iris R. Firstenberg, *Patterns of Problem-Solving*, Prentice Hall, Englewood Cliffs, NJ, 1995.

15. Taguchi, Genichi, *On Robust Technology Development*, ASME Press, NY, 1993.

16. Tribus, M., *Rational Descriptions, Decisions and Designs*, Pergamon Press, Elmsford, NY, 1969.

PROBLEMS

1. Discuss the difference between problem solving and decision making. What are the differences in societal impacts? What are the differences with respect to personal responsibility? What are the differences in consequences?

2. Choose a simple decision that you face. Discuss options, expectations, and values that you hold regarding the decision.

3. Develop a model to estimate the number of vehicles crossing a bridge over the course of a day.

4. Develop a model to estimate the volume of water flowing in the lower Mississippi River.

5. Consider an office building that is 20 stories high and 100 m by 200 m. How many elevators will the building need?

6. What is the maximum distance that a motor vehicle carrying at least one person could travel on level ground while using one liter of gasoline?

7. For problems 3 to 6 above, estimate the uncertainty in your answer.

8. What are the elements of decision making? Choose a typical decision that you might make frequently in your life. Discuss the decision-making elements in the context of this decision.

9. Classify the following research projects by type of research:

a. A scientist examines different material combinations to find new high-temperature superconducting materials.

b. An engineer models the properties of high-temperature superconductor materials in order to design a material production process.
c. A physicist looks for new particles using a high-energy particle accelerator.
d. An engineer studies the transition from laminar to turbulent flow in a boundary layer.
e. Galileo observes Jupiter through his telescope.
f. A chemist measures a reaction rate that is critical to the design of a chemical plant.
g. A physicist models phase transitions in materials at near-absolute-zero temperatures.

10. Consider the bean jar example. Propose three different models for estimating the number of beans in the jar. Discuss the differences between the models.
11. What is the volume of a cow?
12. How many gas stations are there in Los Angeles?

2

Design Options

2.1 THE PURPOSE OF THIS CHAPTER

In the conventional approach to engineering design, we are taught to begin the design process by defining the desired functionality of the system that we seek to design. We are told that form follows from function, which is to say that the design of a system is intended to accomplish its desired function. But the definition of desired functionality can be viewed as a process of imposing constraints on the system, even at the very beginning of the design process. Here, we want instead to instill in the design engineer a sense of freedom, not one of constraint. We want the engineer to think of the design process not as a process of seeking to satisfy a set of constraints at minimum cost but rather as a process of freely taking advantage of an opportunity through the ability to manipulate nature. The opportunity we refer to here is the opportunity to provide benefit through the design of an engineering system. Good engineering designs yield benefits even though they have associated costs. Costs are not all bad, and we need to learn that it can be wise to accept a cost if it yields an adequate benefit. Above all, we want to preserve the ability to make comparisons between design options and make rational choices. We emphasize that engineering design, in its simplest form, can be described as a two-step process: (1) identifying a set of design options or alternatives, and (2) comparing the options and selecting the preferred option from among this set.

In the systems engineering approach, we accept that engineering design is a process of decision making, and that to make a decision we must have options, expectations, and values, then we see that the identification of options is crucial to the success of any engineering design process. Most engineering students are taught to examine options in the design of a physical system. But generally, these are just some of the options available to the engineer. The purpose of this chapter is to create an awareness of the breadth of options available to the design engineer, and to present some approaches to identifying them. In seeking design options, we will allow ourselves to be guided by the opportunity that we choose to pursue through the design process, and we will seek to avoid the imposition of constraints to the extent possible.

2.2 THE ROLE OF OPTIONS IN DECISION MAKING

In the first chapter, we define a decision as an irrevocable allocation of resources. It is also a choice taken from a specific set of options. Without options, there can be no decision. We will refer to the set of all possible options relating to a particular decision as the *option space*. The notion of an option space derives directly from the recognition of systems engineering as a decision-making process. The concept that for every design decision there exists an option space is quite important. If we can define the option space, we will have bounded the design problem. We will know what has to be given consideration and what we can disregard. Blanchard and Fabrycky (1990) note that the systems engineering process begins in conceptual design. Here we will refine this and say that the systems engineering process begins by defining the option space, and this requires two distinctly different activities: *design synthesis* and *parameterization*.

Design synthesis consists of defining feasible system configurations, including both physical and nonphysical elements of a system. In design synthesis, we are not concerned with dimensions of the physical elements of the system, only with the design layout and other configurational decisions. We are concerned, for example, with the configuration of the parts that comprise a system, how they are arranged, and how they are used, but not with their individual specifications. Parameterization consists of identifying the numerical quantities that specify each element in the system and their permissible ranges. A system specification consists of a set of values for the system parameters together with the system configuration. Alternatively, we could view this as the specification of a set of parts and the instructions to assemble them and put them to use.

For some physical elements of a system, however, it can be extremely difficult to provide a specification in terms of a configuration and design parameters. And, even if such a specification were available, still it might be very difficult to manufacture the system using it. A good example is an apple pie. We can think of an apple pie configuration as being crust on the bottom, crust on the top, and apple pie filling in the middle. There are obvious variations on this configuration. But first, let us note that engineering specifications on the crust or filling would be very complex. Second, the specifications would not help much as a guide to making the pie. In this case, what we need to specify is a recipe. So we see that, for some systems, rather than to specify the system itself, we specify the recipe required to make the system. An example of this in engineering is a microelectronic circuit. It can be specified in terms of a set of masks and fabrication steps—a recipe—for its fabrication.

It is also apparent that a system can be used in many different ways, and for entirely different purposes. A pie, for example, can be eaten, it can be thrown, or it can be displayed to entice potential buyers into a shop. It can be eaten immediately after it has baked and cooled, or it can be shipped, stored, and eventually sold to be eaten. All of these different uses of the pie lead to different choices of recipe. For example, if the pie is to be shipped and stored, it might be desirable to include preservatives in the recipe. If the pie is to be thrown, the best option might be to leave off the top crust and use a cream filling. So we see that the use of a system can significantly affect design choices. Although this sounds so obvious that stating it may appear stupid, you will easily find engineering systems where this fact has been neglected.

In the first chapter we presented a very simple example of decision making, guessing how many beans are in a jar. The principal reason why this example is so simple is that the option space is trivial and very easily defined. It is simply the set of real, positive integers. Defining the option space for a typical engineering system can be very difficult, and perhaps even impossible. Think for a moment of defining an option space by a process of elimination rather than a process of building. In the broadest possible sense, even a physical design could be anything. It could be any feasible arrangement of atoms in the universe. Broader still, atoms are not physical entities per se, they are mathematical constructs. So our option space could begin as the set of all possible mathematical constructs. From here, we would get to a workable option space through a process of elimination, discarding those constructs that we no longer wish to consider. Knowledge allows us to discard options. Knowing that something is infeasible or inappropriate, we feel free to discard the corresponding options. Through this process, we begin with an uncountable infinity of options and seek to narrow the set to something workable. So much must be taken into consideration by this approach that it is hard even to think about using it. Still, it gives us a good appreciation for the complexity of a typical option space.

A second approach is to build up the option space by adding options for consideration. This approach works essentially by using existing or emerging technologies to build new systems. We can see how this might work by looking at another example. Suppose we have two parallel shafts, one a driving shaft and the other the driven shaft. They are separated by a specified distance. We want the driving shaft to drive the driven shaft, by means of gears, so that the driven shaft turns at one-half the speed of the driving shaft. What are our options? First, let us observe that the ratio of gear teeth of the driving gear to the driven gear is 1:2, that is, the driven gear must have twice as many teeth as the driving gear. This gives us the desired speed reduction. Now our options include:

- The number of teeth on the driving gear
- The tooth design (straight, helical, etc.)
- The thickness of the gears
- The driving gear material
- The driven gear material
- The method of attachment of the driving gear to its shaft
- The method of attachment of the driven gear to its shaft.

Perhaps we could consider driving gears with from 20 to 50 teeth (31 options). We will consider two standard tooth designs. We will consider each of five materials for the driving gear and the driven gear (25 possible combinations), and three possible attachment methods for each gear (nine possibilities). Gear thickness could vary continuously over some range, yielding an infinite number of possibilities, but let us allow only 10 discrete thicknesses. Our option space now consists of 139,500 possible designs ($31 \times 2 \times 25 \times 9 \times 10$). This is a large number indeed, but it is very much smaller than infinity. Still, this is a very simple example. We have now only to choose the best option among these 139,500 options.

We now see one more reason why systems engineering is not a well-developed discipline. It encompasses a great deal of complexity. The simple fact is that decision making is complex. Face-to-face with such complexity, you might not want to continue. We either will have to discard much of the complexity or find ways of dealing with it. One lone engineer is not going to give explicit consideration to each of 139,500 options. Still, the design of these gears is not an overwhelming task for one engineer. Why is this? Does the engineer have a better way without using a formal approach to systems engineering? Not usually. Normally, an engineer considers only a limited number of options, say 10, and selects the best of these. If the best of these 10 is also the best of the 139,500 options set out previously, the engineer will have found the best available design. If not, and this is quite likely, the engineer will have a design that is not as good as might be possible. We will refer to such designs as *suboptimal.* Most engineered products are the result of suboptimal designs. We use suboptimal designs because they are achievable. We can afford them. Without them we might have nothing. But another thing that suboptimal designs provide is room for improvement. Economic competitiveness may depend on your ability to get closer to the optimal design than your competition.

2.3 SYSTEM OBJECTIVES

We describe the results that we want from a system as a set that we refer to as the system objectives. It is this set of objectives that will drive our selection of system options. Thus, a sensible place to start in the definition of system options is with the definition of system objectives. Some authors adhere to a linear model (by which they mean sequential model) of systems engineering in which system objectives derive from the need for a system, which in turn is driven by human wants or desires. In the linear model, objectives are formulated in a sequence of steps progressing from the first to the last. The model shown in Fig. 2.1 is an example of a linear model. This model can be viewed as a filtering model. A very broad set of needs or desires may be identified. That set is reduced according to the feasibility of obtaining the needs or desires, so that the resulting list is reasonable in terms of extant or projected technology. Finally, the set is further reduced by other factors, such as environmental concerns. In this model, the system objectives seek to satisfy a tempered set of needs or desires.

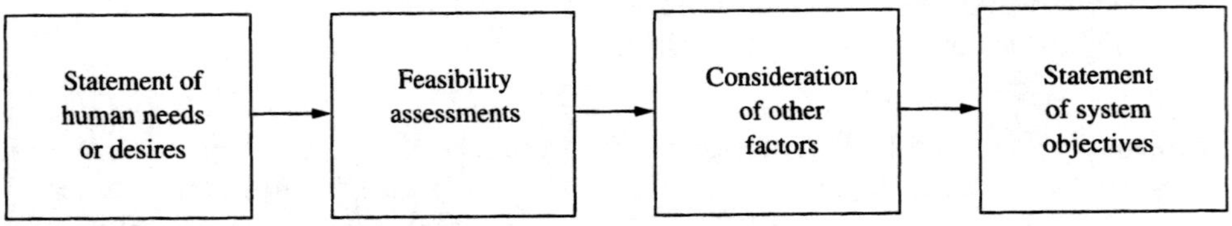

Figure 2.1 A linear model for determining system objectives.

Although human wants or desires may underlie all systems development, the steps by which one converts these to a set of system objectives is not always linear. Figure 2.2 shows a nonlinear model of systems objective determination. In this model there is feed-

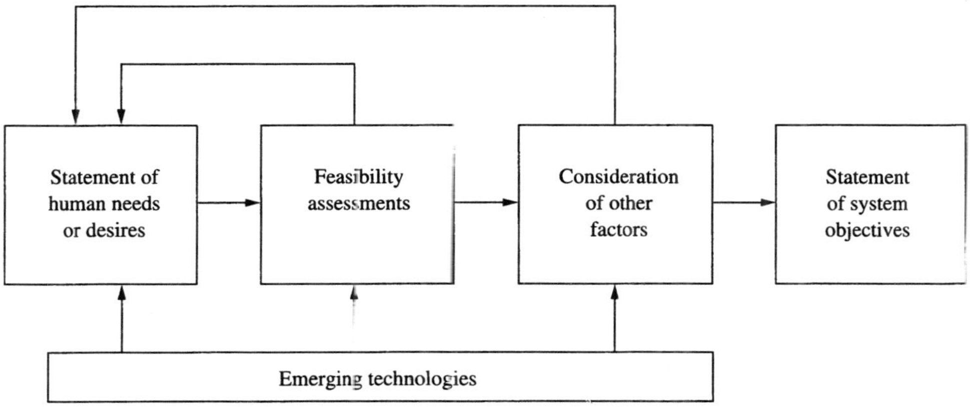

Figure 2.2 A nonlinear model for determining system objectives.

back, so that the needs and desires are themselves modified by both emerging technologies and their prospect.

It is useful here to compare systems engineering with strategic planning. A strategic plan consists of three parts: (1) an honest assessment of where you are now, (2) a statement about where you want to be at some time in the future, and (3) a road map of how to get from here to there. In systems engineering, the extant technology is where we are now. The system objectives are our statement of where we want to be at some time in the future. And the systems engineering process is how we get from here to there.

2.4 OPTIONS IN PHYSICAL DESIGN

When we think of a system, we naturally tend to think of something physical: a car, a computer, a light bulb. A system can be much more than a physical thing, but we will limit our attention for the time to the physical elements of a system. The physical elements of a system are comprised of materials, and these obey the laws of nature. The materials of a system are not always solids. For example, the Atlas rocket launch vehicle consists of a stainless steel balloon, in the shape of a cylinder capped at the ends, and pressurized with nitrogen gas. The gas provides the structural integrity needed to keep the cylindrical structure rigid, much like a modern soda can. Thus, we could say that the nitrogen gas is part of the structure of the vehicle. Take a moment to consider the use of a gas to provide structural strength. Is this lighter than the use of metals? What are desirable properties of the gas?

Together, available materials and the laws of nature comprise the principal limitations on the physical design of a system. Our job is to identify the range of designs that stay within these limitations and yet produce a desired result—the system objectives. We are looking here for alternatives, not optimal designs. Alternatives can be outlined in table form, as shown for major options in a commercial aircraft design in Table 2.1.

TABLE 2.1 MAJOR OPTIONS IN AIRCRAFT DESIGN

Wing		High wing Low wing
Engine	Type	Fan jet Turboprop
	Number	2 3 4
	Location	Under the wing Fuselage mount Tail mount
Tail		Standard T-tail
Fuselage/seating		Narrow body Wide body
Landing gear		Wing mount Fuselage mount
Structural material		Aluminum Composite

It is very clear from this table that we have just begun to outline the possible design options. Despite huge numbers of possible combinations and permutations of design options (Table 2.1 gives 576), we have considered here only the highest level options. One of the things that a rigorous method of design should seek to accomplish is to achieve consistency across the design of the overall system. For example, we do not want one engineer to choose aluminum for one part and another to choose composites for another similar part based on different perceptions of the same attribute of the materials. And one of the problems that we have in accomplishing this is that any large system is designed by tens or hundreds, sometimes even thousands of engineers, each with their own area of expertise, and each working on only a small subsection of some subsystem. In this melee, clear, complete, and concise communication is key. We do not want engineers working on one subsystem to impose major and costly design constraints on engineers working on another subsystem by eliminating design options in an ad hoc way. Thus, one purpose of a table such as this is to put down in writing the major options that management is willing to consider, and communicate those options to others working on the project.

The next step will be to expand this table, providing more detail for a selected set of options. In this way we seek to create a taxonomy of selected alternatives. For example, we might give only very cursory attention to each of the 576 options of Table 2.1, and narrow the range of options to, say, five before proceeding to the next level in the taxonomy. Whenever we eliminate options, however, we should be careful that we are not eliminating potentially good choices.

Packaging Complexity

A significant problem encountered in modern systems is complexity. Many years ago, a single engineer could understand all aspects of a system from design through manufacture to use. Today this is rarely the case. A computer engineer might understand how to create digital circuits but not how to fabricate microelectronic devices. A controls engineer might understand how to design a control system, but not the sensors or actuators that make it work. We cannot understand everything because it is simply too complex. Yet, we must make use of devices that we do not understand in the systems we design. We can do this if the complexity is properly packaged. For example, an engineer can use a light bulb in a control panel design, even though he or she does not understand the design of light bulbs. The light bulb comes in a package, which we can understand. Similarly, we can use a microchip in a circuit, even though we might not understand the detailed design and fabrication of the microchip, because we understand what the microchip does. The complexity of the light bulb or the microchip is not relevant because it has been packaged.

Packaging complexity does two principal things. First, it enables us to make use of advanced technologies that otherwise we could not use. This is because it provides us with ready-made devices that we can incorporate in our designs without a detailed understanding of their design. Second, it reduces the number of possible design choices. This is because, to use a packaged device, we must select it as the package that it is. If we go back to the example of the gear-train design, we might observe that gears are available in standard sizes. Perhaps the selection of two standard gears is acceptable.

When we select standard parts, we get the benefit of the design, development, and testing efforts of the engineers who have been responsible for these parts. The cost of their efforts is amortized over the entire market, not just the parts we might buy, and thus their selection can buy considerable expertise, even though we do not ourselves fully understand the technologies involved.

The keys to using packaged complexity are a good set of catalogs and a telephone. As technologies advance, we can expect to use packaged complexity more and more.

2.5 SYSTEM MANUFACTURE AND DEPLOYMENT

We began our discussion of system options with a discussion of options in the physical elements of the system, and we noted that these must obey the laws of nature. Other elements of the system are not physical, and they do not necessarily obey any such laws. Many elements of an overall system involve human decisions, which obey no known laws of nature. Options in system manufacture and deployment are examples.

Options in manufacturing are of two types: options in physical system design that affect the manufacturing process, and options in the manufacturing process itself. For example, one design might lend itself to automated assembly, while another requires manual assembly. In the latter case, manual assembly is required, but in the former case we have the option of manual or automated assembly. Alternatives in system design that impact manufacturability include, among others, selection of materials, which can affect machining techniques, joining and fastening, and a myriad of other manufacturing activities; shapes of parts, which affect the machining process or part fabrication as well as options for automatic versus manual assembly; and part counts, which affect assembly requirements.

At this point, let us return to the notion of packaged complexity and its impact on the manufacturing process. A good example of packaged complexity is given by the modern hand-held calculator. In the 1950s, a calculator capable of adding, subtracting, multiplying, and dividing was a complex mechanical device with hundreds of parts. The parts were assembled manually. The entire calculator was the size of a typewriter, and may have weighed upwards of 10 kg. Its cost, in the 1950s was as high as $1,000. Modern microelectronics technology packages the equivalent functionality into a single silicon chip, about one centimeter square, plus another chip for the readout display. The microelectronic circuit on the silicon chip may include a million or more individual electronic devices. But they are fabricated and packaged in such a way that they can be counted as a single part in the modern calculator. So, while functionality has increased greatly, the packaging of that functionality actually decreases system complexity. Indeed, you would find it a great deal easier to design a calculator using microelectronics technologies than using the older mechanical technology.

In many cases it is unclear when system manufacture ends and deployment begins. Perhaps we would be wise simply to leave this distinction somewhat unresolved. Consider the construction of a nuclear power plant. It is deployed when its manufacture is complete, including acceptance testing. But there is no clear point of distinction between manufacture and deployment. Another system for which there is no clear distinction between manufacture and deployment is a communication satellite system. Indeed, we could consider deployment of the satellite as a part of the manufacturing process. In both of these examples, design options in the physical elements of the system affect system deployment. A satellite can be designed for deployment on only one specific launch vehicle, or it can be designed to be launched by any of several different launch vehicles. In general, there is a price to be paid for the flexibility of having a choice of launch vehicle.

With respect to the launch of a satellite, there can be many deployment options beyond the selection of a launch vehicle. We can begin to examine these options by drawing simple pictures of the deployment process.

Cartoons

A cartoon is a drawing that by nature is very simple, yet captures some essence of a situation. The use of cartoons to sketch out deployment options can be helpful. We will later expand the use of this tool to the analysis of failures that can occur during system deployment and use. Figure 2.3 shows a cartoon that depicts the deployment of a communications satellite. The case shown depicts an expendable rocket launch, beginning with the first stage and booster liftoff. Strap-on booster rockets are jettisoned after they are expended, and the first stage continues its ascent. First-stage depletion is followed by second-stage ignition and direct injection into geosynchronous transfer orbit. After injection into transfer orbit, the second stage separates from the satellite. The satellite then coasts

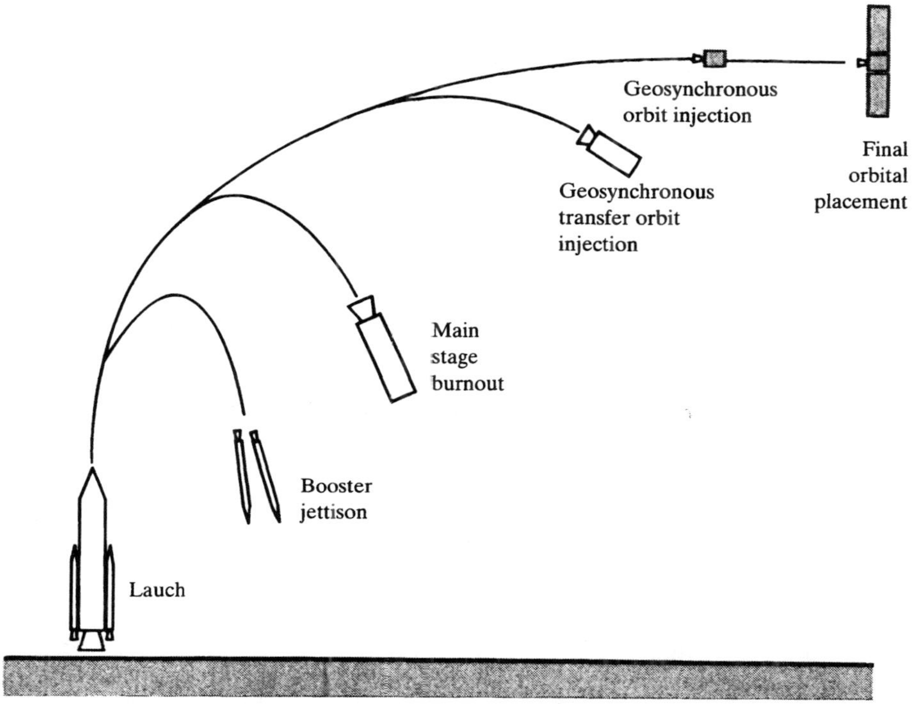

Figure 2.3 A communication satellite launch cartoon.

to geosynchronous orbit altitude, where the final injection burn is accomplished. The satellite's solar panels and antennas are then deployed, and the satellite is moved to its final location in geosynchronous orbit.

We now could begin to examine options using this cartoon as a basis. One option is to coast in low earth orbit before making a second burn with the second-stage rocket, which would place the satellite onto a geosynchronous transfer orbit. This might be done to obtain better tracking data so that the injection into transfer orbit might be made more accurately or to better position the transfer orbit with respect to the final desired satellite position. These options might reduce the propellant required for satellite deployment, leaving more propellant in the satellite after deployment, which in turn will increase the satellite's lifetime on orbit.

Cartoons provide a simple method of depicting deployment options.

2.6 OPTIONS IN SYSTEM OPERATION

The *raison d'être* for a system is to put it to use. Too often, engineers forget this fact in the system design process. You should keep in mind that design affects use, and use affects design. Consider an airplane, for example. If pilots could be relied upon always to make soft landings, the landing gear would not have to be designed to withstand rough landings. If pilots never made mistakes, airplanes would not need safety interlocks to prevent incorrect procedures. Sometimes we think that we are designing a system to a specification, when what we are really doing is designing the system for use where experience has shown the need for certain system attributes. And sometimes it pays to reexamine the specifications. If you could design an airplane that would always make soft landings, perhaps you should reexamine the requirement for strong landing gear.

We will distinguish two types of activity in the system operation phase: normal operation and maintenance and repair. By normal operation here, we do not imply that operation is always normal. Many systems can be operated in a degraded or emergency mode, such as an airplane flying with one engine shut down. Many systems are designed to be operated, even safely, in various degraded modes. When a system breaks or malfunctions or suffers some sort of failure, we have three major options: to fix it, to continue its use in a degraded condition, or to discontinue use of the system and scrap it. Many times the last option is preferred. For example, hand-held calculators are now so inexpensive that, if one malfunctions, it is better to throw it out and get a new one than to try to fix it. This enables the calculator to be designed so that it cannot be taken apart. It means that replacement parts need not be kept in stock, and technicians need not be trained to fix the calculator. Today, many systems and subsystems are designed to be thrown away when they fail, even when the failures would be trivial to fix.

Normal Operation

System operation frequently includes many decisions that can have a significant effect on system lifetime, operating cost, reliability, environmental impact, and other elements of system performance. Sometimes the operation of a system is largely out of the hands of a manufacturer, and different users might use the system in vastly different ways. This is the case of an automobile, which one owner may care for with diligence and another provide no care at all. The control of other systems might be very precise and according to very specific directions. This might be the case of a satellite. Every command issued to the satellite while on orbit is carefully considered. Airplanes fall somewhere between automobiles and satellites.

Expected operation of the system can influence design. For systems such as the automobile, it is important to include in the design features that prevent the operator from destroying the system. Included might be engine overspeed protection, for example.

Product liability is an issue of increasing importance. System design must take a responsible attitude toward system use. Some systems involve inherent risks, and these must be accepted by the user. A chain saw is an example. But other systems are in common use by the general public, and due concern for the user's safety is expected. The automobile is again a good example. The public has come to expect that automobile designs will incorporate a minimum level of safety, giving some protection in the event of an accident or other malfunction. It is also expected that controls are such that even the less capable user is unlikely to do something hazardous. One such interlock on modern automobiles prevents the transmission from being placed in gear unless the operator depresses the brake pedal. Similarly, airplanes have interlocks to prevent the pilot from making certain catastrophic mistakes.

Particularly with respect to well-controlled systems, such as satellites, operational choices can significantly affect system economics. Consider the case of a communication satellite. Communication satellites are generally placed in geosynchronous orbit where, from the earth, they appear to stay in a fixed location in space. Because they remain in a fixed position, earth-based antennas need not track the satellite. Rather they can be rigidly mounted, saving considerable expense. The larger the earth-based antenna, the better the signal-to-noise ratio of the transmitted signals. Thus, larger antennas allow more information to be transmitted through the satellite, enhancing the system value. But larger antennas also need to be pointed very precisely at the satellite. Or put another way, for fixed antennas, the satellite must remain very precisely in its assigned location. Active position control is required to keep the satellite on station, and the amount of propellant required increases as the satellite is required to remain on station within tighter tolerances. Therefore, the more precisely the satellite is maintained in its assigned position, the shorter its lifetime. Its life ends when station-keeping propellant is expended. So we are faced with a tradeoff among satellite lifetime, ground-station cost, and information throughput of the system. These are choices we can make in the normal operation of the system.

Maintenance and Repair

Failures and failure recovery can be a major issue in the design of some systems. We will treat these topics later. For now we will look only at strategies for maintenance and repair. Feasible options vary widely. On the one hand, we could undertake a program of preventive maintenance. At the other extreme, we could fix things only when they break.

Under the preventative maintenance strategy, we try to anticipate failures and replace parts before they fail. This strategy works well if quality control is very good or if we understand failure mechanisms very well and have means for anticipating failures. Sometimes this is possible with amazing precision. Again, on communication satellites that use traveling wave-tube amplifiers (TWTAs), careful analysis of the TWTA signals enables engineers to predict their failure to within days several months before the failure actually occurs. This provides time even for the negotiation of modified contractual arrangements with users.

Some mechanical systems also provide characteristic signals that foretell failure. For example, it is sometimes possible to analyze the noise produced by bearings and detect incipient failures. To do this, however, may require special design considerations. Another approach to the prediction of incipient failure includes the use of various nondestructive evaluation (NDE) methods. NDE is a rapidly developing field that will play an increasing role in systems of the future. Also, sensors are improving rapidly, and these will be employed increasingly for detection of incipient failure.

When failures are not crucial to safety, we can take the fix-it-when-it-breaks approach. Surprisingly, this approach can lead to higher system reliability, particularly if the technicians who fix the system are not highly skilled. Under this approach, we need to give consideration to the stocking of spare parts. A crucial question is: how important is it that the system be available for use at all times? Or, put differently, how impatient are we to get the system fixed when it breaks? If we want the system fixed immediately, then we need rapid access to spare parts. Either the parts must be stocked locally or transported rapidly from a central stocking facility. Which choices are reasonable will depend on many factors, including the demand for the parts. With high demand, local stocking may be economic. With low demand, central stocking may be cost effective. Demand for spare parts is a function of both the number of systems in operation and the reliability of the parts.

Repair options must include consideration of the overall repair strategy. This includes the training of repair technicians. The cost of such training and the skill level required of technicians depends, in turn, on the maintenance strategy. If a design is modularized into subsystems, and if these have a self-test capability, subsystem replacement can be relatively easy, and only minimal technician training will be required. Other systems can be very complex and require extensive technician training.

We should remind ourselves once more that one strategy is to discard the system entirely if it breaks. This strategy has the advantages that no spare parts are needed, no technicians must be trained, and no repair centers are needed.

2.7 END-OF-LIFE OPTIONS

It is becoming increasingly important to consider what happens to a system at the end of its life. In some cases, laws require recycling of materials and essentially mandate a specific end-of-life disposal. Such disposal can impose substantial costs, and its early consideration might lead to decisions to use materials that are less difficult to deal with. Communication satellites pose an interesting problem. Because of the value of geosynchronous orbit, it is required that communication satellites be boosted to a higher orbit at the end of their life. They are essentially parked in a high-tech junkyard where, hopefully, they will stay for eternity.

Part of the problem in even listing end-of-life options is that the time frame of interest could be several years in the future, and it can be very difficult to guess what will be the social and political environment regarding disposal of the system. It may be wise to err on the side of conservatism, considering carefully ecological implications of system disposal and taking an environmentally responsible position with regard to choice of system materials.

Sometimes, if properly planned, disposal of a system can provide benefit to the manufacturer. For example, recycling of certain materials might reduce costs in the long run. You should strive to be aware of such possibilities, and take advantage of opportunities as they arise. Include them in your list of options.

2.8 PARING THE OPTION SPACE

The cost of analyzing alternatives in the option space can be quite high. Therefore, even while we are defining the option space, we are continuously looking for ways of reducing its size. A word of caution is appropriate here: *Do not throw out the baby with the bath water.* By this, we mean be careful in paring options from the option space. History shows that some rather good options have been thrown out at different times in the design process. On the other hand, it is costly to retain for analysis obviously poor options. Sometimes the characteristics of the option space make it possible to pare whole sets of options. This is the case when one choice is independent of the others. Again, in the example of the gears, perhaps the other design parameters are independent of the choice of attaching the driving gear to its shaft. If this is the case, and if the selection of an attachment method is clear, then two-thirds of the options can be eliminated.

When the choice of one system design option does not affect the choice of other options, we can say that that option is *independent* of the others. Unfortunately, it is not always easy to know whether a particular option is independent of other choices and can be analyzed without consideration of them. If it is, however, a choice of the independent option can be made without regard to selection of other options. We will discuss analysis of options in much detail later.

Feasibility Analysis

One approach to paring options is the use of feasibility analyses. Let us go back to the example of guessing how many beans there are in a jar. Our first assessment of the option space was that it consists of the set of real positive integers from one to infinity. But the very simple model given in Chap. 1 gives us great confidence that we can bound the option space to a much smaller set of numbers. Perhaps our first estimate is on the order of 1,000 beans in the jar. We could examine the assumptions we would have to make to change this estimate by, say, an order of magnitude. In many cases, these assumptions will not be reasonable, and we could say with great confidence that the number of beans in the jar is between, say 100 and 10,000. In this case, a simple feasibility analysis has reduced the size of the option space to a rather small set.

One of the problems that we observed in defining the option space is that we can very quickly define far too many possible system configurations to be able to give each careful consideration. Thus, even while we think up such options, we might want to perform simple feasibility analyses to determine if the options are sufficiently competitive that we should keep them in the option space or discard them immediately. Throughout this book, we will be presenting methods of analysis that are applicable to making such choices. But in the end, there is always a lack of precision and completeness in any model. Feasibility models, by the nature of their simplicity, are less precise and complete than more detailed models. Thus, their use is more risky. In light of this aspect of feasibility models, their use actually requires more skill than the use of more detailed models. In the end, despite all our rules, notions, and teachings, the use of feasibility models is largely an art, and it requires practice to apply these models effectively.

2.9 SUMMARY

In this chapter, we have examined the option space. We looked at the creation of options in the physical elements of a system, and in the nonphysical elements. We have taken a total life cycle approach to the development of options, urging consideration of options in all phases of system design, manufacture, deployment, operation, maintenance and repair, and end-of-life disposal. We have shown the use of cartoons as an aid to developing options in the deployment phase of a system. They are useful also in other phases. Finally, we examined the importance of feasibility analysis as a means of paring the option space.

In the end, it is generally impossible to define the entire option space for a given engineering system. And even if we could, we probably could not deal with it. Often, however, we fail to recognize the breadth of options that is available. We particularly fail to see design options in the nonphysical elements of systems. The real lesson in this chapter is that we need to open our eyes to these options, to recognize that they exist, and to take advantage of them.

CLASS PROJECT

The class may be divided into teams of 10 to 15 students each. Each team should pick a system and develop as complete an option space as possible. Students should each work on a different part of the system or system life cycle, contributing their efforts to the total product. The option space should be pared using feasibility analyses as appropriate. Each team should then present their results to the remainder of the class. This assignment should be completed in one week. The time limit emphasizes the level of detail expected.

REFERENCE

1. Blanchard, B. S. and W. J. Fabrycky, *Systems Engineering and Analysis*, 2nd ed., Prentice Hall "International Series in Industrial and Systems Engineering," Englewood Cliffs, NJ, 1990.

PROBLEMS

1. Give examples of five physical system elements and five nonphysical system elements. What is the most significant difference between the physical and the nonphysical system elements?
2. Choose a system and state the objectives of the system.
3. Consider taking a vacation. For your vacation, determine an option space. Do not attempt to obtain a complete option space, but consider how options arise in various phases of the vacation, such as transportation to and from the vacation spot and daily activities. Illustrate your nominal vacation with a cartoon.
4. Choose a simple, convenient, engineering system that is near you at this moment. It could be a pencil, a pair of scissors, a computer diskette, or the chair you are sitting in, for example. Discuss the option space for the system. List some design options that are different from the system you have chosen.

5. Identify five examples of packaged complexity. Try to find examples that you use as a part of a larger system. Can you identify all five examples as being present in the room that you occupy at this moment?
6. You have been called upon to design a system that will provide access to a building for disabled persons. The building in question is comprised of a single story with doors about two meters above the ground level. What are your design options?
7. Consider the bean jar example given in Chap. 1. Perform a feasibility analysis on the option space for this example, and pare the option space in accordance with the results of your feasibility analysis.
8. It is summer, and you are called upon as a consultant to a bathing club. Their pool has developed a leak, and they want advice on what to do. What are the options?
9. You are an engineer for a company that distributes natural gas. It is an unusually cold winter, and your company's distribution system cannot flow enough gas to meet the demand. What are the company's options? Hint: Think about nonphysical options.
10. You are a traffic engineer for the town in which you live. Several complaints have been voiced that rush-hour traffic is too heavy. What actions might the town council take to relieve the problem?
11. Draw cartoons showing the delivery of a product from factory to end user. Consider, for example, a car or a TV, each delivered from a domestic and a foreign supplier. Use your cartoons to identify options.
12. Pick an engineering system that you do not like. Suggest alternatives to the current design.
13. Identify as many nonphysical options in the design of an automobile as you can in 10 minutes. Compare your list to others in your class.
14. Draw a cartoon for getting to work or to an early morning class. Identify in the cartoon things that could go wrong causing you to be late.
15. Many engineering schools conduct an egg-drop contest in which students design devices to protect an egg in a fall of, say, 10 stories onto a pavement or other hard surface. The students then build and test their devices. Develop for this contest a taxonomy of options.

3

Fundamentals of Probability Theory

3.1 THE PURPOSE OF THIS CHAPTER

As soon as we begin to examine systems engineering as a decision-making process, we find that uncertainty plays a major role. Indeed, uncertainty is an inherent part of almost every engineering decision we make. In most of our decisions, we cope with uncertainty intuitively. But rigorous models of physical phenomena must account for uncertainty explicitly. The mathematics that deals with uncertainty is probability theory, and the purpose of this chapter is to introduce some basic concepts of this mathematics. It is not our goal here to provide a rigorous and thorough development of probability theory. Rather, we will discuss only those topics that are important to systems engineering, and we will not be rigorous in their development. If you are interested in a more thorough development of this subject, you are referred to the many excellent texts that are available. On the other hand, we show here the effect of uncertainty on our estimation of parameters relating to engineering design. In particular, we will see that failure to account rigorously for uncertainty can lead to errors of potentially unbounded magnitude.

3.2 THE CONCEPT OF PROBABILITY

To begin our study of probability theory, we need to define the word *set*. A set is simply any collection of things that we choose. For example, we could choose as our set the integers from one to ten. The word set then refers to this collection as we define it, in this case the integers one, two, three, four, five, six, seven, eight, nine, and ten. We say that this set has ten *elements*, which are the integers in the set, and we write the set $\{1, 2, 3, 4, 5, 6, 7, 8, 9, 10\}$. Another set could be all even integers from one to ten. This set has five elements. Sets do not have to be numerical quantities. We could define as our set all of the stations on a particular subway line. Or we could define it as the stations on two intersecting lines. Notice, in the last case, that the number of elements in the set is one less than the sum of the numbers of stations on each of the lines individually. This is because one station is shared by both lines. We also need to define the word *subset.* A subset is any collection of things taken from a set. So, for example, the integers two, four, six, eight, and ten comprise the subset of even integers taken from the set of integers from one to ten. Finally, a *null set*, generally given the symbol Ø, is any set less all of its elements. For example, the set of all integers from one to ten less the set of all odd integers from one to ten and less the set of all even integers from one to ten is the null set. The null set is a set with no elements.

Let us return to the example of Chap. 1, beans in a jar. The number of beans in a jar is an integer. Now, suppose we count the beans in several jars. The number of beans in any particular jar can be either odd or even. We could group the jars according to whether the number of beans in them is odd or even, and we can represent these groupings pictorially in what is called a Venn diagram, as shown in Fig. 3.1. In this case, the set we are

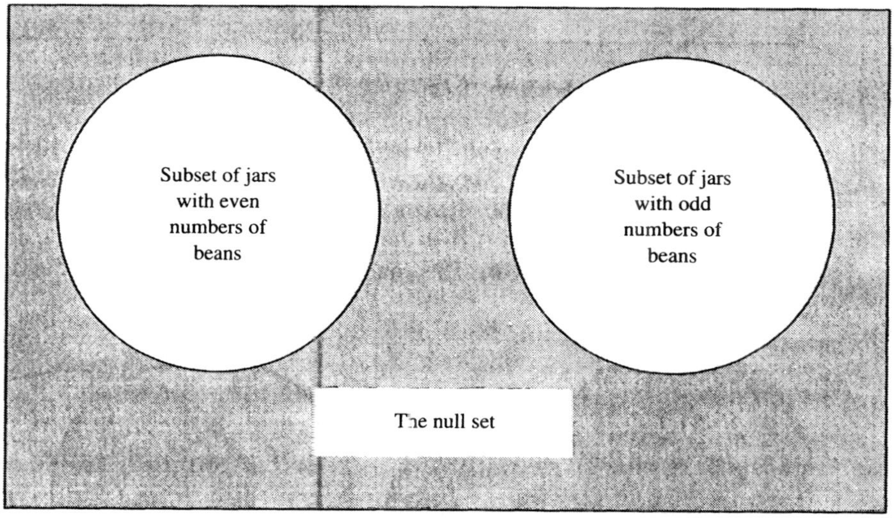

Figure 3.1 A Venn diagram for mutually exclusive sets.

dealing with is the set of positive integers representing the number of beans observed in each of the jars. This set is divided into two subsets, observations that are even numbers and observations that are odd numbers. We can see that each jar has an observed number of beans that is either odd or even, but not both, and thus we refer to the events odd and even as *disjoint events.* Each jar belongs to only one subset. Put differently, no element of one subset is an element of the other subset as well. Therefore, we say that the subsets are *mutually exclusive*. We could, instead, look at two different subsets. Let us take as one subset all jars with an even number of beans, and as the other subset all jars with fewer than 1,000 beans, as shown in Fig. 3.2. Certainly, jars with fewer than 1,000 beans can have an even number of beans. Thus, some elements of each of these subsets are elements of the other subset as well. Therefore, these subsets are not mutually exclusive.

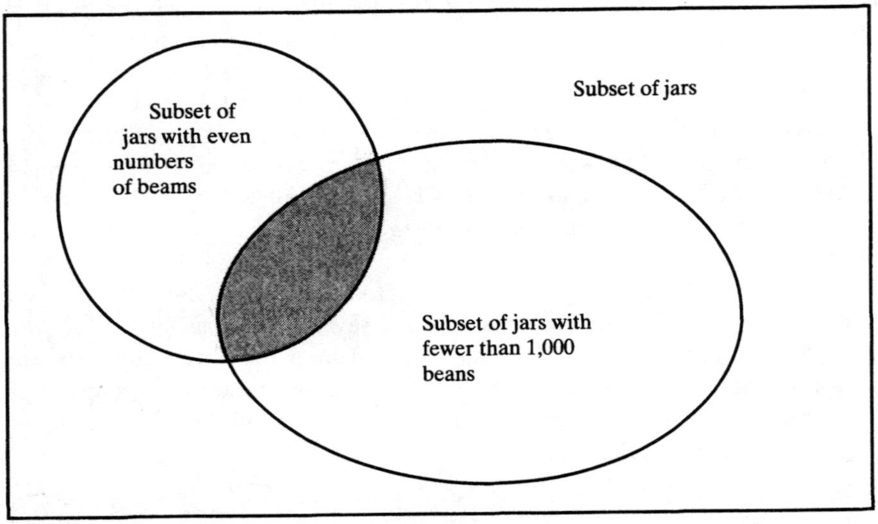

Figure 3.2 A Venn diagram for intersecting sets.

Observing that subsets can be mutually exclusive or not, we now introduce the notion of union and intersection. The *union* of two sets, Set A and Set B, written $A \cup B$, is the set that consists of all of the elements of both sets A and B. The intersection of two sets, Set A and Set B, written $A \cap B$, is the set that consists of all of the elements of Set A that are also elements of Set B. The concepts of union and intersection are shown in Fig. 3.3.

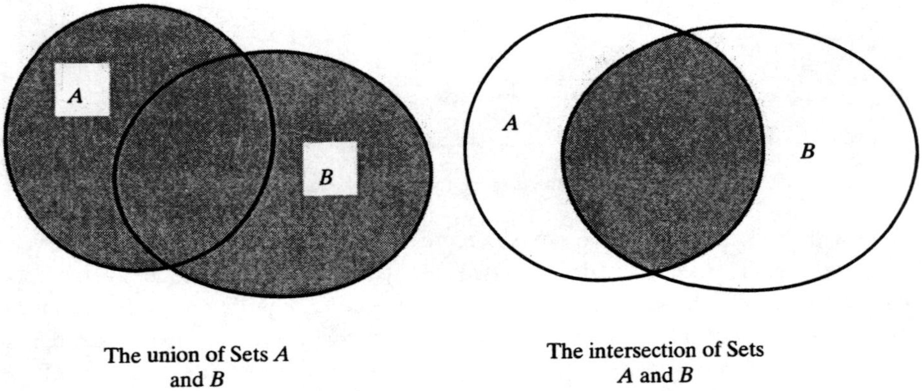

Figure 3.3 Illustration of the union and intersection of sets.

Now we can turn to the concept of probability. To begin, we will define the word *experiment*. An experiment is the process by which we observe or measure something. For example, an experiment might be the toss of a coin, and we observe the outcome of the experiment as heads or tails. Another experiment might be the measurement of the height of your desk using a ruler, and possible outcomes are numerical heights. Key to the concept of an experiment is that we do not know with certainty in advance of the experiment what will be the outcome of the experiment. We refer to the set of all possible outcomes of an experiment as the *sample space*, *S*. Next, we define a *simple event* as the outcome of an experiment that cannot be decomposed. In the case of a coin toss, the simple events that comprise the sample space are heads (H) and tails (T), $S = \{H,T\}$. In the case of two successive coin tosses, the simple events that comprise the sample space are HH, HT, TH, and TT, $S = \{HH, HT, TH, TT\}$. Finally, we define an event as a subset of the sample space. For the case of two successive coin tosses, we could define three events: two heads, one head and one tail, and two tails. Note that the second event is obtained by either of the simple events HT or TH. The concepts of simple event and event are illustrated in Fig. 3.4. We now observe that probability is a nonphysical attribute of the simple events that comprise the sample space of an experiment. As we shall see, it can be interpreted as the relative frequency of occurrence of each element of this space.

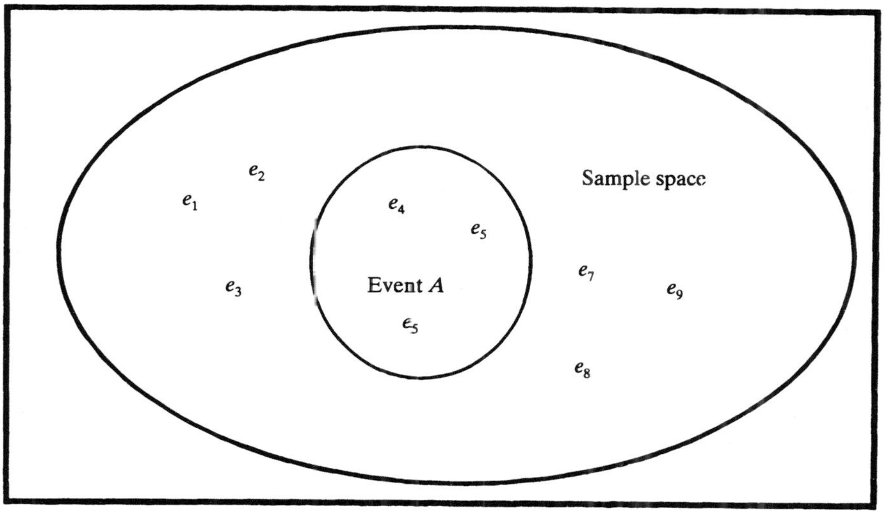

Figure 3.4 Illustration of a sample space showing simple events and an event.

As engineers, we are familiar with many types of experiments. Some examples include the following:

Experiment: Test the tensile strength of a new material to see if it is strong enough for a given purpose.
  Simple event: A measured tensile strength.
  Sample space: Positive real numbers.
  Event: Pass or fail.

Experiment: Conduct an acceptance test of an electronic device.
  Simple event: A condition of the device.

Sample space: All possible conditions of the device.
Event: Pass or fail.

Experiment: Measuring fuel consumption of an engine to determine if it passes a fuel economy test.
Simple event: A measured value of fuel consumption.
Sample space: Positive real numbers.
Event: Pass or fail.

Experiment: Counting the number of beans in a jar to see if yours is the winning guess.
Simple event: A number of beans.
Sample space: Positive integers.
Event: Win or lose.

In any of these cases, if we repeat the experiment, we can get a different result.

Any time that we conduct an experiment for which we cannot predict the outcome, that is, when the sample space contains more than one element with nonzero probability, we say that there is ***uncertainty***.

Uncertainty can be the result of any of the following reasons:

1. We may not understand the experiment, that is, the process we are observing, adequately to be able to predict its outcome. This is to say that our knowledge of the laws of nature may be insufficient. A current example deals with our ability to predict the properties of new, high temperature, superconducting materials.

2. Even though we may understand the process or the laws of nature that apply with sufficient detail, we may not have the computational capabilities needed to accurately assess the outcome of the experiment. Prediction of the airflow over the surface of an airplane in various flow regimes is an example of this deficiency. Another is the computation of the characteristics of an electric power system grid.

3. We may not be able to measure the initial conditions, the boundary conditions, or the inputs to the process with sufficient accuracy, or we may not be able to replicate the experimental conditions with sufficient precision. A coin toss is a typical example. If we could flip the coin exactly the same each time, we would expect it to land on the same side every time. But very small changes in the inputs to this experiment cause the outcome to be uncertain. Chaos theory deals with problems of this sort.

4. Some processes may actually be inherently random. Although some scientists argue that this may appear to be the case whenever we lack an understanding of nature, it appears that a fundamental nature of some systems is their randomness. This is the subject of quantum mechanics, for example.

These are the causes of uncertainty with which an engineer must contend.

Now we can define the word *probability*. A probability is a number, $P(e_i)$, associated with each element e_i of a sample space, that obeys the following conditions:

$$0 \le P(e_i) \le 1$$

$$\sum_{i=1}^{m} P(e_i) = 1$$

where m is the number of elements in the sample space, and

$$P(A) = \sum_{i \in A} P(e_i)$$

In plain language, a probability is a number between zero and one, such that the sum of the probabilities for all elements in a sample space equals one and such that the probability of any subset, $P(A)$, of the sample space is the sum of the probabilities of the elements in the subset. Taking as our set the events associated with an experiment, the probability of an event is a measure of the likelihood of the occurrence of the event. $P(e_i) = 0$ infers that the event will not occur. $P(e_i) = 1$ infers that the event is certain to occur, and the condition that the sum of the probabilities of all elements in the sample space equals one assures that an experiment must result in an outcome that is contained in the sample space. So, for example, if we say that the sample space for a coin toss consists of H and T, we deny the possibility that the coin can land on its edge. Note that the numerical limits, 0 and 1, that we place on probabilities are arbitrarily chosen. The advantage of 0 and 1 is that they provide an easy interpretation of probability numbers.

To be sure, any experiment we might conduct can be conducted only once. And the outcome of that experiment is unique. But, for many things, we feel that we understand nature sufficiently to repeat an experiment, isolating the successive experiments from influences that would cause them to differ structurally. This is the essence of the hard sciences. So, for example, we might flip a coin ten times, with the expectation that each of the flips is a similar experiment. It is not at all obvious that this can be done. And indeed, researchers have conducted experiments that they thought were similar, but in fact were not. In the coin-flip example, we would not think the experiments were similar if we used bent coins and if we used a different coin each time. This is because we sense that the bend in the coin will somehow influence the outcome of the flip. Hidden factors can always influence an experiment, and if these factors differ from experiment to experiment, then the experiments are not similar.

Our ability to repeat an experiment depends on our knowledge of nature to assure that the successive experiments are indeed similar, that is, that hidden factors do not influence the outcomes.

Given that we can conduct a series of similar experiments, we can now refer to two different types of probability: *objective* and *subjective*. An objective probability is one that is measured in terms of the outcomes of a series of similar experiments. The probability of heads in a coin flip is an example. We could flip a coin 1,000 times and count the number, k, of times heads appears. Then we could use as an estimate of the probability of heads on the next toss, $P(H)$, the ratio $k/1{,}000$. We have measured $P(H)$ by means of repeated experiments. But it is important to keep in mind that this is only a measurement or estimate; it is not precisely the probability of heads on the next toss. For example, k could be 487 for one set of 1,000 flips and 516 for another.

Some experiments simply cannot be repeated. Consider the question, what will be the average price of 87 octane gasoline ten years from today? To be sure, we may know a lot about the history of gasoline prices. But we cannot predict with certainty the price of gasoline ten years in advance. Furthermore, the day in question is one particular day. We can only wait until that day arrives and observe the price of gasoline to know it. At that point, we cannot go back and conduct the experiment over. Yet, we might make an assessment of the probabilities of various prices based on our experience. This we will call a subjective probability. Another example of a subjective probability is the probability assigned to a particular outcome of a scientific experiment or measurement never before conducted.

Subjective probabilities derive from our intuition. Objective probabilities are based on the notion that the future will be just like the past, that is that measurements made in past experiments can be applied to the future.

We now return to the Venn diagrams of the above figures to explore more properties of probabilities. Consider two events, A and B, that are mutually exclusive. Then the probability that one or the other event occurs is given by

$$P(A \cup B) = P(A) + P(B)$$

In the case that the events are not mutually exclusive, the probability that both A and B occur is given by

$$P(A \cap B) = P(A\,B)$$

This then leads to the probability that either event A or event B or both occur in the general case

$$P(A \cup B) = P(A) + P(B) - P(A\,B)$$

Another important case deals with *conditional probabilities*. We ask the question, what is the probability that event A occurs given that event B occurs, namely $P(A|B)$, which is read the probability of A conditional upon B, or A given B. In answer to this question, we introduce the notion of independence. We say that events A and B are *independent* if and only if their probability of joint occurrence $P(AB)$ is given by

$$P(A\,B) = P(A)\,P(B)$$

In general, however,

$$P(A\,B) = P(A \mid B)\,P(B)$$

which can be written in the form

$$P(A \mid B) = \frac{P(A\,B)}{P(B)}$$

Note that, in the case that A and B are independent.

$$P(A \mid B) = \frac{P(AB)}{P(B)} = \frac{P(A)\,P(B)}{P(B)} = P(A)$$

3.3 PROPERTIES OF A RANDOM VARIABLE

We return once more to the example of estimating the number of beans in a jar. We defined in Chap. 1 the variable n to be an estimate of the true number of beans, N. In fact n is a set of numbers, consisting of positive integers, each representing a guess at the true number N, and each with an assigned probability. We can refer to the variable n as a quantitative variable. Some events are not quantitative. Take as an example the weather in Washington, D.C. Suppose we were concerned whether the sun is shining or not. The events *sunny* or *cloudy* are not quantitative. We refer to them as qualitative. Yet we can define these events in a quantitative manner if we define a variable x such that $x = 1$ if *sunny* occurs, and $x = 2$ if *cloudy* occurs. In these cases we refer to the variables n and x as random variables.

> *A* ***random variable*** *is a numerically valued function representing the set of events that can result from an experiment.*

There are two types of random variables, *discrete* and *continuous*. A discrete random variable is one that is countable. That is, there is a one-to-one mapping between the number set 1, 2, 3, 4, ... and the sequential values that the random variable can assume. A continuous random variable can assume an infinite number of values over a given range, such as the set of positive numbers from one to ten, and is thus uncountable.

Associated with each value of a discrete random variable, or with each range of a continuous random variable, is a probability. The set of probabilities for all possible values (or ranges of values) of the random variable give the likelihoods of each value (or each range of values) occurring as the result of an experiment. With this notion of probability, we can now define the concept of a probability distribution. We will begin with discrete random variables. Consider the toss of a fair coin. The possible events are heads and tails. Each has a probability of 0.5. If we define the random variable, x, which can assume two values, 1 representing heads and 2 representing tails, we can then draw the probability distribution of x as shown in Fig. 3.5. Next, we can draw a cumulative distribution function (cdf) by asking the question, what is the probability that an outcome X, drawn from x is less than or equal to some number, x, namely, what is $P(X \le x)$? We draw the cdf as shown in Fig. 3.6. This cdf tells us that there is zero probability that $X < 1$, probability 0.5 that $X < 2$, and probability 1 that $X \le 2$. We will later make considerable use of cdfs in system simulation.

Now consider an experiment that is not limited to a finite number of possible outcomes. Throwing darts at a target is an example. In this case, let X be the distance by which a dart misses the center of the target. The random variable x is a continuous random variable. Now we must ask the question, what is the probability that X will lie between a lower and upper bound, x and $x + \delta x$? We can look at this question as δx approaches zero:

Figure 3.5 A discrete distribution.

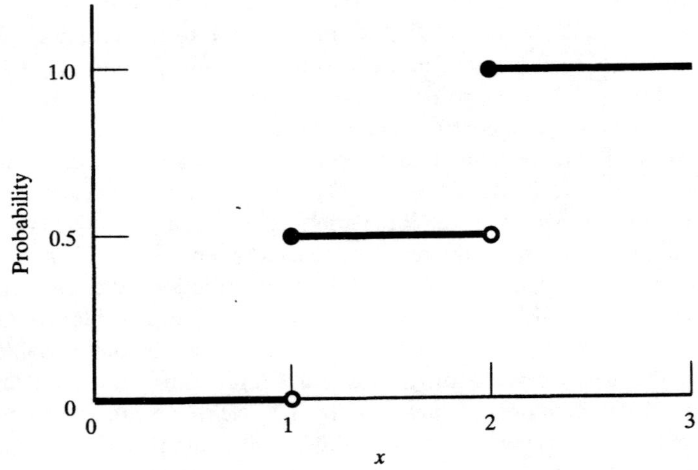

Figure 3.6 A cdf for a discrete random variable.

$$f(x) = \lim \frac{P(X \leq x + \delta x) - P(X \leq x)}{\delta x}$$

We refer to $f(x)$ as the probability density function, pdf, of x. The pdf is the derivative of the cdf, and again, it is customary to write it in terms of the random variable. Thus the integral of the pdf over the full range of the random variable must be unity, assuring that the random variable takes on values always within its allowable range. We will illustrate several common pdfs later in this chapter.

Probability distributions have several properties that are of importance. We define a few of them here:

Moments of a distribution—The first moment of a pdf is given by the equation

$$E\{x\} = \int_{-\infty}^{+\infty} x\, f(x)\, dx = m$$

The first moment of a pdf is called the *mean* of the distribution, and is generally given the symbol m. It is also sometimes referred to as the expected value and written $E\{x\}$. The jth moment of a distribution, that is, the expected value of x^j, is given by

$$E\{x^j\} = \int_{-\infty}^{+\infty} x^j f(x)\, dx$$

The jth *central moment* of a pdf, the moment about the mean, for $j \geq 2$ is given by the equation

$$M_j = \int_{-\infty}^{+\infty} (x - m)^j f(x)\, dx$$

The second central moment, $j = 2$, is called the *variance*, $V\{x\}$. It measures the extent to which the random variable tends to vary from the mean. The square root of the variance is called the *standard deviation*, σ. If σ is zero, there is no randomness in the random variable, and the only value that it takes on is the mean. The third moment, $j = 3$, is called the *kurtosis*, and it reflects the degree of asymmetry in the pdf. A pdf that is symmetrical about its mean has a kurtosis of zero, but a kurtosis of zero does not guarantee symmetry.

Moments of a discrete distribution are similarly defined. The first moment is given by

$$E\{x\} = \sum_i P(X_i)X_i = m$$

summed over all i. This quantity is referred to as the mean or average value of x. Likewise, the higher central moments are given by

$$M_j = \sum_i P(x_i)\,(x_i - m)^j$$

and from here the definitions of variance, standard deviation, and kurtosis are the same as before.

A convenient relationship exists between $E\{x^2\}$ and $V\{x\}$. Note that

$$V\{x\} = \int_{-\infty}^{+\infty} (x - m)^2 f(x)\, dx$$

$$v\{x\} = \int_{-\infty}^{+\infty} (x^2 - 2xm + m^2) f(x)\, dx$$

But

$$E\{x^2\} = \int_{-\infty}^{+\infty} x^2 f(x)\, dx$$

and

$$m = \int_{-\infty}^{+\infty} x f(x)\, dx$$

Thus,

$$V\{x\} = E\{x^2\} - m^2$$

One more property of probability distributions is of interest, the *most likely* value. The most likely value of a discrete random variable is the value of the random variable that has the highest associated probability. For a continuous random variable, the *most likely* value is the value of the random variable for which the pdf is a maximum. *Note that, in general, the mean or average value of a random variable is not the same as the most likely value.*

Finally, we should recognize the meaning of the area under a section of a pdf. This area represents the probability of the random variable lying within the bounds of the area:

$$P(x_a \leq x \leq x_b) = \int_{x_a}^{x_b} f(x)\, dx$$

3.4 THE MATHEMATICS OF RANDOM VARIABLES

Engineers frequently develop and use mathematical models that involve parameters that are measured or estimated. It is important to recognize that these parameters are, in reality, random variables. Failure to treat them as such is mathematically incorrect, and it can lead to serious modeling errors. Therefore, we will take the time here to develop the arithmetic of random variables.

Let x and y be independent continuous random variables with pdfs $f(x)$ and $g(y)$. We will seek first to determine the sum $x + y$. Because both x and y are random variables, we must realize that their sum is not a unique number, rather the sum is also a random variable. So we must think about what we mean when we ask for the sum of two random variables. To be complete, we would obtain the pdf of the sum. But let us start with only the mean or expected value of the sum. The expected value of the sum of two independent random variables is given by

$$E\{x + y\} = \int_{-\infty}^{+\infty}\int_{-\infty}^{+\infty} (x + y) f(x)\, g(y)\, dx\, dy$$

$$= \int_{-\infty}^{+\infty}\int_{-\infty}^{+\infty} xf(x)\, g(y)\, dx\, dy + \int_{-\infty}^{+\infty}\int_{-\infty}^{+\infty} yf(x)\, g(y)\, dx\, dy$$

We have, for example,

$$\int_{-\infty}^{+\infty} xf(x)\, g(y)\, dx\, dy = \int_{-\infty}^{+\infty} xf(x)\, dx \int_{-\infty}^{+\infty} g(y)\, dy$$

$$= E\{x\} \times 1$$

recognizing that the first integral on the right-hand side of the above equation is the expected value of x, and the second integral is one. It follows that

$$E\{x + y\} = E\{x\} + E\{y\}$$

We emphasize here that the expected value of the sum of random numbers is the sum of the expected values of the random numbers. It is **not** in general the sum of the most likely values. The misuse of most likely values, typically given as best guesses, can result in serious errors in estimating the expected value of a sum.

Substituting $-y$ in place of y in the above equations leads to the result that

$$E\{x - y\} = E\{x\} - E\{y\}$$

And again we warn you to be careful not to use most likely values in place of expected values in these equations.

We can use a similar procedure to derive expressions for the multiplication and division of two random variables. For multiplication

$$E\{xy\} = \int_{-\infty}^{+\infty}\int_{-\infty}^{+\infty} xyf(x)\, g(y)\, dx\, dy$$

$$= \int_{-\infty}^{+\infty} xf(x)\, dx \int_{-\infty}^{+\infty} yg(y)\, dy$$

$$= E\{x\}\, E\{y\}$$

And, finally, for division, we substitute $1/y$ in place of y, and we obtain

$$E\left\{\frac{x}{y}\right\} = E\{x\}\, E\left\{\frac{1}{y}\right\}$$

Now we see a potentially serious problem. The expected value of $1/y$ is not the same as $1/E\{y\}$. Thus, it is not true that the expected value of x divided by y is the expected value of x divided by the expected value of y. Consider the simple example where y is a random variable with $g(y) = 1$ over the range 0 to 1, and zero elsewhere. It is easy to show that the expected value of y is 0.5.

$$
\begin{aligned}
E\{y\} &= \int_0^1 y\,dy \\
&= \left[\frac{y^2}{2}\right]_0^1 \\
&= 0.5
\end{aligned}
$$

But the expected value of $1/y$ for the same pdf is quite different:

$$
\begin{aligned}
E\left\{\frac{1}{y}\right\} &= \int_0^1 \frac{1}{y}\,dy \\
&= [\ln y]_0^1 \\
&= \infty
\end{aligned}
$$

This result should give us considerable cause for concern. In our everyday work as engineers, we frequently use models that include division by a random variable. $F = ma$ is an example, written to solve for a

$$a = \frac{F}{m}$$

any time that we do not know m precisely. Consider the model we used in Chap. 1 to calculate the number of beans in a jar. It has in the denominator the estimated volume of a bean. As you looked at this model in Chap. 1, you probably were not concerned about the fact that you would be using the model incorrectly by inserting best guesses into it for V_j, V_b, and f, or that this error could result in an estimate of n that could be far from the correct best guess. It is only by good fortune that we get reasonable answers from such models when we neglect uncertainty. But, at any time, errors can creep in that destroy any semblance of accuracy. ***To be assured that a model involving random variables gives reasonably accurate results, we must take into account randomness.*** The forthcoming chapters of this book will discuss how to do this.

An exercise for the student is to extend the above results to the higher-order moments. Using the same approach, and for the same arithmetic operations, derive the following equations for the variance of the arithmetical operations:

$$
\begin{aligned}
V\{x + y\} &= V\{x\} + V\{y\} \\
V\{x - y\} &= V\{x\} + V\{y\}
\end{aligned}
$$

$$V\{xy\} = V\{x\}\,V\{y\} + E^2\{x\}\,V\{y\} + E^2\{y\}\,V\{x\}$$

$$V\left\{\frac{x}{y}\right\} = V\{x\}\,V\left\{\frac{1}{y}\right\} + E^2\{x\}\,V\left\{\frac{1}{y}\right\} + E^2\left\{\frac{1}{y}\right\}V\{x\}$$

In the above, we have considered only independent random variables, and we have examined only two properties of arithmetic operations on them, expected value and variance. Yet, even in this case, we see cause for alarm and need for extreme caution whenever we use deterministic models, that is, models that do not account for randomness. In the case that the random variables of a model are not independent, or for more complex functions of the random variables, the mathematics is far more difficult. For this reason, we will not continue to develop the formal mathematics of random variables. In the next chapter, however, we will develop a general methodology for the solution, by means of computer simulation, of expressions involving random variables.

Two final properties of random variables are worthy of mention. The first is known as the *strong law of large numbers*. We take a sequence of random variables selected from a common distribution, $f(x)$. Let $X_1, X_2, X_3, \ldots, X_n$ denote the variables, and let m denote the mean of the distribution. Then, the strong law of large numbers states that

$$\frac{\sum_{i=1}^{n} X_i}{n} = m \qquad \text{as } n \rightarrow \infty$$

This law says that the mean value of a sequence of random variables selected from a distribution tends precisely to the mean of the distribution as the number of random variables becomes large.

The second property involves the *central limit theorem*. The central limit theorem states, for the same sequence of random variables,

$$P\left\{\frac{\sum_{i=1}^{n} Xi - nm}{\sigma\sqrt{n}} \leq a\right\} \rightarrow \frac{1}{\sqrt{2\pi}}\int_{-\infty}^{a} e^{-x^2/2}\,dx$$

This is such an interesting result because the form of $f(x)$ from which the X_i were taken was not specified. It says that the distribution of the sum of a sequence of random variables tends toward a particular form regardless of the form of the original distribution. This form we shall soon recognize as the very important standard normal distribution. The central limit theorem explains why so many random variables found in nature exhibit a normal distribution.

The strong law of large numbers and the central limit theorem can be extremely useful in obtaining estimates of the sums or differences of random variables. Consider a typical engineering example. We seek to estimate the cost or weight of a system that we hope to build. We divide the system into subsystems, and estimate the cost or weight of each subsystem. Then we add these to get the total system cost or weight. Let C_i be the cost of the ith subsystem. Then we take the total cost, C, to be the sum of the C_i.

$$C = \sum_i C_i$$

In general, when we estimate costs, we make best guesses. Typically, we know that our guesses can be in error. It is not likely that our guesses would be high by very much. But, they could be low by a substantial amount. Suppose we bound the possible error on each guess by saying that we could be high by only as much as 10 percent, but we could be low by as much as a factor of 2. These are not uncommon numbers. Then, as you shall see later, the expected value of each C_i will exceed your best guess by about 30 percent. From the central limit theorem, we know that, independent of the distributional forms of our knowledge on each of the C_i, the distributional form of C will approach a normal distribution, which is symmetrical and for which the expected value and the best guess are the same. Thus, in this case, the sum of the best guesses of the C_i will result in an estimate of the expected value or best guess of C that is low by 30 percent. Experienced engineers know that a good rule of thumb is to make a best estimate using the above model (albeit neglecting uncertainty) and then double the result. This common factor-of-two error is explained here as the neglect of uncertainty and the mathematical errors that result.

3.5 SOME IMPORTANT DISTRIBUTIONS

According to our definitions of probabilities, density functions, and cumulative distribution functions, there are rather few restrictions on the forms these quantities may take. Probabilities are real numbers that can range from zero to one, such that the sum of probabilities for all elements in the sample space is one. For a continuous random variable, a probability density function is a real, positive function of the random variable, defined over a range such that its integral over that range is one. And a cumulative distribution function is a function of a random variable, defined over the range of the random variable, such that it is a monotonically increasing function of the random variable, with a value of zero corresponding to the minimum value of the random variable, and a value of one corresponding to the maximum value of the random variable. Aside from these restrictions, there are no limitations on the form that distributions can take. There are, however, a few very important kinds of distributions. They are important because they describe important classes of observed phenomena. We shall discuss some of these below. We shall group these distributions according to whether they are discrete or continuous.

Discrete Distributions

Uniform—The discrete uniform distribution is perhaps the simplest distribution that you will encounter. It applies to the case where there are n possible events, each of which can occur with the same probability, $1/n$. Figures 3.5 and 3.6 illustrate for $n = 2$ the discrete uniform distribution. Examples of where the discrete uniform distribution applies include the toss of a fair coin and the roll of a fair die.

Binomial—The binomial distribution applies to the case when we have repeated experiments. Consider the case of a coin toss, which has two and only two possible events, heads and tails. Now we repeat the toss n times, and count the number of, say, heads, H. H is a binomially distributed random variable. We can examine the distribution of H using a tool that we call the event tree. An event tree for the case of two successive tosses of a fair coin is shown in Fig. 3.7. Each outcome of the tree has probability 0.25. Then we can construct the following table of probabilities:

Number of heads	Probability
0	0.25
1	0.50
2	0.25

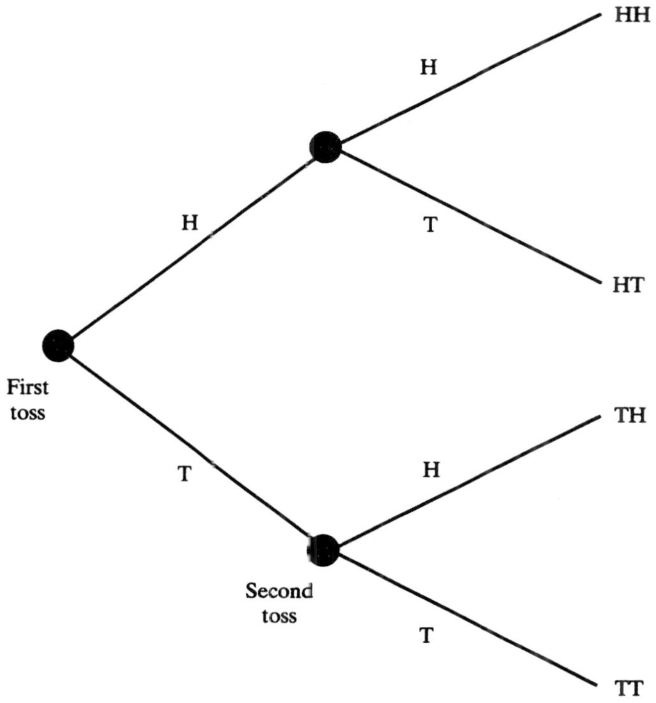

Figure 3.7 An event tree for two successive coin tosses.

Extending this to the general case, we get the binomial distribution:

$$P(x,n) = \frac{n!}{x!(n-x)!} p^x(1-p)^{n-x}$$

where p is the probability of success on any given trial, the probability of failure is $1 - p$, and x is the number of successful outcomes out of n trials. An example binomial distribution is shown in Fig. 3.8. The concept of a binomial distribution can be extended to the case where more than two outcomes on each trial are possible.

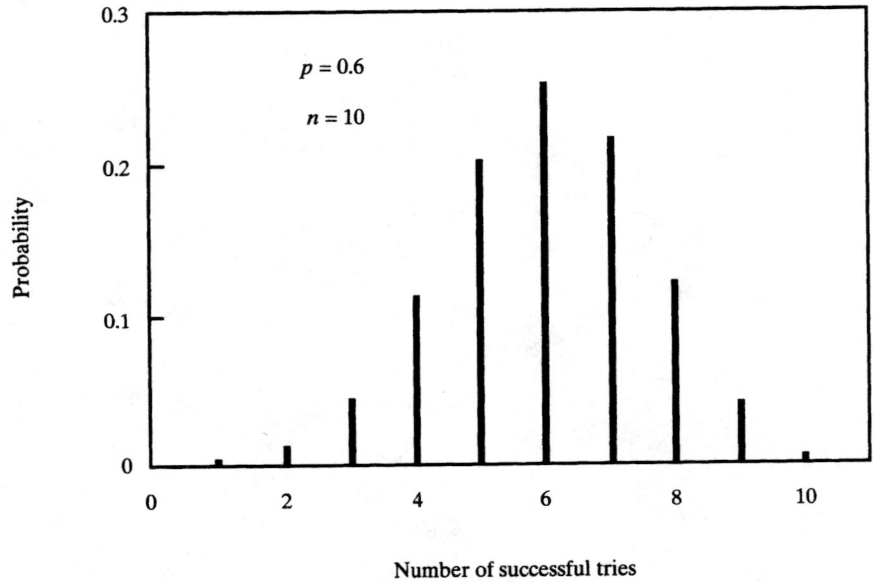

Figure 3.8 An example of a binomial distribution.

Note that, for large values of n, the factorial term in the above equation can be rather difficult to evaluate. An approximate form is available that makes evaluation possible, however. For n large, we can write

$$P(a \le x \le b) = \sum_{x=a}^{b} \frac{n!}{x!(n-x)!} p^x(1-p)^{n-x}$$

$$\sim \Phi(\beta) - \Phi(\alpha)$$

where

$$\alpha = \frac{a - np - 0.5}{\sqrt{np(1-p)}}$$

$$\beta = \frac{b - np + 0.5}{\sqrt{np(1-p)}}$$

and the function $\Phi(\gamma)$ is defined as the cumulative distribution function of the standard normal distribution.

$$\Phi(\gamma) = \frac{1}{\sqrt{2\pi}} \int_{-\infty}^{\gamma} e^{-\mu^2/2}\, d\mu$$

This integral has no convenient closed form solution. However, it can be evaluated with reasonable accuracy using the following polynomial (Abramowitz and Stegun, 1972):

$$\Phi(\gamma) = 1 - \frac{1}{2}(1 + d_1\gamma + d_2\gamma^2 - d_3\gamma^3 - d_4\gamma^4 + d_5\gamma^5 + d_6\gamma^6)^{-16} + \varepsilon(\gamma)$$

where

$$|\varepsilon(\gamma)| < 1.5 \times 10^{-7}$$

and

$$\begin{aligned} d_1 &= 0.04986\ 73470 \\ d_2 &= 0.02114\ 10061 \\ d_3 &= 0.00327\ 76263 \\ d_4 &= 0.00003\ 80036 \\ d_5 &= 0.00004\ 88906 \\ d_6 &= 0.00000\ 53830 \end{aligned}$$

Geometric—The geometric distribution also relates to repeated experiments. It addresses the question: how many trials of a random process must be attempted to obtain success? For example, how many times must we flip a coin until heads occurs? Again, we could draw an event tree such as that in Fig. 3.7, stopping the extension of branches each time we encounter a branch that has the outcome heads. From this, we obtain the equation

$$P(n) = (1 - p)^{n-1}p$$

where p is the probability of success on any given trial, and n is the number of trials until success occurs. A geometric distribution is shown in Fig. 3.9.

Poisson—The Poisson distribution is important for two reasons. First, it represents many processes that occur in nature. Second, it provides a convenient approximation to the binomial distribution for large values of n. A Poisson process can be viewed as a process that occurs over time, but with no memory. Thus, the occurrence of the next event in a Poisson process does not depend on when the last event occurred. The Poisson distribution is given by

$$P(i)\ \frac{\mu^i}{i!}\ e^{-\mu}$$

where i is the number of events occurring during a unit of time, and m is the average rate of event occurrence, as shown in Fig. 3.10.

The Poisson distribution provides a good approximation to binomial probabilities when n is large and $\mu = np$ is small. For good results, we need $np < 7$.

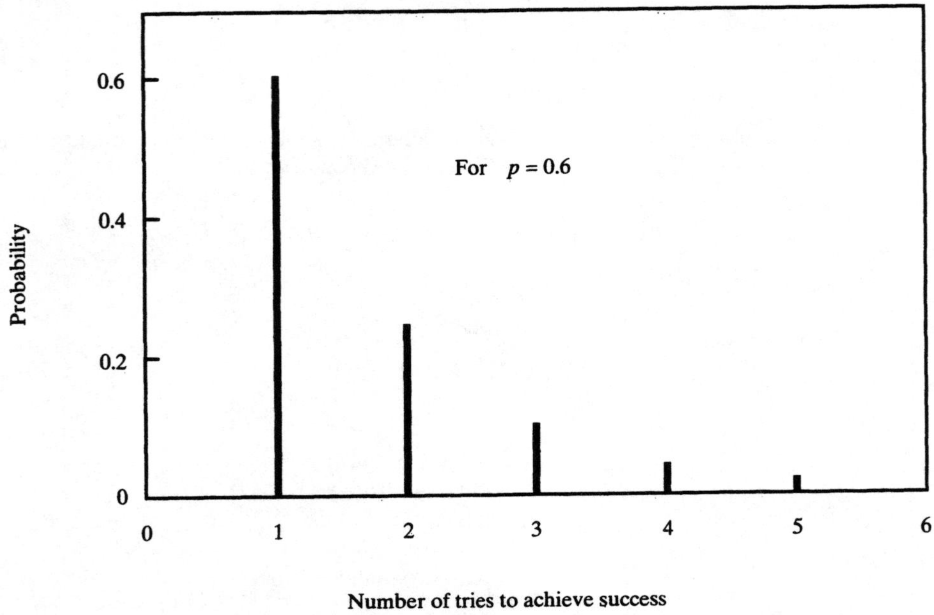

Figure 3.9 An example of a geometric distribution.

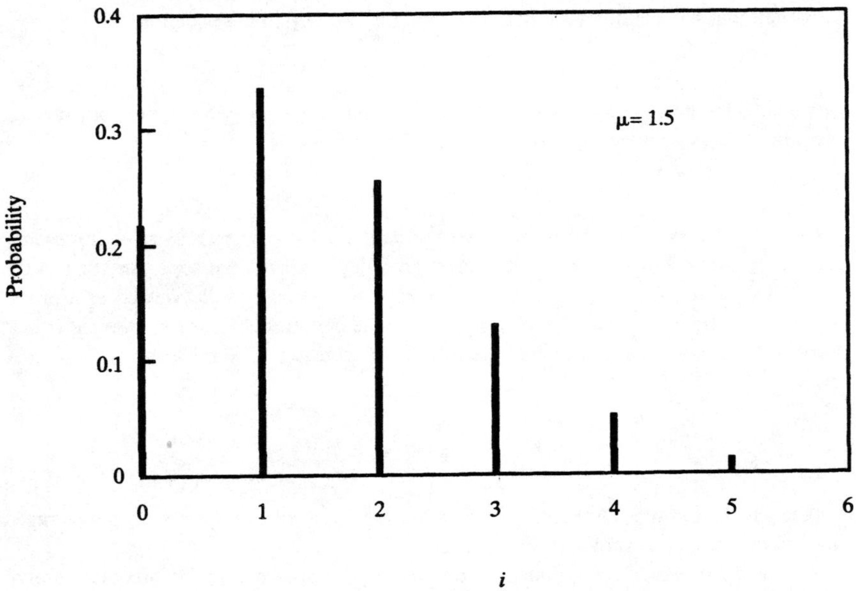

Figure 3.10 An example of a Poisson distribution.

Continuous Distributions

Uniform—The continuous uniform distribution has much the same form as the discrete uniform distribution. It is defined over an interval, a to b, such that any value of the random variable over that range is equally likely. The pdf is given by

$$f(x) = \frac{1}{b - a} \quad \text{for} \quad a \leq x \leq b$$

$$f(x) = 0 \quad \text{for} \quad x < a \text{ or } x > b$$

and the cdf is a straight line that increases from 0 to 1 over the same interval. The continuous uniform distribution is illustrated in Figs. 3.11 and 3.12.

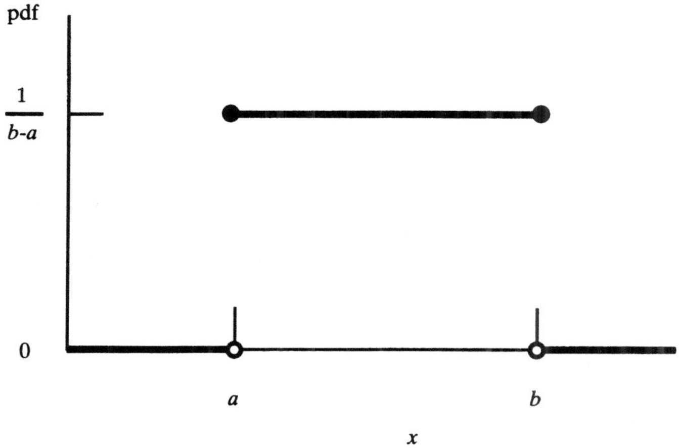

Figure 3.11 A pdf for a uniform distribution.

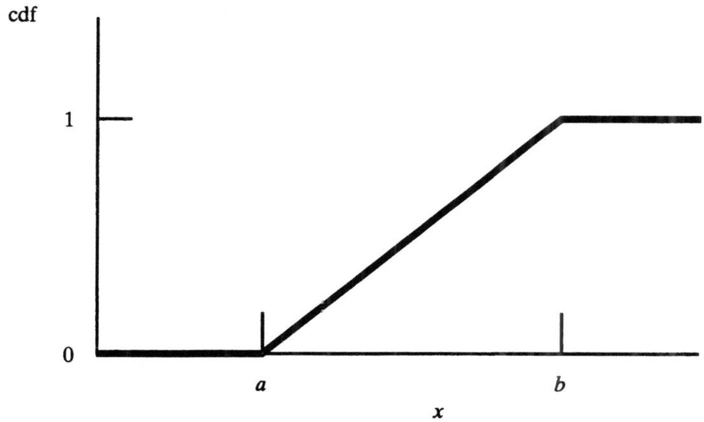

Figure 3.12 A cdf for a uniform distribution.

Normal or Gaussian—The normal distribution, also called the Gaussian distribution, is perhaps the most important of all continuous distributions, for the reason explained by the central limit theorem. Many natural phenomena tend to be described by a normal distribution. The pdf for the normal distribution has the form

$$f(x) = \frac{1}{\sqrt{\sigma 2\pi}} e^{-(x - m)^2/2\sigma^2} \text{ for } -\infty < x < \infty$$

where m is the distribution mean and σ is the standard deviation. A closed-form integral for this pdf does not exist, and thus the cdf has a very complex mathematical form. If we define

$$z = \frac{x - m}{\sigma}$$

the pdf of the normal distribution can be written in the form

$$f(z) = \frac{1}{\sqrt{2\pi}} e^{-z^2/2}$$

This distribution we call the *standard normal distribution* and an approximation to the cdf of this distribution is given above. A standard normal distribution is illustrated in Fig. 3.13. Notice that the normal distribution is rather narrow measured in terms of standard deviations. Indeed, it is practically zero for values of the random variable that are more than three standard deviations from the mean. Nonetheless, strictly speaking, the random variable spans the range from $-\infty$ to $+\infty$. This is a characteristic of the normal distribution that limits its application, for example, in cases where the random variable cannot be less than zero.

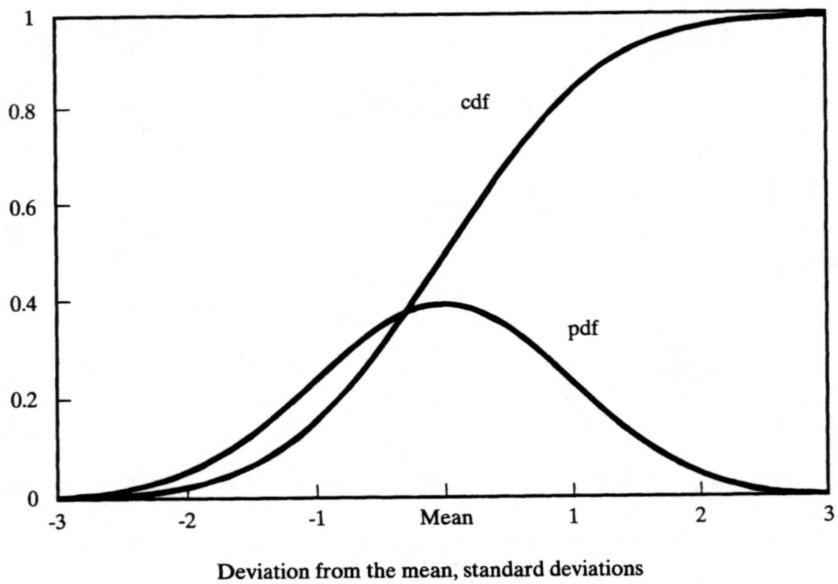

Figure 3.13 The pdf and cdf for a normal distribution.

Gamma—The normal distribution has an exponential form with the variable, x, scaled by the standard deviation, σ, and appearing in the exponent squared. This gives the normal distribution its symmetry, and also causes it to decay very rapidly as x deviates from the mean. The gamma distribution also has an exponential form, but the variable, x, appears only to the first power. The gamma distribution is defined only for positive x. Also, by the simple substitution

$$y - a = x$$

we obtain a form of the gamma distribution that is defined only for $y \geq a$. This is useful for creating distributions that represent things such as schedules, which may have zero probability up to a minimum estimated time for completion. The form of the gamma distribution is

$$f(x) = \frac{1}{\mu\Gamma(\beta)} \left(\frac{x}{\mu}\right)^{\beta - 1} e^{-x/\mu}$$

where μ is the scaling parameter for x, and β is another parameter of the distribution. The function $\Gamma(\beta)$ is called the gamma function, and is related to the factorial function.

$$\Gamma(\beta) = \int_0^\infty e^{-t}\, t^{\beta - 1}\, dt$$

For an integer, i,

$$\Gamma(i) = (i - 1)\,!$$

An example of the gamma distribution for two values of β is shown in Fig. 3.14.

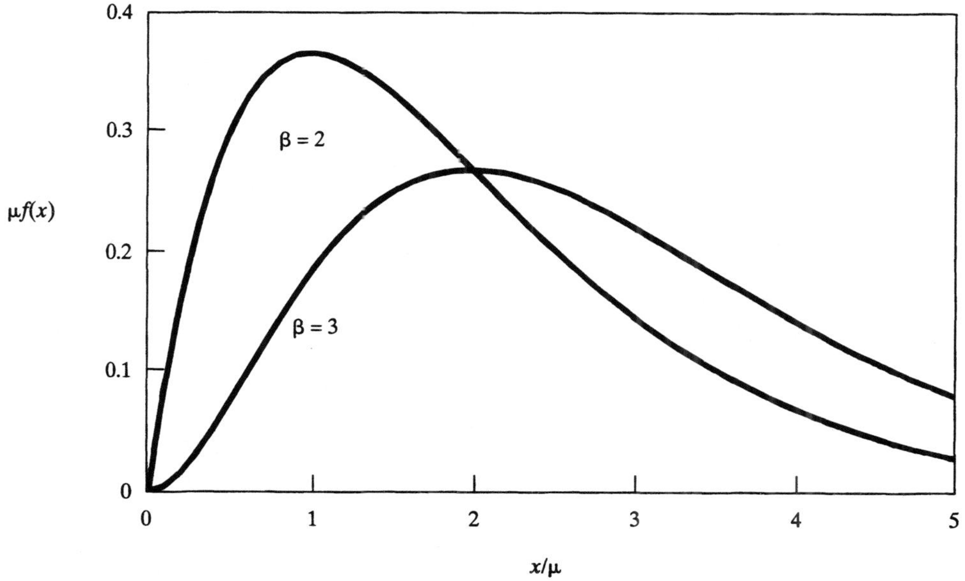

Figure 3.14 An example of gamma distributions.

A variation of the gamma distribution occurs for a value of β equal to one. The form of the distribution then reduces to

$$f(x) = \frac{1}{\mu} e^{-x/y}$$

This important form is known as the **exponential** distribution. It is closely related to the Poisson distribution, and its importance lies in the fact that it, too, is a memoryless distribution. One can interpret x as the waiting time for the occurrence of an event whose probability of occurrence is constant over all time. The next distribution, the Weibull distribution, also reduces to the exponential distribution for a value of b equal to one, thus Fig. 3.15 includes an illustration of the exponential distribution.

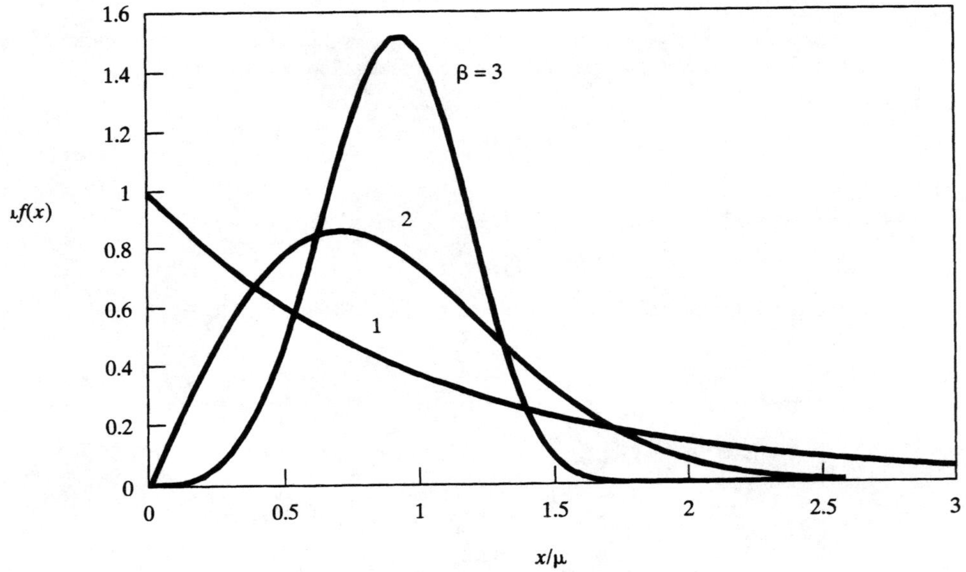

Figure 3.15 An example of Weibull distributions.

Weibull—The Weibull distribution is another exponential distribution, but it includes provision for varying the power of the parameter x in the exponent. The Weibull distribution is important because it appears to provide a good representation of the frequency distribution of failures for many types of equipment. It is particularly popular for failure modeling in the electric power industry. The Weibull distribution is defined for positive x, and has the form

$$f(x) = \frac{\beta}{\mu}\left(\frac{x}{\mu}\right)^{\beta - 1} e^{-(x/\mu)\beta}$$

where μ and β are parameters similar to those in the gamma distribution. It is easy to see that this form reduces to the exponential distribution for the case of β equal to one. It also bears considerable resemblance to the normal distribution for β equal to two, except that it is nonzero only for positive x. An illustration of the Weibull distribution is given in Fig. 3.15.

3.6 BAYES' FORMULA

An important case arises when we wish to determine the probability of an event B, given that event A has already occurred. In engineering, this case relates to the use of test data, for example. Suppose that we have an event, B, that initially we believe occurs with probability $P(B)$. We refer to $P(B)$ as an *a priori* probability—before the test. Then we conduct a test for which event A occurs. The question is, now what is our estimate of $P(B \mid A)$, that is, what is the *a posteriori* probability of B—after the test? We determine this probability using Bayes' formula.

Consider an experiment in which we sample one of k outcome spaces, which we refer to here as states of nature given by $S_1, S_2, S_3, \ldots, S_k$. We further take these outcome spaces to be mutually exclusive and to be the only possible states of nature. In addition, consider an experiment that results in event A, such that $P(AS_i) > 0$. A Venn diagram for this is shown in Fig. 3.16. Because the states of nature are mutually exclusive

$$P(A) = P(A\,S_1) + P(A\,S_2) + P(A\,S_3) + \ldots + P(A\,S_k)$$

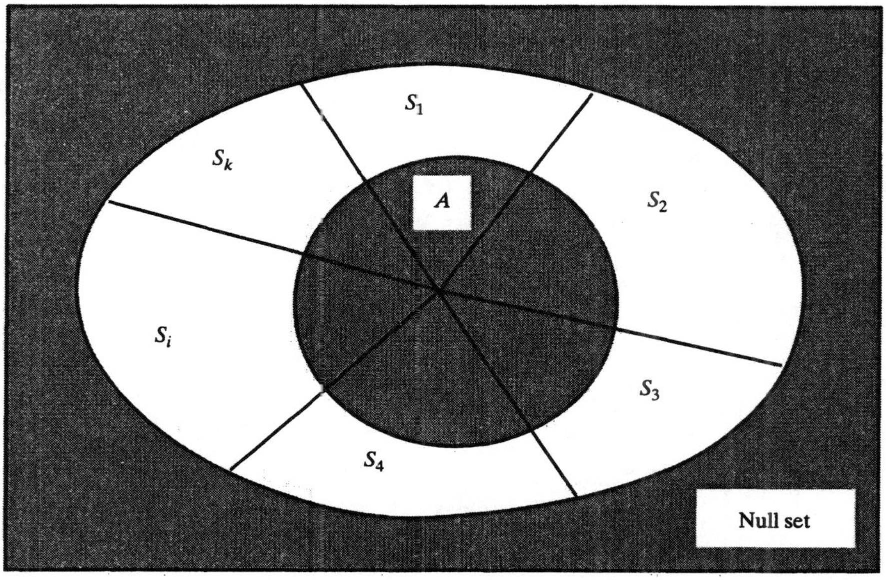

Figure 3.16 A Venn diagram for mutually exclusive states of nature.

We can then compute the probability that the sample came from state of nature i as

$$P(S_i \mid A) = \frac{P(A\,S_i)}{P(A)}$$

$$= \frac{P(S_i)P(A \mid S_i)}{\sum_{j=1}^{k} P(AS_j)}$$

$$= \frac{P(S_i)P(A \mid S_i)}{\sum_{j=1}^{k} P(S_j)P(A \mid S_j)}$$

The importance of this equation can be seen by means of an example. Let us go back to our example of guessing how many beans are in a jar. Suppose we feel that our best guess is 1,000 beans, but that the actual number could be as few as half this number or as great as twice this number. A very simple representation of this information is given by the dotted line in Fig. 3.17. Here we have simply represented our state of information as a triangular probability distribution, and to further simplify the example we have taken units of 50 beans as the smallest division of a guess.

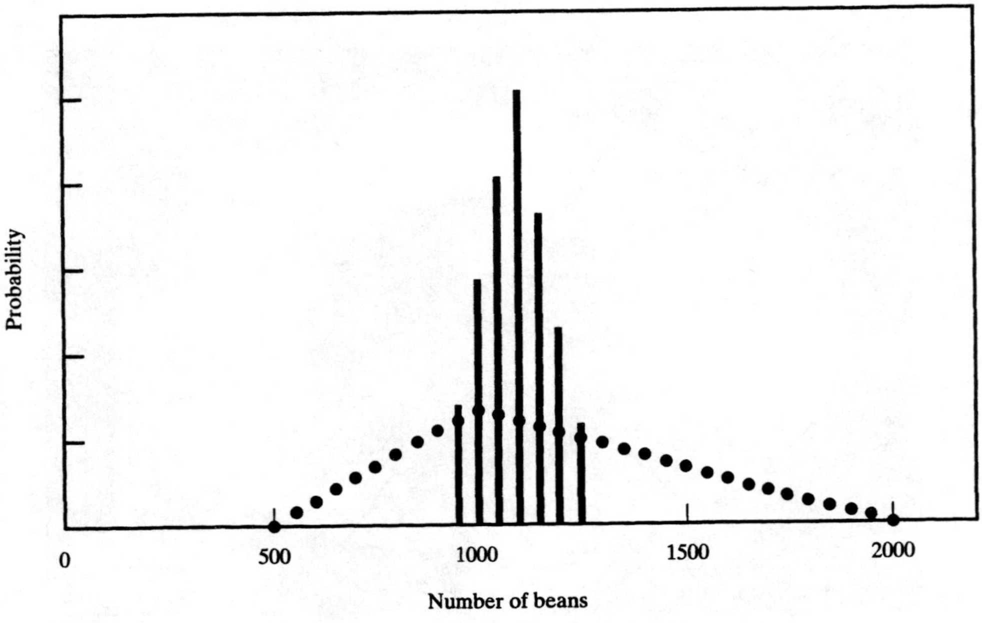

Figure 3.17 *A priori* and *a posteriori* distributions of beans in a jar.

Now suppose we can conduct a test. For example, suppose we have a jar that is one-tenth the volume of the jar in our example, and suppose we can fill it with beans and count them. A simple model for the number of beans in the large jar is 10 times the number of beans in the test jar. But we must recognize that this is not a perfect model. For example, edge effects, which vary with volume, can affect the number of beans that will fit in a unit volume, causing error in our model. And still there is randomness in how the beans will pack each time they are put into the jar. Now suppose that our test, event A as we call it above, gives a best estimate of 1,100 beans, and we feel that this could be in error by up to 150 beans. We will represent our state of information on A given i as a symmetrical, triangular distribution, also simplified to units of 50 beans. Using Bayes' formula to update our *a priori* distribution—the dotted line in Fig. 3.17—we get the distribution represented by the bars in this figure.

Two things are immediately evident by studying the *a priori* and *a posteriori* curves in Fig. 3.17. First, the result of the test is to give an *a posteriori* distribution that is much more peaked than the *a priori* distribution. This tells us that our state of information after the test is likely to be much better than the state of information before the test. Obviously, after the test, we are much more confident in our guess than we were before the test. The second thing we see is that the *a posteriori* distribution is neither symmetrical nor triangular, even though the distribution we took on $P(A \mid S_i)$ was both symmetrical and triangular. In using Bayes' formula, we integrate our test results with our prior information to obtain the *a posteriori* state of information. We do not throw out our *a priori* information. If we simply used the test result as though it were all the information that we have, we would have an *a posteriori* distribution that is both symmetrical and triangular. But this requires that we assume that any guess is as good as any other guess—that is, that our *a priori* distribution is a uniform distribution. The fact is we know more than this, and Bayes' formula enables us to include our prior information in our final result.

In a sense, Bayes' formula represents our implicit treatment of information. To use Bayes' formula, we apply data obtained through experimentation to our *a priori* model of nature, and we get an updated model of nature. Bayes' formula applies whether we use it or not. So, the insight we obtain here is that we cannot interpret data exclusive of an *a priori* model. Or conversely, the information that we get from data depends on our *a priori* knowledge of nature. This is really quite profound. It tells us that, if we are not ready to accept data in terms of our extant knowledge base, we will not be able to properly interpret them. And it explains why science is a process of carefully and methodically building knowledge bit by bit on an existing base. A word of humility is called for here. This certainly is a very broad interpretation of Bayes' formula, and one that he is not likely to have held.

3.7 HYPOTHESIS TESTING

In the final section of this chapter we discuss hypothesis testing. Hypothesis testing is important to a systems engineer because engineering design is decision making, and this involves the choice between alternatives with the goal of obtaining the best outcome. We can view the process of choosing an option, say A versus B, as equivalent to testing the hypothesis that option A will produce an outcome that is preferred to the outcome that option B will produce. Uncertainty frequently provides the situation where we cannot be sure which option will provide the most desired outcome. When this is the case, we say that the decision involves *risk*.

A hypothesis test begins with the creation of an *alternative hypothesis* and a *null hypothesis*. The alternative hypothesis is the hypothesis that we seek to prove. The null hypothesis is a contradiction of the alternative hypothesis such that, if the null hypothesis is false, the alternative hypothesis must be true. In engineering design, an alternative hypothesis could be stated as follows: *Option A will produce an outcome that is at least as good as option B.* The null hypothesis can be stated as: Option A will not produce an outcome that is preferred to option B. Later we will discuss preferences in detail. For now, however, we will simplify this hypothesis by assuming that the outcomes of options A and B can be represented by real numbers, V_A and V_B, such that the preferred outcome has the higher number. Now our alternative hypothesis is simply $V_A \geq V_B$, and we can test the hypothesis by evaluating the probability that $V_A \geq V_B$. To begin, we need to recognize that there are five possible outcomes of such a hypothesis test:

1. The null hypothesis is true, and we conclude that it is true.
2. The null hypothesis is true, but we conclude that it is false.
3. The null hypothesis is false, but we conclude that it is true.
4. The null hypothesis is false, and we conclude that it is false.
5. We conclude that there is insufficient evidence to determine whether the alternative hypothesis is true or false.

Outcomes 1 and 4 are correct conclusions. Outcomes 2 and 3 are incorrect. We give the errors of outcomes 2 and 3 special names:

> We say that a *type I* error occurs when the null hypothesis is true, but we conclude that it is false.
>
> We say that a *type II* error occurs when the null hypothesis is false, but we conclude that it is true.

It is important to realize that the probability of type I and type II errors will generally differ for the same hypothesis test, so that it is important to check them both. We generally denote by the symbol α the probability of making a type I error, and by the symbol β the probability of making a type II error. Finally, it is important to realize that data obtained by means of an experiment can be insufficient to support either the null or the alternative hypotheses. Thus, data do not always support one choice or another.

In our example, we will check the validity of the alternative hypothesis by asking, what is the probability that it is true? Consider distributions for V_A and V_B as shown in

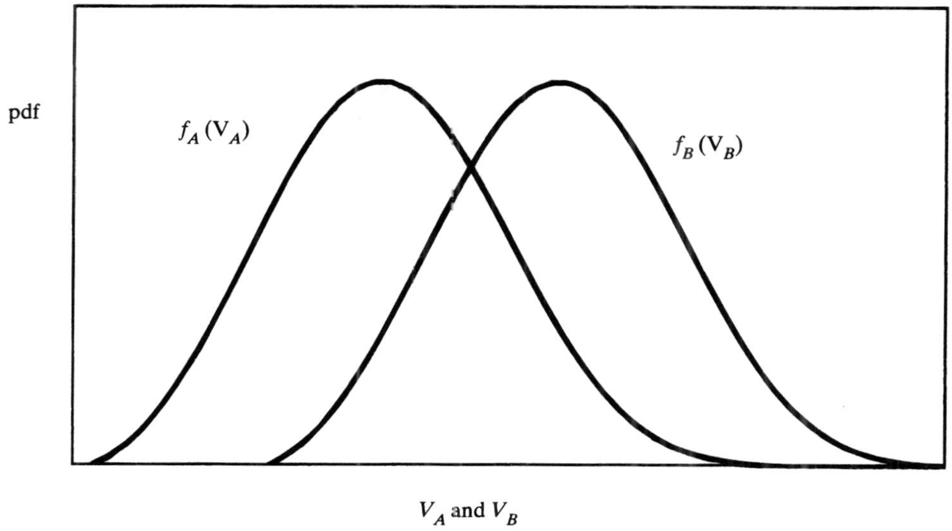

Figure 3.18 Distributions on the merit of options *A* and *B*.

Fig. 3.18. This question then translates into the following equation:

$$P(V_A \geq V_B) = \int_{-\infty}^{+\infty} f_A(V_A) \int_{-\infty}^{V_A} f_B(V_B)\, dV_B\, dV_A$$

Later, we will show that it is important to use value functions that are constructed to be valid under conditions of uncertainty.

3.8 SUMMARY

In this chapter we began to examine engineering models from the more rigorous point of view that they are probabilistic models, rather than deterministic models. This led us to a number of observations. First, we found that deterministic models discard the informational content of the modeling process and obscure the meaning of the outputs of the models. Second, we tried treating deterministic models as models of expected values, but we found that this could lead to serious errors as the expected value of an inverse is not equal to the inverse of an expected value. We defined probability distributions for both discrete and continuous variables, and then discussed several important distributions. We found that the central limit theorem explains the importance of the normal distribution. Toward the end of the chapter we discussed Bayes' formula, and we found that test results do not supplant prior knowledge, rather they update prior knowledge. Lastly, we discussed decision making as a process of hypothesis testing, and we derived an approach to test the hypothesis that one engineering option is at least as good as others.

At this point, we see a growing richness of analytical techniques that emerge from consideration of engineering design as a matter of decision making under conditions of uncertainty. But at the same time, we should be developing an awareness that failure to consider uncertainty formally and explicitly in all engineering analyses can lead to serious mathematical errors.

REFERENCES

1. Abramowitz, M. and I. A. Stegun, *Handbook of Mathematical Functions*, Dover Publications, New York, 1972.
2. Çinlar, E., *Introduction to Stochastic Processes*, Prentice Hall, Englewood Cliffs, NJ, 1975.
3. Mendenhall, W. and R. J. Beaver, Introduction to Probability and Statistics, PWS-Kent Publishing Company, Boston, MA, 1991.
4. Ross, S. M., *Introduction to Probability Models*, Academic Press, NY, 1972.
5. Wonnacott, R. J. and T. H. Wonnacott, *Introductory Statistics*, Wiley Series in Probability and Mathematical Statistics, John Wiley, New York, 1985.

PROBLEMS

1. Suppose a fair coin will be flipped until it comes up heads or until it has been flipped five times, whichever comes first. What is the probability that a head will be obtained? Suppose that the first three flips are tails. For this series of flips, what is the probability that a head will be obtained?

2. Derive expressions for the variance and standard deviation of $x + y$, $x - y$, xy, and x/y, where x and y are independent random variables.

3. Derive expressions for the expected value and variance of y, where

$$y = \sum_{i=1}^{n} x_i$$

and where the x_i are independent random variables. Simplify this result assuming that the x_i all have the same mean and standard deviation.

4. Derive expressions for the expected value and variance of y, where

$$y = \sum_{i=1}^{n} x = nx$$

Hint: This is not the same as the case of problem three because there is only one random variable x, so the assumption of independence does not apply.

5. Prove the law of large numbers.

6. A thumbtack is flipped until it lands point down. What is the expected number of flips if the probability of point down on any given flip is 0.3? Draw an event tree for this process.

7. You are given 20 non-Gaussian random variables, x_i, with means m_i and standard deviations σ_i. Write an approximate probability density function for the sum of these random variables.

8. A Japanese restaurant serves soup from a large kettle. At the end of each day, the owner of the restaurant saves the remaining soup and adds new ingredients to refill the kettle. This kettle has been kept hot for over 100 years, and the owner of the restaurant advertises that, when you eat his soup, part of your meal is over 100 years old. Is this true? Assume that the kettle is half emptied each day.

9. A machine tends to fail, on average, four times per year. What is the probability that it will fail five times in a one-year period? Six times? Three times?

10. Based on a computer model, you compute that the probability of a thumbtack landing point up on any given flip is 0.65. By changing the assumptions in the model, within reason, you are able to get different results, ranging from 0.5 to 0.75. Now you take the thumbtack and flip it 10 times. It lands point up seven times. What is your estimate of the probability that it will land point up on the next flip?

11. Let x and y be uniformly distributed, continuous random variables such that $0 \le x \le 2$, and $1 \le y \le 2$. Given X, drawn from x, and Y, drawn from y, what is the probability that $X \le Y$?

12. Your company manufactures widgets whose nominal lifetime is 1,000 hours, and whose standard deviation of lifetime is 100 hours. Your laboratory produces a new widget designed to last 30 percent longer. One is tested, and it lasts 1,250 hours. Do you believe that they have succeeded?

13. Your company is establishing a communications satellite network consisting of two satellites. They first must be placed in orbit. They can be launched separately or together on the same launch vehicle. In either case, the probability of a successful launch is 0.9. The cost to launch a single satellite is \$50 million, and both can be launched together for a cost of \$80 million. Each satellite costs \$75 million. Your company will self insure for the cost of the satellite, and the launch vehicle company will provide another launch at no cost if a launch attempt results in the loss of the satellite or failure to place it properly in orbit. Which launch option has the lower expected total cost?

14. Suppose in Problem 13 that, in order to launch both satellites together, they must be designed for the more complex launch, and this increases the cost of each satellite by \$5 million. Now which satellite design option, with or without the ability to make a dual launch, has the lower expected total cost?

15. You are asked to estimate the weight of an engineering system. You divide the system into 10 subsystems and estimate the weight of each subsystem independently. You then add these weights to get the total system weight. Later, you think about the uncertainty in your estimate, and you conclude that you should revise your estimate to account explicitly for uncertainty. You decide that your uncertainty on each estimate is represented well by a gamma distribution with $\beta = 2$. At most, each system could be 10 percent lighter than your initial estimate, and you take each deterministic subsystem weight estimate to be the most likely value of the corresponding gamma distribution. What is the expected error in your initial, deterministic estimate?

16. Derive equations for the mean and standard deviation of the exponential, gamma, and Weibull distributions.

17. You are asked to estimate the time it will take to complete the assembly of an engineering system. You divide the overall assembly into ten sequential subsystem assembly tasks and estimate the time required for each as follows:

Subtask	Distributional form	Mean	Std. deviation
1	Exponential	4 hours	*
2	Gamma (β = 2)	7 hours	*
3	Exponential	3 hours	*
4	Weibull (β = 3)	9 hours	*
5	Normal	12 hours	1 hour
6	Normal	8 hours	1 hour
7	Gamma (β = 3)	3 hours	*
8	Weibull (β = 2)	5 hours	*
9	Uniform	4 hours	1 hour
10	Gamma (β = 4)	13 hours	*
* Determine from equations derived in Problem 16.			

Estimate the pdf of the time to complete the entire assembly.

18. What is the probability that a car can be driven across the United States without needing repair?

19. You are asked to estimate the number of beans in a jar. $V_j = 1.5\ l \pm 5\%$, $V_b = 0.12$ cc $\pm$ 20%, $\mu = 0.6 \pm 0.2$. What is the expected value of the number of beans? How does this compare to a deterministic estimate? What is the standard deviation of the estimated number of beans Assume uniform distributions.

20. You are asked to estimate the volume of a spherical tank. You estimate the diameter of the tank to be 1.0 m ± 0.05 m. What is the expected value of the volume? What is the standard deviation of your estimate?

21. You have developed a process to produce small spheres. The diameter of the spheres ranges from 75 to 125 microns with a target diameter of 100 microns. The specific gravity of the material is 2. How many spheres would you expect to find in 1 kg of spheres?

22. You pull into the parking lot of a small shopping area. The lot has 25 spaces, but they are full. You are the only person waiting to park. How long will you have to wait?

23. You are betting heads on the flip of a coin. Before the flip, however, the coin gets bent so that the head side is concave. Did your chance of winning increase, decrease, or stay the same as a result of the coin being bent?

24. Driving to class, you get into a minor accident. Going home you should feel that your chance of an accident is less likely, more likely, or the same as it was when you left home?

4

Monte Carlo Modeling

4.1 THE PURPOSE OF THIS CHAPTER

In the previous chapters we found that deterministic modeling of engineering systems does not carry the notion that modeling is performed to provide information for a decision-making process, and it can lead to serious mathematical errors. Then we found that the mathematics of the modeling process, including uncertainty, can be quite complex. In this chapter, we present a general method for analyzing engineering systems that makes substantial use of computers and, in so doing, gets around much of the mathematical difficulty in the explicit treatment of uncertainty.

4.2 THE NOTION OF A MONTE CARLO MODEL

The notion of Monte Carlo modeling comes from the gaming tables at the casinos of Monte Carlo. Here, gamblers play games of chance, such as dice, roulette, and a variety of card games. These games work to create both fun and excitement for the players, and profit for the casinos, because minute differences in the spin of a roulette wheel, the toss of a die, or the shuffle of a deck of cards create complete unpredictability in the result. (In engineering, we call such systems chaotic.) If we record, for example, the sequence of numbers obtained on successive spins of the roulette wheel, even for extremely long number sequences, we would see no repeating pattern of numbers, and we would see no individual numbers appearing more often than others. And in the roll of a fair die, the numbers one through six appear with equal probability and randomness.

This notion of randomness can be used together with the mathematics of probability theory and the power of the modern computer to provide a computational procedure that is rich in its provision of information. The method carries the name Monte Carlo. In concept, the Monte Carlo method is quite simple. Consider, for example, that you want to compute the area of an irregular shape drawn in a square, as shown in Fig. 4.1. It is easy to compute the area of the square,

$$A_s = s^2$$

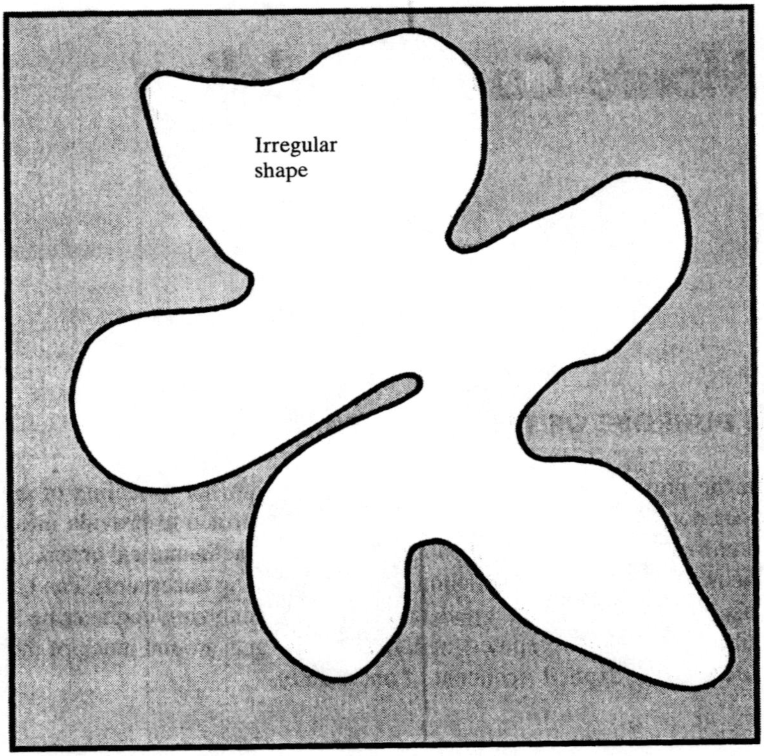

Square, side length s

Figure 4.1 A model for the computation of an irregular area by the Monte Carlo method.

as a function of the dimension of a side, s. It is much more difficult to compute the area of the irregular shape. But, clearly, the area of the irregular shape is a fraction of the area of the square.

$$A_i = \mu A_s = \mu s^2$$

Where μ is a number, $0 \leq \mu \leq 1$. And it would be easy to evaluate A_i if we could evaluate μ. Now consider the following process. Suppose the irregular shape drawn inside the square were hung on the wall, and suppose you were positioned about 3 meters from the wall and blindfolded. Now, suppose someone spun you around and then aimed you in the general direction of the wall and handed you several darts to throw at the wall. If the darts are thrown in random directions, then you would expect μ to be the ratio of the number of darts that land inside the irregular area divided by the total number of darts to land inside the square (including the irregular area). Indeed, according to the law of large numbers, if random, this process converges to the correct value of μ as the number of darts thrown becomes large. This is the essence of the Monte Carlo method, but we replace the physical process of throwing darts with the computer process of selecting and using random numbers.

Suppose we can have a computer choose uniformly distributed random numbers on the interval from zero to one. Let the number represent the fraction of the distance along a side, and let two successive numbers give a horizontal-vertical position on the square. Then we merely check to see whether the position is inside the irregular shape or not, and we keep count of the total number of tries and the number of tries that land inside the irregular shape. The ratio of the latter number to the former is μ. The advantage of the computer is that it can make many tries in very little time and keep track of the statistics at the same time. Indeed, a computer may be capable of performing as many as 1,000 tries per second.

So we see that Monte Carlo modeling relies on random numbers generated by computer to drive a computer simulation of a system. The computer also provides the mechanism for doing the associated accounting needed to obtain the desired statistics. Our study of the Monte Carlo method will follow this guide. In solving for the area of the irregular shape or to determine the number of beans in a jar, we must take five steps:

1. Generate random numbers, R_i.
2. Convert the R_i to variables of interest, such as an x or y coordinate or the volume of the jar or a bean.
3. Compute the variable of interest, such as in-the-irregular-shape or not-in-the-irregular-shape, or the number of beans in the jar.
4. Repeat steps 1 through 3 and collect data.
5. Analyze the data to obtain statistics of interest.

The remainder of this chapter will discuss these steps in detail.

4.3 RANDOM NUMBER GENERATORS

The first problem that we have in performing Monte Carlo modeling is to obtain a very large set of random numbers. Actually, we would like to have several very large sets of random numbers so that we can do successive simulations with different random numbers. One approach to obtaining a large set of random numbers is to perform a physical, random process, record the results, and publish a table. But this is cumbersome and requires considerable computer data storage space. The storage requirements will become clearer as we go on. It might be necessary to use several million or even perhaps a billion random numbers in a single simulation. So a better approach is to find an algorithm by which random numbers can be generated directly as needed in the computer. This is not easy to do, because of all the things that we want computers to be, random is not one of them. So we must use a deterministic machine to produce a random output, to the extent possible. Considerable work has been done on this problem, and many algorithms exist. They differ widely in how well they achieve randomness.

Randomness in a random number generator is measured by several statistics. First, we desire a random number generator that produces a uniformly distributed set of random numbers. No specific numbers or ranges of numbers should be favored over others on the range of zero to one. But, in addition to a uniform distribution, number sequences should not repeat. Stronger still, the random number set should exhibit no periodicities. In short, there should be no predictability of the random number over the range of zero to one. Still, random number algorithms generate a sequence of numbers. That is, one number is used as a "seed" to generate the next number. So you might expect that some predictability would exist. Getting rid of such predictability is the real difficulty.

A class of random number generators in common use are the linear congruential generators [Thesen, 1989]. Taking X_i to be sequential numbers drawn from the random numbers x, the X_i are given by

$$X_{i+1} = (A * X_i + C)\text{Mod}(M)$$

where A and C are constants, and Mod(M) refers to the modulo operation. The modulo operator takes the calculated quantity, divides by the integer M, and selects the remainder. (23 modulo 7 equals 2, as 7 divides into 23 three times with a remainder of 2.) Consider the following case:

$$X_{i+1} = (5 * X_i + 1)\text{Mod}(16)$$

This random number generator produces the sequence of random numbers shown in the following table. The numbers X_i can then be converted to a sequence of random variables on the interval from zero to one by dividing by M, namely 16.

There are several tests of goodness of a random number generator. First, we ask, does it produce a uniformly distributed random variable? It is easy to look at the above table and determine that each number from 0 to 15 appears exactly once in the sequence. But in a very long string of random numbers, a more convenient form of this test is needed. In this case we look at histograms. A histogram is a graphing technique that collects occurrences of events in bins. For example, we could define as our bins the number ranges $0.00 \le X_i < 0.01$, $0.01 \le X_i < 0.02$, $0.02 \le X_i < 0.03$, and so on. Then we count the number of random variable samples that lie within each bin, and look to see how uniform are these counts. The above algorithm passes this test, within the granularity of the number.

Another test examines the predictability of the random numbers by looking at the correlation between the ith value and the $(i + 1)$th value. To do this, we can construct a plot of the X_{i+1} versus the X_i, as shown for the above algorithm in Fig. 4.2. In this figure, we see a clear correlation between successive numbers. This is not a good random number generator, and its use in a Monte Carlo simulation could produce results that are misleading. Of course, we could also plot X_{i+2} versus X_i or, in general, X_{i+n} versus X_i and look for similar results.

i	X_i	$5 * X_i + 1$	X_{i+1}
1	1	6	6
2	6	31	15
3	15	76	12
4	12	61	13
5	13	66	2
6	2	11	11
7	11	56	8
8	8	41	9
9	9	46	14
10	14	71	7
11	7	36	4
12	4	21	5
13	5	26	10
14	10	51	3
15	3	16	0
16	0	1	1

We will take a look at two random number algorithms in common use on PCs. The first algorithm is called Randu. The *C* code for this algorithm is as follows:

```
float randu( ) {
   static long seed = 12345;
   seed = 65539*seed;
   seed = seed & 0 × 7fffffff;
   return(seed/2147483647.);}
```

Note that the number 2147483647 is equal to $2^{31} - 1$, which is the largest integer that can be stored in a 32-bit computer. Some researchers report that successive values of the random numbers generated by this algorithm demonstrate the type of correlation described above.

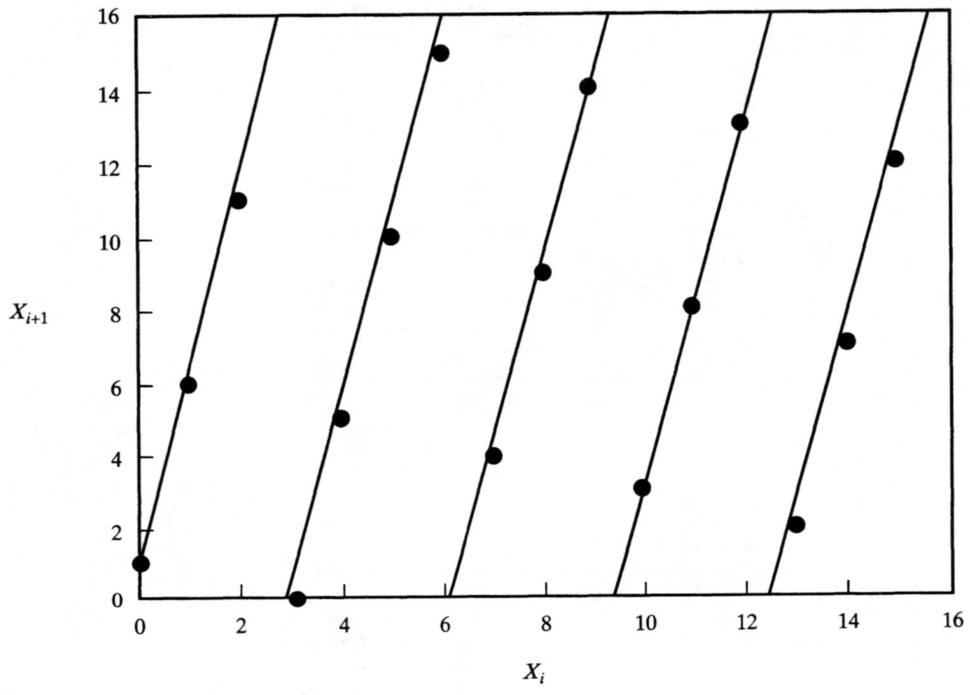

Figure 4.2 Correlation between successive random variables.

The second algorithm is called the Lewis algorithm. The C code for this algorithm is as follows:

```
float lewis ( ) {
    static long double seed = 12345;
    seed = fmod (seed*16807.,2147483647.);
    return(seed/2147483647.);}
```

This algorithm passes both the histogram test and the correlation test. A plot of X_{i+1} versus the X_i for this algorithm is shown in Fig. 4.3.

We will not discuss all the aspects of random number generators here. But we will close with some words of advice.

1. Do not use unknown random number generators.
2. Check the generators you use very carefully, using several different types of tests.
3. Do not use ad hoc methods to improve a random number generator.

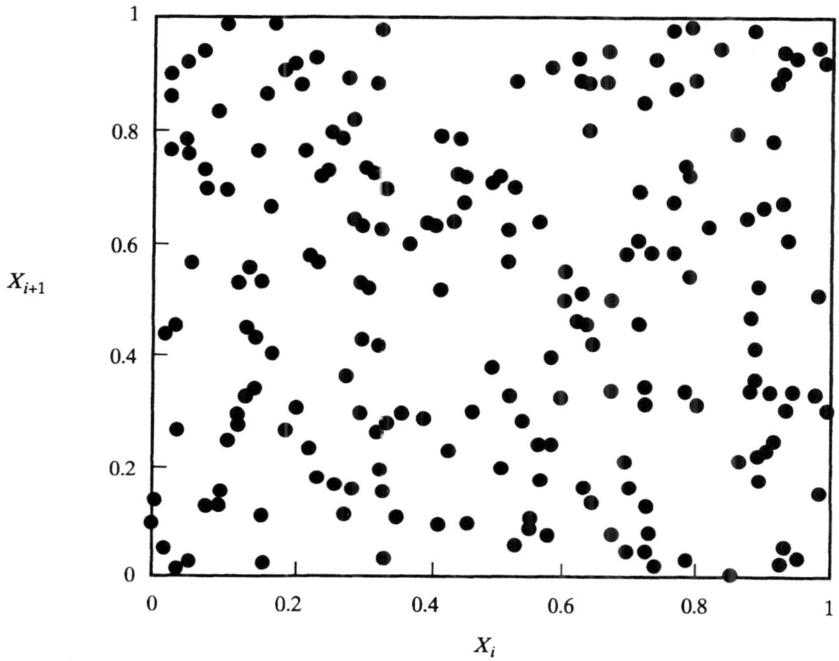

Figure 4.3 Lack of correlation between successive random variables obtained via the Lewis method.

4.4 SAMPLING A DISTRIBUTION

Once we have the ability to generate a uniformly distributed random variable between zero and one, we can proceed with a wide range of simulation tasks. First, however, we have to generalize our ability to obtain a uniformly distributed random variable, and extend this capability to any general form of distribution. A procedure for doing this is illustrated in Fig. 4.4. We begin with a uniformly distributed random variable on the range zero to one, and enter that variable on the ordinate of a cdf for our general distribution. We then trace horizontally to the cdf curve and then vertically to the abscissa. The value on the abscissa is the sampled value of our random variable for the case of a general distribution.

There is one problem with this procedure that is worthy of note. That is, in order to perform this operation efficiently, we need a convenient mathematical form of the cdf that we can solve easily for x given the random number. Unfortunately, many distributions do not possess such convenient mathematical forms. In particular, the pdf of the normal distribution cannot be integrated in closed form, and the mathematical form of the cdf is quite messy. Therefore, we often resort to approximations for the form of the cdf. Keep in mind that Monte Carlo simulation models use highly repetitive calculations, and a goal

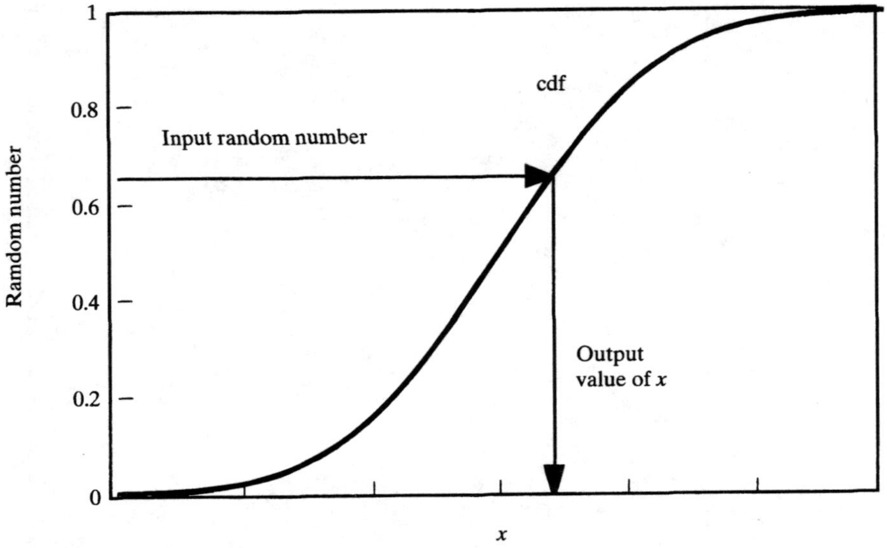

Figure 4.4 Illustration of the procedure for sampling a general distribution using a uniformly distributed random variable.

often is to perform as many simulations as practical under constraints of available computational power and time. Therefore, we are almost always seeking the highest possible efficiency in our computational procedures. It might disturb you that we now turn to approximations in the simulation, having done so much to tear apart the approximation of deterministic modeling. But all simulation involves approximation, and we have made a vast improvement in our modeling process merely by incorporating uncertainty explicitly in the model. Now we must give up a little accuracy in terms of the form of the cdf in order to gain accuracy by virtue of increasing the number of simulations that we can perform.

Because of the difficulties associated with many distributional forms, it is not inappropriate to turn to the use of some rather simple forms. One such form is the triangular distribution, such as that of the *a priori* distribution shown in Fig. 3.17. The questions that need to be answered to provide a triangular distribution are very simple:

1. What is your best guess of the variable (β)?
2. What value of the variable are you confident comprises the lower bound (α)?
3. What value of the variable are you confident comprises the upper bound (γ)?

Figure 4.5 shows how the answers to these three simple questions can be used to construct a pdf and cdf, both with simple mathematical forms that lend themselves well to simulation. Furthermore, note that the triangular distribution provides considerable flexibility to capture a good deal of the essence and form of the uncertainty that lies in many physical systems. It will not be often that you will feel compelled to abandon the use of triangular distributions for more appropriate distributional forms. However, one such time occurs when the pdf you seek to describe has very long tails. The peak of the distribution occurs at the point of the answer to question 1, and the minimum and maximum values of the random variable x correspond to the answers to questions 2 and 3. The height of the peak of the pdf is easily determined from the fact that the area under the distribution must be equal to one. The pdf of the triangular distribution is given by:

$$f(x) = 0 \qquad \text{for} \quad x < \alpha$$

$$f(x) = h\frac{x - \alpha}{\beta - \alpha} \qquad \text{for} \quad \alpha \leq x \leq \beta$$

$$f(x) = h\frac{\gamma - x}{\gamma - \beta} \qquad \text{for} \quad \beta \leq x \leq \gamma$$

$$f(x) = 0 \qquad \text{for} \quad x > \gamma$$

where

$$h = \frac{2}{\gamma - \alpha}$$

It can be shown that the mean and variance of the triangular distribution are given by the simple relationships

$$m = \frac{\alpha + \beta + \gamma}{3}$$

$$V\{x\} = \frac{(\beta - \alpha)^3 + (\gamma - \beta)^3}{6(\gamma - \alpha)} - (m - \beta)^2$$

And the cumulative distribution function is given by

$$P\{X \leq x\} = 0 \qquad \text{for} \quad x < \alpha$$

$$P\{X \leq x\} = \frac{(x - \alpha)^2}{(\beta - \alpha)(\gamma - \alpha)} \qquad \text{for} \quad \alpha \leq x \leq \beta$$

$$P\{X \leq x\} = 1 - \frac{(\gamma - x)^2}{(\gamma - \beta)(\gamma - \alpha)} \qquad \text{for} \quad \beta \leq x \leq \gamma$$

$$P\{X \leq x\} = 1 \qquad \text{for} \quad x > \gamma$$

4.5 FORMULATING MONTE CARLO MODELS

We are now ready to discuss the layout of a Monte Carlo model. All Monte Carlo models essentially follow the logic shown in Fig. 4.5. The computer code must allocate considerable space for the storage of results. Let us consider a very simple example, that of

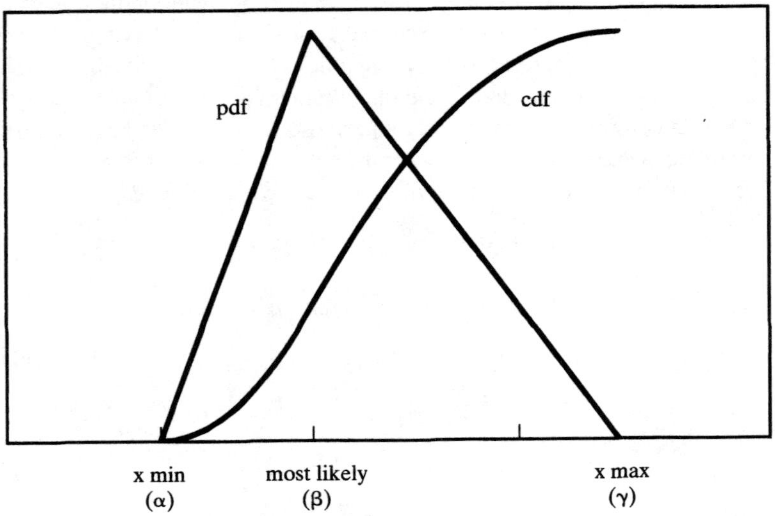

Figure 4.5 Construction of a pdf and cdf for a triangular distribution.

beans in a jar. A deterministic model was given for determining the number of beans in a jar. It is repeated here.

$$n = \frac{3V_j f}{4\pi r_b^3}$$

where V_j is the volume of the jar, r_b is the radius of a bean, and f is the fraction of the jar that is filled with beans rather than air.

What we want to do using a Monte Carlo simulation model is to construct our state of information on n. This is represented as a probability distribution on this variable. Because of the randomness associated with each individual simulation in a Monte Carlo model, we also want to know that we have conducted a sufficient number of simulations that the model provides a stable, converged solution. We could use the results of the simulations to compute the parameters of the distribution on n, namely the mean and standard deviation. But we may also want more details on the shape of the distribution on n. To get more detail, we collect data on n in the form of a histogram, as explained in Sec. 4.3. Suppose we expect n to vary over the range of 100 to 2,000. We could set up bins with a width of 10 beans each, so that the bins would be for values of n between 100 and 109, 110 and 119, 120 and 129, and so on. For each bin, we need a counter. Let us denote the bin counter by the symbol B_i, i for the ith bin. Then, as we later perform simulations, each time we obtain an answer for n, we will increment the counter in the bin corresponding to the value of n. We begin by setting the B_i all to zero. So, for example, if on the first simulation we obtain $n = 112$, we would then increment B_2 from zero to one. We will also keep a running total of all the successive ns obtained so that we can later divide by the number of simulations performed to determine the mean value of n. And we could similarly keep data to compute the higher-order moments.

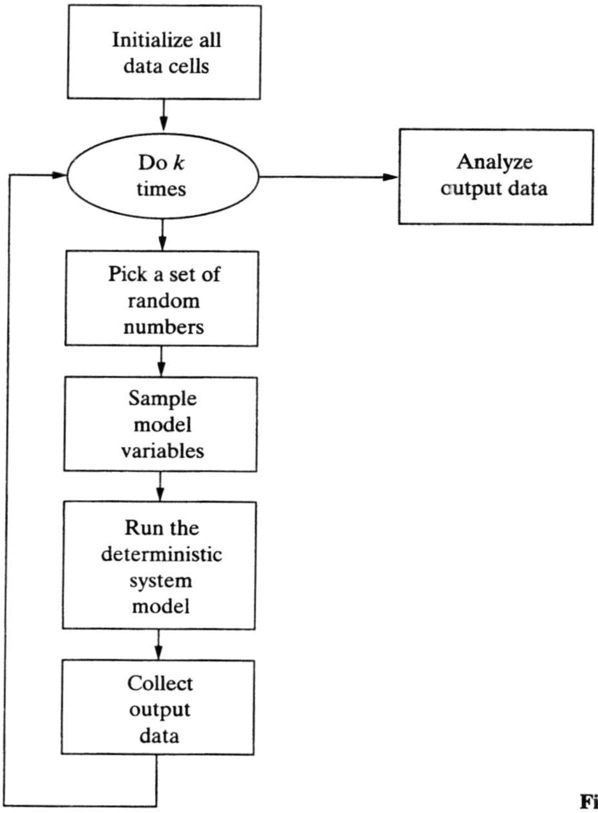

Figure 4.6 The logic flow of a typical Monte Carlo model.

Once we have set up the appropriate counters, we are ready to begin the Monte Carlo simulations. We next choose the number of simulations that we wish to perform. On the one hand, we want to perform a sufficient number of simulations to obtain useful output data. But each simulation requires some amount of computing time, and in the end we are limited always by computing time. A typical starting number for the number of simulations is 1,000. But this number must be carefully chosen in light of the complexity of the calculations to be performed. For a relatively simple set of calculations, which could be performed at the rate of, say, 100 per second, 1,000 simulations will require only about 10 seconds. But for much more complex calculations, 1,000 simulations could be prohibitively time consuming. Fewer than 1,000 simulations can give reasonable results for some problems, but especially if the number of variables in the model is large, you should take care that the number of simulations you perform is sufficient so that you do not obtain misleading results.

Our bean-jar model contains three variables, V_j, r_b, and f. So, for each simulation, we must obtain values for each of these three variables. To do this, at the beginning of each simulation we pick three random numbers, R, uniformly distributed on the interval 0 to 1, and, using the method shown in Fig. 4.3, convert these random numbers into sampled variables. Again, it is recommended that you use triangular distributions, as these are easy to deal with and easy to construct from data. For the triangular distribution, we obtain

$$x = \alpha + \sqrt{R(\beta - \alpha)(\gamma - \alpha)} \qquad \text{for} \quad 0 \le R \le \frac{\beta - \alpha}{\gamma - \alpha}$$

$$x = \gamma - \sqrt{(1 - R)(\gamma - \beta)(\gamma - \alpha)} \qquad \text{for} \quad \frac{\beta - \alpha}{\gamma - \alpha} \le R \le 1$$

where x is the variable of interest, such as V_j, r_b, or f, and R is a random number uniformly distributed on the interval zero to one. It is assumed at this point in the model that you have constructed these distributions for each variable. Once you have values for each of the three variables, you simply put these values into the equation for n and solve it. You then collect the data, which in this case is merely n, and repeat the process. When you have completed the desired number of simulations, you are ready to analyze the data that you have collected.

There are a number of variations on the above model that will become obvious as the need for them arises. For example, sometimes Monte Carlo models involve logic and branching such that you will not be able to determine how many random numbers you will need for an individual simulation at the beginning of the simulation. In this case, the obvious solution is to generate random numbers and sample variables as needed in the model.

4.6 ANALYSIS OF MONTE CARLO MODEL RESULTS

Generally, the first things we want to know are the mean and standard deviation of the variables of interest in our simulation. In the bean-jar example, this is n. These properties of the distribution on n can be calculated directly from the output data. We also generally like to see the pdf and cdf of the distribution. These can be constructed from the histogram data. To construct the cdf, we simply begin at the first bin and keep a running cumulative sum of all counts collected. This sum divided by the number of simulations is the value of the cdf that may be plotted at the upper limit of the bin. The pdf, which may be plotted at the center of the bin, is simply the count in each bin divided by the total number of simulations and normalized by the width of the bin. An example of this is shown in the following table, constructed for 1,000 simulations.

i	Bin	B_i	ΣB_i	pdf	cdf
1	100–109	1	1	0.0001	0.001
2	110–119	0	1	0.0000	0.001
3	120–129	6	7	0.0006	0.007
4	130–139	9	16	0.0009	0.016
5	140–149	7	23	0.0007	0.023

Now we see something of a problem immediately by examining the table. That is, because of our finite sample size, there is *noise* in the computation of the pdf and cdf. It shows up more vividly in the pdf, which is the derivative of the cdf. If you were to plot these data directly, the noise would be seen on the graph. If the noise is excessive, it is a sign that you might not have performed a sufficient number of simulations or that you have chosen a bin size that is too small (too narrow). But some noise will always be present in Monte Carlo model output, and you can accept it. You can also smooth the pdf and cdf curves to hide the noise, provided that it does not concern you. By increasing the bin size, for example, by a factor of two or three, you can smooth the noise, but do not hide the noise if it is a concern. Resolve your anxieties by running more simulations and by further checking the model, and perhaps your random number generator.

Often you need more assurance that a model is providing valid results. One approach is to collect data in a still more dispersed manner. Instead of having a single array of bins in which to collect the histogram data, we could define, say 10 sets of bins. Then, we can collect the results from, say the first 100 simulations in the first set of bins, the results from the second 100 simulations in the second set of bins, and so on. (Be sure that you use the same number of simulations for each set of bins.) From these data, we are able to calculate 10 means, and we can estimate the standard deviation of the means. We could then estimate the standard deviation of the mean of the means. The standard deviation of the mean of the means would be the standard deviation of the means divided by the square root of the number of means that were used to compute the mean of the means, in this case the square root of 10. This procedure is recommended as routine practice in Monte Carlo model output analysis.

Now we see how much data we may have to collect to thoroughly analyze the output of a Monte Carlo model. In the case presented, where we took n to vary from 100 to 2,000 (actually 1,999) and a bin size of ten beans, leading to 190 bins, and then we repeated this 10 times, we obtain a total of 1,900 bins. If our model had 10 output variables of interest, you can see that the number of bins needed would be getting quite large.

4.7 AN ALTERNATIVE ANALYSIS OF MONTE CARLO RESULTS

An alternative to the method of histograms presented in the previous section is to keep all of the model output data, and analyze it upon completion of the Monte Carlo simulations. Again, consider the example of the bean jar. The result of each simulation is a computation of n. Every computed value of n can be saved in a table. Upon completion of the Monte Carlo simulations, this table might have $k = 1{,}000$ values of n. Next, sort the table so that the values of n appear in ascending numerical order. Each value of n then adds $1/k$ to the cumulative distribution function, and this value is added to the cdf at the value of n. The process is illustrated in the following table.

The only problem with this method is its requirement for data storage. On older computers, where storage space was limited, approaches such as the histogram method were used to reduce storage requirements. Modern PCs, however, generally have sufficient storage capability that model outputs can all be stored for later analysis.

n sorted	cdf (k = 1,000)
<276	0.000
276	0.001
281	0.002
285	0.003
289	0.004
289	0.005

4.8 MONTE CARLO MODELS WITH EMBEDDED DECISIONS

One of the major strengths of Monte Carlo models is their ability to incorporate simulations of a wide range of system elements as they are not limited to simulation of only the physical elements of a system. For example, we can simulate human decision processes. We can incorporate in a simulation design choices that would be taken differently depending on the outcome of an earlier part of the simulation, and we can include a certain amount of randomness in these choices as well. Indeed, in later chapters of this book, we will see that the Monte Carlo method is exceptionally powerful, and it can be used to simulate a very wide range of real-life systems and situations. One activity that can be modelled using the Monte Carlo method is the research process itself.

4.9 SUMMARY

In the earlier chapters, we discussed some of the problems associated with deterministic modeling. We showed the need for probabilistic modeling of engineering systems, but also pointed to the complexity of probabilistic models. In this chapter we developed a method that addresses these issues, the method of Monte Carlo simulation. We presented the notion of a Monte Carlo model that is based on the properties of random numbers and their use for certain types of calculations. We then discussed random number generators for use in digital computers. As noted, there are several potential problems with random number generators, and we presented two algorithms, one of which we recommend over the other. Several cautions are provided relating to the use of random numbers. The random number generators discussed provide uniformly distributed random numbers on the interval from zero to one. Next an algorithm for generalizing these distributions is presented, but we noted that, in the interest of computational speed, we need to keep simple the form of the distributions that we use. We showed the triangular distribution and how it is constructed from three questions, and noted that it provides a convenient computational tool. Finally, we looked at the logic of Monte Carlo simulation and the treatment of simulation results.

REFERENCE

1. Thesen, A. and L. E. Travis, *Simulation of Industrial Systems*, Classnotes for IE 442, University of Wisconsin, Madison, WI, Spring 1989.

PROBLEMS

1. Write a computer program to generate random numbers. Use your program to generate, say, 100,000 random numbers. How long did the computer take to generate the random numbers? Roughly how long does it take for the computer to generate a single random number?
2. Using a sample of the random numbers generated in problem 1, construct a plot similar to that in Fig. 4.2. Does your random number generator appear to produce random numbers?
3. Compute the mean and standard deviation of the random numbers generated in problem 1. How do they compare with the mean and standard deviation of a uniformly distributed variable on the range zero to one?
4. In problem 1, count the number of random numbers that lie within the ranges 0.00 to 0.01, 0.10 to 0.11, 0.20 to 0.21, 0.30 to 0.31, and so on to 0.90 to 0.91. Compare these numbers. Does your random number generator appear to produce a uniformly distributed random variable?
5. Write a Monte Carlo simulation program to compute the area of a rectangle embedded in a square. Let the rectangle have dimensions 0.7 by 0.4. Perform the simulation 10 successive times without restarting the random number generator (that is, use a different random number set each time). Compare the ten answers. What is the standard deviation of your result?
6. Write a Monte Carlo simulation program to determine the volume of a sphere of unit radius.
7. Consider a random variable x that has a triangular probability density function with a minimum value of 1, a most likely value of 2, and a maximum value of 4. Construct the cumulative distribution function of x and sample the cdf using uniformly distributed random numbers. From the samples, reconstruct the probability density function. Does the reconstructed pdf compare well to the original pdf?
8. Write a Monte Carlo simulation model for the bean-jar example. Choose a deterministic system model, and use this model in your simulation. If possible, get an actual jar and fill it with beans. First, guess at the number of beans in the jar, n. Second, use your model in a deterministic way. Guess at the parameters of your model and calculate the number of beans in the jar. Then use your Monte Carlo model to obtain the pdf and cdf of n, using triangularly distributed random variables for each of the model parameters. Also compute the mean and standard deviation of n. Compare your results for each of the three methods, and comment on your feelings regarding the accuracy and information content in each. Finally, count the beans in your jar and compare the actual count to your model and simulation results.

5

Optimization

5.1 THE PURPOSE OF THIS CHAPTER

We shall soon see that there are very strong similarities between decision making and optimization. Indeed, much of engineering design is turning to optimization implemented on digital computers as an aid to achieving better designs. In fact, optimization treats engineering design as a decision-making process. The purpose of this chapter is to introduce the basic concepts and methods of optimization as an approach to both the study and conduct of decision making. A broad range of techniques are introduced, and an understanding of the basic concepts behind the techniques is presented. They are presented largely in vector notation, which enormously simplifies their presentation. All boldface characters are vectors. If you are not familiar with vector notation, see App. A. These concepts are important for an understanding of the chapters that follow and they provide deeper insight into the process of decision making. The subject of optimization is quite important and extensive. This chapter provides only the briefest of introductions. There exist many substantial textbooks on the subject, and the interested reader is referred to these.

5.2 THE NEED FOR OPTIMIZATION IN SYSTEMS ENGINEERING

We have defined systems engineering to be a process of decision making as opposed to problem solving. Decision making is the taking of choices from sets of options in order to obtain the most desired outcomes. This is precisely the process of optimization. Indeed, we can write the general optimization problem in the form

$$\text{maximize} \quad J = f(\boldsymbol{x},\boldsymbol{u})$$

$$\text{subject to constraints} \quad \boldsymbol{g}(\boldsymbol{x},\boldsymbol{u}) = \boldsymbol{0}$$

$$\boldsymbol{u}_{\min} \leq \boldsymbol{u} \leq \boldsymbol{u}_{\max}$$

where J is the objective or value measure, which is a function of state variables $\boldsymbol{x}$ and control variables $\boldsymbol{u}$. The variables $\boldsymbol{x}$ derive from physical constraints and other considerations, and the variables $\boldsymbol{u}$ are free, within bounds, to be chosen in order to achieve the highest possible real value of J. Recall that we identified three things necessary for decision making: options, which here are given by the $\boldsymbol{u}$, expectations, which derive from the constraint equations, and values, given here by $f(\boldsymbol{x},\boldsymbol{u})$. But more than a model of decision making for study, optimization is a rigorous mathematical procedure for dealing with large option spaces. We obtain methods for both writing option spaces in an abbreviated form and selecting the best choice from the option space.

We can see the power of the optimization formulation of options rather easily. Consider the problem

$$\text{maximize} \quad J = 2 - u^2 \qquad \text{for} \quad -1 \leq u \leq 1$$

The object here is to select the value of the scalar u on the interval from -1 to $+1$ that maximizes J. There are an infinite number of values of u on this interval, and thus an infinite number of options. But this is understood in the above expression without further elaboration. One approach to finding the value of u that maximizes J is to examine all values of u, and simply choose the best value. But, since there are an infinite number of values of u, this is not feasible. The mathematics of optimization makes it possible to select the best value of u without having to search through an infinity of possibilities. Any capable student of elementary calculus can solve for the optimal value of u in the above equation.

In some cases, optimization is an important part of our ability to make a choice. For example, suppose that we are trying to choose between two different materials for a part. We cannot simply analyze the same part given each of the two materials. We must reoptimize the part for each material before making the comparison on the basis of materials. A design may be good for one material and bad for another, but a different design to incorporate the same functionality might alter the choice of materials. Consider an example in which you are asked to select the function, $g(u)$ or $h(u)$, that has the highest maximum. Obviously, you cannot make this choice until you have determined the maxima for both functions. Optimization is needed to do this. The same is true any time that a choice is to be made in an engineering system. Optimization is needed before making the choice.

Later in this chapter, we will see that some decision-making situations are so complex that it is very difficult merely to obtain a feasible selection of choices. Here again optimization techniques will come to our aid.

5.3 NOTIONS OF MINIMUM AND MAXIMUM

Optimization is the process of choosing values of control or decision variables to minimize or maximize a value function. So, we will begin by defining exactly what we mean by the words minimum or maximum. Also, we can note that a minimum of a function, J, occurs at the same values of the control variables as a maximum of the negative of the same function. Thus, we need only discuss the process of maximization, as everything for this case applies to the case of minimization with the sign of J changed. The concept of a maximum point is illustrated in Fig. 5.1 for a function of a single control variable. A maximum point on a function is any locally highest point on the function. From a maximum point, the function does not go up in any direction. Relate this to climbing a hill. You are on a peak any time that there is no direction that you can go to get to a higher elevation without having to descend first. Still, there may be a higher peak than the one you are on. To get to it, you must first descend from the peak you are on and move toward the higher peak. In general, a hilly region will contain several peaks, but only one of them will be the highest. This peak we refer to as the *globally* highest. The other peaks we refer to as *local* peaks. A local peak is the highest point in the immediate area, but not necessarily the highest peak in the entire region. And we note that the global peak is also a local peak.

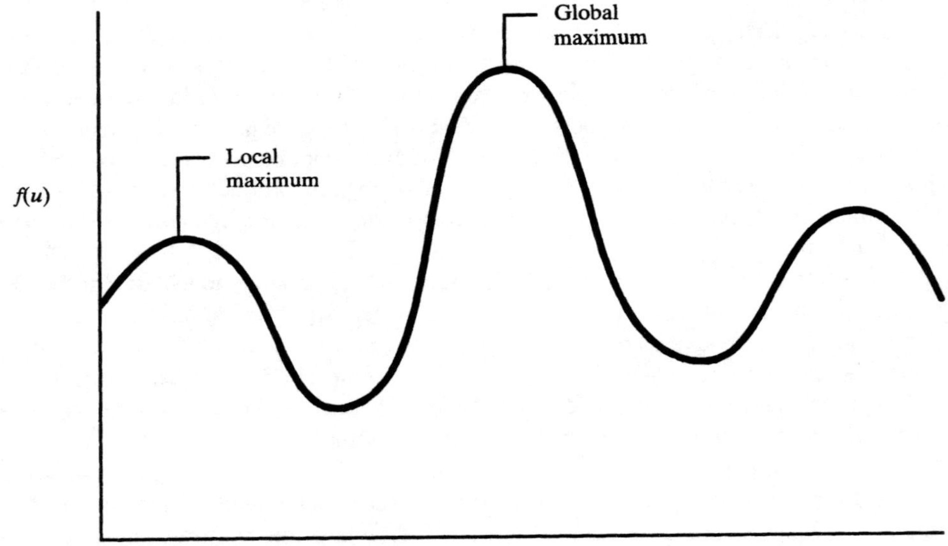

Figure 5.1 Illustration of local and global maxima.

When we deal with real, scalar functions, we also refer to maximum points as local maxima or global maxima. A local maximum point on a function is any point from which there is no small movement, taken in any allowable direction, that leads to a point on the function that has a higher value. And the point of global maximum of a function is the point for which the value of the function is at its overall highest within the allowable region of the control variables. It is this notion that is the basis of all mathematical optimization.

Not all functions are comprised of smooth curves. The function shown in Fig. 5.2 is an example of a function with discontinuous derivatives. The concepts of minimum and maximum apply equally but, as we shall see, the mathematics of optimizing these two types of functions differ.

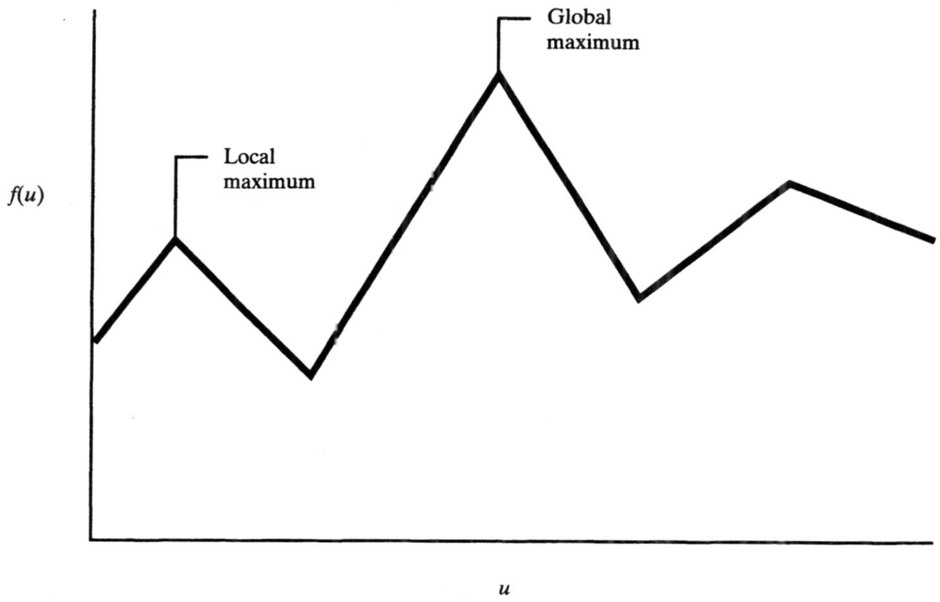

Figure 5.2 Illustration of maxima and minima of a function that has discontinuous derivatives.

5.4 MAXIMIZATION OF A FUNCTION

We will begin our mathematical development of optimization techniques by examining a scalar function of a single control variable, u. Our problem can be stated

$$\text{maximize} \quad J = f(u)$$

By the notion of a maximum as described above, if the point u_0 is a maximum point, then it must be true that, for small δu

$$f(u_0 + \delta u) - f(u_0) \leq 0$$

This condition states simply that the function does not go uphill in any direction from u_0.

Now, if all the derivatives of f exist and are finite, we can expand f in a Taylor series to obtain

$$\begin{aligned} f(u_0 + \delta u) &= f(u_0) + \left.\frac{df}{du}\right|_{u_0} \delta u + \frac{1}{2} \left.\frac{d^2f}{du^2}\right|_{u_0} \delta u^2 \\ &\quad + \frac{1}{6} \left.\frac{d^3f}{du^3}\right|_{u_0} \delta u^3 + \ldots \\ &= f(u_0) + \sum_{n=1}^{\infty} \frac{1}{n!} \left.\frac{d^nf}{du^n}\right|_{u_0} \delta u^n \end{aligned}$$

Thus, the condition for a maximum becomes

$$\sum_{n=1}^{\infty} \frac{1}{n!} \frac{df^n}{du^n}\bigg|_{u_0} \delta u^n \leq 0$$

for all small δu. In other words, a local maximum occurs at a value of u for which u cannot be changed by a small amount, in any direction, such that the function becomes larger. This means that we can look at the above condition as δu approaches zero, and consider either positive or negative values of δu. Because we can take δu as small as we wish, and because one condition for validity of a Taylor series is that all derivatives must exist and be finite, we always can choose δu small enough that the term of the lowest order derivative dominates, unless that derivative is exactly zero. Thus, unless

$$\frac{df}{du}\bigg|_{u_0} = 0$$

this term dominates the Taylor series expansion in the region near u_0. But if it is non-zero, we can then choose δu either positive or negative as needed to make the function grow in the vicinity of u_0. Consequently, a *necessary condition* for a maximum of a function whose derivatives all exist and are finite in the region of the maximum point is that the first derivative of the function must be zero at the maximum point. When this condition is satisfied, the higher-order derivatives dominate the expansion. So next we look at the second term. The second term contains δu raised to the second power. Thus, no matter whether δu is taken positive or negative, δu^2 cannot be negative. It follows that a *sufficient condition* for a maximum is that

$$\frac{d^2f}{du^2}\bigg|_{u_0} < 0$$

This condition assures that the function goes downhill in all directions from u_0.

If the second derivative is also zero at the same point where the first derivative is zero, the third-order term will dominate unless it is zero. Then we could apply the same logic as above to show that a necessary condition is that the third derivative must be zero, and a negative fourth derivative is a sufficient condition. We continue this logic until we reach a non-zero even-order derivative. But remember that these conditions apply only in the case that $f(u)$ can be expanded in a Taylor series. This requirement demands that $f(u)$ and **all** its derivatives exist and be finite. This requirement is not met by a function such as that shown in Fig. 5.2, and these conditions **do not apply** to such a function.

We derived the above conditions for the case that f is a scalar function of a single real control variable u. These conditions are easily extended to the case where f is a scalar function and $\boldsymbol{u}$ is a vector of arbitrary dimension. The vector form of the Taylor series has very much the same appearance as the scalar form.

$$f(\boldsymbol{u}_0 + \delta\boldsymbol{u}) = f(\boldsymbol{u}_0) + \frac{df}{d\boldsymbol{u}}\bigg|_{\boldsymbol{u}_0} \delta\boldsymbol{u} + \frac{1}{2}\,\delta\boldsymbol{u}^T \frac{d^2f}{d\boldsymbol{u}^2}\bigg|_{\boldsymbol{u}_0} \delta\boldsymbol{u} + \ldots$$

Here, the superscript T refers to the vector transpose. The first-order term is a dot product, leading to the necessary condition that

$$\left.\frac{\partial f}{\partial \boldsymbol{u}}\right|_{\boldsymbol{u}_0} = \mathbf{0}^T$$

If $\boldsymbol{u}$ is a vector of dimension k, this condition is k equations. The second-order term, which leads to the following sufficient condition, is a little more complex.

$$\delta \boldsymbol{u}^T \left.\frac{\partial^2 f}{\partial \boldsymbol{u}^2}\right|_{\boldsymbol{u}_0} \delta \boldsymbol{u} < 0$$

The second partial of f with respect to $\boldsymbol{u}$ is a square matrix of dimension $k \times k$. In order for this condition to be met for all arbitrary $\delta \boldsymbol{u} \neq \mathbf{0}$, the matrix of partial derivatives must be *negative definite.* A matrix is negative definite if and only if its eigenvalues are negative.

5.5 SEARCH METHODS

The necessary conditions for a maximum point given above tell us what the characteristics of a function will be at a maximum point of a smooth function, that is, a function all of whose derivatives exist and are finite, provided that the maximum point is not on a boundary of the control. However, they do not tell us how to find such points. Generally, we apply the necessary condition to obtain a set of equations that we can then solve to find the maxima of a function. For example, consider the equation

$$f = 2 + u - u^2$$

Applying the necessary condition, we get

$$\frac{df}{du} = 1 - 2u = 0$$

which is easily solved to yield $u = 0.5$. Now we look at the sufficient condition at this value of u

$$\frac{d^2 f}{du^2} = -2$$

which satisfies the sufficient condition for a maximum at $u = 0.5$. In this case, we could solve easily for the optimal value of u. This does not always occur. Consider the function

$$f(u) = u^2 \sin u$$

This function has the form shown in Fig. 5.3 for the range $0 \leq u \leq 10$. It has multiple maxima. The necessary condition for a maximum of this function is

$$\frac{df}{du} = u(u \cos u + 2 \sin u) = 0$$

which cannot be solved directly for u. We say that this equation is *transcendental* in u.

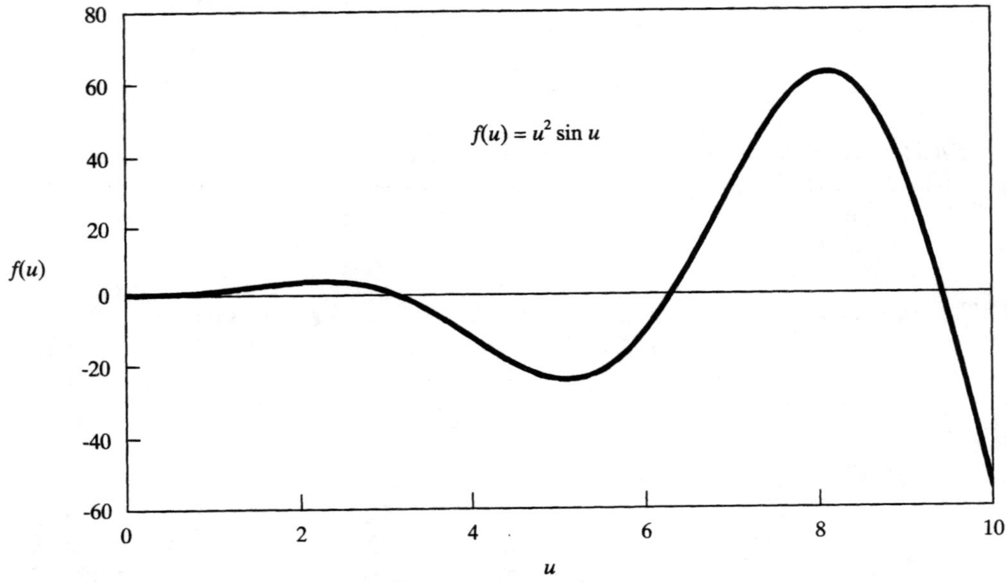

Figure 5.3 A function whose first derivative produces a transcendental function.

which means that it cannot be solved in closed form. To obtain solutions for u, we need an *iterative* solution procedure. Iterative solution procedures begin by taking a guess at the answer, and then by some logic they try a succession of guesses that approach the answer closer and closer. The process is stopped when it has *converged* to a value of u such that the necessary conditions are satisfied to a sufficient degree of accuracy.

Iterative procedures can be broadly classified according to the degree to which they approximate the function being optimized. In essence, you can think of a Taylor series expansion of the function around the optimal point, taken to some order, and that is the order of the convergence of the iterative procedure. A zeroth-order search procedure merely scans the function and heads in the direction of larger values. A first-order method uses the first derivative to point the way to larger values of the function. A second-order method uses both the first and second derivatives to both point the way toward the maximum and estimate the distance. We will briefly discuss methods that fit into these three classes. There are methods of still higher order, but these can be difficult to use and are of lesser importance.

Zeroth-Order Methods

Conceptually, the simplest iterative method for solving a transcendental equation is to try every possible value of the variable in the allowable range and see which values correspond to solutions. This is the method of *global search*. While conceptually simple, however, the method is difficult to apply because you simply cannot try every possible value if, as typically is the case, there is an infinite or near-infinite number of possible values. So the first step toward resolving this difficulty is to divide the region of interest of the variable into a finite and reasonably small number of values. In the case of Fig. 5.3, for example, we could divide the region from 0 to 10 into 101 values, 0, 0.1, 0.2, and so on. This is what was done to plot the curve in the figure. (Note that it is important to divide up the region of interest sufficiently to capture the basic structure of the function.) Simply selecting the best of these values gives a fair approximation to the optimal value. But sometimes we want a more precise result. In this case, one approach is to locate the highest point of those computed and the point to either side of that point, and divide this area into, say, 100 intervals and repeat the above process. We could repeat these steps until we find the maximum point with any desired accuracy.

There are, obviously, more efficient approaches. Instead of dividing the selected interval between the highest and two adjacent points into 100 intervals, we could simply divide each interval in half, compute two new points, choose the highest of these five points and its two adjacent points, and then repeat this process until the desired level of convergence is obtained. This process is depicted in Fig. 5.4. There are even more efficient methods of dividing the intervals for successive iterations, but we will not cover them here.

The advantages of zeroth-order iterative procedures include their conceptual simplicity, ease of programming, and inherent stability. These methods require calculation only of the value of the function itself at each point. As we shall soon see, higher-order methods require more. This makes programming easier. And, although it is possible to construct pathological cases that can give these methods convergence problems, the ability to find the maxima of nearly any function, whether it or its derivatives exist and are finite or not, is essentially assured. This is not the case in many other techniques.

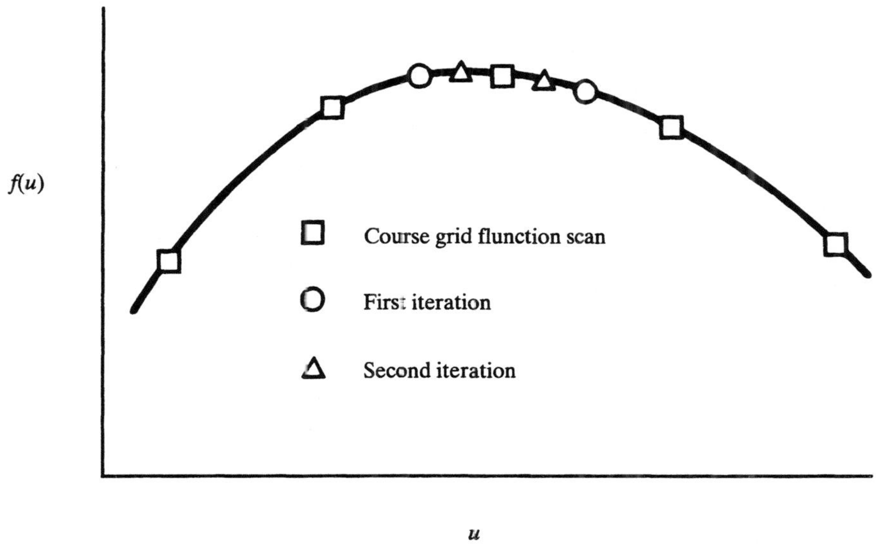

Figure 5.4 Illustration of a zeroth-order search.

The disadvantage of zeroth-order methods is that they tend to be slow and, for problems with more than just one or two control variables, they can become overwhelming. For example, suppose you have a problem containing ten control variables, and you feel it necessary to consider just five values of each variable to start the iteration. The starting iteration then requires 10^5 evaluations of the function.

First-Order Methods

The basic idea behind first-order iteration methods is this: if you are trying to climb a mountain as quickly as possible, you can't go very wrong by taking the steepest direction. This approach does not always lead to a global maximum, but it does lead at least to a local peak. Although first-order methods can be applied to searches in one dimension, their power becomes apparent when they are applied to a search in many dimensions. There are many different first-order methods, but they are typified by the method of steepest ascent, which we shall outline here. This method is depicted in Fig. 5.5. First, we choose a starting point. This is obtained simply as a guess. Next, we determine the direction of the steepest upward slope. The first-order term in a Taylor series expansion around the point of the initial guess is

$$\frac{\partial f}{\partial \boldsymbol{u}}\,\delta\boldsymbol{u}$$

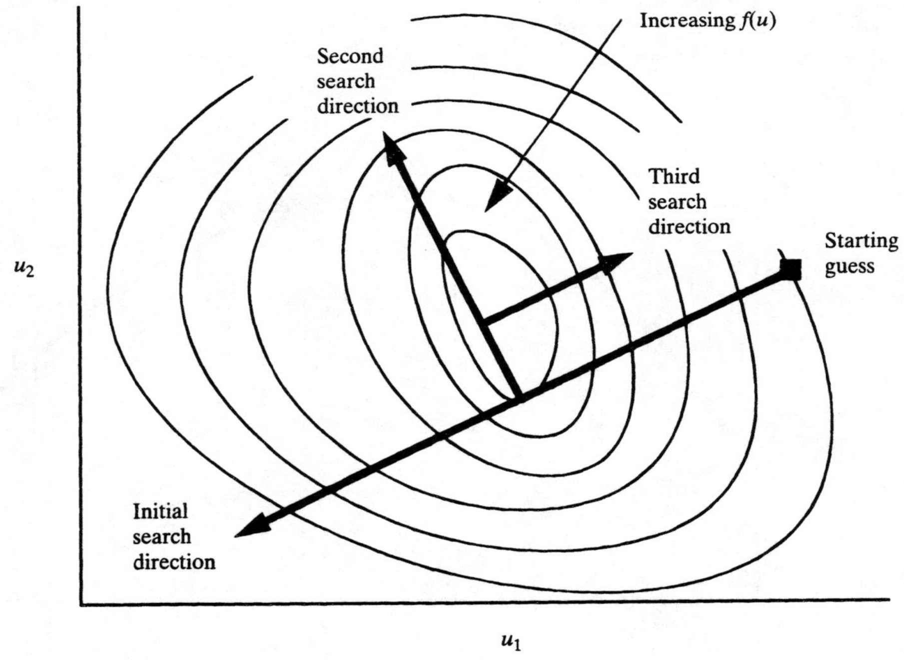

Figure 5.5 Illustration of a steepest ascent search in two dimensions.

and the function is locally steepest upward when $\delta \boldsymbol{u}$ is chosen to be in the same direction as $\partial f/\partial \boldsymbol{u}$, which is the gradient of f. This provides a direction in which to search for the maximum. Now, a one-dimensional search is conducted in this direction to find the maximum point on the line defined by the starting guess and the direction $\delta \boldsymbol{u}$. The search on the line could be a zeroth-order search, or it could be a second-order search. Once the maximum point on the line is found, this provides a new starting guess, and the process is repeated until there is no longer any direction in which to go to make the function larger. That is, when the necessary condition

$$\frac{\partial f}{\partial \boldsymbol{u}} = \mathbf{0}$$

is met to a sufficient degree of accuracy. Note that each new one-dimensional search is orthogonal to the last.

There are two problems that we must face in applying a first-order method. First, we must obtain the partial derivatives $\partial f/\partial \boldsymbol{u}$. Second, we must decide when to stop the iteration process. There are two fundamentally different approaches to determining the partial derivatives, analytical and numerical. In the analytical approach, we take the mathematical partial derivatives of the function with respect to the control variables, and program these so that the derivatives can be computed directly at each iteration point on the function. In the numerical method, we begin by computing the function at the iteration point, $\boldsymbol{u}_0$, and then move a small distance from this point in each direction, to the points $\boldsymbol{u}_0 + \Delta \boldsymbol{u}$. At each of these points we recompute the function, and the partial derivatives are then given by

$$\frac{\partial f}{\partial u_i} = \frac{f_{u_i+\Delta u_i} - f_{u_i}}{2\ \Delta u_i}$$

This is called the method of forward differences. Immediately, we see that one advantage of using numerically obtained derivatives is that we do not have to derive and program each of the partial derivatives, a time-consuming process in which we could easily make errors. (This is a significant advantage because the derivatives of most functions are more complex than the functions themselves.) But, on the other hand, we encounter two problems. First, we need to select the Δu_i. It is desirable to keep them small, as the definition of a derivative is that it is the above expression in the limit as $\Delta u_i \rightarrow 0$. The problem with choosing Δu_i too small is that the numerator of the above equation becomes very difficult to calculate with sufficient accuracy, and the derivatives become corrupted with computational noise. The second problem is that the above expression does not compute the value of the partial derivatives at the point $\boldsymbol{u}_0$, but rather at some point between $\boldsymbol{u}_0$ and $\boldsymbol{u}_0 + \Delta \boldsymbol{u}$.

A better approach to the numerical computation of derivatives is the method of forward and backward differences. In this method, we compute values of the function at the points $\boldsymbol{u}_0 + \Delta \boldsymbol{u}$ and $\boldsymbol{u}_0 - \Delta \boldsymbol{u}$, and then compute the partial derivatives using the equations

$$\frac{\partial f}{\partial u_i} = \frac{f_{u_i+\Delta u_i} - f_{u_i-\Delta u_i}}{2\ \Delta u_i}$$

This form has the advantages that it uses larger differences in $\boldsymbol{u}$, which reduce computational noise, and it centers the evaluation of the derivatives on the point $\boldsymbol{u}_0$. Its disadvantage is obvious, for $\boldsymbol{u}$ of large dimension, the function must be evaluated roughly twice as many times as in the case of forward differences. But experience shows that accurate evaluation of the derivatives is a very important element in obtaining rapid convergence of any iterative technique, and the added computation may be well worth the effort. Although the method of forward differences sometimes works, I would never use it.

The second problem we noted above was deciding when to stop the iteration. Clearly, convergence is obtained when $\partial f/\partial \boldsymbol{u} = \boldsymbol{0}$. But in practice, we rarely reach this point. We have to decide what is close enough. Is it 0.000001? 0.0001? 0.01? There is really no way to tell, and it differs for each problem. If you are familiar with the problem, you may be able to pick a reasonable number for convergence. But we often do not have this luxury. First, we usually take the dot product of the gradient with itself, namely $(\partial f/\partial \boldsymbol{u})(\partial f/\partial \boldsymbol{u})^T$, and apply a scalar stopping rule. But this works well only if all variables are roughly of the same order of magnitude. If they are not, additional scaling should be applied to assure that each variable is equally important in the stopping criterion. Second, we can look at the value of this scalar for the starting iteration, and ask that it be reduced, through the iteration process, by, say, three orders of magnitude. So, if the starting dot product is 1.473, we would be happy stopping the iteration when it is reduced to 0.001473 or less. When convergence is easy, we might demand greater accuracy. But frequently we are satisfied with the improvement obtained in the function without demanding extremely accurate convergence.

You may also note that similar stopping rules must apply to the one-dimensional search that the steepest ascent method uses. The maximum point on the one-dimensional search line occurs when the dot product $(\partial f/\partial \boldsymbol{u})\delta \boldsymbol{u} = 0$. Generally, you want to converge the one-dimensional search to an accuracy of about one order of magnitude more than you are seeking from overall search. This prevents the overall optimum point from being missed because of sloppy convergence on the one-dimensional search.

Second-Order Methods

Second-order methods make use of both the first derivative and second derivative terms in the Taylor series. For example, we can expand the first derivative around the guess point and invoke the necessary condition to obtain

$$\left.\frac{\partial f}{\partial \boldsymbol{u}}\right|_{\boldsymbol{u}+\delta\boldsymbol{u}} \approx \left.\frac{\partial f}{\partial \boldsymbol{u}}\right|_{\boldsymbol{u}_0} + \delta\boldsymbol{u}^T \left.\frac{\partial^2 f}{\partial \boldsymbol{u}^2}\right|_{\boldsymbol{u}_0} = \boldsymbol{0}^T$$

We then solve this equation for $\delta\boldsymbol{u}$.

$$\delta\boldsymbol{u}^T = -\left.\frac{\partial f}{\partial \boldsymbol{u}}\right|_{\boldsymbol{u}_0} \left[\left.\frac{\partial^2 f}{\partial \boldsymbol{u}^2}\right|_{\boldsymbol{u}_0}\right]^{-1}$$

This formulation is known as Newton's method. Two advantages of a second-order method appear immediately from this equation. First, each step in the iteration process can go in a new direction, determined by both the first and second derivatives, so that convergence can be much quicker. And, second, $\delta\boldsymbol{u}$ is fully determined, so that you no longer perform a one-dimensional search. Again, this enables much more rapid convergence. Indeed, for a quadratic function, this procedure converges in a single iteration. Thus, we say that second-order methods have *quadratic convergence.*

The major problem with this method is in determining the matrix of partial derivatives, $\partial^2 f/\partial \boldsymbol{u}^2$. Again, the elements of this matrix can be determined either analytically or numerically. And again, the problems with each method are the same as they were for the first derivatives. The equations for numerical computation of the matrix of second partials are

$$\frac{\partial^2 f}{\partial u_i \partial u_j} = \frac{f_{+\Delta u_i,+\Delta u_j} - f_{+\Delta u_i,-\Delta u_j} - f_{-\Delta u_i,+\Delta u_j} + f_{-\Delta u_i,-\Delta u_j}}{4\ \Delta u_i\, \Delta u_j} \qquad \text{for} \quad i \neq j$$

$$\frac{\partial^2 f}{\partial u_i^2} = \frac{f_{+\Delta u_i} - 2f + f_{-\Delta u_i}}{\Delta u_i^2} \qquad \text{for} \quad i = j$$

where the subscript notation used in these equations has, for example, the following meaning

$$f_{+\Delta u_i,-\Delta u_j} = f(u_i + \Delta u_i,\ u_j - \Delta u_j)$$

These equations can be derived by simply taking numerical derivatives of the numerically determined first derivatives. Whatever problems may be associated with numerical computation of first derivatives, they are much more difficult to deal with in the computation of the second derivatives. This is because the numerator contains a difference, and the denominator contains a small number squared. You can expect problems with numerical noise in the calculation of these derivatives, and the best advice that can be offered is to pay particular attention to maintaining as much computational accuracy as possible. Second-order methods frequently fail to work. That is, they frequently fail to converge to a solution. The main reason for their failure is the failure to accurately compute the first and second derivatives.

Conjugate Gradient Methods

There are a few extremely powerful techniques that are variations on the above methods. We shall discuss one here, the Fletcher-Powell method. The Fletcher-Powell method is constructed in such a way that it begins as a steepest ascent method but, if the function to be maximized is quadratic in k variables, it converges to the solution in exactly k iterations. It also provides strong convergence for functions that are not quadratic. The Fletcher-Powell method belongs to a class of methods referred to as *conjugate gradient* methods.

The general quadratic form in k variables is given by

$$f = a + \boldsymbol{b}^T\boldsymbol{u} + \boldsymbol{u}^T\boldsymbol{C}\boldsymbol{u}$$

Differentiating, the necessary condition is

$$\frac{\partial f}{\partial \boldsymbol{u}} = \boldsymbol{b}^T + 2\boldsymbol{u}^T C = \boldsymbol{0}^T$$

We can solve this for the optimum $\boldsymbol{u}$

$$\boldsymbol{u} = -\frac{1}{2}C^{-1}\boldsymbol{b}$$

The problems are that, in general, we do not have a quadratic form nor do we have the matrix C^{-1}. But, even though the general optimization problem is not quadratic, if all the derivatives of the function exist and are finite, that is, if a Taylor series expansion of the function is valid, then the function can be approximated well by a quadratic in the vicinity of the optimum point. Thus, quadratic methods have the potential to work very well as they approach the optimum point.

On the other hand, we observed above that one of the major problems with quadratic methods is that we must obtain a matrix of the sort C^{-1} and, if we use finite difference methods, this process is prone to error from computational noise. If the matrix of partial derivatives is much in error, and this may be the case more often than you might like, the iteration process may be slowed considerably or even diverge. That is, it may fail to converge altogether.

The Fletcher-Powell method begins as a gradient method, and uses only first derivatives. As with other methods, it begins with an initial guess. Next, the gradient is determined, and a one-dimensional search is conducted to find the maximum point in the gradient direction. At this point a new search direction is computed, and a new one-dimensional search initiated. And the process repeats until convergence is obtained. At each maximum point, $\boldsymbol{u}_s$, on a one-dimensional search, a new direction is computed using the equation

$$\boldsymbol{r}_s = -H_s\left[\frac{\partial f}{\partial \boldsymbol{u}}\bigg|_{\boldsymbol{u}_s}\right]^T$$

The matrix H_s is referred to as the Hessian matrix. The process is started by choosing an arbitrary negative definite matrix, H_0, which guarantees that the search will head uphill so long as the gradient is computed with reasonable accuracy. A convenient initial Hessian matrix is the negative identity matrix

$$H_0 = -I$$

Now, the Fletcher-Powell method iteratively constructs updates of the Hessian matrix such that it always remains negative definite. It is this property of the Hessian matrix that gives the Fletcher-Powell method its great stability in seeking a maximum, while at the same time enabling quadratic convergence as the optimum point is approached.

The Fletcher-Powell method takes the following steps:

1. Make an initial guess at the optimum point, $\boldsymbol{u}_0$. Initialize the Hessian matrix to the negative identity matrix as shown above.
2. Compute the initial search direction $\boldsymbol{r}_0$.

$$\boldsymbol{r}_0 = -H_0\left[\frac{\partial f}{\partial \boldsymbol{u}_0}\bigg|_{\boldsymbol{u}_0}\right]^T$$

3. On the line $\boldsymbol{u}_0+\lambda\boldsymbol{r}_0$, search for the value, λ_0, that maximizes f.
4. Define $\beta_0 = \lambda_0\boldsymbol{r}_0$. Compute

$$\gamma_0^T = \frac{\partial f}{\partial \boldsymbol{u}}\bigg|_{\boldsymbol{u}_1} - \frac{\partial f}{\partial \boldsymbol{u}}\bigg|_{\boldsymbol{u}_0}$$

5. Update the Hessian matrix as follows:

$$H_1 = H_0 + A_0 + B_0$$

where the matrices A_0 and B_0 are given by

$$A_0 = \frac{\beta_0\beta_0^T}{\beta_0^T\gamma_0}$$

and

$$B_0 = -\frac{H_0\gamma_0\gamma_0^T H_0}{\gamma_0^T H_0\gamma_0}$$

6. Repeat this process until the necessary condition

$$\frac{\partial f}{\partial \boldsymbol{u}} = \boldsymbol{0}^T$$

is met with sufficient accuracy.

Experience shows that the Fletcher-Powell method works well provided that you take care to obtain accurate numerical derivatives, and provided that you take care to assure that for each one-dimensional search, λ is determined with precision. Again, it pays to seek convergence on λ to a level of accuracy that is significantly better than that that you are seeking for the overall function.

5.6 CONSTRAINED OPTIMIZATION

We frequently encounter problems in optimization of the form,

$$\text{maximize} \quad J = f(\boldsymbol{x},\boldsymbol{u})$$

$$\text{subject to} \quad \boldsymbol{g}(\boldsymbol{x},\boldsymbol{u}) = \boldsymbol{0}$$

In this formulation, the constraints, $\boldsymbol{g}(\boldsymbol{x},\boldsymbol{u}) = \boldsymbol{0}$, comprise a vector set of equations. For example, consider the following problem. Maximize

$$J = -u_1^2 - u_2^2$$

subject to the constraint

$$u_1 + u_2 - 1 = 0$$

We can interpret this problem graphically as shown in Fig. 5.6. We are looking for the highest value of J that lies on the line given by the constraint.

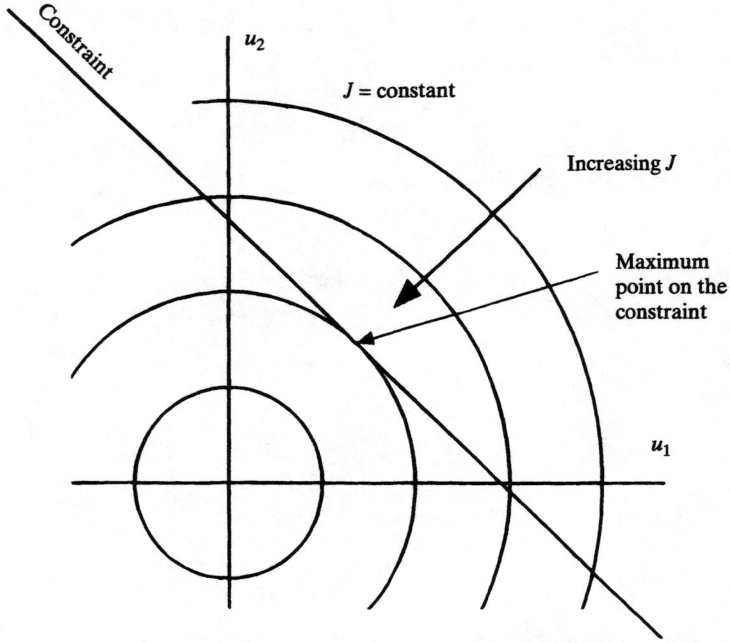

Figure 5.6 Geometry of a constrained maximization.

There are two approaches to the solution of this problem. The first is to assure that the constraint is satisfied for every permissible solution in J, and then seek the maximum of the constrained J. To begin, we solve for u_1:

$$u_1 = 1 - u_2$$

and then substitute this into the equation for J:

$$J = -(1 - u_2)^2 - u_2^2$$

Now we treat J as unconstrained and solve for u_2. The necessary condition gives

$$\frac{dJ}{du_2} = 2(1 - u_2) - 2u_2 = 0$$

which gives

$$u_2 = \frac{1}{2}$$

We can now use the constraint equation to solve for u_1.

$$u_1 = \frac{1}{2}$$

Although this method is simple and easy to understand, it does not always work. It fails when we cannot solve the constraint equations for the components of $\boldsymbol{u}$ in order to substitute them into J, that is, when the constraint equations are transcendental in $\boldsymbol{u}$. Unfortunately, this happens rather frequently in real engineering problems.

The second method gets around the problem posed when the $\boldsymbol{g}$ are transcendental in $\boldsymbol{u}$. It is called the *adjoint* or *Lagrangian* method. In this method, we adjoin the constraint to the criterion function using a multiplier called a *Lagrangian multiplier* or *adjoint variable*. In so doing, we create the augmented or adjoined function, $f^*(\boldsymbol{x},\boldsymbol{u})$.

$$f^*(\boldsymbol{x},\boldsymbol{u}) = f(\boldsymbol{x},\boldsymbol{u}) + \lambda \cdot \boldsymbol{g}(\boldsymbol{x},\boldsymbol{u})$$

Notice that, if the constraint is satisfied, that is, if $\boldsymbol{g} = \boldsymbol{0}$, then quantitatively $f^*(\boldsymbol{x},\boldsymbol{u}) = f(\boldsymbol{x},\boldsymbol{u})$. In the example given above,

$$f^*(\boldsymbol{x},\boldsymbol{u}) = -u_1^2 - u_2^2 + \lambda(u_1 + u_2 - 1)$$

Now we apply the necessary condition

$$\frac{df^*(\boldsymbol{x},\boldsymbol{u})}{d\boldsymbol{u}} = \boldsymbol{0}^T$$

and we enforce the constraints

$$\boldsymbol{g}(\boldsymbol{x},\boldsymbol{u}) = \boldsymbol{0}$$

If the dimensionality of $\boldsymbol{u}$ is n, and the dimensionality of $\boldsymbol{g}$ is m, this gives us $n + m$ equations. For our example, we get

$$-2u_1 + \lambda = 0$$

$$-2u_2 + \lambda = 0$$

$$u_1 + u_2 = 1$$

This is easily solved to obtain

$$\lambda = 1$$

$$u_1 = \frac{1}{2}$$

$$u_2 = \frac{1}{2}$$

Clearly, both methods give the same answer. The adjoint method gives added insight, however. Note that, in the adjoint method,

$$J = f^*(\boldsymbol{x},\boldsymbol{u}) = f(\boldsymbol{x},\boldsymbol{u}) + \lambda \cdot \boldsymbol{g}(\boldsymbol{x},\boldsymbol{u})$$

If we had written $\boldsymbol{g}$ in the form

$$\boldsymbol{g}(\boldsymbol{x},\boldsymbol{u}) = \boldsymbol{g}^*(\boldsymbol{x},\boldsymbol{u}) - \boldsymbol{c} = \boldsymbol{0}$$

we would see that

$$\frac{\partial J}{\partial c} = -\lambda$$

where we observe that λ is a row vector, not a column vector. Now we see that the interpretation of λ is that it is the cost of the constraints, $\boldsymbol{c}$. This is a very important result the meaning of which will become clear as we get into more advanced topics.

5.7 LINEAR PROGRAMMING

A very important class of optimization problems take the form

$$J = \boldsymbol{c}^T\boldsymbol{u}$$

where we wish to maximize J subject to constraints of the form

$$A\boldsymbol{u} = \boldsymbol{b}$$

$$\boldsymbol{u} \geq \boldsymbol{0}$$

A, $\boldsymbol{b}$, and $\boldsymbol{c}$ are constants. A is an $m \times n$ matrix, $\boldsymbol{b}$ is an m-vector, and $\boldsymbol{c}$ is an n-vector. These equations are linear in the variables $\boldsymbol{u}$. The above form is the standard format of a linear programming problem. Note that constraints expressed in the form

$$A\boldsymbol{u} \leq \boldsymbol{b}$$

can be transformed into the standard form by the introduction of the slack variables, $\boldsymbol{s}$, giving

$$A\boldsymbol{u} + \boldsymbol{s} = \boldsymbol{b}$$

$$\boldsymbol{s} \geq \boldsymbol{0}$$

Also note that an unrestricted variable, u_i, can be replaced by the difference between two variables, $u_i = u_i^+ - u_i^-$, where $u_i^+, u_i^- \geq 0$.

The solution to problems of this sort is obtained by various methods referred to as *linear programming*. It is easy to visualize this problem in two dimensions, as shown in Fig. 5.7. The arrow indicates the direction of increasing J. Because of the form of J, the value space is linear. Further, if the constraints are specified as given above and in a way that closes the set of feasible solutions, namely the set of solutions that satisfy the constraints, then it can be shown that the set of feasible solutions is convex. Convexity means that a line that connects any two points that lie within the set of feasible solutions is itself entirely within the set of feasible solutions. Given the condition of convexity, it can be further shown that, for a closed set, the optimal solution lies at an extreme point of the convex set of feasible solutions. Now, two possibilities exist. Either the optimal solution lies at a unique point in the set of feasible solutions as it does in Fig. 5.7, or it lies along a surface as shown in Fig. 5.8 and, hence, is not unique. We say that the solution in the latter case is *degenerate*. In the degenerate case, an infinity of optimal solutions exist. A final possibility is shown in Fig. 5.9. Here, the constraints leave the set open in such a way that the solution is unbounded. No finite solution exists in this case.

Knowing the above characteristics of a linear programming problem enables the development of some extraordinarily rapid and strongly convergent iteration techniques.

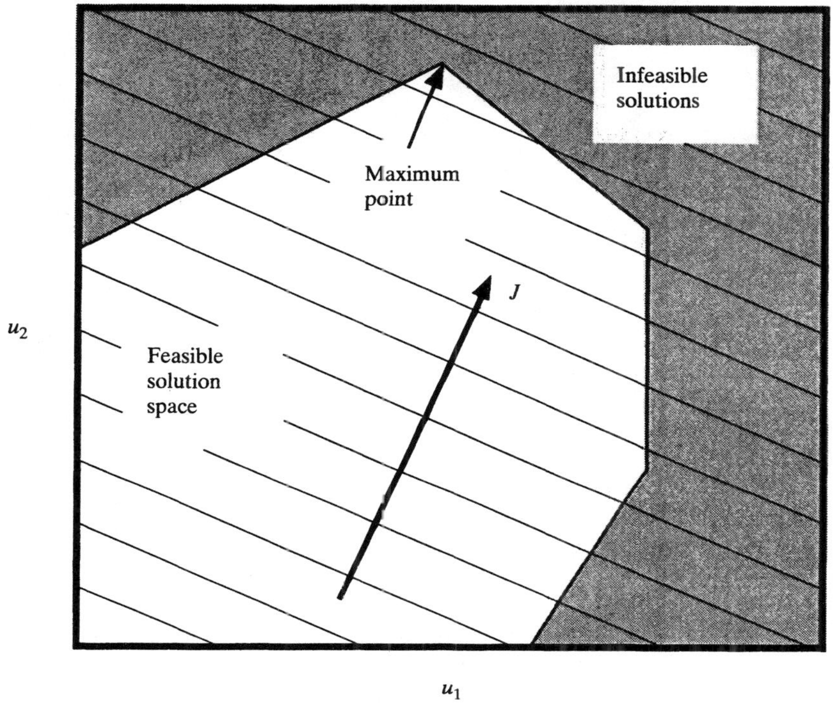

Figure 5.7 Illustration of a linear programming problem in two dimensions.

These techniques enable optimization of problems involving even thousands of variables. We would never, given the current state of the art, consider problems of this scope in the case of nonlinear optimization.

Perhaps the best-known linear programming method is the simplex method. It is comparatively easy to understand and computationally effective. But there now exist much more powerful methods. The most important new method was developed by Narendra Karmarkar in 1984, and it is called the *interior-point algorithm.* Still, the simplex method retains considerable importance, and we will explain it here. The simplex method begins by finding an extreme point in the convex set of feasible solutions, and then moves from extreme point to extreme point in a direction such that J is always increasing, until an extreme point is reached from which there is no subsequent extreme point whose value of J is larger. This is the optimal point. Since this is a solution to the degenerate case as well, this is taken as the optimal solution.

We begin the simplex formulation by observing that the matrix A can be decomposed into two matrices

$$A = [B , N]$$

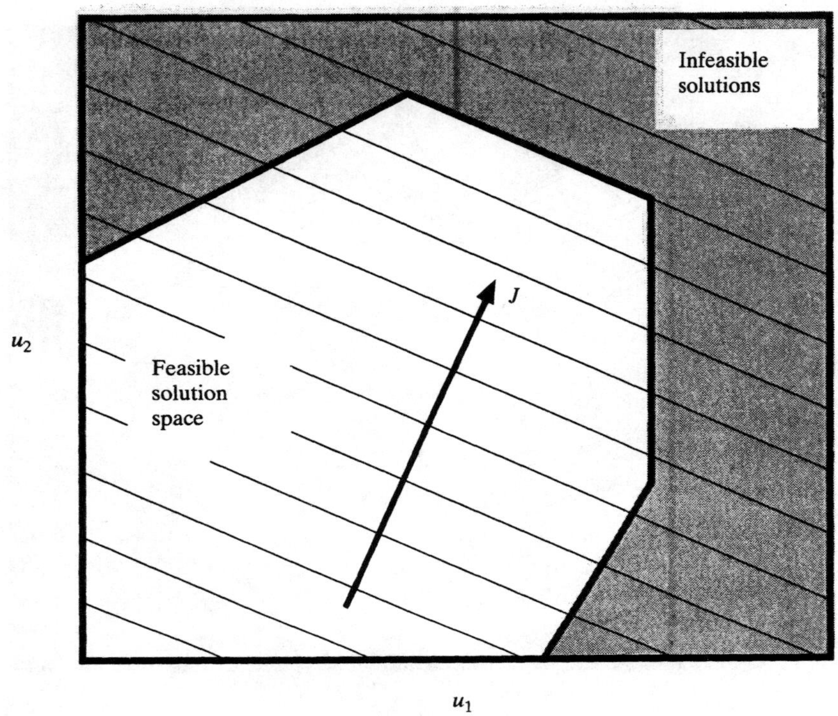

Figure 5.8 Illustration of a linear programming problem in two dimensions with a degenerate solution.

where B is an $m \times m$ matrix of full rank and consisting of the vectors

$$B = [\boldsymbol{a}_{B_1}, \ . \ . \ . \ , \boldsymbol{a}_{B_m}]$$

The vectors comprising B are linearly independent, and B is called the basis. N is an $m \times (n - m)$ matrix consisting of the vectors $\boldsymbol{a}_j$. Likewise, $\boldsymbol{u}$ and $\boldsymbol{c}$ can be decomposed

$$\boldsymbol{u} = \begin{bmatrix} \boldsymbol{u}_B \\ \boldsymbol{u}_N \end{bmatrix} = \begin{bmatrix} \bar{\boldsymbol{b}} \\ 0 \end{bmatrix} \qquad \boldsymbol{c} = \begin{bmatrix} \boldsymbol{c}_B \\ \boldsymbol{c}_N \end{bmatrix}$$

where

$$\bar{\boldsymbol{b}} = B^{-1}\boldsymbol{b} \geq \mathbf{0}$$

We call the variables $\boldsymbol{u}_B$ basic variables, and the variables $\boldsymbol{u}_N$ nonbasic variables. We can now describe the simplex algorithm as follows:

1. Find a starting guess $\boldsymbol{u}$, with a basis B, that is an extreme point in the set of feasible solutions. This could be easy to do, such as the case when $\boldsymbol{u} = \mathbf{0}$ is a feasible solu-

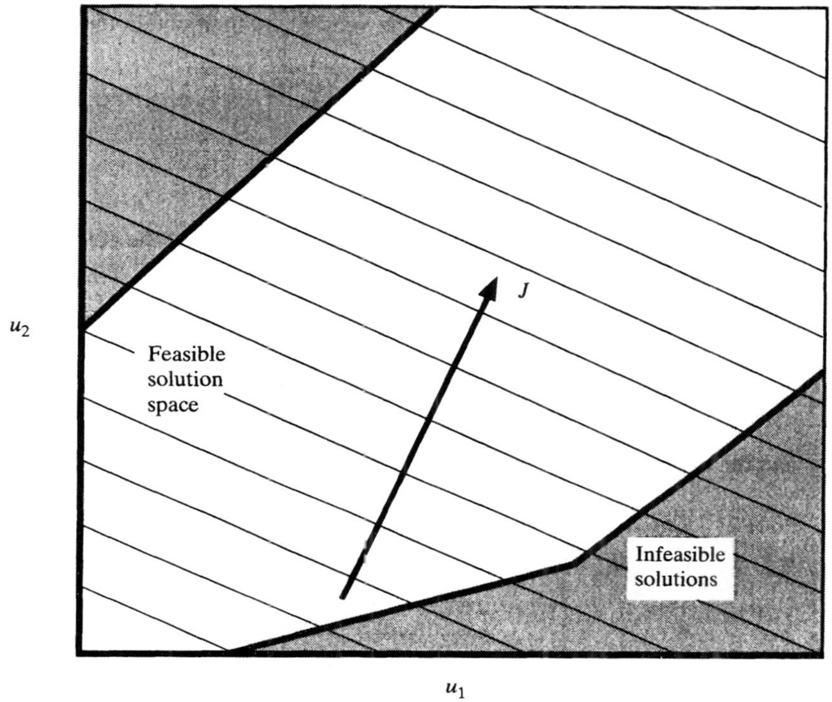

Figure 5.9 Illustration of an unbounded feasible space in two dimensions.

tion, or it may be difficult to find. In the latter case, methods exist for the determination of $\boldsymbol{u}$.

2. Calculate

$$\boldsymbol{c}_B^T \boldsymbol{B}^{-1} N - \boldsymbol{c}_N^T$$

If all the elements of this vector are non-negative, the guess $\boldsymbol{u}$ is an optimal extreme point. If at least one element of this vector is negative, pick the most negative element. If, for this element,

$$B^{-1}\boldsymbol{a}_j \leq \boldsymbol{0}$$

then the function is unbounded, and no finite optimal solution exists.

3. Compute the index r

$$\underset{1 \leq i \leq m}{\text{minimum}} \left\{ \frac{\bar{b}_i}{y_{ij}} : y_{ij} > 0 \right\} = \frac{\bar{b}_r}{y_{rj}} \geq 0$$

$$\boldsymbol{y}_j = B^{-1}\boldsymbol{a}_j$$

where y_{ij} is the ith element of the vector, and then obtain the new extreme point

$$u_{B_i} = \bar{b}_i - \frac{\bar{b}_r}{y_{rj}} y_{ij} \qquad \text{for} \quad i = 1, \ldots, m$$

$$u_j = \frac{\bar{b}_r}{y_{rj}}$$

and all other $u_i = 0$. Form the new basis by deleting the column $\boldsymbol{a}_B$ and adding the column $\boldsymbol{a}_j$. Then go back to step one taking as $\boldsymbol{u}$, $\boldsymbol{u}_j$.

This procedure can be implemented in a tableau format, which shows the simplicity of the computations. Recognizing that the objective function can be written in the form

$$J - \boldsymbol{c}_B^T \boldsymbol{u}_B - \boldsymbol{c}_N^T \boldsymbol{u}_N = 0$$

and the constraints can be written in the form

$$B\,\boldsymbol{u}_B + N\,\boldsymbol{u}_N = \boldsymbol{b}$$

we obtain the following tableau. RHS refers to the right hand side of the above equations.

J	$\boldsymbol{u}_B^T$	$\boldsymbol{u}_N^T$	RHS
1	$-\boldsymbol{c}_B^T$	$-\boldsymbol{c}_N^T$	0
$\mathbf{0}$	B	N	$\boldsymbol{b}$

The first row, corresponding to the objective function, is updated by adding to it the $\boldsymbol{c}_B^T$ times the new constraint rows.

	J	$\boldsymbol{u}_B^T$	$\boldsymbol{u}_N^T$	RHS
J	1	$\mathbf{0}^T$	$\boldsymbol{c}_B^T B^{-1} N - \boldsymbol{c}_N^T$	$\boldsymbol{c}_B^T \bar{\boldsymbol{b}}$
$\boldsymbol{x}_B$	$\mathbf{0}$	I	$B^{-1}N$	$\bar{\boldsymbol{b}}$

Despite the complex appearance of this tableau, it provides a very convenient format for storing the data necessary to conduct a linear programming optimization. In particular, it provides a convenient format for performing step 2 outlined above. If we complete step 2 and find that

$$\boldsymbol{c}_B^T B^{-1} N - \boldsymbol{c}_N^T \geq \mathbf{0}$$

then we have found the maximum extreme point and we can stop. If instead, we find

$$B^{-1}\boldsymbol{a}_j \leq \mathbf{0}$$

then the optimal solution is unbounded, and again we can stop. On the other hand, if this is not the case, we update the tableau by removing the basic variable u_B, corresponding to the index r determined previously, and introducing the variable u_j. This can be done by the following set of steps:

1. Divide the rth row, corresponding to u_B, by y_{rj}.
2. Multiply the resulting rth row by y_{ij} and subtract it from the ith constraint row, for $i = 1, \ldots, m, i \neq r$.
3. Multiply the resulting rth row by

$$c_B^T B^{-1} a_j - c_j$$

and subtract the result from the objective row.

This procedure has been implemented by many people, and a number of fine computer software packages are available for linear programming optimization. Many of these packages will handle problems of very large dimensionality. Indeed, one of the great virtues of the linear programming problem lies in its rapid and highly successful convergence. Although there are convergence problems that you can encounter, these are rare, and steps can be taken to overcome them. For these reasons, there is a strong tendency among those who are familiar with optimization problems to cast a problem in the linear programming format whenever possible.

5.8 INTEGER PROGRAMMING

A variation occurs on the linear programming problem when, in addition to the constraints given above, the solution is constrained to integer values of $\boldsymbol{u}$. Several engineering optimization problems can be formulated in the integer programming format. The research and development portfolio problem is an example. Suppose you need information on a design, and that information can be obtained through each of 10 research projects. But there are resources for only five of the projects. Which are the best five to fund? This problem can be put in the integer programming format if the benefits of each project are independent of the other projects. Other examples occur in design and manufacturing where we are dealing with things that come in units of integers, such as machines, fasteners, and people. We sometimes have the urge to assume that these things that come only in units of integers can be approximated by continuous variables, and we then find the maximum of the continuous function and round it to an integer solution. But it is not difficult to show that this approximation can yield results that are sometimes substantially worse than the best possible solution.

While it is easy to visualize the integer programming problem, as shown in Fig. 5.10, the mathematics of the optimization are greatly complicated because continuum mathematics no longer applies, and thus the conditions of optimality derived for cases such as the linear programming problem are not valid. The allowable solutions in Fig. 5.10 are only those points marked by the squares. The maximum point is denoted by a

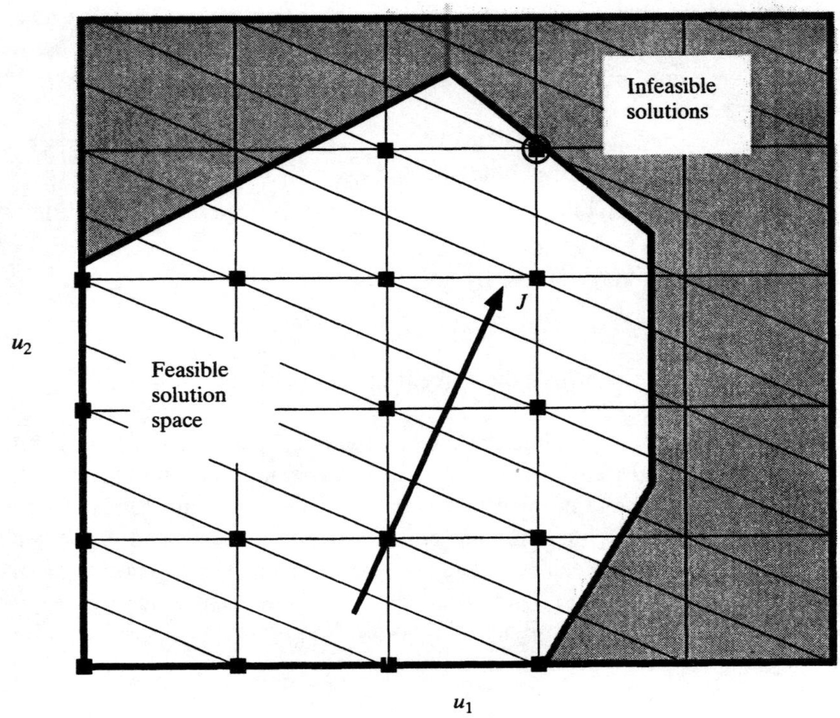

Figure 5.10 Illustration of an integer programming problem in two dimensions.

circle drawn around the square corresponding to the optimal integer values of $\boldsymbol{u}$. In a closed set, these are countable, and you could seek the optimal solution by enumeration, that is, by trying every possible feasible solution. But this can be difficult because there can be a very large number of feasible solutions and they can be difficult to find. A solution technique in common use makes use of repeated application of linear programming optimization methods. It is referred to as *Gomory's algorithm.* The procedure is essentially the following:

1. Solve the corresponding linear programming problem neglecting the constraint that the solution is constrained to integer values of $\boldsymbol{u}$. If this solution provides values of $\boldsymbol{u}$ that are all integers, then it is the solution to the integer programming problem. If not, continue.

2. If the solution to the linear programming problem produces values of $\boldsymbol{u}$ that are not all integers, introduce a new constraint into the formulation of the linear programming problem that excludes the solution from the set of feasible solutions but which does not remove any of the feasible, integer solutions $\boldsymbol{u}$, that is for example, any of the points marked by the squares in Fig. 5.10. Then go back to step 1.

The Assignment Problem

The assignment problem is an example of the application of integer programming. It relates to the assignment of resources to accomplish some end purpose. For example, suppose we desire to produce a range of products, each with demand d_j, which can be manufactured by each of various machines. Suppose further that the cost incurred by machine i in producing product j is c_{ij}, and let u_{ij} be the quantity of product j produced by machine i. We can now formulate the assignment problem as follows:

$$\text{minimize} \quad J = \sum_i \sum_j c_{ij} u_{ij}$$

$$\text{subject to} \quad \sum_i u_{ij} \geq d_j \quad \text{for all } j$$

$$u_{ij} \geq 0 \quad \text{and} \quad u_{ij} \quad \text{are integers}$$

5.9 NETWORK FLOW OPTIMIZATION

Network flow problems relate to the flow of physical things or substances through a system that consists of nodes and branches as shown in Fig. 5.11. Examples of network flows include the distribution of a liquid or gas through a system of pipes, the flow of traffic on roads, the flow of information over a telecommunications network, and the distribution of electric power over a network of wires. An example of a problem in systems engineering that could be addressed using network flow optimization is the following: Given an electric power distribution grid, what additions to the existing grid could be made to increase the grid capacity in an effective manner? Another problem is, given a road system, which roads should be improved to effectively increase the flow of traffic? In these cases, flow is a real, non-negative quantity, and we are interested in finding the maximum capacity of the system.

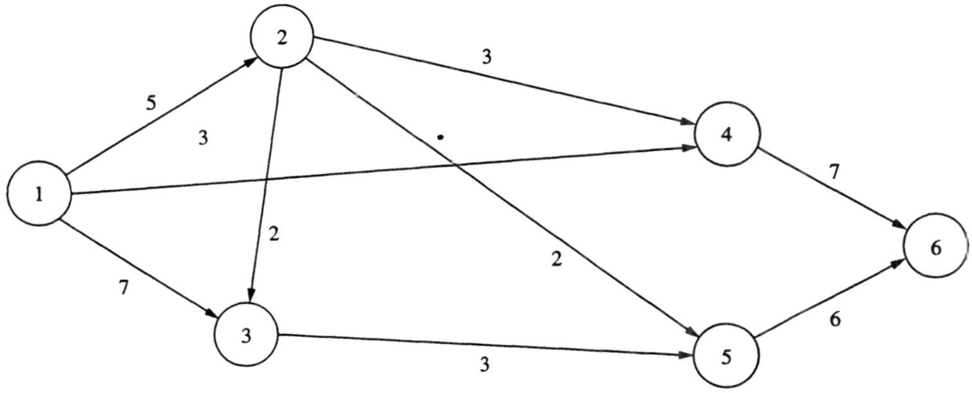

Figure 5.11 Illustration of a network flow maximization problem.

The flow in a network emanates from nodes referred to as sources, and disappears into nodes referred to as sinks. Specifically, we shall define a source as a node such that the flows on all branches emanating from the node are outward from the node. And we shall define a sink as a node such that the flows on all branches converging on the node are into it. Note that we can always construct a network with sources and sinks defined this way. The flow from a source enters the branches emanating from the source, and transfers to other branches at intermediate nodes. Conservation laws apply at the intermediate nodes, assuring that the inflow and outflow are equal at these nodes. Referring to Fig. 5.11, we define the quantity x_{ij} to be the flow from the ith node to the jth node. Further, for now, we will restrict our discussion to the case where there is a single source node and a single sink node. In general, there can be multiple sources and sinks, but again, we can reduce the problem to the case of a single source and a single sink by applying proper capacity constraints. We shall designate the source node as node 1 and the sink node as node N. The capacity of the system is then equal to the flow out of the source,

$$J = \sum_j x_{1j}$$

or, equivalently, the flow into the sink,

$$J = \sum_i x_{iN}$$

subject to constraints of the form

$$0 \le x_{ij} \le c_{ij} \qquad \text{for all } i, j$$

and

$$\sum_i x_{ik} - \sum_j x_{kj} = 0 \qquad \text{for} \quad k = 2, 3, \ldots, N-1$$

In this formulation, we refer to the c_{ij} as the branch capacities. Note that the system considered above is fully specified. Therefore, the optimization process does not involve control of the system; rather, it results in a determination of the capacity of the system. Given that we can determine the capacity of the system, then we can look at options, such as increasing the capacity of certain branches or the addition of new branches so as to increase the capacity of the system. Thus, by repeated application of the optimization process, we are able to evaluate design options.

Consider the following approach to the solution of the network flow problem as stated above. Pick any path from the source node to the sink node whose branches all have a positive flow capacity. Now the branch with the lowest capacity will limit the flow along this path. If we assume that this path carries the limiting flow, and we subtract that flow from the capacities of all the branches, the remaining capacities of all the branches will be reduced, and the remaining capacity of the limiting branch will be zero. Then we could pick another path and repeat the process, and so on until no paths from source to sink remain that have a positive flow. This process is shown in the table below. We are then inclined to think that the network capacity is the sum of the flows, F_i, that we found for the individual paths in the figure. As shown here, the total flow is 11.

Path	F_i	c_{12}	c_{13}	c_{14}	c_{23}	c_{24}	c_{25}	c_{35}	c_{46}	c_{56}
Init. cond.	—	5	7	3	2	3	2	3	7	6
1,2,4,6	3	2	7	3	2	0	2	3	4	6
1,4,6	3	2	7	0	2	0	2	3	1	6
1,2,5,6	2	0	7	0	2	0	0	3	1	4
1,3,5,6	3	0	4	0	2	0	0	0	1	1

Unfortunately, this simple approach does not always provide the system capacity as is demonstrated in the following table that solves the same problem but in a different order, yielding a flow of 9. We see that the method is sensitive to the order in which the paths are chosen and, hence, cannot be used in the form described above. It is possible, however, to modify this approach to obtain a viable optimization algorithm. The resulting, more systematic approach, is referred to as the *labeling process* of Ford and Fulkerson.

Path	F_i	c_{12}	c_{13}	c_{14}	c_{23}	c_{24}	c_{25}	c_{35}	c_{46}	c_{56}
Init. cond.	—	5	7	3	2	3	2	3	7	6
1,2,5,6	2	3	7	3	2	3	0	3	7	4
1,2,3,5,6	2	1	7	3	0	3	0	1	7	2
1,4,6	3	1	7	0	0	3	0	1	4	2
1,2,4,6	1	0	7	0	0	2	0	1	3	2
1,3,5,6	1	0	6	0	0	2	0	0	3	1

The labeling process method, instead of examining paths one at a time, essentially fans out to examine all paths simultaneously. We first define quantities that we call *excess capacities*, d_{ij}, for each branch. These are defined so that excess capacities in both directions are permitted. However, if reverse flow is not permitted, we invoke the constraint $c_{ji} = 0$.

$$d_{ij} = c_{ij} - x_{ij} + x_{ji} \qquad \text{for all } i, j$$
$$d_{ji} = c_{ji} - x_{ji} + x_{ij}$$

We start the procedure with all flows equal to zero, and the capacities given by the excess capacities. We now label the nodes as in the following table, corresponding to Fig. 5.11. We begin with the source node, node 1, and label all the nodes that can be reached from this node with positive flow. The label given to the target node is (α,β), where α is the excess capacity of the flow to the target node, and β is the name of the node from which we came. If we labeled the source node, we stop the labeling and continue with the algorithm. If we did not label the source node, we continue to fill out the table. In the example shown, we then trace from the nodes just labeled to new, unlabeled nodes. We label each of these as before, (α,β), where β is the name of the node from which we came to the present node, and α is now the minimum of the excess capacity on the branch from the previous node to the current node or the value of α for the preceding node. Thus, α is the minimum excess capacity for the current path. We continue this process as shown in the table until we reach the sink node. In the table shown, we see that we can increase the flow by the flow that is given in the label for the sink node, in this case, 3. We can also determine the path that this flow takes, as it is given by the indices β. If, in this labeling process, we cannot reach the sink node on a path that has positive excess capacity, we have reached the solution, and we can stop the process.

From node	Target node	Label (α, β)
1	2	5,1
1	3	7,1
1	4	3,1
2	5	2,2
4	6	3,4

Assuming that we reached the sink node with positive excess capacity, we now go back to the network, and update the excess capacities by the added flow into the sink node, α_N.

$$\hat{d}_{uv} = d_{uv} - \alpha_N$$

$$\hat{d}_{vu} = d_{vu} + \alpha_N$$

$$\hat{d}_{ij} = d_{ij} \qquad \text{for} \quad i, j \text{ not on the labeling path}$$

Next, we reconstruct the above table using the new excess capacities. And we repeat this process until we can no longer reach the sink node on a path with positive excess capacity. When this occurs, we have obtained a solution.

5.10 THE TRANSPORTATION PROBLEM

The transportation problem is a broad class of the liner programming problem that is so named because of the context within which it was first formulated. Suppose that a product is to be distributed from m distribution centers or sources to n destinations. Denote by a_i the amount of the product available at the source i, and by b_j the amount of the product desired at destination j. Let c_{ij} be the cost of shipping the product from source i to destination j, and let u_{ij} be the amount of product shipped from source i to destination j. We also take shipping cost to be proportional to the amount shipped. Then the form of the transportation problem is as follows:

$$\text{minimize} \quad J = \sum_{i=1}^{m} \sum_{j=1}^{n} c_{ij}\, u_{ij}$$

subject to the constraints

$$\sum_{j=1}^{n} u_{ij} = a_i \qquad \text{for} \quad i = 1, 2, \ldots, m \quad a_i > 0$$

$$\sum_{i=1}^{m} u_{ij} = b_j \qquad \text{for} \quad j = 1, 2, \ldots, n \quad b_j > 0$$

$$u_{ij} \geq 0 \qquad \text{for all } i, j$$

This is the standard form of the linear programming problem. Indeed, the transportation problem always takes this form, which is considerably simpler than the general form of the linear programming problem. The solution of this problem is straightforward and can be accomplished using a linear programming algorithm.

5.11 CALCULUS OF VARIATIONS

The calculus of variations is an optimization technique that deals, in essence, with an infinity of control variables in the sense that we are seeking not the optimal values of $\boldsymbol{u}$, but rather an optimal path $\boldsymbol{u}(t)$. It is a very powerful technique in its ability to represent engineering options and find optimal solutions. The general optimization problem in the calculus of variations may be written in the form

$$j = \int_{t_a}^{t_b} f(\boldsymbol{x}, \dot{\boldsymbol{x}}, t)\, dt$$

subject to

$$\dot{\boldsymbol{x}} = \boldsymbol{g}(\boldsymbol{x}, \boldsymbol{u}, t)$$

$$\boldsymbol{x}(t_a) = \boldsymbol{x}_a$$

$$\boldsymbol{x}(t_b) = \boldsymbol{x}_b$$

where $\dot{\boldsymbol{x}} = d\boldsymbol{x}/dt$. We refer to the right hand side of J as a *functional*, and the constraints $\boldsymbol{g}$ are in differential form. The calculus of variations is particularly important in the use phase of a product's life cycle, and it often deals with nonphysical aspects of a system. For example,

1. What path should an airplane or ship take between two points to minimize time enroute or fuel consumption?

2. What path from Earth to Mars enables a spacecraft to traverse the space using a minimum amount of propellant?

We also use the calculus of variations as an aid to making design decisions such as the following:

3. What two-dimensional shape encloses the maximum area with the minimum perimeter?
4. For a ball rolling downhill, starting from rest at point A and ending at point B, what is the shape of the slope that enables the ball to traverse from A to B in the minimum time?
5. For a cantilever beam, what thickness as a function of length maximizes load carrying capability for a specified weight?
6. What is the optimum price as a function of time to place on a limited resource?

In the preceding problem formulation, note that J is a scaler to be extremized over the interval t_a to t_b. t is the independent variable. Suppose we want to prove that a straight line provides the shortest path between two points. From Fig. 5.12, we see that the distance between a and b is given by the integral of ds over a to b. Since

$$ds = \sqrt{dt^2 + dx^2} = \sqrt{1 + (dx/dt)^2}\, dt$$

we obtain

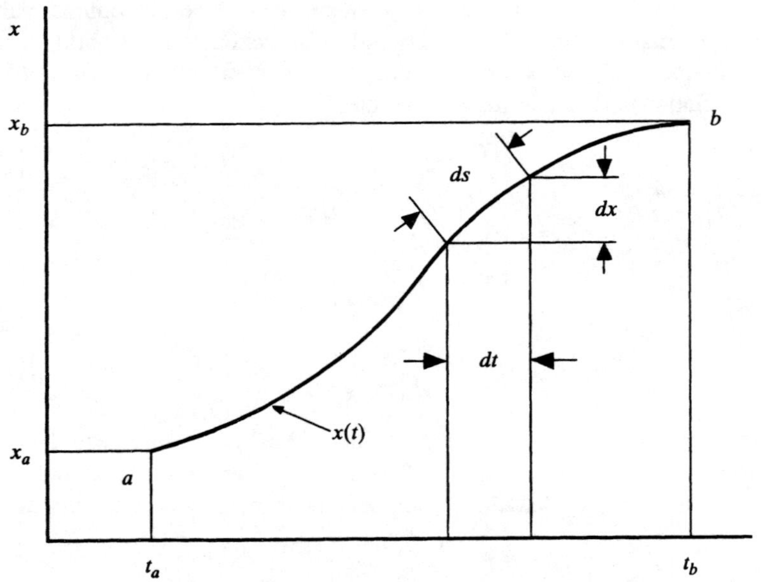

Figure 5.12 The distance between two points.

$$J = \int_{t_a}^{t_b} \sqrt{1 + \left(\frac{dx}{dt}\right)^2}\, dt = \int_{t_a}^{t_b} \sqrt{1 + \dot{x}^2}\, dt$$

Here, dx/dt is not specified in any way. In fact, the problem is to determine dx/dt as a function of t. One approach would be to break up the interval between a and b into some large but finite number of intervals, and determine the optimal value of dx/dt on each interval using calculus. But, no matter how many intervals we choose, the fact that we use only a finite number of intervals imposes the constraint that dx/dt is limited to that number of values, and we would always expect that relaxing the constraint, that is, increasing the number of intervals might lead to a better result. To obtain the true minimum of J, we must allow for continuous variation in dx/dt. In essence then, J is a function of an infinite number of variables.

To solve a problem of this sort, we proceed much the same as we did in the case of a function of a finite number of variables. We form the variation of J, δJ and determine the appropriate necessary conditions. For our example case, the minimum distance between two points,

$$J = \int_{t_a}^{t_b} f(\dot{x})\, dt = \int_{t_a}^{t_b} \sqrt{1 + \dot{x}^2}\, dt$$

and

$$\delta J = J[f(\dot{x} + \delta\dot{x})] - J[f(\dot{x})]$$

where

$$f(\dot{x} + \delta\dot{x}) = \sqrt{1 + (\dot{x} + \delta\dot{x})^2}$$

Expanding this,

$$f(\dot{x} + \delta\dot{x}) = \sqrt{1 + \dot{x}^2 + 2\dot{x}\delta\dot{x} + \delta\dot{x}^2}$$

$$= \sqrt{(1 + \dot{x}^2)\left(1 + \frac{2\dot{x}\delta\dot{x}}{1 + \dot{x}^2} + \frac{\delta\dot{x}^2}{1 + \dot{x}^2}\right)}$$

Now, we will compare a candidate optimal function $\dot{x}(t)$ to a neighboring function $\dot{x}'(t)$

$$\delta\dot{x}(t) = \dot{x}'(t) - \dot{x}(t)$$

where we will assume that $\delta\dot{x}$ is small. Note, however, from Fig. 5.13, that simply because δx is small, it does not follow that $\delta\dot{x}$ is small. The sawtooth curve shown in this figure is a variation for which δx can be made arbitrarily small, while $\delta\dot{x}$ remains large. We refer to δx and $\delta\dot{x}$ as the *variations* of $x(t)$ and $\dot{x}(t)$, respectively. When $\delta\dot{x}(t)$ becomes small as δx is made small, we say that $\delta\dot{x}(t)$ is a *weak variation*. Restricting ourselves to weak variations for the time being, we may take $\delta\dot{x}$ to be small so that $\delta\dot{x}^2$ is of second order and can be neglected, giving

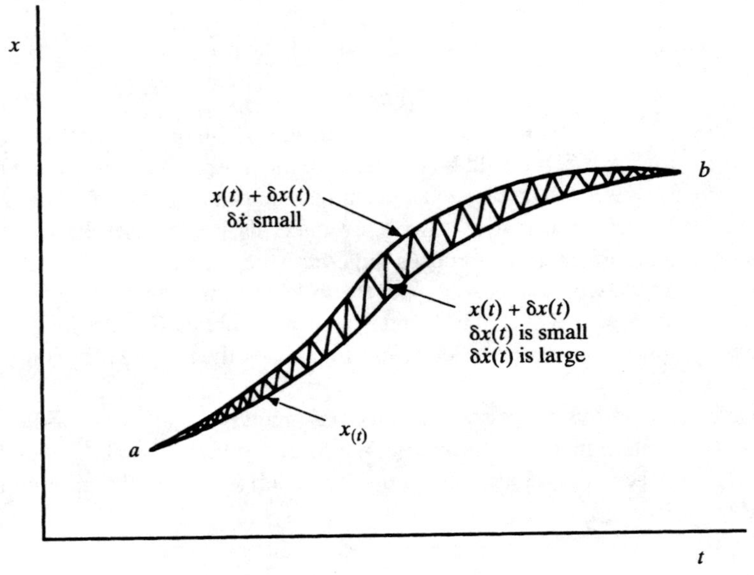

Figure 5.13 Variations in $\delta\dot{x}(t)$ with $\delta x(t)$ small.

$$f(\dot{x} + \delta\dot{x}) = \sqrt{1 + \dot{x}^2}\left(1 + \frac{\dot{x}\delta\dot{x}}{1 + \dot{x}^2}\right)$$

$$= \sqrt{1 + \dot{x}^2} + \frac{\dot{x}\delta\dot{x}}{\sqrt{1 + \dot{x}^2}}$$

Now we can write the variation in J as follows:

$$\delta J = \int_{t_a}^{t_b} \left[\sqrt{1 + \dot{x}^2} + \frac{\dot{x}\delta\dot{x}}{\sqrt{1 + \dot{x}^2}}\right] dt - \int_{t_a}^{t_b} \sqrt{1 + \dot{x}^2}\, dt$$

$$= \int_{t_a}^{t_b} \frac{\dot{x}}{\sqrt{1 + \dot{x}^2}}\, \delta\dot{x}\, dt$$

Clearly, if the path $\dot{x}(t)$ minimizes J, then $\delta J \geq 0$ for all $\dot{x}(t) + \delta x(t)$. But, in the above equation, we see that the sign of δJ depends on the sign of $\delta\dot{x}(t)$ if $\delta x(t)$ appears to first order. Thus, to first order, for a minimum, we must have

$$\delta J = 0$$

We note further that $\delta\dot{x}(t)$ is an arbitrary function within limits. The limits are that $\delta x(t)$ and $\delta\dot{x}(t)$ must both be small. But merely keeping $\delta\dot{x}(t)$ small does not assure that $\delta x(t)$

will remain small. Thus, we must express δJ in terms of $\delta x(t)$ rather than $\delta \dot{x}(t)$. Since t is the independent parameter,

$$\delta \dot{x} = \frac{d}{dt}\,\delta x(t)$$

Substituting this into the above equation for δJ gives

$$\delta J = \int_{t_a}^{t_b} \frac{\dot{x}}{\sqrt{1+\dot{x}^2}}\,\frac{d}{dt}\,\delta x\,dt$$

This equation can be integrated by parts.

$$\delta J = \frac{\dot{x}}{\sqrt{1+\dot{x}^2}}\,\delta x\,\Big|_{t_a}^{t_b} - \int_{t_a}^{t_b} \frac{d}{dt}\left(\frac{\dot{x}}{\sqrt{1+\dot{x}^2}}\right)\delta x\,dt = 0$$

Since the points a and b are given,

$$x(t_a) = x_a$$

$$x(t_b) = x_b$$

$$\delta x(t_a) = 0$$

$$\delta x(t_b) = 0$$

Recall now that $\delta x(t)$ must remain small everywhere, but it is otherwise arbitrary. Thus, it is allowable for $\delta x(t)$ to be zero everywhere except in one small region. It follows that, if we are to enforce the condition $\delta J = 0$, then we must have

$$\frac{d}{dt}\left(\frac{\dot{x}}{\sqrt{1+\dot{x}^2}}\right) = 0$$

for all t on the interval $t_a < t < t_b$. If this condition is not met, we could choose a $\delta x(t)$ that would make $\delta J < 0$. The above equation is

$$\ddot{x}\left[\frac{1}{\sqrt{1+\dot{x}^2}}\left(1 + \frac{\dot{x}^2}{1+\dot{x}^2}\right)\right] = 0$$

But the bracketed part of this equation is non-zero positive for all finite $\dot{x}$. Thus, we obtain

$$\ddot{x} = 0$$

which is the equation of a straight line. And thus we have shown that the path of minimum distance between two points is a straight line.

Necessary Conditions

Now let us turn to a more formal derivation of the necessary conditions for the extremization of a functional. Using the more general form of J, we obtain

$$\delta J = \int_{t_a}^{t_b} f(\boldsymbol{x} + \delta\boldsymbol{x}, \dot{\boldsymbol{x}} + \delta\dot{\boldsymbol{x}}, t)\, dt - \int_{t_a}^{t_b} f(\boldsymbol{x}, \dot{\boldsymbol{x}}, t)\, dt$$

But, to first order,

$$f(\boldsymbol{x} + \delta\boldsymbol{x}, \dot{\boldsymbol{x}} + \delta\dot{\boldsymbol{x}}, t) = f(\boldsymbol{x}, \dot{\boldsymbol{x}}, t) + \frac{\partial f}{\partial \boldsymbol{x}}\, \delta\boldsymbol{x} + \frac{\partial f}{\partial \dot{\boldsymbol{x}}}\, \delta\dot{\boldsymbol{x}}$$

Hence, δJ can be written

$$\delta J = \int_{t_a}^{t_b} \left[\frac{\partial f}{\partial \boldsymbol{x}}\, \delta\boldsymbol{x} + \frac{\partial f}{\partial \dot{\boldsymbol{x}}}\, \delta\dot{\boldsymbol{x}} \right] dt$$

which, in turn, can be written

$$\delta J = \int_{t_a}^{t_b} \left[\frac{\partial f}{\partial \boldsymbol{x}}\, \delta\boldsymbol{x} + \frac{\partial f}{\partial \dot{\boldsymbol{x}}} \frac{d}{dt}\, \delta\boldsymbol{x} \right] dt$$

Integrating the second term in this equation by parts,

$$\int_{t_a}^{t_b} \frac{\partial f}{\partial \dot{\boldsymbol{x}}} \frac{d}{dt}\, \delta\boldsymbol{x}\, dt = \frac{\partial f}{\partial \dot{\boldsymbol{x}}}\, \delta\boldsymbol{x} \Big|_{t_a}^{t_b} - \int_{t_a}^{t_b} \frac{d}{dt} \left(\frac{\partial f}{\partial \dot{\boldsymbol{x}}} \right) \delta\boldsymbol{x}\, dt$$

Thus, δJ becomes

$$\delta J = \frac{\partial f}{\partial \dot{\boldsymbol{x}}}\, \delta\boldsymbol{x} \Big|_{t_a}^{t_b} + \int_{t_a}^{t_b} \left[\frac{\partial f}{\partial \boldsymbol{x}} - \frac{d}{dt} \left(\frac{\partial f}{\partial \dot{\boldsymbol{x}}} \right) \right] \delta\boldsymbol{x}\, dt$$

Again, $\delta\boldsymbol{x}(t)$ is an arbitrary function of t, subject only to the restriction that it remain small. Thus, in order that $\delta J \geq 0$ in the case of minimization, or $\delta J \leq 0$ in the case of maximization, for all admissible $\delta\boldsymbol{x}(t)$, we must take $\delta J = 0$ everywhere on the interval $t_a < t < t_b$. This gives us the condition

$$\frac{\partial f}{\partial \boldsymbol{x}} - \frac{d}{dt} \left(\frac{\partial f}{\partial \dot{\boldsymbol{x}}} \right) = 0$$

This result is known as the *Euler-Lagrange equation.* It is a necessary condition for an extremum of J. Independently, at the end points, we obtain

$$\frac{\partial f}{\partial \dot{\boldsymbol{x}}}\, \delta\boldsymbol{x} \Big|_{t_a}^{t_b} = 0$$

This equation is called the *transversality condition.* Its name comes from the fact that it demands that the vectors $\partial f/\partial \dot{\boldsymbol{x}}$ and $\delta\boldsymbol{x}$, if they are both non-zero, be orthogonal or transverse at the times t_a and t_b. The transversality condition is also a necessary condition for an extremum of J, and both the transversality condition and the Euler-Lagrange equations must be satisfied on an optimal arc, $\boldsymbol{x}(t)$.

Problem Formulation in the Calculus of Variations

There are three basic extremization formulations in the calculus of variations. These are the problems of Lagrange, Mayer, and Bolza, and they take the following forms, where it is desired to extremize J:

The problem of Lagrange—

$$J = \int_{t_a}^{t_b} f(\boldsymbol{x}, \dot{\boldsymbol{x}}, t)\, dt$$

The problem of Mayer—

$$J = g(\boldsymbol{x}_i, t_i)$$

The problem of Bolza—

$$J = g(\boldsymbol{x}_i, t_i) + \int_{t_a}^{t_b} f(\boldsymbol{x}, \dot{\boldsymbol{x}}, t)\, dt$$

We can show that the problems of Lagrange and Bolza are equivalent to the problem of Mayer by increasing the dimensionality of $\boldsymbol{x}$ from n to $n + 1$, where the $(n + 1)$th component is given by

$$\dot{x}_{n+1} = f(\boldsymbol{x}, \dot{\boldsymbol{x}}, t)$$

Then

$$x_{n+1}(t_b) = \int_{t_a}^{t_b} f(\boldsymbol{x}, \dot{\boldsymbol{x}}, t)\, dt$$

and so the problem of Lagrange can be written

$$J = x_{n+1}(t_b)$$

Similarly, we can write the problem of Bolza

$$J = g(\boldsymbol{x}_i, t_i) + x_{n+1}(t_b) = g * (\boldsymbol{x}_i, t_i)$$

Thus, we see that the three problem formulations are equivalent, and all are subject to the necessary conditions given above.

The Open Time Case

There are occasions when we do not want to constrain the times t_a and t_b; rather, we wish to extremize J with respect to either or both of these quantities as well. For example, we will consider the case in which t_b is free to be optimized. Now, in obtaining the variation in J, we must allow also for variations in the upper limit of integration. These variations are as shown in Fig. 5.14. From the figure, we see the first-order relationship

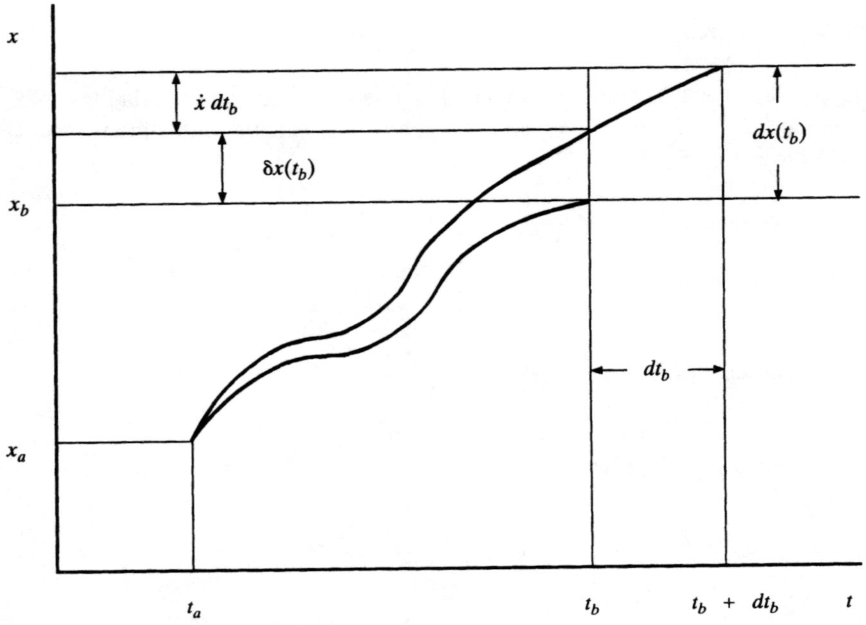

Figure 5.14 Variations in x at the final time for the case where the final time is free to be optimized.

$$dx_b = \delta x_b + \dot{x}_b\, dt_b$$

which we can rearrange to give

$$\delta x_b = dx_b - \dot{x}_b\, dt_b$$

Now taking the Bolza formulation, the variation of J becomes

$$\delta J = \frac{\partial g}{\partial x_b}\, dx_b + \frac{\partial g}{\partial t_b}\, dt_b + f(t_b)\, dt_b + \frac{\partial f}{\partial \dot{x}}\, \delta x \Big|_{t_a}^{t_b} + \int_{t_a}^{t_b} \left[\frac{\partial f}{\partial x} - \frac{d}{dt}\left(\frac{\partial f}{\partial \dot{x}} \right) \right] \delta x\, dt$$

where the term in $f(t_b)$ comes from differentiating with respect to the limits of integration. Combining the above two equations,

$$\delta J = \frac{\partial g}{\partial x_b}\, dx_b + \frac{\partial g}{\partial t_b}\, dt_b + f(t_b)\, dt_b + \frac{\partial f}{\partial \dot{x}}\, dx_b - \frac{\partial f}{\partial \dot{x}}\, \dot{x}_b\, dt_b + \int_{t_a}^{t_b} \left[\frac{\partial f}{\partial x} - \frac{d}{dt}\left(\frac{\partial f}{\partial \dot{x}} \right) \right] \delta x\, dt$$

Grouping the terms in dt_b (note that all other terms are the same whether or not we fix t_b) and assuring that $\delta J = 0$

$$\frac{\partial g}{\partial t_b} + f(t_b) - \frac{\partial f}{\partial \dot{x}_b} \dot{x}_b = 0$$

By a similar process, we can extend this to the case where x is an n-dimensional vector and to the initial time, t_a. We obtain the necessary conditions

$$\frac{\partial g}{\partial t_a} - f(t_a) + \frac{\partial f}{\partial \dot{\mathbf{x}}_a} \dot{\mathbf{x}}_a = 0$$

$$\frac{\partial g}{\partial t_b} + f(t_b) - \frac{\partial f}{\partial \dot{\mathbf{x}}_b} \dot{\mathbf{x}}_b = 0$$

In these equations, it is common to group the terms in f, and define the Hamiltonian, H:

$$H = \frac{\partial f}{\partial \dot{\mathbf{x}}} \dot{\mathbf{x}} - f$$

Thus, if g is autonomous, that is, if it does not contain terms in which t appears explicitly, then

$$\frac{\partial g}{\partial t} = 0$$

and we must have $H = 0$ at t_a and t_b if they are optimally chosen. Furthermore, we see that for autonomous g,

$$H = -\left.\frac{\partial J}{\partial t}\right|_{t=t_b}$$

Suppose, on the other hand, our problem is to minimize t_b. Then

$$g = t_b$$

and

$$\left.\frac{\partial g}{\partial t}\right|_{t_b} = 1$$

and we must have $H(t_b) = 1$.

The Maximum Principle

Thus far we have written J, in the Lagrange formulation, as

$$J = \int_{t_a}^{t_b} f(\mathbf{x}, \dot{\mathbf{x}}, t)\, dt$$

and we have expanded f to obtain the variation of J as follows:

$$f(x + \delta x, \dot{x} + \delta\dot{x}, t) = f(x,\dot{x},t) + \frac{\partial f}{\partial x}\,\delta x + \frac{\partial f}{\partial \dot{x}}\,\delta\dot{x}$$

where δx and $\delta\dot{x}$ are variations in x and $\dot{x}$, respectively. In order that this expansion of f be valid, $\partial f/\partial x$ and $\partial f/\partial \dot{x}$ and all higher order derivatives must exist and be finite and δx and $\delta\dot{x}$ must be small. We referred to this as a weak variation. However, we saw in Fig. 5.13 that it is possible for δx to be small while $\delta\dot{x}$ is large beyond all bounds. Variations that allow $\delta\dot{x}$ to be large are called *strong variations.* They arise in problems of optimal control. The necessary conditions that we derived for the case of weak variations apply also in the case of strong variations, but we have an additional condition that applies as well. This condition is referred to as the *Pontryagin maximum principle* or the *Weierstrass condition.*

Suppose that $x(t)$ minimizes J with respect to strong variations in x. Then, for all admissible strong variations $x^*(t)$

$$J \le J^*$$

In other terms,

$$\int_{t_a}^{t_b} f(x,\dot{x},t)\,dt \le \int_{t_a}^{t_b} f(x^*,\dot{x}^*,t)\,dt$$

For maximization, we would simply reverse all inequalities involving J. By $x^*(t)$, we imply that we have kept

$$|\delta x| = |x^* - x| < \varepsilon$$

where ε is an arbitrarily small positive number. But

$$|\delta\dot{x}| = |\dot{x}^* - \dot{x}|$$

is not necessarily small, and it does not approach zero as $|\delta x|$ does. We will derive a necessary condition by constructing a strong variation and requiring that the above condition on f be satisfied.

Let $Q(x_Q,t_Q)$ and $P(x_P,t_P)$ be points along a minimizing arc $x(t)$ without corners, that is, without strong variations in $\dot{x}$, and with $t_Q > t_P$ as shown in Fig. 5.15. Let x^* be a comparison curve where

$$x^*(t) = \Gamma(t) \qquad \text{for} \quad t_P \le t \le \alpha$$

$$x^*(t) = \Delta(t,\alpha) \qquad \text{for} \quad \alpha \le t \le t_Q$$

and x^* is an admissible variation such that:

1. $t_P \le \alpha \le t_Q$
2. $\Delta(t,\alpha)$ is an arbitrarily chosen n-parameter family (x is an n-vector) of arcs connecting the movable point R on $\Gamma(t)$ to point Q, constructed such that $\Delta(t,\alpha)$ will coincide with $x(t)$ as $\alpha \to t_P$, that is, as $R \to P$. For example, $\Delta(t,\alpha)$ could be

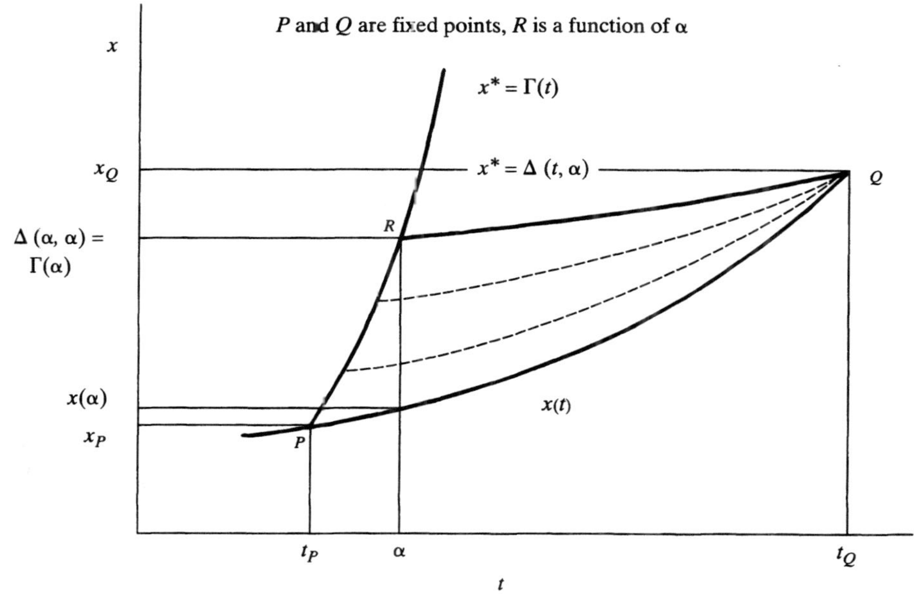

Figure 5.15 A strong variation.

$$\Delta(t,\alpha) = x(t) + \frac{\Gamma(\alpha) - x(\alpha)}{t_Q - \alpha}(t_Q - t)$$

It follows that

$$\Delta(t_Q, \alpha) = x_Q$$

$$\Delta(t, \alpha)\big|_{\alpha=t_P} = \Delta(t, t_P) = x(t)$$

and

$$x_P = x(t_P) = \Gamma(t_P) = \Delta(t_P, t_P)$$

Now note that, while $|\delta x| < \varepsilon$ at all points between t_P and t_Q as $\alpha \to t_P$, the same is not true of $|\delta \dot{x}|$. Along $\Delta(t,\alpha)$, $|\delta \dot{x}| < \varepsilon$ as $\alpha \to t_P$; therefore $\Delta(t,\alpha)$ is a weak variation from $x(t)$. But along $\Gamma(t)$, $|\delta \dot{x}|$ does not go to zero as $\alpha \to t_P$. Along $\Gamma(t)$, at point P,

$$|\delta \dot{x}|_P = |\dot{\Gamma}(t_P) - \dot{x}(t_P)| \neq 0$$

Thus we have constructed a strong variation from $x(t)$ at the point P. On the arc segment $\Gamma(t)$, $|\delta \dot{x}|_P$ is arbitrary.

As we showed above, if $x(t)$ is to minimize J with respect to strong variations in $x(t)$, then

$$J(x) \leq J(x^*)$$

Hence

$$\int_{t_P \text{ along } x}^{t_Q} f(x, \dot{x}, t)\, dt \leq \int_{t_P \text{ along } \Gamma}^{\alpha} f(\Gamma, \dot{\Gamma}, t)\, dt + \int_{\alpha \text{ along } \Delta}^{t_Q} f(\Delta, \dot{\Delta}, t)\, dt$$

In the limit as $\alpha \rightarrow t_P$ we define

$$\alpha = t_P + \delta t$$

Then

$$\int_{t_P}^{t_P+\delta t} f(\Gamma, \dot{\Gamma}, t)\, dt + \int_{t_P+\delta t}^{t_Q} f(\Delta, \dot{\Delta}, t)\, dt - \int_{t_P}^{t_Q} f(x, \dot{x}, t)\, dt \geq 0$$

Our task now is to evaluate this inequality. We proceed as follows.

We have already examined the variation of J resulting from weak variations, for example, from $x(t)$ to $\Delta(t,\alpha)$. Including end-point variations, in the Lagrange form, δJ can be written

$$\begin{aligned}\delta J &= f(t)\, dt \Big|_{t_a}^{t_b} + \frac{\partial f}{\partial \dot{x}}\, dx \Big|_{t_a}^{t_b} - \frac{\partial f}{\partial \dot{x}}\, d\dot{x}\, dt \Big|_{t_a}^{t_b} \\ &\quad + \int_{t_a}^{t_b} \left[\frac{\partial f}{\partial x} - \frac{d}{dt}\left(\frac{\partial f}{\partial \dot{x}} \right) \right] \delta x\, dt \\ &= \left[-H\, dT + \frac{\partial f}{\partial \dot{x}}\, dX \right]_{t_a}^{t_b} + \int_{t_a}^{t_b} \left[\frac{\partial f}{\partial x} - \frac{d}{dt}\left(\frac{\partial f}{\partial \dot{x}} \right) \right] \delta x\, dt\end{aligned}$$

where dT is the variation in t_a and t_b and dX is the variation in x_a and x_b with t_a, t_b, and δx at t_a and t_b. Thus, dX/dT is the slope of the boundary curves at the end points t_a and t_b. Expanding the first term in this equation,

$$\begin{aligned}\left[-H\, dT + \frac{\partial f}{\partial \dot{x}}\, dX \right]_{t_a}^{t_b} &= \left[\left(f - \frac{\partial f}{\partial \dot{x}}\, \dot{x} \right) dT + \frac{\partial f}{\partial \dot{x}}\, dX \right]_{t_a}^{t_b} \\ &= \left[\left(f - \frac{\partial f}{\partial \dot{x}}\, \dot{x} + \frac{\partial f}{\partial \dot{x}}\, d\dot{X} \right) dT \right]_{t_a}^{t_b}\end{aligned}$$

We must be careful to note that $dx/dt = \dot{x}$ is the slope of the extremal curve $x(t)$; dT is the change in t at the boundary. For the case at hand, $\Delta(t,\alpha)$ is a weak variation from $x(t)$ as $\alpha \rightarrow t_P$. The end points of the $\Delta(t,\alpha)$ curve are R on the curve $\Gamma(t)$ and the fixed point Q on $x(t)$. We can now evaluate the variation of J along $\Delta(t,\alpha)$.

$$\begin{aligned}&\int_{t_P+\delta t}^{t_Q} f(\Delta,\dot{\Delta},t)\, dt - \int_{t_P}^{t_Q} f(x,\dot{x},t)\, dt \\ &\quad = -\left[f(x,\dot{x},t) - \frac{\partial f(x,\dot{x},t)}{\partial \dot{x}}\, \dot{x} + \frac{\partial f(x,\dot{x},t)}{\partial \dot{x}}\, \dot{X} \right] dT \Big|_{t=t_P}\end{aligned}$$

The minus sign in front of the bracketed term obtains because this term comes from the lower limit of the above equation.

We have one integral yet to evaluate in the preceeding inequality. This integral can be obtained using the mean value theorem, which states

$$\int_{a}^{a+\delta t} f(t)\, dt = f(a + b\ \delta t)\ \delta t$$

where $f(t)$ is continuous in $(a,a + \delta t)$, and b is a number, $0 \le b \le 1$. Using this theorem,

$$\int_{t_P}^{t_P+\delta t} f(\Gamma,\dot{\Gamma},t)\, dt = f[\Gamma(t_P + b\ \delta t), \dot{\Gamma}(t_P + b\ \delta t), t_P + b\ \delta t]\ dt$$

where $0 \le b \le 1$.

We now obtain, for the inequality,

$$f[\Gamma(t_P + b\ \delta t),\Gamma(t_P + b\ \delta t),t_P + b\ \delta t]\ \delta t - \left[f(\boldsymbol{x},\dot{\boldsymbol{x}},t) - \frac{\partial f(\boldsymbol{x},\dot{\boldsymbol{x}},t)}{\partial \dot{\boldsymbol{x}}}\ \dot{\boldsymbol{x}} + \frac{\partial f(\boldsymbol{x},\dot{\boldsymbol{x}},t)}{\partial \dot{\boldsymbol{x}}}\ \dot{X}\right] dT\ \Bigg|_{t=t_P} \ge 0$$

Since, as $\delta t \to 0$, $b\ \delta t \to 0$,

$$\Gamma(t_P + b\ \delta t) \to \Gamma(t_P) = \boldsymbol{x}(t_P)$$

$$\dot{\Gamma}(t_P + b\ \delta t) \to \dot{\Gamma}(t_P) = \dot{X}(t_P)$$

and noting that

$$\dot{X} = \text{slope of } \Gamma(t) \text{ at } t = t_P$$

Thus, the above inequality becomes

$$f(\boldsymbol{x},\dot{X},t)\Big|_{t_P} - \left[f(\boldsymbol{x},\dot{\boldsymbol{x}},t) - \frac{\partial f(\boldsymbol{x},\dot{\boldsymbol{x}},t)}{\partial \dot{\boldsymbol{x}}}\ \dot{\boldsymbol{x}} + \frac{\partial f(\boldsymbol{x},\dot{\boldsymbol{x}},t)}{\partial \dot{\boldsymbol{x}}}\ \dot{X}\right]\Bigg|_{t_P} \ge 0$$

This inequality holds for any arbitrary point P along $\boldsymbol{x}(t)$. Thus, a necessary condition for $\boldsymbol{x}(t)$ to furnish a strong minimum to J is that, at every point on $\boldsymbol{x}(t)$

$$E(\boldsymbol{x},\dot{\boldsymbol{x}},\dot{X},t) = f(\boldsymbol{x},\dot{X},t) - f(\boldsymbol{x},\dot{\boldsymbol{x}},t) - \frac{\partial f(\boldsymbol{x},\dot{\boldsymbol{x}},t)}{\partial \dot{\boldsymbol{x}}}\ (\dot{X} - \dot{\boldsymbol{x}}) \ge 0$$

E is called the Weierstrass E function. It is interesting to note that Weierstrass was not a well-known mathematician; rather, he was a high school teacher at the time that he derived this condition. We note also that

$$H(\boldsymbol{x},\dot{\boldsymbol{x}},t) = -f(\boldsymbol{x},\dot{\boldsymbol{x}},t) + \frac{\partial f(\boldsymbol{x},\dot{\boldsymbol{x}},t)}{\partial \dot{\boldsymbol{x}}}\ \dot{\boldsymbol{x}}$$

and

$$H(\boldsymbol{x},\dot{\boldsymbol{x}}^*,t) = -f(\boldsymbol{x},\dot{X},t) + \frac{\partial f(\boldsymbol{x},\dot{\boldsymbol{x}},t)}{\partial \dot{\boldsymbol{x}}}\ \dot{X}$$

Thus, the Weierstrass condition can be written in terms of the Hamiltonian as follows:

$$H(\boldsymbol{x},\dot{\boldsymbol{x}},t) \geq H(\boldsymbol{x},\dot{\boldsymbol{x}}^*,t)$$

in order to minimize J. To maximize J

$$H(\boldsymbol{x},\dot{\boldsymbol{x}},t) \leq H(\boldsymbol{x},\dot{\boldsymbol{x}}^*,t)$$

These two inequalities are referred to as the Pontryagin maximum principle. Rozonoér (1960) shows that, for systems in which the control is linear, the maximum condition is a sufficient condition.

Example

The conditions derived above will become a bit clearer if we look at an example. To be sure, most problems in the calculus of variations are too difficult to solve analytically, and the example we will present is quite simple. But the application of the necessary conditions is not the issue. Rather the difficulty rests instead in the solution of the resulting equations, not in their derivation. Problems of optimal control posed in the calculus of variations invariably result in a *two-point boundary value problem.* A two-point boundary value problem results when a set of differential equations is solved, requiring boundary conditions that can be obtained only at two different points in time. This always occurs here because the differential equations that govern the system appear in pairs, consisting of both the constraint equations and the adjoint equations, and the boundary conditions always apply at both end points. This does not pose a problem if the equations are linear, as the principle of superposition may then be used to obtain a solution. This principle states that, if $\mu(t)$ and $\nu(t)$ are independent solutions of a differential equations, then

$$x(t) = \alpha\mu(t) + \beta\nu(t)$$

is also a solution, where α and β are constants. But it does cause substantial problems in the more common case when the system equations are nonlinear.

With this said, we will examine a problem that consists of a linear set of equations. The problem is illustrated in Fig. 5.16. A ferryboat seeks to cross a river, from point A and B, in minimum time. The river has a current of constant velocity, $\boldsymbol{v}_s$, everywhere. The question is: at what angle should the boat be steered, and how fast should the boat be driven.

Let v_f denote the speed of the boat, such that

$$0 \leq v_f \leq v_{f\max}$$

and let $\boldsymbol{i}$ be a unit vector in the direction in which the boat is steered. The motion of the ferryboat is given by

$$\dot{\boldsymbol{x}} = \boldsymbol{v}_s + v_f\boldsymbol{i}$$

We refer to this as the *equation of state*. And we can now write the objective functional as follows:

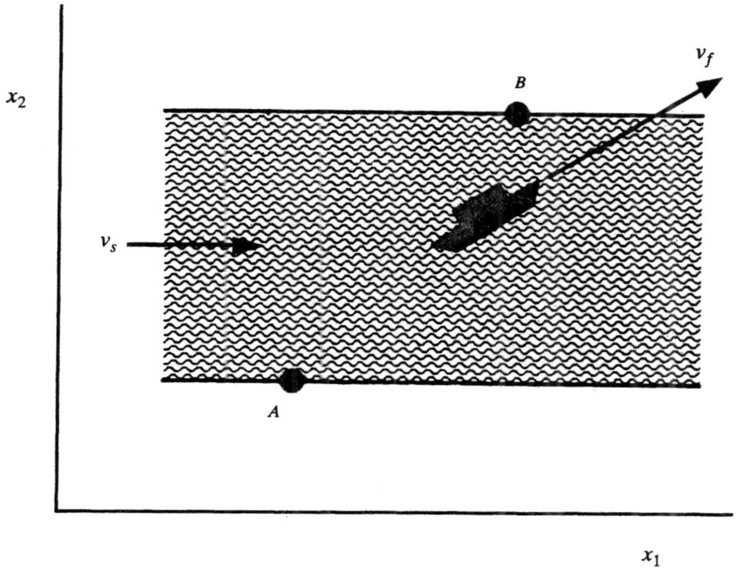

Figure 5.16 Illustration of the ferryboat problem.

$$J = \int_{t_A}^{t_B} \lambda(\dot{\boldsymbol{x}} - \boldsymbol{v}_s - v_f\boldsymbol{i})\,dt$$

where we wish to minimize t_B. We begin by applying the Euler-Lagrange condition. This leads to the equation

$$\dot{\lambda} = \mathbf{0}^T$$

which gives the result that λ is a constant. Next, we determine the Hamiltonian.

$$\begin{aligned} H &= \lambda\dot{\boldsymbol{x}} - \lambda\dot{\boldsymbol{x}} + \lambda(\boldsymbol{v}_s + v_f\boldsymbol{i}) \\ &= \lambda(\boldsymbol{v}_s + v_f\boldsymbol{i}) \end{aligned}$$

Now, to obtain the optimal control, that is, to minimize t_B, we maximize the Hamiltonian with respect to the control. $\boldsymbol{v}_s$ is a constant. Therefore, to maximize the Hamiltonian with respect to the control, we maximize the term $v_f\lambda\boldsymbol{i}$. This term is maximized by making v_f as large as possible, that is, by sailing at top speed, and by making $\boldsymbol{i}$ parallel to λ, that is, by steering in the direction of the adjoint vector. Also, because we wish to minimize t_B, we have the condition that $H_B = 0$. Since λ is a constant, $\boldsymbol{i}$ is also a constant, and the equation of state becomes

$$\dot{\boldsymbol{x}} = \boldsymbol{v}_s + v_f\boldsymbol{i}$$

which integrates to give

$$\boldsymbol{x}_B - \boldsymbol{x}_A = (\boldsymbol{v}_x + v_f \boldsymbol{i})t \Big|_{t_A}^{t_B}$$

This equation can be solved to determine $\boldsymbol{i}$, which also provides the direction of λ. The magnitude of λ derives from the condition on the Hamiltonian, namely

$$\lambda \boldsymbol{v}_s + v_f \lambda \boldsymbol{i} = 0$$

Keep in mind that λ is a row vector, and thus the above is a scalar equation. We also know that the directions of λ and $\boldsymbol{i}$ are the same, that is

$$\lambda = |\lambda| \boldsymbol{i}$$

which enables us to solve for the magnitude of λ.

$$|\lambda| = -v_f \frac{\boldsymbol{i} \cdot \boldsymbol{i}}{\boldsymbol{i} \cdot \boldsymbol{v}_s} = -\frac{v_f}{\boldsymbol{i} \cdot \boldsymbol{v}_s}$$

This completes the solution of the ferryboat problem.

As noted above, this problem contains the essentials of all optimal control problems. However, in general, it is much more difficult to obtain solutions because the constraint and adjoint equations are nonlinear and cannot be integrated in closed form. A solution approach exists, but it is rather cumbersome and it can take considerable computer time. Since analytical solutions to the resulting system equations cannot be obtained, we integrate them numerically. This would be easy if we had all the necessary boundary conditions at one point in time. Unfortunately, as in the ferryboat problem for which boundary conditions were given at points A and B, we do not. So we must guess at the missing boundary conditions in order to numerically integrate the system equations. Then we compare the resulting final state of the numerically integrated equations to the desired state and adjust our guessed boundary conditions accordingly.

5.12 SUMMARY

In this chapter we have taken a brief look at the meaning and methods of optimization. We see that the essentials of optimization are the same as those for a decision-making process. We identify options, which are control variables, expectations, which are governed by the constraint equations, and a value system, which is the objective function. We also see that many cases of decision making that can be expressed in an optimization form involve such complexity or such a large number of options that we could never enumerate them explicitly. For example, a linear programming problem in a hundred variables has a 100-dimension infinity of possible options. If for each control variable we chose only two options, we would still have 2^{100} possible options to consider. This is far more options than we could begin to examine one by one. Yet, we can examine them and find the optimum case using the simplex method. The same is true for other optimization techniques. The optimization methods enable us to consider broad ranges of system options and to choose optimal system designs without the complexity of having to consider every possible option individually. Viewed in this light, optimization is a technique for greatly reducing the work of considering and choosing among system options. In the end, however, optimization depends on proper formulation of the objective function. In other words, an optimization is only as good as the objective function for which the maximum or minimum is sought. Thus, we must turn next to the proper formulation of objective functions.

REFERENCES

1. Bazaraa, M. S. and C. M. Shetty, *Nonlinear Programming, Theory and Algorithms,* John Wiley, New York, 1979.
2. Cooper, L. and D. Steinberg, *Introduction to Methods of Optimization,* W. B. Saunders Company, Philadelphia, PA, 1970.
3. Courant, R. and F. John, *Introduction to Calculus and Analysis,* John Wiley, New York, 1974.
4. Gillett, B. E., *Introduction to Operations Research: A Computer-Oriented Algorithmic Approach,* McGraw-Hill Series in Industrial Engineering and Management Science, 1976.
5. Hildebrand, F. B., *Advanced Calculus for Applications,* Prentice Hall, Englewood Cliffs, NJ, 1976.
6. Hillier, F. S. and G. J. Lieberman, *Introduction to Operations Research,* McGraw-Hill, New York, 1990.
7. Rozonoér, L. I., L. S. Pontryagin's Maximum Principle in the Theory of Optimum Systems, I, II, III, Automat. i Telemeh. 20, 1320–1334. 1441–1458, 1561–1578, Oct., Nov., Dec., 1959, *Automation and Remote Control,* 20, pp 1288–1302, 1405–1421, 1517–1532, June, July, August, 1960 (Translation).
8. Takayama, A., *Mathematical Economics,* The Dryden Press, Hinsdale, IL, 1974.

PROBLEMS

1. Draw a function that cannot be expanded in a Taylor series.

2. Find the extreme points of

$$y = x^4 - x^2 + 1$$

Identify them as minima or maxima.

3. Find the maximum of

$$y = 2x + 1$$

$$\text{subject to} \quad 0 \leq x \leq 1$$

4. Using a zeroth-order technique, find the maximum of

$$y = \sin x \ln x$$

$$\text{subject to} \quad 0 \leq x \leq \pi$$

5. Solve problem 4 using the gradient method.
6. Solve problem 4 using Newton's method.
7. Find the maximum of

$$f = -2x^2 + xy - 3y^2$$
$$\text{subject to} \quad x + 2y = 2$$

by the method of substitution.

8. Find the maximum of

$$f = -2x^2 + xy - 3y^2$$
$$\text{subject to} \quad x + 2y = 2$$

by the adjoint method. Suppose the constraint were written in the form

$$x + 2y = c$$

Write an expression for f as a function of c in the vicinity of $c = 2$.

9. Using the calculus of variations, determine the shape that encloses the maximum area with a given perimeter.
10. You are in charge of a scientific satellite with instruments I_i, $i = 1, 2, \ldots, I$. Scientists have proposed experiments E_j, $j = 1, 2, \ldots, J$, each with value V_j. Each experiment requires time T_j on a subset of the instruments. The total time during which experiments may be performed is T_m. Which experiments do you approve? Formulate the solution as an LP problem.
11. You are climbing a mountain when suddenly you come to a deep depression that lies between you and the peak. Since climbing requires a lot of effort, you think it might be possible to walk around the depression. Are you guaranteed that this is possible, and that you can continue to ascend the mountain without first having to descend?
12. You are driving home from class each day, taking the same route. One day you discover a variation of your route by which you can reduce the time between two points, *A* and *B*, that lie along the original route. The rest of the route remains the same. If you take this alternate route, can you be sure that you will save time? Suppose later you find an even better route that is different between class and point *A* and between point *B* and home, but the same as your original route between points *A* and *B*. Would your new route be better still if you add to it the variation in the route between points *A* and *B*?
13. You are designing a car for a 1/4-mile drag race. You face a tradeoff in which you can gain low-speed acceleration at the expense of high-speed acceleration. Does this tradeoff sound good?
14. You enter a game in which you are given a loop of string (a string tied together at the ends). You then lay the string on the ground, go to a window four stories above the ground, and throw 100 pennies down trying to get as many as possible inside the loop of string. In what shape should you lay the string?
15. You have designed an airplane with which you hope to capture a closed-course speed record. Should you try to capture the speed record on a windy day such that during at least one leg of the flight the airplane will have a tail wind, or should you fly on a calm day?
16. You are designing a bucket to hold 5 liters of water. The water will drain from the bucket through a 2 mm hole in the bottom of the bucket. What should be the shape of the bucket if you want to minimize the time for the bucket to drain?

17. You are designing an elevator system for a building. Elevators cost a lot to buy, operate, and maintain, and they occupy space that could otherwise be put to productive use. What would you do to minimize the number of elevators needed?

18. You are designing an antiaircraft gun to be used with a radar tracking and aiming system. What should be the repeatability of your gun? That is, if you aim the gun precisely at the center of a target, what dispersion of hits around the center should you seek to obtain? Hint: The radar is not perfect.

19. An electric vehicle services a manufacturing operation. Its value is a function of its speed (S, m/s) and range (R, km) according to

$$V = S + 0.007\,R$$

Range is a function of speed such that

$$R \leq 8 - S$$

The minimum speed needed to perform required functions is 3 m/s and, for safety considerations, the maximum speed is 6 m/s. Solve graphically for the optimal speed and range.

20. You are asked to design a cantilever beam of rectangular cross section with width w and thickness $t(x)$, where x is distance along the beam. The beam is subject to a single load placed at the end of the beam of weight W. (a) Determine $t(x)$ to minimize the weight of the beam. (b) Derive a calculus of variations formulation for $t(x)$ to minimize weight subject to a constraint on the maximum displacement of the end of the beam. Do not solve the equations of part (b).

6

Engineering Microeconomics

6.1 THE PURPOSE OF THIS CHAPTER

Our view of systems engineering as a decision-making process leads us to the recognition that, in order to obtain a good system design, we must understand the purpose of the system. That is, we must define the values against which the system will be judged. More often than not, these values involve economic measures. Therefore, it is important that we be sufficiently grounded in the theory of microeconomics that we can properly compare system alternatives. The material presented in this chapter is not intended to comprise a thorough discourse on microeconomics. Rather it touches only on those topics that are of importance to the systems engineer, and it covers these only to a relatively shallow depth. An understanding of these topics should help to keep you from making many of the errors we commonly see in the design process that relate to poor formulation of values. The reader who is interested in pursuing this topic to a greater depth is referred to the many fine textbooks that are available. The presentation of microeconomics here is somewhat different from that in other textbooks, however. While other textbooks typically have the goal of presenting microeconomics as a subject in its own right, we present it here as a tool of systems engineering, with very specific applications in mind.

6.2 THE CONCEPT OF PREFERENCE

The Nobel economist Paul Samuelson (1973) defines economics as follows:

> *Economics is the study of how men and society end up choosing, with or without the use of money, to employ scarce productive resources that could have alternative uses, to produce various commodities and distribute them for consumption, now or in the future, among vari-*

ous people and groups in society. It analyzes the costs and benefits of improving patterns of resource allocation.

This definition of economics ties the subject very closely to our concept of systems engineering. We define systems engineering as the treatment of the engineering design process as a decision-making process. Samuelson defines economics essentially as the study of that process. He incorporates in his definition the allocation of scarce resources, the production of commodities and their distribution for consumption, and the efficiency of this process. Certainly, this is what systems engineering is all about. We should have no further doubts about the relevance of microeconomics to systems engineering, and the need for systems engineers to understand some basic principles of the discipline and their applications to systems engineering.

Basic Human Needs and Wants

Recall that we defined engineering to involve the allocation of scarce resources of nature to *benefit at least some segment of mankind.* The notion of benefit included in this definition is very important. It infers that the purpose of engineering, and indeed the reason engineers get paid for their work, is because they provide things that are of value to society. Things derive value because people need or want them. So we will begin our study of microeconomics by examining basic human needs and wants. It is possible to list a hierarchy of human needs and wants. This list, shown in Table 6.1, is surprisingly short and simple. We begin the list with the basic needs that sustain life and move on to the wants that make life more comfortable and enjoyable. Everything we do as engineers is aimed at satisfying something or some combination of things on this list.

TABLE 6.1 BASIC HUMAN NEEDS AND WANTS

Basic human needs
Air, water, food
Protection from the elements, warmth
Sleep
Security
Reproductive capability
Contact with other humans
Human wants
Clothing
Health care
Light
Education
Entertainment
Transportation
Commerce
Communication
Comfort

There is a general tendency for people to strive to fulfill their needs in an order that reflects the importance of each need to the sustenance of life. The most immediate need is for air, without which a person dies rather rapidly. Following this need comes the needs for water, food, shelter from the elements, and sleep. Fulfillment of each of these needs in turn sustains life a bit longer. As we continue down the list, we begin to encounter needs that sustain the species as well as the life of an individual. In general, humans give higher priority to their needs than to their wants. As a result, we begin to fulfill our wants only after our needs are adequately satisfied. This explains, for example, why education is more effective when students are well fed and in a safe environment. If we as engineers recognize this hierarchy of needs and wants, we will better understand the value that engineering systems provide to society.

The notion that people tend to satisfy their needs and wants in a hierarchical manner has been formalized through the conduct of literally thousands of economic studies of the subject. The results of these studies are summarized by the Engel curve shown in Fig. 6.1. This curve clearly shows that people tend to their needs first, and then expand their expenditures into their wants.

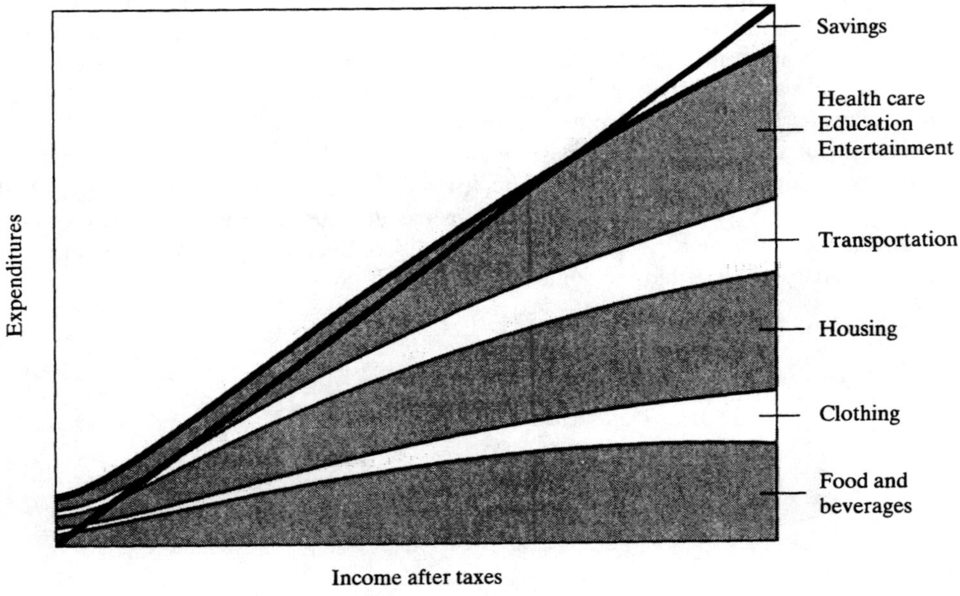

Figure 6.1 Income after taxes. (from *Economics* by Paul A. Samuelson, McGraw Hill, 1973. Reprinted with permission.)

The Concept of Value

Value is an attribute of something. We say that something has value to an individual if, under some circumstance, the individual would choose to consume it, that is, if the individual would choose this thing over other options. Value is implied by choice. Thus, if at a particular moment, an individual is presented with two options, A and B and chooses A, we would say that, to this individual, A has more value than B. It follows that value is expressed by individuals through their preferences. Value is relative and hard to quantify. Economists refer to value as *utility*, and give it units of utiles. There is a well-developed theory of utility, and we will give considerable attention to it in the next chapter.

Some authors use the word *value* in reference to preferences under conditions of certainty, and *utility* in reference to preferences under conditions of risk. We will not make this distinction here.

Indifference Curves

For an individual, relative values can be expressed in the form of *indifference curves*. To understand indifference curves, let us begin by observing that not every unit of a good has the same value. For example, perhaps you like strawberry ice cream, and you would like to eat some. The first scoop you eat provides considerable satisfaction. It has a relatively high value. Now, if you choose to have a second scoop, it too provides some satisfaction. It has value. But if you are forced to continue eating scoop after scoop, there comes a point where you really don't want more. The nth scoop has no value. This is the concept of *decreasing marginal value*. The added value obtained by each additional scoop of ice cream is less than the added value of the previous scoop. Thus, if we plot conceptually the total value of the ice cream as a function of the number of scoops consumed, the curve would have the form shown in Fig. 6.2. Of course, if you consumed enough ice cream, you might get sick, and we might even note that the value of added scoops beyond this point is negative. But we will neglect this effect here and assume that the marginal value remains non-negative.

Now let us consider two different types of goods, such as food and clothing. Each has decreasing marginal value. Given this, we can consider the tradeoffs a person makes in choosing between the two. If a person has lots of food, but little clothing, the marginal value of clothing will be higher than the marginal value of food. The person would be willing to trade food for clothing. On the other hand, if the person has lots of clothing, but little food, the marginal value of food will be higher. Then the person would seek to trade clothing for food. We can plot consumption choices between food and clothing as shown in Fig. 6.3. A person who has the combination of food and clothing denoted by point A would be willing to trade food for clothing in any way that is beneficial to his or her overall well being. That is, the person would trade in order to increase the total value of food and clothing. Presented with a set of possible trades, the person would prefer to make some of the trades and prefer not to make other trades. We could picture the region in which trades are preferred and the region in which they are not preferred on Fig. 6.3. Separating these two regions is a line along which the person is indifferent to the trade. This line is called an *indifference curve*. All combinations of food and clothing along a given indifference curve have the same total value to the person.

If we repeat the process by which we defined the above indifference curve, starting at different combinations of food and clothing, we could map out the indifference curves of the person. These curves would depict the trading preferences of the person, and indi-

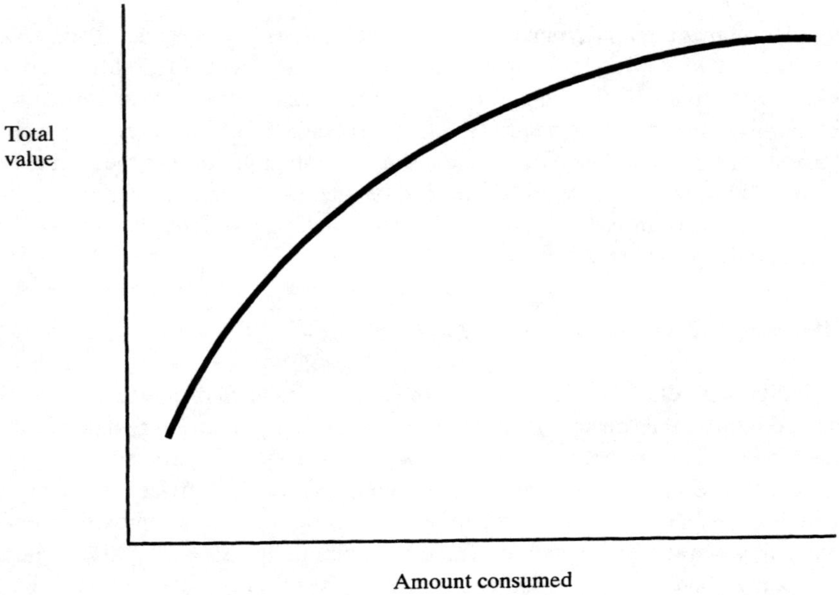

Figure 6.2 The concept of decreasing marginal value with amount of a commodity consumed.

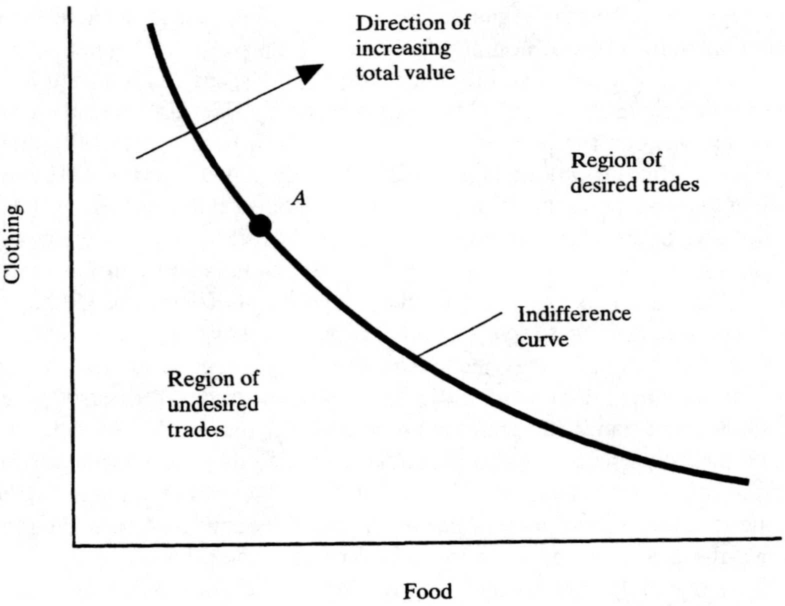

Figure 6.3 The concept of an indifference curve.

cate that overall the person typically would prefer increased quantities of both commodities. This set of indifference curves defines the person's relative values for food and clothing. We say that this is the person's *value space*.

Preferences Under Conditions of Limited Resources

We will continue with the example of food and clothing. Suppose that the person in this example has a budget for the purchase of food and clothing, and suppose that budget is limited. The choice that the person can make is how much food to buy versus how much clothing. We denote the person's budget by B, and the price per unit of food by P_f and clothing by P_c. Let q_f represent the quantity of food purchased and q_c the quantity of clothing. Purchases of food and clothing are then limited by the resource constraint

$$B = P_c q_c + P_f q_f$$

This resource constraint can be plotted on the value space shown in Fig. 6.3 as shown in Fig. 6.4. We can now examine the choices that the person will make. We can assume that the objective of the choice is to expend the budget in such a way as to maximize total value. This is done by choosing an amount of food and an amount of clothing subject to the above constraint that provides the highest possible value. This choice occurs at point A in the figure, the point of tangency between the budget line and the indifference curve of maximum obtainable value. This is, in general, the process by which tradeoffs are made under conditions of resource limitations.

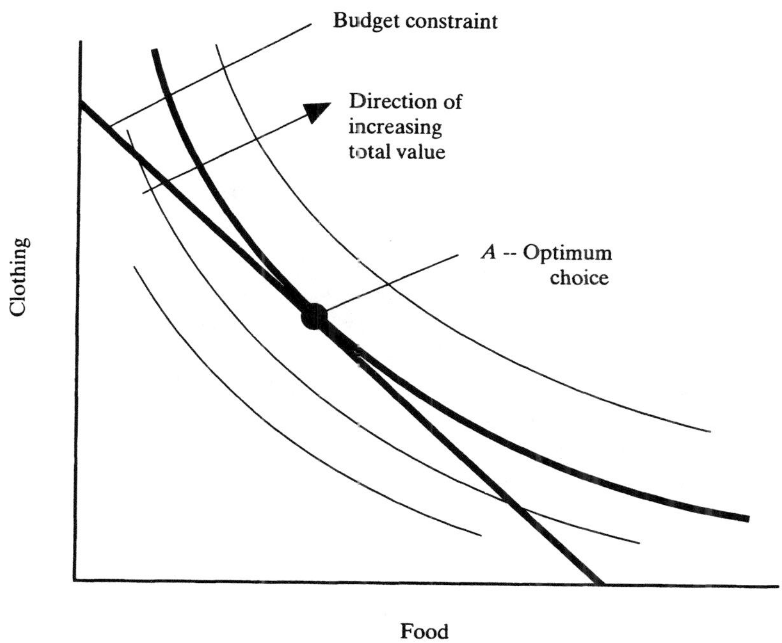

Figure 6.4 Tradeoffs between commodities purchased subject to a budget constraint.

An Engineering Example

We return again to the example of guessing how many beans are in a jar. Let us assume that we are in a game against nature, but with the nonsymmetrical payoff function of Fig. 1.1. A game against nature is one that we are in alone. That is, there are no other players. This greatly simplifies the mathematics. We presented in Chap. 4 a methodology for quantifying our state of information relative to the number of beans in the jar, and we obtained a probability distribution. We can now use that distribution to examine possible guesses.

Combining the payoff function given in Fig. 1.1 with our computed distribution on the number of beans in the jar, $f(n)$, we can calculate the expected value of any guess, u.

$$E\{u\} = \int_0^\infty f(n)P(n - u)\, dn - C$$

where $P(n - u)$ is the value of the prize, which derives from Fig. 1.1, and C is the cost of making a guess. If you base your decisions on expected values, you might be inclined to select the value of u that maximizes $E\{u\}$. But most of us do not base our decisions on expected values. We tend to be somewhat risk averse. That is, if possible, we tend to avoid risk. Risk here might be measured by our chance of losing money on the game. Conversely, we might be interested in our chance of making money. Recognizing that we win if our guess is such that

$$u \le N \le u + 200(1 - C/250)$$

where N is the actual number of beans in the jar. Our chance of making money by playing this game is estimated by $R\{u\}$, where

$$R\{u\} = \int_u^{u+200(1-C/250)} f(n)\, dn$$

In general, $E\{u\}$ and $R\{u\}$ will be functions of u that will each have a maximum value for a different value of u. This is shown in Fig. 6.5. If now we wish to select a value of u, we need to trade off our preferences between $E\{u\}$ and $R\{u\}$ in order to select a value of u that gives us the desired combination of both commodities. We do this by adding to Fig. 6.5 our indifference curves or value space. The optimal tradeoff and selection of u is also shown in Fig. 6.5. Our best choice of u maximizes the value of the combination of $E\{u\}$ and $R\{u\}$ that we obtain with our guess. So we see that proper expression of our preferences is highly relevant to engineering decisions.

Preferences Involving Time

The above discussion treats preference between two different commodities. In the engineering example, these commodities happened to be attributes of our guess. Another commodity that we need to be aware of is time. Suppose that we are offered commodity

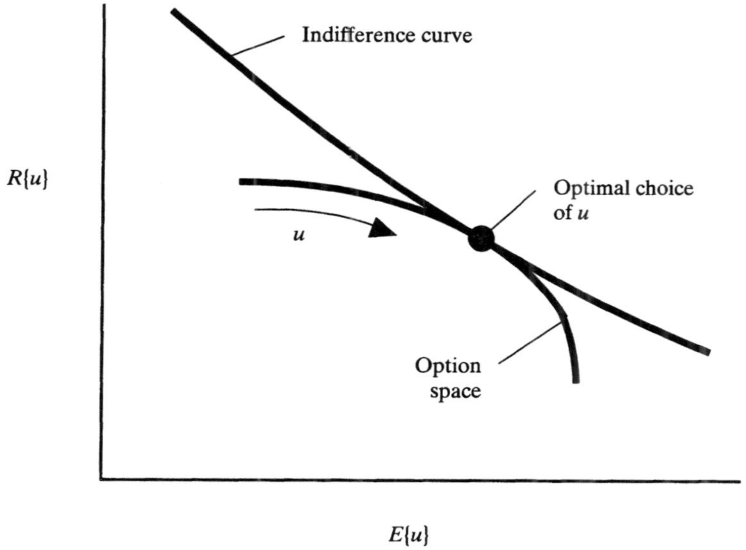

Figure 6.5 The optimum guess of how many beans are in the jar trading off risk of losing versus expected value of the prize.

X in quantity q_0, which will be delivered to us at time t_0. If we value the commodity, and we look forward to receiving it, then we would prefer to have the commodity at time t_0 rather than at a later time t_1. With this concept in mind, we can draw an indifference curve in the space of t and q, and it will appear as shown in Fig. 6.6. It is common to assume that the cost of a delay in access to the commodity is independent of when the delay occurs. This is the basis of the notion of a social discount rate. The existence of a non-zero discount rate infers that a commodity received at some future date has less value than a commodity received today. Note that this is not the same concept as interest. We will return to this subject shortly.

6.3 EQUILIBRIUM ECONOMICS

Now that we understand some of the basic concepts behind preferences, we can begin to examine the aggregate behavior of producers and consumers. It is important that we gain this understanding because it will later form much of the basis for the analysis of costs and benefits, both of which are important in the formulation of values for engineering decision making. The title of this section is "Equilibrium Economics" to emphasize that much of the theory developed here applies only in the case of a steady state economy.

Demand Curves

Demand curves represent the aggregated decision making of a set of consumers. From our discussion of preferences and the notion of an indifference curve as shown in Fig. 6.3,

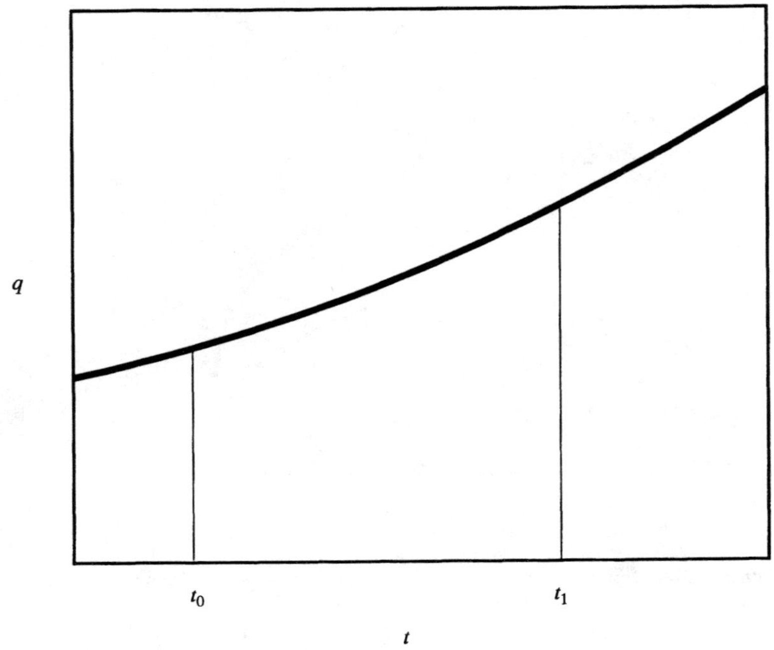

Figure 6.6 An indifference curve in time.

we see that for a given commodity each individual has a region of desired trades. If instead of food and clothing, we examine trades between money and some commodity, X, we conclude that the individual will purchase the commodity if it is in the region of desired trades. That is, the consumer will choose to consume the commodity if consumption raises the consumer's total value or, we could say, *welfare*. In other words, a consumer will purchase a unit of a commodity if the price of the unit is less than the value that its purchase gives to the consumer.

Consumers are different. Therefore, one consumer may place a high value on a commodity and be willing to pay a high price per unit, while another places a lower value on the commodity and will purchase it only at a lower price. It is also the case, because of decreasing marginal value of a commodity as shown in Fig. 6.2, that a given consumer will purchase more of a given commodity as its price is lowered. Demand curves represent the aggregate of these decisions. At a sufficiently high price, no one will choose to consume a given commodity. As the price is lowered, those consumers who value the commodity the most will begin to purchase it. As the price is lowered still more, additional consumers will purchase the commodity, and those consumers who have already purchased some of the commodity may increase their consumption. As the price of the commodity continues to decrease, more and more of it will be purchased. This is shown in Fig. 6.7.

The negative of the slope of the demand curve at a point A, corresponding to a price P and a demand q, is an important parameter of the demand curve. It is called the *demand*

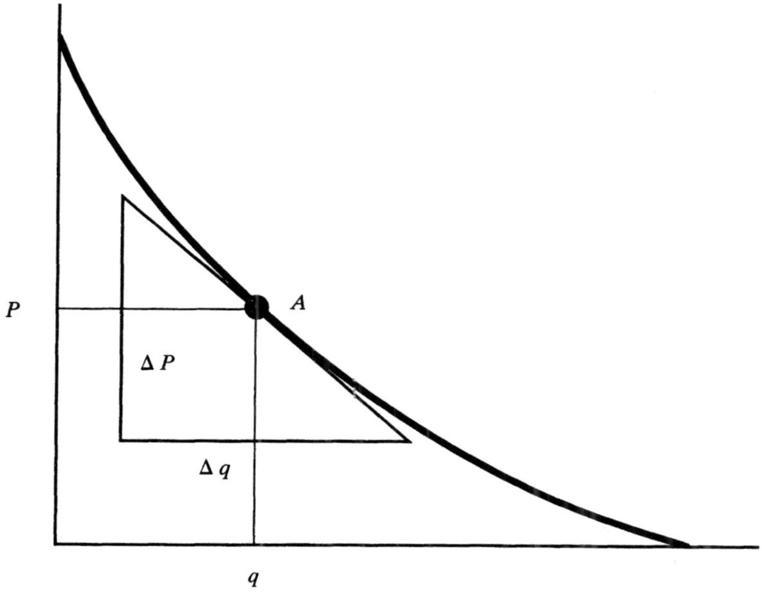

Figure 6.7 Illustration of demand elasticity.

elasticity, e. It determines, on the margin, the additional consumption of a commodity given a small decrease in its price.

$$e = -\frac{\Delta q/q}{\Delta P/P}$$

where q is the rate of consumption in units per unit time, and P is the price or amount charged per unit. Here, e is expressed as the ratio of the fractional increase in consumption to the fractional decrease in price. The demand for a commodity whose elasticity is much less than one, say 0.1, is said to be inelastic. The consumption of such a commodity is very insensitive to price. The demand for a commodity whose elasticity is one or more is said to be elastic. Consumption of this commodity will vary substantially with changes in price.

A very special demand curve occurs for the case when $e = 1$ everywhere. This is called a unity elasticity demand curve. Expressing the equation for e in differential form, we get

$$-\frac{dq}{dP}\frac{P}{q} = 1$$

which we can integrate to obtain

$$\ln q = -\ln P + \ln c$$

We then put this equation in the form

$$Pq = c$$

where c is a constant. Thus, for this demand curve, regardless of price, the total expenditure on the commodity by consumers is the same.

Despite the fact that we are discussing equilibrium economics, and we have to be very careful about misusing demand curves, economists also refer to short-term and long-term elasticities. Short-term elasticities refer to behavior of consumers to a price change in the short term—immediately. Long-term elasticities refer to behavior of consumers after a sufficient time has passed for them to adjust their purchasing behavior to a price change. In general, short-term elasticities tend to be low, that is, the short-term demand for commodities tends to be inelastic. Long-term demand tends to be more elastic and, in fact, there is some tendency of long-term elasticities toward unity. In essence, people tend, in the long term, to spend in accordance with the Engel curve of Fig. 6.1, and they will adjust their market behavior to achieve this, given time.

So far we see that demand curves provide us with information on the market behavior of consumers. They do more, however. They also provide us with quantitative information on the value of the commodity to the consumers. Let us consider a commodity with the demand curve shown in Fig. 6.8, whose price is set at P_1. The corresponding demand is q_1. Now let's look at the demand curve at $q = 0$. Above the price P_0, no one will purchase the commodity. That is because, to all potential consumers, its value for the first unit is less than P_0. However, the first unit is purchased as soon as the price is lowered below P_0. This price is then a measure of the value of the first unit of the commodity to the consumer who purchases it. As the price is lowered another small increment, another unit is purchased. The value of this unit is equal to the price at which the decision to purchase it would be made. Thus, the total value to all purchasing consumers of the commodity, given the demand curve $P(q)$, is

$$V_c = \int_0^{q_1} P(q)\, dq$$

But, together, the consumers spend $P_1 q_1$ on their purchase of the commodity. This sum detracts from the value that they gain from the purchase of the commodity. Therefore, the net value obtained by all purchasing consumers is V_c less this expenditure.

$$S_c = \int_0^{q_1} P(q)\, dq - P_1 q_1$$

This area, shown in Fig. 6.8, is called the *consumer surplus*. It is the surplus value, or the value net of cost, that the consumers receive as a result of their consumption of the commodity. Clearly, the lower the price of a commodity, the higher the consumer surplus.

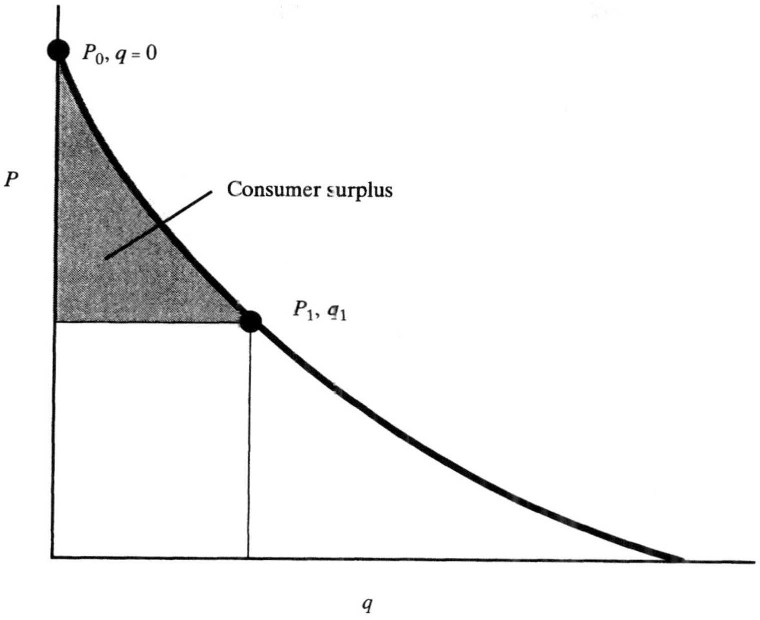

Figure 6.8 Consumer surplus.

Supply Curves

Supply curves are the suppliers' equivalent of demand curves. A supply curve depicts the willingness of suppliers to provide a commodity as a function of its market price. The notion behind a supply curve is that suppliers are willing to supply only a small amount of a commodity at a low price, and that amount increases as the price is increased. A typical supply curve is shown in Fig. 6.9. To be sure, there exist economies of scale that cloud this concept of supply. Indeed, supply curves could have more complex shapes. But let us keep the discussion simple and consider only supply curves of the general shape shown.

We can go through a similar logic for supply curves that we went through for demand curves. Suppliers, on the margin, decide whether to add to the supply of a commodity based on whether or not it benefits them to do so. The supply curve is thus a measure of the marginal cost of supplying the commodity. If the price is set at P_1, the producers then gain a net benefit from supplying the commodity that is measured by

$$S_p = P_1 q_1 - \int_0^{q_1} C(q)\, dq$$

where $C(q)$ is the producers' marginal cost of production. The benefit given by S_p is referred to as the *producer surplus*.

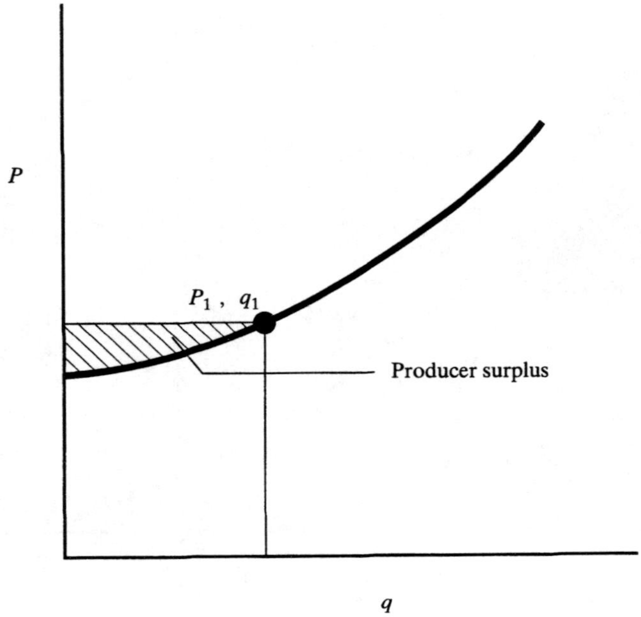

Figure 6.9 Producer surplus.

Pricing Theory

We can plot the demand and supply curves for a given commodity together on the same set of axes as shown in Fig. 6.10, and then use this plot to discuss the effects of various pricing strategies. First, we note that, in general, the demand and supply curves will intersect each other as shown in the figure. In a market with perfect competition, the equilibrium price will be set at the price corresponding to this point of intersection. The logic is this. If the price were set higher, consumers would not purchase the total supply, and the suppliers would lose on the resulting excess production. On the other hand, if the price were set lower, the producers would not be willing to meet the demand of the consumers. Further, if the price is set higher than the marginal cost of supply, at least one supplier can benefit by increasing production and, in a competitive market, would choose to do so. The resulting pricing policy provides the total net benefit $S_c + S_p$, and this is the maximum possible total net benefit from production and consumption of the commodity under static equilibrium conditions. This condition does not, however, determine price in a dynamic or changing market, as would be obtained during the introduction of a new technology.

In the case where there is not perfect competition in the marketplace, the price might not be set as above. In general, if a producer has some control over the marketplace, it is possible to set a price such that the producer surplus can be increased, albeit at a loss in overall benefit. When a single producer has complete control over the marketplace we say

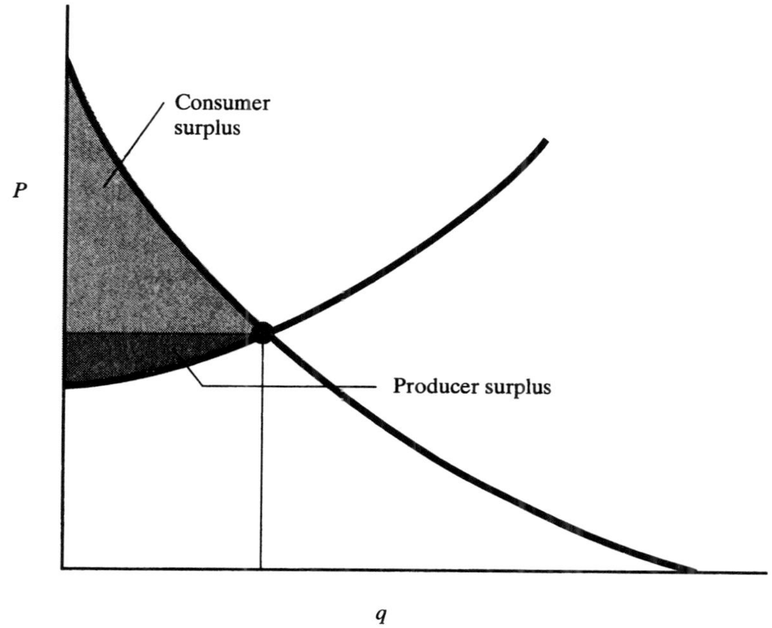

Figure 6.10 Price in a market with perfect competition.

that the producer has a monopoly. In this case the producer is free to set the price so as to reap the largest possible benefit for himself or herself. In this case the price would be set according to the algorithm:

$$\underset{\text{with respect to } P_1}{\text{maximize}} \quad S_p = P_1 q_1 - \int_0^{q_1(P_1)} C(q)\, dq$$

When the price is set according to this algorithm, the producer collects a surplus that includes a *monopoly rent*. This is shown in Fig. 6.11.

It is important to recognize that there are different pricing schemes, and that they result in different benefits to both consumers and producers. Furthermore, it is important to keep in mind that the illustrations of the pricing schemes given above are for the case of equilibrium only.

In closing this discussion of pricing theory, we will note that market prices do more than set demand and supply. They represent an information system. When the price of a commodity begins to rise or fall sharply, information regarding the supply and/or demand of the commodity is being transmitted to both consumers and producers. Conversely, people with special or inside information regarding the market for a particular commodity cannot use that information for their own gain without transmitting it through the price structure of the marketplace to others. Herein lies a major problem of centrally planned economies or regulated economies in which prices are not allowed to adjust in response

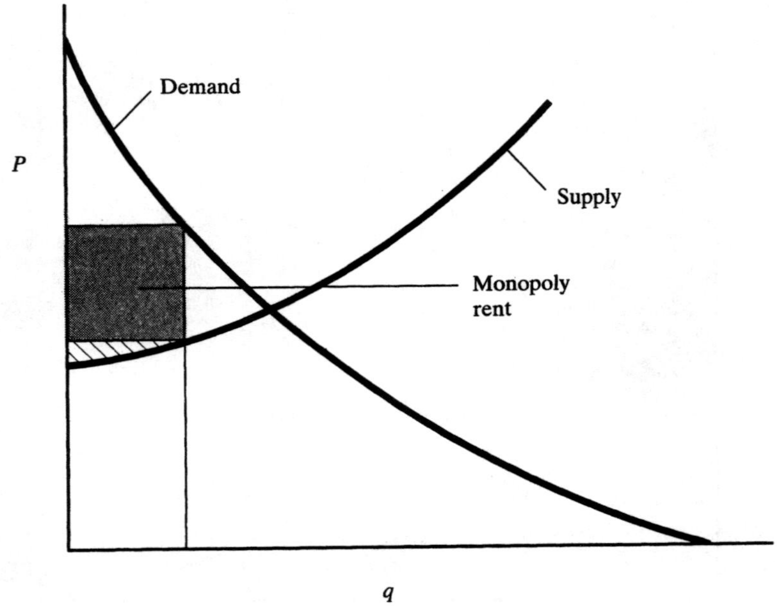

Figure 6.11 Monopoly pricing.

to market forces. In this case, the information that would normally be transmitted to both consumers and producers through the pricing system is thwarted. Unless this loss of information is replaced by some other mechanism, both consumers and producers will find it difficult to adjust their behavior to achieve optimal production and consumption patterns.

The Effect of Changing the Supply Curve

With the above understanding of demand and supply, we can begin to examine changes in equilibrium conditions of demand and supply as a function of improvements in product design or production. We will consider production first. It may be possible to improve the production of a product by redesign or by investment in production facilities or by research and development on the production process. We assume here, however, that all the attributes of the product that are of relevance to the consumers remain unchanged. Then, the effect of improving product production is to lower the supply curve, from S_1 to S_2 as shown in Fig. 6.12. In general, improvements in product production both lower and flatten the supply curve, although this need not always be the case. But it is easy to see that the main beneficiaries of this change in a competitive market are the consumers. The benefit to suppliers is not easily explained in the context of equilibrium economics. Herein lies a limitation of equilibrium economics. Do not try to explain the benefits to suppliers of research and development on the production process using equilibrium eco-

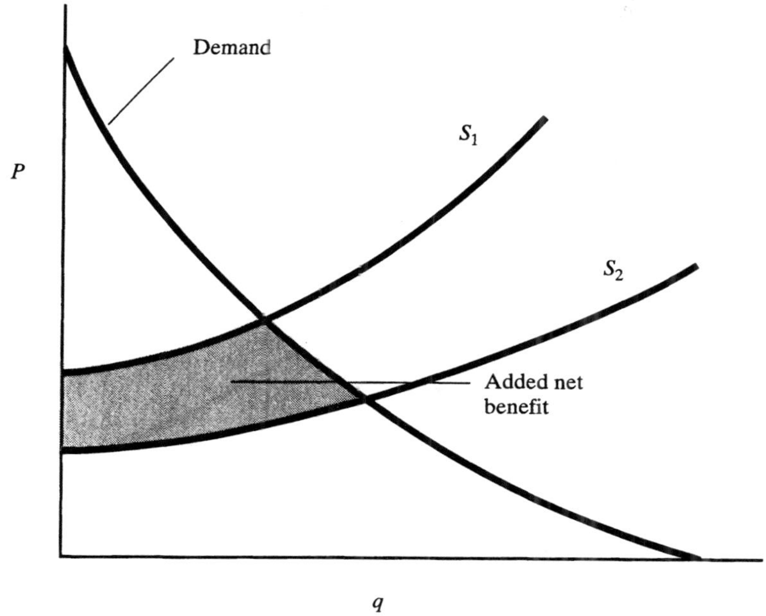

Figure 6.12 The effect of improving the supply curve.

nomics. Rather, the benefits lie mainly in the transients that such benefits cause in the marketplace. In a competitive marketplace, a supplier can capture a larger market share, at least until the competition catches up. In some cases, the improvement can be sufficiently significant that a producer obtains some monopoly powers, and can set prices at an advantageously high level. In these cases, the benefits to the producer of such improvements can be large.

In the long term, we should keep in mind that the purpose of improving our production process is to remain competitive. But sometimes, remaining competitive requires that we do things that lower our net producer surplus measured in equilibrium terms.

The Effect of Product Research and Development

The effect of product research and development is to enable us to produce new or improved products. New products produce entirely new areas for economic activity, and the effect of improved products is to shift the demand curve as shown in Fig. 6.13. As products are improved, the demand for them increases, and again consumers benefit. But it is not a foregone conclusion that the suppliers who perform the research also benefit. It depends on what happens to the supply curve as a result of the improvements and on the transients that the introduction of the improved products create.

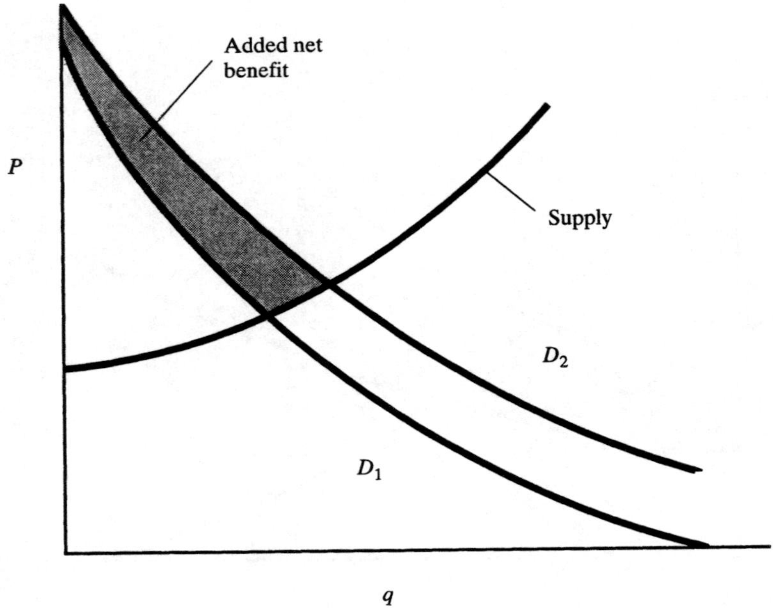

Figure 6.13 The effect of product research and development.

Regulation

Sometimes we desire to have suppliers provide society with a commodity in a monopolistic marketplace. We make this choice because we perceive that, overall, we will benefit from it. An example is the provision of electric power. An electric power company operates typically in a region as the only supplier of electricity for the region. Also, many of these companies are privately held, and operate to their own benefit. To prevent abuses to the society they serve, these monopolies are regulated, and we call them *regulated monopolies*. The objective of regulation is to prevent the assessment by the supplier of a monopoly rent that is unfair to the consumer. In essence, society tries to capture the benefit that accrues because of the economies of having only a single supplier, while not allowing that supplier to allocate those benefits to itself. Regulation, therefore, seeks to keep the price at a level that is at or near the competitive market price.

The Effect of Taxes on Prices

Taxes, in the form of value-added or sales taxes levied against consumers, or other taxes levied against producers, impose a cost that is in addition to the cost represented by the supply curve. The effect of taxation is illustrated in Fig. 6.14. In a competitive market, taxes reduce consumption and reduce both consumer and producer surplus. The taxing

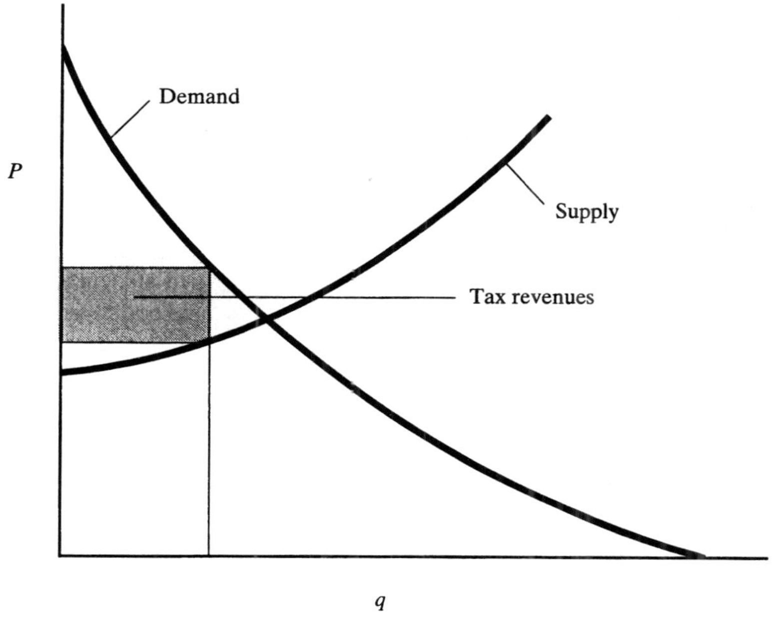

Figure 6.14 The effect of taxes on prices.

entity, typically a local or regional government, allocates some of the benefit of supply and consumption to itself. To the extent that this government represents the interests of its society, it can return the lost benefit in the form of other services.

6.4 DISCOUNTING AND PRESENT VALUE

The notion of the time value of things was introduced earlier. Here we will explore that notion and its mathematics to a greater extent.

The Social Rate of Discount

We observed earlier that, in general, we prefer to receive things that we value sooner rather than later. Perhaps you desire a new car or a new television set. You would use these things now, if you had them, to gain satisfaction or enjoyment. The new car might improve your quality of life by permitting you to take a trip that otherwise you could not. The same is true of larger items that are consumed by society as a whole, such as a new highway system or a new communication system. We express our preference for having something today as opposed to, say, a year from today by our willingness to pay a premium to get the item today. Suppose that you would be willing to pay a 20 percent premium on the price of a new car in order to obtain it today as opposed to a year from today. What you are saying is that the use of the car for the coming year has a value that is 20 percent of the purchase price of the car. To put it in different terms, the worth to you of a car received one year from today would be $1/1.2 = 0.83$ times its value received today. We say that the value of the car received in the future has been *discounted* to the present. It is common to assume that discounting applies to such personal decisions at a rate that remains constant into the future. Applied to decisions of an individual, we say that this rate is the person's discount rate. Applied to decisions of a society, we say that this rate is the *social rate of discount*. Discounting provides a method to assess the time value of things. It should not be confused with interest.

The social rate of discount is important because we need it to compare social alternatives. For example, suppose we plan to build a bridge. In comparing two designs we find that one costs more but could be completed one year sooner. Which is the better? To answer this question, we must determine the value that society places on the earlier completion date versus the latter date. Clearly, we would prefer the more expensive design if the added value of the earlier opening date is greater than the added cost.

Typically accepted values of the social rate of discount range from 5 to 10 percent.

Present Value

In order to compare alternatives, we must reduce benefits received from a project or activity over a period of time to a common measure. Generally, we choose to measure these benefits in terms of their value to us today. Let B_i be a sequence of benefits received in year i from a project, such as the construction of a bridge. We might be inclined simply to add these benefits to obtain the total benefit. But since they are received in different years, they are not expressed in common units and, hence, cannot be added. For example, a benefit of \$1 million received in the year 2000 is not the same as a benefit of \$1 million received in the year 2001. Thus, if we receive a benefit of \$1 million in each of these years, we cannot simply say that our total benefit received is \$2 million. We must first convert both amounts to a common base. To do this, we discount each benefit according to when it is received. Then we can add the discounted amounts, as they are benefits measured at a common time. We frequently choose to discount to the present benefits that we expect to receive in the future. We then refer to these discounted benefits as the present value of benefits, and we compute them according to the following equation:

$$PV(B) = \sum_{i=0}^{n} (1 + r)^{-i} B_i$$

where r is the discount rate and n is the number of time periods that the benefit continues, not counting the period for $i = 0$.

Likewise, costs that are incurred in different periods of time cannot be compared directly. They too must be reduced to a common base, and we refer to them as the present value of costs. If we incur a cost stream, C_i, the present value of that cost stream is given by a similar equation,

$$PV(C) = \sum_{i=0}^{n} (1 + r)^{-i} C_i$$

Again you are reminded that the discount rate used in the determination of present values is not the same as the interest rate that might be applied for other purposes, although in some cases we might choose them both to have the same numerical value.

A mathematical identity is sometimes useful in the computation of present values. Consider the present value of a stream of benefits, B, that is constant in every year and continues forever:

$$PV(B) = B \sum_{i=0}^{\infty} (1 + r)^{-i}$$

If we take $(1 + r)^{-1} = \gamma$, the series in this equation can be written

$$\sum_{i=0}^{\infty} \gamma^{i} = \frac{1}{1 - \gamma} = 1 + \frac{1}{r}$$

Thus, for a constant stream of benefits with an infinite horizon,

$$PV(B) = B\left(1 + \frac{1}{r}\right)$$

This identity frequently is useful in estimating the present value of a benefit or cost stream.

6.5 INTEREST AND ANNUITIES

Traditional economic theory arbitrarily divides all productive factors into three categories: natural resources, labor, and capital. Natural resources are provided by nature, and economists generally consider them to be fixed in supply—they can neither be created nor destroyed. Thus, we say that their supply is inelastic. The return on this factor is called *rent*. Labor is taken to be the human factor in production, and the return on it is called *wages*. Capital consists of goods produced by the economic system itself to be used as inputs for further production or services. Capital goods are both outputs and inputs of the economic process. They can be short-lived or long-lived. The yield of capital is called *interest*. A discussion on the interplay between these factors in the economic production system is beyond the scope of this book. But it is interesting to note that John von Neumann, a distinguished mathematician who made a few significant contributions to economic theory, showed, subject to a rather strong set of assumptions, that the maximum real rate of interest that can be sustained in an economy is equal to the rate of technological innovation as measured by productivity. The real rate of interest is the rate of interest corrected for inflation or the changing value of money.

Despite the fact that interest and present value are different concepts, their mathematics is similar. The interest rate is expressed as a fraction of capital, i. Thus the annual interest on an amount of capital, C, is Ci. In the design and analysis of engineering systems, interest is generally a cost that must be accounted for.

Annuities

An *annuity* is a series of payments made over a period of time in order to retire a debt. The debt may be incurred before, during, or after the series of payments is made. The concept of an annuity is that the present value of the series of payments that comprise the annuity equals the present value of the debt. Within the constraints of this concept, there are no limits on how you construct the series of payments, and this can lead to considerable confusion.

A common form of annuity is a set of payments that are all equal in current-year monetary units. So, for example, suppose a debt of \$1 million is incurred today, and it is to be repaid by 20 equal payments made once a year beginning one year from today. The present value of the debt is obviously \$1 million. This must equal the present value of the payments. Thus,

$$\frac{P}{r}\,[1 - (1 + r)^{-20}] = \$1 \text{ million}$$

where P is the annual payment. This is how banks typically finance debt, where r is the interest rate that they charge (although banks usually require monthly payments). The problem with using this approach for economic analysis of engineering alternatives is that, because of inflation, the payment stream is not constant over time. To obtain a payment stream that is constant over time, we first would have to compute a *real* interest rate by reducing r according to the expected rate of inflation, ι, according to the equation

$$1 + r = (1 + r_{\text{real}})\,(1 + \iota)$$

We would then replace r in the above equation by r_{real}. Next, we compute P, which is now a value of P expressed in today's dollars. That value must then be inflated by the inflation rate $(1 + \iota)$ to compute the current-year payment in the annuity to be made at each year.

Failure to adjust annuity payments as described above can distort the economics of an engineering system as perceived by the public or others. For example, electric utility rates can be strongly affected by a new generating facility that enters the rate base. The rate base is the value of the capital investment upon which the utility company determines a capital recovery payment that is entered as an expense to be recovered by charging its customers. Because of distortions caused by inflation over the decades lifetime of utility equipment, the value of a single generating plant can comprise as much as half the book value of a utility system at the time that the plant enters service. What we mean here is that a system whose components may have a lifetime of, say, 50 years, but that are depreciated by a method that does not account for inflation, may have a book value that is much lower than its actual, current dollar worth. But charges are based on book value, not worth, so that, when a major new component, such as a generating plant, is added to the system, the new component dominates the book value of the system, and causes an immediate and large increase in charges. This increase in charges then turns customers against the new component—it is to blame for their increased rate. Also, failure to adjust the repayment annuity for expected inflation on the new system component loads the repayment to the front end of the repayment period, further increasing customers' rates. Adjustment of the repayment schedule to account for expected inflation can significantly reduce the cost impact to customers of new capital equipment added to the system.

Another problem that arises with the determination of annuities, appears when a capital item, such as a production tool, is used in the production of several different products. To determine the cost of each product, the company must allocate the cost of the tool across the products produced. Now, suppose the tool has a lifetime of 20 years, but the products change over this period. So, in order to allocate the cost of the tool to each product, the cost must first be allocated over time. In one approach, the company could choose to allocate the cost of the machine mainly over the first few years of its life. Then the products produced during the early years of the tools life would be expensive, and those produced later would be cheap. Alternatively, the company could reverse this policy, and reverse the cost pattern of the products. What we begin to see is that the cost of a product is largely arbitrary, particularly if the product's production is capital intensive. Indeed, the cost of the product can be almost anything that the company's accountants want it to be, of course within the limits of the applicable laws. But, generally, these leave considerable flexibility.

The problems of cost allocation become even more difficult to resolve when a particular tool is used in the manufacture of several different products at any given time, as well as having a changing product mix over time. Now different rates can be charged to the different products as well as the time-value distortions referred to earlier. For products that are capital intensive, reasonable determinations of cost may vary over more than an order of magnitude.

As a systems engineer, you should understand the problems inherent in the determination and use of annuities, and you should be aware of how this will affect the price and, hence, net value of the system you are designing.

6.6 INFLATION AND DEFLATION

One mistake that engineers frequently make in formulating and analyzing values that involve money is to assume that units of money are constants of nature, like the speed of light, the wavelength of light emitted by the sodium atom, or the mass of a proton. They are not. Units of money are whatever the society using them happens to decide they are at the time. Thus, a dollar spent in 1960 is not the same as a dollar spent in 1990. So if we try to measure, say, the cost of a system by observing how much is spent on it over an extended period of time, perhaps 20 years, we cannot determine the cost by simply adding the yearly costs together. This would be much like trying to estimate the weight of a system by adding together the weights of the system components, one measured in kilograms, one measured in pounds, one in stone, and so on, with no attempt to convert these to a common unit. In fact, if we were given component weights each in a different unit, we would have little trouble converting all the weights to a common unit for taking their sum. This is because there exist well-defined conversions between the different units. Unfortunately, no such conversions exist with money.

The price of gasoline is an interesting topic. Some researchers believe that the reason for improving fuel efficiency in modern automobiles and their downsizing is the steadily increasing price of gasoline. But is the price of gasoline really increasing? How might we measure it? I can remember buying gasoline in the United States in the late 1950s. The price was about $0.25 per gallon. In 1994, I paid about $1.15 per gallon, a factor of 4.6 times more, if we don't worry about the changing value of the dollar. But in the 1950s, I had to work about 15 minutes to earn the money to buy a gallon of gasoline. Today, I work less than 2 minutes. And today's gasoline is not the same as the gasoline I purchased in the 1950s. It is higher octane, cleaner burning, better for the automobile engine, and it contains little or no lead. In short, it is a vastly improved product. By this measure, we might conclude that gasoline is today more than ten times less expensive than it was 35 years ago. But this comparison is not fair. My time is worth more today. I have much more education and experience, and I am paid for that. So perhaps the difference is not so large.

In economics, we try to smooth out all these differences, and we create inflation indices to measure the aggregate effects of inflation. We then apply such an index to adjust for the changing value of money by reducing monies spent in different years to units of a common year. When we take a stream of monies spent over a series of years and reduce them all to a common year, we refer to the resulting set of values as *deflated* or in *constant-year dollars*. This stream of monies, now all in common units, can be added to obtain a total.

This may sound like a good procedure, but the problem is that prices of different commodities do not change together in proportion to each other. Some prices may go up, and others down. Some may go up quite a lot, and others not so much. Thus, to obtain a price index, we must weight the relative importance of different commodities whose prices are reflected in the index. As a result, there are many different inflation indices, and it may be hard to decide which is the best to use. To name a few, there is the consumer price index (CPI), the wholesale price index (WPI), and the Handy-Whitman index, which is a construction price index. Each of these is an index of a weighted set of commodities in a marketbasket of goods.

The problem of creating a price index is still more complex. Some goods enter the marketbasket over time, and others leave. For example, microelectronics are important in today's marketplace. They did not exist 50 years ago. So, in comparing monies spent over a wide span of time, there may be very little that remains constant in the comparison. Still, failing to make some correction for the varying value of money is worse than making a correction that is not precise.

6.7 THE VALUE OF A FORECAST

We can now begin to apply the above economic concepts to the engineering design of systems. Suppose we wish to design a system for measuring and predicting wheat crop yields. Such systems exist, and they make use of measurements taken by researchers in the wheat fields, aerial photography, satellite imagery, and survey data. To design an optimal system, we need to know what is the value of improved forecasts. Value of better forecasts accrues to many individuals involved in the supply of wheat, as they are better able to plan their activities. But value of better forecasts also accrues to the consumers, even though better forecasts in themselves do not increase the supply of wheat. We can begin to understand the value of a forecast to consumers by reference to a two-period consumption model, such as the Hyami-Peterson model.

Consider the consumption of a good, X, over the course of two successive periods in time. Let X_1 be consumption in time period 1 and X_2 be consumption in time period 2, subject to the constraint that total consumption is limited.

$$X_1 + X_2 \leq X_T$$

We can picture the value received from various consumption patterns as shown in Fig. 6.15. The consumption pattern that maximizes total value is given as point A. Now suppose that, at the time that X_1 is consumed, the consumer believes that, instead of a total available supply of X_T, there is only a supply $X_T - \Delta$. The consumer believes this because the forecast is in error by amount Δ. Then, for period 1, the consumer will shift his or her consumption so that, at the start of period 1, the expected consumption pattern is given as point A'. In accordance with this shift, the consumer reduces consumption for period 1 to the amount corresponding to point A', a bit less than that corresponding to point A. But,

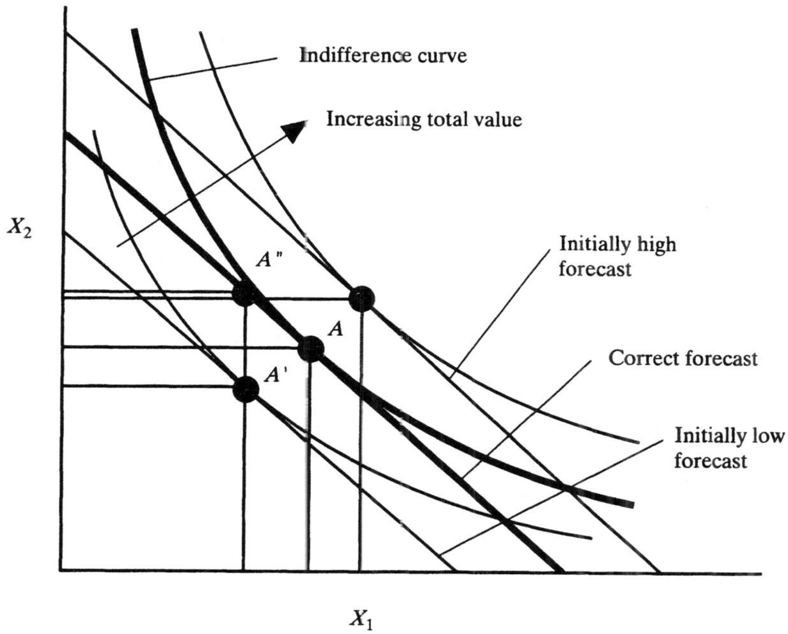

Figure 6.15 A two-period model showing the value lost to consumers as a result of inaccurate forecasts.

at the end of period 1, the consumer becomes aware of the true supply and takes the best possible choice for period 2, given the consumption of period 1. This leads to the final consumption pattern denoted by point A''. It is easy to see that there is value lost to the consumer because poor information leads to a consumption pattern that is less than optimal under conditions of perfect information. That is, point A'' is on an indifference curve of lower value than point A. It is easy to repeat this process with an initial forecast that is high, and show that this forecast also results in a loss in total value. Since forecasts that are both high and low result in lost value to consumers, it follows that inaccuracies in forecasts or expectations of consumers do create economic loss.

The above arguments can be extended to the case of multiperiod consumption patterns with time-varying forecast accuracy, but this is beyond the scope of the present discussion.

An engineering example of the importance of forecast errors is the following. Consider a manufacturing facility that has several machines that produce a number of products, and where there is some flexibility in the assignment of machines to tasks. Nominally, you might think that the best assignment of machines would be that assignment that would minimize production cost. But this is not necessarily the case. First, we have the problem of time value of money. And so, we would want to consider the present value of costs. This leads to an issue of what time horizon to consider in the analysis. But minimizing cost is not likely to be our objective anyway. More likely, it is maximizing some function of the present value of net revenues or profits. Cost, after all, is minimized by producing nothing, which also leads to no sales and no revenues. Maximizing some measure of profits instead emphasizes the desire to increase production as a means to increased profitability. If there were total flexibility between machines at no cost, then a maximization such as this could be carried out on a daily basis, and tasks assigned to machines accordingly. Usually, however, there is a nonrecoverable setup cost each time tasks are reassigned. Because of this cost, forecasting the demand for tasks becomes important, and errors in the forecast will result in an economic loss to the company.

6.8 RESOURCE ECONOMICS

Another problem that arises in engineering that we can address using economics involves the allocation of limited resources such as oil. This problem was originally formulated by Hotelling (1931). Suppose we are designing a system that will use energy. A building is a good example. A question we must address is: how much should we invest to reduce the system's future energy consumption? In a building, we can install added insulation, use energy-conserving windows, and other such energy-conserving measures, but these are expensive alternatives, and we want to know that the savings will be worth the expense. To answer this question, we need a long-term forecast of energy prices, one that looks out over the next 50 years or so. Surprisingly, there is a method for making such forecasts, although any forecast is subject to error. We can look at the problem from several different points of view, but the mathematics is much the same.

For the example we will consider here, we will assume that all oil resources are held by a single entity (such as the OPEC—Organization of Petroleum Exporting Countries—

nations might be if they included all oil-producing nations and if they were in complete agreement on a production strategy, which is far from the actual case). It would then be reasonable to assume that this entity would establish a pricing policy that would maximize the value of the oil resource to itself. Working as a unit, this entity would control the market as would a monopolist and set price as a function of time to maximize the net present value of the resource. In mathematical terms, the suppliers would seek to maximize the functional

$$J = \int_0^\infty e^{-rt}[P(t) - C(Q,q,t)]q(P,t)\, dt$$

where r is the discount rate of the oil producers, $P(t)$ is the price of oil set as a function of time, $q(P,t)$ is demand as a function of price and time (it could also be a function of previous demand), and $C(Q,q,t)$ is the cost of supply. Q is the amount of oil remaining in the reserve stock. We maximize this functional with respect to the control variable $P(t)$, subject to the constraint.

$$\frac{dQ}{dt} = \dot{Q} = -q$$

The resource limitation can be specified also as constraints of the form

$$Q(t) \geq 0 \quad \text{for all } t, \text{ and} \quad q(P,t) \geq 0$$

or, alternatively, simply by choosing a cost function $C(Q,q,t)$ that becomes very large as Q approaches a limiting total amount. Taking the above constraints, we can write the boundary conditions

$$Q(0) = Q_0$$

$$Q(T) = 0$$

where T is the time at which the resource runs out.

As we see from Chap. 5, this is a problem in optimization, and specifically in the calculus of variations. If we choose simple forms for C and q, it is not difficult to solve. We will assume a linear demand curve given by

$$P = P_0 - aq$$

where P_0 and a are constants. This leads to the following functional:

$$J^* = \int_0^T [e^{-rt}(P_0 - aq - C)q + \lambda(\dot{Q} + q)]\, dt$$

$$= \int_0^T \{e^{-rt}[(P_0 - C)q - aq^2] + \lambda(\dot{Q} + q)\}\, dt$$

For this functional, the Hamiltonian is

$$H = -e^{-rt}[(P_0 - C)q - aq^2] + \lambda q$$

The necessary conditions follow. From the Euler-Lagrange equation,

$$\dot{\lambda} = 0$$

which integrates to yield

$$\lambda = \text{constant}$$

Allowing the time T to be unconstrained gives a condition on the Hamiltonian,

$$H(T) = 0$$

This condition can be met either by (Condition A) setting

$$q(T) = 0$$

or by (Condition B) satisfying

$$q(T) = \frac{P_0 - C + \lambda e^{rT}}{a}$$

We now turn to the Weierstrass condition. In order to maximize J^*, we minimize the Hamiltonian with respect to q.

$$\frac{\partial H}{\partial q} = -e^{-rt}[(P_0 - C) - 2\,aq] - \lambda = 0$$

Solving this equation for q,

$$q(t) = \frac{P_0 - C + \lambda e^{rt}}{2\,a}$$

This equation must apply at time T. But, at time T, it is clearly in conflict with Condition B above. Therefore, Condition A must apply, and at time T we must have $q(T) = 0$. Since at time T, $q(T) = 0$, we can solve the above equation for λ.

$$\lambda = -(P_0 - C)e^{-rT}$$

Combining the above two equations, the optimal rate of consumption is given by

$$q(t) = \left(\frac{P_0 - C}{2\,a}\right)(1 - e^{-r(T-t)})$$

All that remains is to determine T. To do this, we use the constraint equation

$$\dot{Q} = -q(t) = -\left(\frac{P_0 - C}{2\,a}\right)(1 - e^{-r(T-t)})$$

Integrating between the limits of 0 and T,

$$Q_0 = \left(\frac{P_0 - C}{2\,a}\right)\left(T - \frac{1}{r} + \frac{e^{-rT}}{r}\right)$$

which can be solved for T.

By substituting the optimal consumption rate into the demand equation, we can solve for price as a function of time.

$$P(t) - C = \frac{P_0 - C}{2}\,(1 + e^{-r(T-t)})$$

Indeed, this is a relatively modest rate of price increase. But, in reality, we do not see even this rate of price increase. The reason is because there is technological innovation. Technological innovation enters this problem formulation in two ways. First, it can be embedded in the cost of production. We could expect that the cost of production will decrease with time as a result of technological innovation. This trend of cost reduction will compete with increasing costs resulting from depletion in the function that we specify for C. But still another factor enters to further reduce price. That is the expectation resulting from research on alternative energy sources that the future demand for the resource, here oil, will decrease as alternative sources become available. The problem as formulated above is deterministic. Incorporating expectations on the future requires a probabilistic formulation that is much more difficult to solve. Yet for sufficiently simple forms for C, q, and the expectation variable, the problem can be solved. The solution shows that, indeed, there is a measurable price decrease over all future time resulting from the credible pursuit of alternative energy sources.

The planning for a wide range of engineering systems requires long-range forecasts of prices. The mathematics of resource economics is an important tool in obtaining such forecasts.

.9 SHADOW PRICES

Consider an economic activity that involves the production of goods j in quantities q_j. Each good produces a profit p_j per unit of production. In addition, each good, j, requires the consumption of resources i in quantities a_{ij}. And, finally, let the resources be constrained by limits b_i. To maximize profits, we can formulate this as an optimization problem in the linear programming format.

$$\begin{aligned} &\text{maximize} && J = \boldsymbol{p}^T\boldsymbol{q} \\ &\text{subject to} && A\,\boldsymbol{q} \le \boldsymbol{b} \\ & && \boldsymbol{q} \ge \boldsymbol{0} \end{aligned}$$

This problem can be solved as shown in Chap. 5. Now, if we vary $\boldsymbol{b}$ and resolve this problem, we can determine the sensitivity of J, the maximum total profits, with respect to changes in the constraints.

$$\frac{\partial J}{\partial \boldsymbol{b}} = \boldsymbol{y}$$

In other words, $\boldsymbol{y}$ is the marginal value of the resources. We could also write

$$J = \boldsymbol{y}A\boldsymbol{i}$$

where $\boldsymbol{i}$ is a unit vector,

$$i = \begin{bmatrix} 1 \\ 1 \\ 1 \\ \cdot \\ \cdot \end{bmatrix}$$

Thus, $\boldsymbol{y}$ is the value to this economic activity of each of the resources. Determined in this manner, we refer to the elements of $\boldsymbol{y}$ as *shadow prices.*

6.10 SUMMARY

In this chapter, we examined a number of basic concepts in microeconomics that are relevant to systems engineering. We began with a brief look at human needs and wants, which comprise the source of demand for systems engineering, and we saw that the set of basic needs and wants is short and simple. Next, we saw, through the Engel curve, how these needs and wants drive priorities in human preferences. We extended the concept of preferences to consider choices between two or more commodities, and introduced the notion of value. We saw that value is demonstrated through human choices. We extended the notion of preferences to cover consumption of a commodity over multiple time periods, and discussed the time value of commodities.

Our discussion turned next to equilibrium economics. We defined demand and supply curves in terms of choices made by consumers and producers, respectively, based on perceived values. The demand curve then led to the notion of elasticity. Next we talked about consumer and producer surplus, which comprise the net benefits to consumers and producers of economic activities in equilibrium markets. Then we used these concepts to discuss pricing theory in competitive and monopolistic markets. Finally, we looked at the benefits of research and development on the manufacturing process and on the product or commodity.

The next topic we examined dealt with discounting and present value. We found that we cannot directly compare costs or benefits that occur at different points in time. Before we make comparisons, we must reduce costs and benefits to a common point in time by obtaining their present values. The key parameter in this process is the social rate of discount. We found that the mathematics of discounting is very similar to the mathematics of interest, but recognized that the two are different concepts. Next, we discussed the effects of inflation and deflation on economic analysis of costs and benefits. We observed that units of money are not fixed in time, and that monetary amounts need to be corrected to a common basis before adding together costs and benefits accrued over significant periods of time.

Finally, we turned to activities related to systems engineering. We looked at the value of information in intertemporal consumption decisions, and examined one method for developing long-term forecasts of limited resource prices. In closing this chapter, we considered the concept of shadow prices as a means of measuring the marginal cost of resource constraints.

REFERENCES

1. Hazelrigg, G. A. and K. Lietzke, Research and Development and the Price of Oil, *Journal of Energy*, vol. 3, no. 6, November-December 1979, pp. 349–356.
2. Hillier, F. S. and G. J. Lieberman, *Introduction to Operations Research*, Holden-Day, San Francisco, CA, 1974.
3. Hotelling, H., The Economics of Exhaustible Resources, *The Journal of Political Economy*, vol. 39, no. 2, April 1931, pp. 137–175.
4. Samuelson, P. A., *Economics*, McGraw-Hill Book Company, New York, 1973.
5. Takayama, A., *Mathematical Economics*, The Dryden Press, United States, 1974.

PROBLEMS

1. Pick 10 common items. What needs or wants do they satisfy?

2. Pick two items from the above list. Determine an indifference curve for these items.

3. For problem 2, assume that you are given a fixed budget to spend on these items. How much of each would you buy?

4. You are a monopolist supplier for a product whose perfect competition equilibrium price is P_0 at quantity q_0. Assume that the demand curve is linear with elasticity e, and supply is linear with supply elasticity s. Derive an expression for your optimal price.

5. Consider problem 4 with a 10 percent sales tax. Now what is your optimal price?

6. You borrow $100,000 to be repaid by equal monthly payments over a period of 20 years at an annual interest rate of 8 percent. What is your monthly payment?

7. Consider problem 6. Suppose that inflation over the 20-year repayment period is constant at 6 percent. Plot the monthly payment in constant dollars. Suppose instead of a fixed dollar payment over the 20-year repayment period, repayment is made in payments of equal value. Plot these payments.

8. You are designing a building that will have a lifetime of 50 years. At a cost of $100,000, you could add insulation that would save $7,000 per year in heating cost. Is adding the insulation a good choice?
9. Your company produces a car whose average annual repairs cost the company $600. The car is warranted for 4 years against all repairs. A design change would reduce annual repair costs by $100. How much could this change add to the cost of the car without increasing the total cost of the car and its warranty?
10. You are designing a boiler for a system that must operate 40 years. There are two materials you can choose from, A and B. A will cost $1 million and last four years. B will cost $1.5 million and last 8 years. Then each must be replaced. Which is the better choice?
11. You are designing a refrigerator. You have two options to consider: A, with a compressor that gives the refrigerator a lifetime of 20 years and a cost of $900; and B, with a compressor that gives the refrigerator a lifetime of 10 years and a cost of $500. Which should the educated consumer prefer?
12. Draw a figure similar to Fig. 6.15 that shows that consumers lose as a result of initially high forecasts.
13. An old cigarette advertisement included the phrase, "I'd walk a mile for a Camel." What was the ad trying to imply?
14. Many states conduct super lotteries with multimillion dollar prizes. For example, you might hear, "Our lottery is worth $20 million, paid at the rate of $1 million per year for 20 years." Is this false advertising?
15. A Congressman, discussing truth-in-lending legislation, once stated, "How many people realize that 1½ percent per month is 18 percent per year?" What do you think?
16. True or false: Sales taxes reduce the profitability of business.
17. True or false: Consumers buy more of a product if it is produced by a monopolist rather than under competition.
18. True or false: A dollar received today is worth the same as a dollar received 2 years from today if the future dollar is adjusted for inflation?
19. What is the value of money?
20. You sell a car to a person who agrees to pay you $250 per month for 36 months. How much did you sell it for?
21. Which is less expensive, buying a $100,000 house or renting an apartment for $900 per month?
22. Consider the following information services:
    - Weather prediction
    - Position/location determination
    - Communications
    - Crop forecasts
    - Stock market price forecasts.

    How does each service provide value to its consumers? If you were selling the service, what would be the major issues with which you would have to deal?
23. You are designing a refrigerator. How might you use the following:
    - Interest rate projections
    - Electricity price forecasts
    - Forecasts of consumer purchasing preferences
    - Forecasts of your company's cost of capital.

7

Utility Theory

7.1 THE PURPOSE OF THIS CHAPTER

In the above chapters, we have seen the importance of properly formulated objective functions in the decision-making process. The previous chapter examines microeconomic theory as it contributes to proper formulation of objective functions. We began that chapter with a discussion of personal preferences, but then turned to the aggregate behavior of groups. In this chapter we will return to the preferences of an individual, and we will examine these in some detail, including preferences under conditions of risk. We will see that there exists a rigorous mathematical framework within which we can examine the preferences of an individual. But we will also see that it has some limitations. Then we will use this framework to discuss some problems of importance to systems engineering. We will discover that engineers make common use of techniques that have a rigorous mathematical basis, but frequently they ignore that basis. By the end of this chapter, you should have a new appreciation of the formulation and use of objective functions in the engineering design process.

7.2 RATIONALITY

Never try to explain with a rational model that which is perfectly well explained by stupidity.

To be sure, people can act in ways that we say are irrational or sometimes even stupid. But we cannot judge the rationality or stupidity of a person's actions unless we first can make some statements regarding their values or objective function. We must first respect the individuality of the person or decision maker, and seek to determine their preferences before we make judgments. This point needs further emphasis: *All judgmental statements require as their basis some assumption regarding the objective of that which is being judged.* To say something is good or bad, we must first determine or assume its objective. It may be bad in our eyes, but good in the eyes of the design engineer. The difference may lie totally in the fact that we assume its purpose to be different than the purpose chosen by the designer. With this said, we will begin this chapter with a discussion of rationality.

Let us begin by defining a terminology for preferences. Assume that a person, whose current status we shall refer to as his or her status quo, is presented with a set of options, which we shall denote by A, B, C, D, and so on. Let these options denote the entire changes in the person's status quo, if they are selected. For example, option A may comprise seeing a movie at a cost of \$5. The change in the person's status quo includes both seeing the movie and paying for it. Suppose further that, out of this set the person is given only two options, say A and B, and asked to choose one. The choice taken reveals the person's preference between the options. Let us assume that A is chosen over B. Then we would say that the person prefers (at this moment in time) A to B. When we say that the person prefers A to B, we mean that the person prefers to receive the outcome of option A rather than the outcome of option B. We will express this preference mathematically as follows:

$$A > B$$

Here the symbol $>$ is read, "is preferred to," and the above expression is read, A is preferred to B. If instead we had written

$$A < B$$

we would read, "B is preferred to A." It may also turn out that the person cannot decide between A and B, he or she is indifferent to which option is received. Then we write

$$A \sim B$$

and read as, "the person is indifferent between A and B." Two additional symbols are used to indicate preferences:

$$A \geq B$$

$$A \leq B$$

which are read, "B is not preferred to A," and "A is not preferred to B," respectively.

We will refer to the statement "A is preferred to B" as a *strong preference* for A, and the statement "B is not preferred to A" as a *weak preference* for A. The strong preference for A indicates that A is valued over B by some amount. This preference can be interpreted to mean that the cost of A could be increased by at least some small increment, and the person would still choose A over B. Now let us present the person with three options, A, B, and C, and ask the person to rank the options in pairs. That is, choose between A and

B, then B and C, then A and C. Suppose the person responds with the rankings

$$A > B$$

$$B > C$$

$$C > A$$

According to this, the person has a strong preference for A over B, a strong preference for B over C, and a strong preference for C over A. Perhaps A is an apple, B is a pear, and C is an orange. This preference ordering is shown in Fig. 7.1. Suppose now we give this

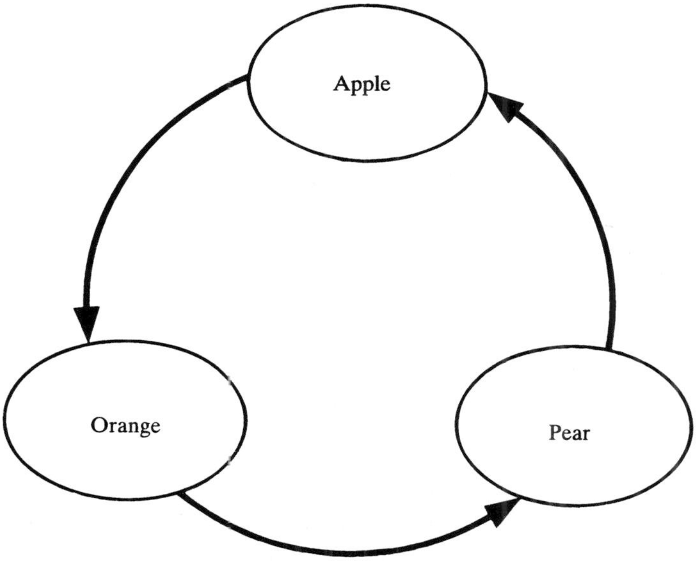

Figure 7.1 An intransitive preference ordering.

person C. But then, before the orange is eaten, we offer a trade for B. Since the person has a strong preference for B over C, he or she would be willing to accept the trade and, in fact, pay to obtain it. Say \$0.01 is the payment. Now the person has B. But, before eating the pear, we offer the person a chance to trade for A, plus, of course, a side payment of another \$0.01. Since the person has a strong preference for A, the trade will be made. Again, before eating the apple, we offer a trade for C, plus the side payment. And again, the person's strong preference dictates that the trade will be made. But now we are right back where we started from, except that we are \$0.03 richer and the person with whom we are conducting this experiment is \$0.03 poorer. As long as this preference ordering holds, we can continue in the above trading loop, simply draining the person's bank account. Obviously this preference ordering, if exercised in the manner shown, would lead most people to a status that is not preferred to the original status quo. Thus, we say that such a preference ordering is *irrational* or *intransitive*.

More precisely, we shall say that a preference ordering is rational if it is transitive, that is, if it satisfies the following condition:

$$\text{if} \quad A > B > C$$

$$\text{then} \quad A > C$$

And we shall say that a person is rational if his or her preference ordering is transitive. The remainder of this chapter will assume that we are dealing with rational individuals.

7.3 THE NOTION OF UTILITY

We turn now to the notion of utility. We will seek to assign to each option, i, a real number, u_i, such that

$$\text{if} \quad A > B$$

$$\text{then} \quad u_A > u_B$$

and

$$\text{if} \quad A \sim B$$

$$\text{then} \quad u_A = u_B$$

We will call the u_i the *utility* of option i, and give it units of *utiles*. If this can be done, then a comparison of the real numbers u_i will reveal the decision maker's set of preferences.

In some cases it is easy to see that utilities can be assigned to options in such a manner. For example, suppose the objective of a process is to earn a profit, and the larger the profit the better. We could simply take u_i to be the profit, and this utility would certainly reflect our preference ordering for the options. In many cases, where options are considered in an economic context, utility can be expressed in terms of profit and loss or benefits and costs. In these cases, utility is expressible as a function of a single economic parameter. This will be the case much of the time. But, when there is no single attribute of the system in terms of which utility can be determined, it is not so clear that such a utility can be determined.

Consider your preferences for music. Could you determine a utility function of the attributes of a set of songs that would predict your preference for which ones you would like to hear? Could you even identify the attributes of importance? One approach is to observe your choices, and then to assign numbers representing utilities to each song such that these numbers give the desired preference ordering. Indeed, such numbers could be interpreted as utilities, and they would be reflective of your preferences. But their use is limited to establishing your preference ordering. Utilities that provide only preference orderings are referred to as *ordinal utilities*. Ordinal utilities cannot be added or subtracted, and they do not measure strength of preference. They only can be compared. Thus, if the ordinal utility of A is 15, and the ordinal utility of B is 30, you can say only that B is preferred to A. You cannot say that B is twice as preferred as A. Nor could you compare the combination of A and B to C, which might have a utility of, say 33.

Utilities that provide both preference ordering and strength of preference are called *cardinal utilities*. Unfortunately, for many cases, it may be difficult or impossible to specify cardinal utilities. This has been the subject of much study in economics, and some experts assert that cardinal utilities do not exist. However, we do not ascribe to this point of view, and much of what follows uses cardinal utilities.

One thing might seem a bit strange regarding utilities. Consider the case where $A \succ B \succ C$. This preference ordering is specified by the utilities $u_A = 7$, $u_B = 5$, $u_C = 3$. But it is equally well specified by the utilities $u_A = 27.54$, $u_B = 19.75$, $u_C = 13.86$. The fact is that there is no unique set of numbers that specify utilities for any set of preferences. The preference orderings given by a utility function are indifferent to a positive, affine transformation of the sort

$$\hat{u} = \alpha u + \beta$$

where α and β are constants, and α is positive. We call this property *scale indifference* of utility preferences.

The fact is that there do exist preferences that cannot be expressed in conventional utilities as we present them here. But a discussion of them is beyond the scope of this book. The interested reader is referred to the extensive literature on the subject, for example, see Chipman et al (1971).

The Law of Diminishing Marginal Utility

The law of diminishing marginal utility states that *as the amount of a good consumed increases,* the *marginal utility of the good decreases.* This concept was present in terms of value in Chap. 6, and illustrated in Fig. 6.2. We illustrate this concept for utility in Fig. 7.2. The slope of the curve at any point is referred to as the marginal utility. Also, if above

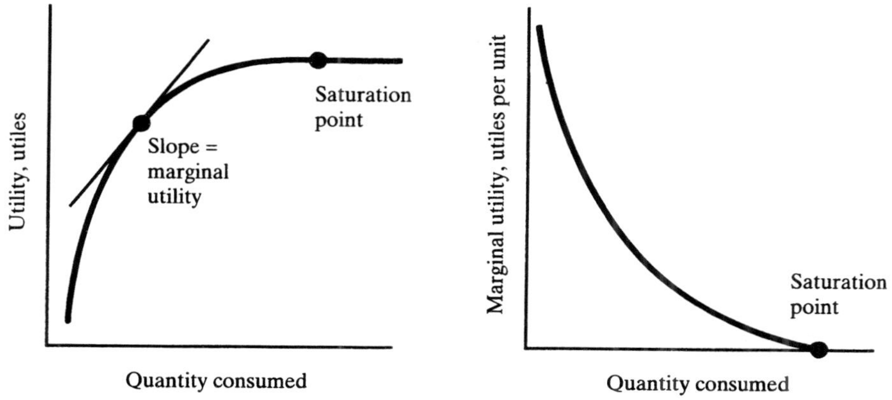

Figure 7.2 The concepts of utility and marginal utility.

some level of consumption the utility remains constant, we refer to that level as the saturation point.

The Law of Equal Marginal Utilities

The law of equal marginal utilities states that *each good is demanded up to the point where the marginal utility per dollar spent on it is exactly equal to the marginal utility of a dollar spent on any other good.* This law says that the value of a unit of currency lies in its purchasing power for alternative goods. The law of equal marginal utilities is illustrated in Fig. 7.3 for three goods. The marginal utility achieved is subject to a budgetary constraint of the form

$$\beta = P_A q_A + P_B q_B + P_C q_C$$

which is depicted as the horizontal line in the figure. P is price and q is quantity.

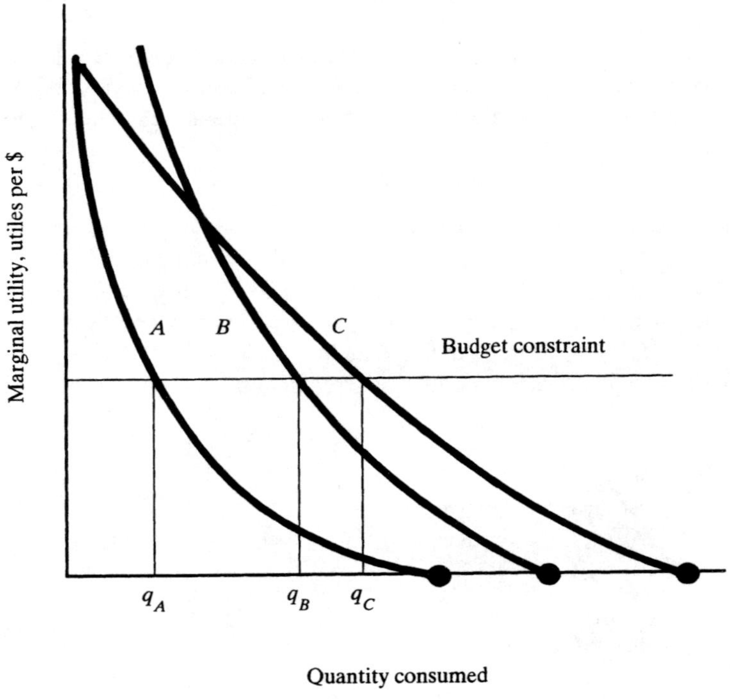

Figure 7.3 The concept of equal marginal utilities in a bundle of goods.

Measuring Utility

The preceding laws imply a method for measuring utility. First, we see that utility is measured in terms of alternatives. Indeed, we determine utility by posing alternatives and asking for or observing preferences. Also, keep in mind that utility preference orderings are insensitive to a positive, affine transformation of the form

$$\hat{u} = \alpha u + \beta \qquad \text{where} \quad \alpha > 0$$

Therefore, in a set of utility functions, one of them is largely arbitrary if only it satisfies the conditions of a utility function on preference ordering. Consider the utility for commodity A, received in quantity q_A, with saturation point q_{As}. Thus, for quantities received of this commodity between zero and the saturation quantity, utility is increasing with quantity received. Above the saturation quantity, there is no increase in utility. These considerations dictate that the utility function be monotonically increasing up to q_{As}, and constant above this quantity. We will find it useful, however, to include the restriction that the utility function exhibits decreasing marginal utility, with marginal utility becoming zero at the saturation point. These restrictions lead to utility functions of the general form shown in Fig. 7.2.

Given a utility function for one commodity in a bundle of commodities, we can now turn to the law of equal marginal utilities for the determination of utility functions for each of the other commodities in the bundle. For example, we could choose a commodity bundle consisting of two commodities, A and B. Then, we could pose several different budget constraints and determine for each the commodity mix, q_A versus q_B. We know that, at this mix, the marginal utilities of A and B per dollar spent on each are equal. This is to say that we know

$$\frac{1}{P_A}\frac{du_A}{dq_A} = \frac{1}{P_B}\frac{du_B}{dq_B}$$

we can rearrange this equation and write it in the form

$$\left.\frac{du_A}{du_B}\right|_\beta = \left(\frac{P_A}{P_B}\right)\left.\frac{dq_A}{dq_B}\right|_\beta$$

Then, determination of the right hand side of this equation enables us to obtain a utility function for commodity B, given that we have the utility function for commodity A. Note that the utility function for commodity B still contains an arbitrary constant. To obtain a value for this constant, we need to make a comparison between two alternatives. For example, we could be at the status quo where we have none of either commodity. We could then pose two alternatives: (1) receive n units of commodity A, where n is a fixed number, and (2) receive m units of commodity B. We seek to determine m such that we are indifferent between the two alternatives. This implies the condition

$$u_A(n) = u_B(m)$$

This completes the determination of u_B.

7.4 MULTIATTRIBUTE UTILITY FUNCTIONS

Most goods can be described in terms of a set of descriptors that we call attributes. Examples of attributes include size, shape, color, taste, smell, speed, and reliability. So it is natural to think in terms of expressing utility in terms of the attributes of the good.

Indeed, economists sometimes use this approach to predict the demand for an entirely new product or service. To do this, economists describe the new product or service in terms of attributes that occur in existing systems. They then examine demand as a function of the attributes of the existing products or services rather than for the aggregate product or service. Finally, they create a new demand function by lumping attributes in such a way that they describe the new product or service, and then compute demand as a function of the demand for this bundle of attributes.

Engineers frequently like to create a system utility function by listing the attributes of the system and then combining these by forming a weighted sum of measures of the attributes. A utility function formed this way is called a linearly additive utility function. While this may be done, the procedure is valid only under rather restrictive conditions, and it frequently leads to an inappropriate utility function, that is, one that does not represent the preference ordering of whom it represents. For example, in an airplane, we might be inclined to treat as attributes cost, speed, range, payload, takeoff distance, landing distance, reliability, maintenance cost per flight hour, and so on. We might then take as our utility function, a weighted sum of measures of each of these attributes. Engineers do this because it is easier than developing a correct utility function, or because they do not understand utility theory.

A better approach is to recognize that the airplane is to be placed in commercial service (if that is the case), that the purpose of this service is for the airline to make money, and to then compute the profitability of the company as a function of the attributes of the airplane. In so doing, we will obtain a properly formulated utility function that expresses utility as a function of many system attributes through their effect on system economics. However, we still have to account for preferences under conditions of uncertainty and risk.

Suppose, instead of the airline, we are the manufacturer of the airplane, and we are interested in optimizing its design. What we may mean by optimizing here is to design the airplane in such a way as to maximize our company's profits as a function of the design of the airplane. Again, although it may be rather complex, we can express company profits as a function of the airplane's design. This would require, among other things, a market model that would predict sales as a function of performance and cost parameters, and a cost model that would predict design, test and qualification, and manufacturing cost as a function of design parameters. We will return to the development of such models later in this book. For now, however, we will simply emphasize the need to develop system utility functions that truly represent the worth of the system to the decision maker.

Occasionally, it will be appropriate to treat the attributes of a system as utility independent. When this occurs, we may find a utility measure for the system using a *multiattribute utility function*. We will review a few commonly used multiattribute utility functions and comment on their validity. The functional forms given below work only when the system utility, u_s, can be decomposed into component utilities representing the utilities of the individual attributes.

$$u_s = f\,[u_1(x_1), u_2(x_2), u_3(x_3), \ldots]$$

where u_1, u_2, u_3, etcetera are the utilities of attributes X_1, X_2, X_3, etcetera, obtained in quantities x_1, x_2, x_3, and so on. We will refer to attributes that satisfy this condition as *independent attributes*. For independent attributes, a change in the quantity received of one attribute does not affect the utility of other attributes. Whenever possible, you will find it useful to characterize a system in terms of a set of attributes that are independent. A test for the utility independence of attributes is given in App. B.

Linearly Additive Utility

The additive utility function is perhaps the multiattribute utility form most commonly used by engineers. For a utility-independent system that may be described in terms of n attributes, the system utility is given by

$$u_s = \sum_{i=1}^{n} a_i u_i(x_i)$$

where u_i is the utility of attribute i occurring in quantity x_i and a_i is a weighting factor. Although, strictly speaking, the u_i are utilities, it is not uncommon for the attribute itself to be used in an expression of this form. For example, for an airplane, an attribute might be speed, and this might be used as a measure of utility directly. But this need not be the case, and instead we might first convert the attribute into a utility using a transformation curve such as that shown in Fig. 7.2. Failure to convert the attribute to a utility is a common error made by engineers. If we choose to substitute the attributes themselves in place of their utilities in the above equation, the resulting utility function, u_s, might be valid only in a small region around nominal values of the attributes. That is, in general, the utility function will not apply for the case of large changes in the attributes.

A big advantage of the linearly additive utility function is that it is conceptually simple and easy to use. A big disadvantage is that it is frequently used incorrectly. And for many of those cases where it is used correctly, it could have been reduced to an economic measure, and that would have been a better approach. A method for determining the weighting factors, a_i, is given in App. C.

Multiplicative Utility

The multiplicative utility form is given by

$$u_s = \alpha \prod_{i=1}^{n} u_i(x_i)$$

where again the u_i are the utilities of each attribute, and α is an arbitrary positive constant. $\alpha = 1$ is perfectly acceptable. The multiplicative utility form often enables a simpler form for each of the component utilities than does the linearly additive form.

Log-Linear Utility

The log-linear utility form is given by

$$u_s = \sum_{i=1}^{n} a_i \log u_i(x_i)$$

$$u_{i,\text{ log form}} = \log\,(u_{i,\text{ linear form}})$$

This form bears resemblance to both the linearly additive form and the multiplicative form. If we allow the transformation we see that these forms are actually identical, given the proper choice of u_i. Also, we see that

$$\sum_{i=1}^{n} a_i \log u_i = \log\left[\prod_{i=1}^{n} u_i^{a_i}\right]$$

Thus, the log-linear and the multiplicative forms are identical, and hence all three forms are equivalent, given the proper choice of the u_i.

Lexicographic Preference Ordering

The lexicographic form is not equivalent to the above forms. In fact, it does not always provide a preference ordering. It takes the form

$$X \geq Y \qquad \text{if and only if}$$

$$[u_1(x_1),\, u_2(x_2),\, u_3(x_3),..]\; L \geq [u_1(y_1),\, u_2(y_2),\, u_3(y_3),..]$$

which means that, for all i,

$$u_i(x_i) \geq u_i(y_i)$$

The notation $L \geq$ means lexicographically greater than or equal to. This says that in order that X be preferred or equivalent to Y, each and every attribute of X must be better than the corresponding attribute of Y. When X is lexicographically preferred to Y, we say that X dominates Y, because it is better in every way. Note that, if some attributes of X are better than the corresponding attributes of Y, but others are not, the lexicographic utility does not indicate a preference between X and Y.

7.5 ARROW'S IMPOSSIBILITY THEOREM

A camel is a horse designed by a committee.

As you delve deeper into utility theory, you begin to see the richness of the concept as a means of predicting market behavior and, thus, as a means of valuing engineering systems. Then you may become tempted to think about replacing the economic theory of the previous chapter with a utility-based theory of value. Indeed, economists have pursued this approach with considerable effort. But, in 1951, Kenneth Arrow, who later won the Nobel prize in economics, presented the results that today we refer to as *Arrow's impossibility theorem.*

Consider a process in which there are three voters, I, II, and III, each of whom casts a single vote, and all votes are of equal importance. In addition, there are three candidates running for election, A, B, and C, such that the voters each have the following rational preferences:

Voter	Preferences
I	A > B > C, A > C
II	B > C > A, B > A
III	C > A > B, C > B

Now consider the choices of these three voters in each of the following three elections:

	Elections		
Voters	A versus B	B versus C	A versus C
I	A	B	A
II	B	B	C
III	A	C	C
Result	A > B	B > C	C > A

Thus, the group preference is A > B > C > A. We recognize immediately that this preference ordering is intransitive or, we could say, irrational. Indeed, the decision making of a group can appear to be irrational. Arrow's impossibility theorem shows how this is possible, even though each individual member of the group exhibits rational decision making.

The impossibility theorem does more than show the irrationality of group decision making, however. It makes an important additional statement: *in general, we cannot write a utility function for a group.* Indeed, there are special conditions under which a utility function does exist for a group. But these conditions are special, and we cannot trust that they will be met by chance. Thus, any time we have a group decision-making process, we must take into account the fact that the group is likely to exhibit intransitive preferences. Think what this means to systems engineering. Suppose 1,000 engineers are working on the design of a complex system: an airplane, a car, a bridge, a computer. Each engineer makes dozens of design decisions. In making these design decisions, each engineer makes use of a system utility function. Each engineer, acting as an individual, may be exhibiting perfectly rational decision-making behavior. Yet, the decision making of the group, taken as a whole, might appear utterly irrational. To prevent this, it is necessary that all engineers working on a system be using, if not the same utility function, a set of functions designed such that they can at least be combined into a single group utility function.

One way to achieve a single utility function for an entire design is for management to impose it on all engineers who are working on a project. In fact, there could be advantages to the expert development of a design utility function by properly trained systems engineers, economists, operations researchers, managers, and others. Such a utility function could incorporate measures of the market value of a system, design risk, legal risks, corporate profitability or other return-on-investment measures, economic risk, and corporate image. Any such top level goals will be better achieved through a process of rational decision making than through a process that allows intransitive preferences exhibited at the group level.

7.6 PARETO OPTIMALITY

Arrow's impossibility theorem raises some interesting questions. For example, in the above election, who wins? Each person favors a different candidate, so election of all three candidates at once does not provide a definitive result. Election of candidates through a process of successive elections between pairs of candidates leads to an intransitive preference ordering. Note here that the preference ordering is intransitive, not indifferent. So, how could such an election be resolved? One notion is to count votes using unequal weighting. This process could lead to a clear decision, but whose votes are weighted higher?

Perhaps the same three voters go out for pizza after the election. Together they want to order a large pie, but each person wants different toppings. Can this be resolved? If the voters can agree to weight one person's preference higher, then the issue can be resolved. But why would the others give in to allowing one of them to get his or her way? This question is, in essence, asking whose preferences are the strongest? The problem can be resolved if we allow for the possibility of side payments. Perhaps you and I are two of the voters at the pizza table, and you really prefer pepperoni, but I want mushrooms. So you turn to me and say, I'll give you a dollar if you vote for pepperoni. And you do the same with the other person. If we agree, then you get your choice, and we are satisfied. We could have bargained over the amounts of the side payments, with each person seeking to maximize his or her own utility, and in the process, we will have measured the strength of our preferences. In fact, allowing side payments to be made enables the formulation of a group utility function.

In general, we have *n* people involved in a decision-making process, and their individual preferences are such that no group utility function exists. Then, by means of side payments, coalitia are formed such that the group can exhibit a set of preferences. Each person who joins a coalition does so because he or she is no worse off for joining than they would have been acting as an individual. The side payments act as transferrals of utility between the individuals in a coalition. Thus, an individual who prefers option A to option B may instead prefer option B plus \$$X$ to option A. If, by forming this coalition, none of the members of the coalition are worse off than they would have been acting alone, and at least one person is better off, then there is a net economic gain that results from formation of the coalition. The group as a whole achieves maximum gain when they form one grand coalition, and maximum gain is achieved when it is no longer possible for any member of the coalition to renegotiate side payments to improve his or her status. This condition is referred to as *Pareto optimality*.

7.7 DECISION MAKING IN THE PRESENCE OF RISK

The subject of decision making in the presence of risk has been a topic of research for a very long time. Over 200 years ago, Bernoulli posed a problem that is now called Bernoulli's paradox. The problem is this: what would you pay to enter a game wherein a fair coin is flipped until, on the nth flip, it lands heads. You then win a prize of \$$2^n$. Since n must be at least one, and with probability 0.5 will be larger than one, most people are willing to pay something more than \$2 to play the game. Perhaps you would be willing to pay \$3 or \$4 or even \$5 to play the game. The paradox, however, is this. The expected value of the prize, P, is given by

$$P = \sum_{n=1}^{\infty} (\text{probability of } n)(\text{prize given } n)$$

$$= \sum_{n=1}^{\infty} \left(\frac{1}{2^n}\right) 2^n$$

$$= \sum_{n=1}^{\infty} 1 = \infty$$

That is, the expected value of the prize is infinite. But despite this fact, very few people would pay more than a few dollars to enter the game. The search for answers to questions such as those raised by Bernoulli's paradox has led to an extension of utility theory to the case of decision making in the presence of risk.

Von Neumann–Morgenstern Lotteries

Bernoulli himself proposed a solution to his paradox. He suggested that the value of the prize, \$$2^n$, be replaced by $\log 2^n$. By so doing, he replaced the monetary prize with a utility that has decreasing marginal value. Now the expected utility of Bernoulli's game is

$$E\{u\} = \sum_{n=1}^{\infty} \left(\frac{1}{2^n}\right) \log 2^n$$

which can be shown to be finite. It is this notion of expected utility that von Neumann and Morgenstern in the 1940s expanded upon.

Consider an action that has two possible outcomes, X and Y. X occurs with probability $P(X)$, and Y occurs with probability $1 - P(X)$. Von Neumann and Morgenstern argued that if $P(X)$ were very close to one, the decision maker should behave as though X is a certain event. And likewise, if $P(X)$ were very close to 0, the decision maker should behave as though Y is a certain event. This logic led to the notion of expected utility in the form

$$\begin{aligned} E\{u\} &= u[P,X; (1 - P),Y] \\ &= pu(X) + (1 - p)u(Y) \end{aligned}$$

where we have written $P(X)$ simply as P. Here, p is a real number on the range $0 \leq p \leq 1$, and may be interpreted as a probability. In fact, we will show that we can take $p = P(X)$. This is an exceptionally powerful result as it extends the theory of utility to the case where outcomes of actions or decisions can be expressed only in probabilistic terms, and it does so in a very simple form that is easy to implement.

Basically, what von Neumann and Morgenstern have shown is that, if an individual can express a transitive set of preferences between every possible pair of gambles, taken over some set of alternatives, then it is possible to associate with each alternative a utility such that, if the individual's decisions are guided strictly by expected utilities, the resulting decisions will be in accord with the individual's preferences. When we evaluate an individual's risk preferences using a gamble as noted above, we refer to the gamble as a *von Neumann–Morgenstern lottery*. Here, a lottery is not something that continues over time. It happens once and only once, and it is played by a rational individual or a group that acts transitively. The general form of the expected utility is

$$E\{u\}\sum_{i=1}^{n} p_i u(X_i)$$

where the X_i comprise the full set of possible outcomes, and

$$\sum_{i=1}^{n} p_i = 1$$

Using the concept of the von Neumann–Morgenstern lottery, and taking the p_i to be the probabilities of the various independent and only events, we can take another look at Bernoulli's paradox, this time using the logarithmic utilities proposed by Bernoulli. The expected utility of the game is

$$E\{u\} = \sum_{n=1}^{\infty} \left(\frac{1}{2^n}\right) \log 2^n = \log 4$$

A person with this utility for money would be willing to pay \$4 to enter this game. To derive this result, recall that we chose $u(\text{prize}) = \log(\text{prize})$. So we see that, in this example, the application of von Neumann-Morgenstern utility leads to a very reasonable answer. Here we see quite clearly that the *utility of money is not, in general, a linear function of the amount of money.*

The Axioms of Von Neumann–Morgenstern Utility

We will now examine briefly the assumptions upon which von Neumann–Morgenstern utility theory is based. These assumptions are the following [Luce and Raiffa, 1957]:

1. All outcomes of a von Neumann-Morgenstern lottery can be ordered in terms of the decision maker's preferences or indifferences, and that ordering is transitive.
2. Any compound lottery, that is, any lottery that has as an outcome another lottery, can be reduced to a simple lottery that has among its outcomes all the outcomes of the compound lottery with the associated probabilities of their ultimate occurrence, and the decision maker will be indifferent between the original compound lottery and the equivalent simple lottery.

In interpreting this assumption, you must be careful to observe that the probabilities of occurrence of the outcomes in the later stages of a compound lottery may be conditional upon the outcomes of the earlier stages. With the probabilities properly computed, indifference between the compound and simple lotteries may seem obvious, but this assumption nonetheless removes from one's utility of a compound lottery any joy or anticipation associated with the gambling process itself.

3. Let the set of all outcomes from the compound lottery be designated by A_1, A_2, ... , A_r , ordered such that

$$A_1 > A_2 > A_3 > \ldots > A_r$$

Then there exists a number, p, such that

$$A_i \sim [p,A_1; (1 - p),A_r]$$

where p is a real number on the interval $0 \leq p \leq 1$. This assumption assures the continuity of preferences between outcomes A_1 and A_r.

4. Let B_i be given by

$$B_i = [p,A_1; (1 - p),A_r]$$

Then, in any lottery, B_i can be substituted for A_i and the preferences of the decision maker will remain unchanged.

This assumption means that we can construct an equivalent lottery that contains only the outcomes A_1 and A_r .

5. The decision makers preferences and indifferences among lotteries is transitive.

6. Given two lotteries, L_1 and L_2, where

$$L_1 = [p_1, A_1; (1 - p_1), A_r]$$

$$L_2 = [p_2, A_1; (1 - p_2), A_r]$$

L_1 is preferred or indifferent to L_2, if and only if $p_1 \geq p_2$.

If the above assumptions hold, then there exist a set of numbers, u_i, associated with the outcomes, A_i, such that for two lotteries, L_1 and L_2, the expected values

$$q_{1,1}u(A_1) + q_{1,2}u(A_2) + q_{1,3}u(A_3) + \ldots + q_{1,r}u(A_r)$$

and

$$q_{2,1}u(A_1) + q_{2,2}u(A_2) + q_{2,3}u(A_3) + \ldots + q_{2,r}u(A_r)$$

reflect the preferences between the lotteries, where the q_{ij} are the probabilities of the jth outcome in the ith lottery. Further, if for the decision maker transitive preferences exist over the set of lotteries such that we can assign numbers $u_i(L_i)$ to each lottery with the condition that $u_i(L_i) \geq u_j(L_j)$ if and only if $L_i \succeq L_j$, we say that there exists a utility function over the lotteries. If this utility function also has the property that

$$u[q, L_1; (1 - q), L_2] = qu(L_1) + (1 - q)u(L_2)$$

then we say that the utility function is linear.

Utility theory extended to the von Neumann–Morgenstern lottery is an extremely powerful method for creating a value space. It is limited to a decision maker or group that has a set of preferences on the outcomes of the lottery that is transitive. Figure 7.4 shows typical utility curves for three individuals, one of whom is risk neutral, one who is a risk

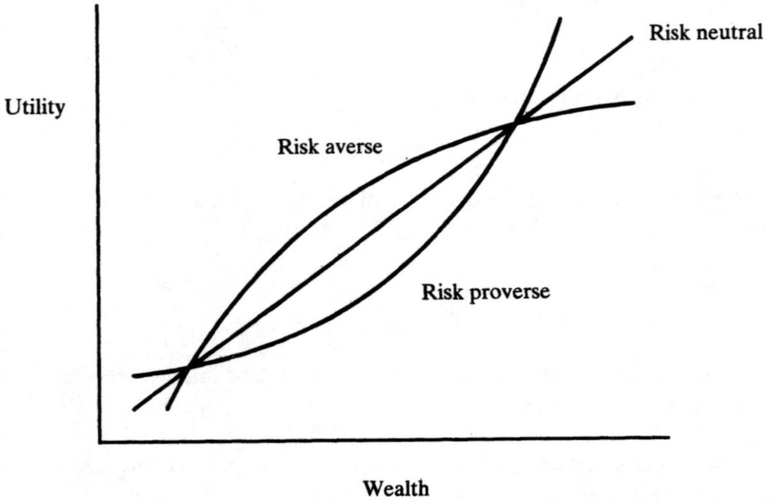

Figure 7.4 Utility curves for various risk preferences.

taker or who is risk proverse, and one who is risk averse or who prefers to avoid risks. In Chap. 12, we will outline a method for determining utility curves under conditions of risk.

7.8 THE VALUE OF BETTER INFORMATION

Suppose we are presented with a set of n lotteries, $L_1, L_2, \ldots, L_n$, each of the form

$$L_i = [q_{i,1}, A_1;\ q_{i,2}, A_2;\ \ldots;\ q_{i,r}, A_r]$$

and we are asked to choose between them. This is precisely the problem of engineering design. We are given a number of design options, the lotteries, each with a set of possible outcomes and each outcome with an associated probability, and our job is to choose the option that produces the most desired outcome. The best choice is the lottery that has the highest expected utility.

Now suppose we can purchase information to aid us in our decision. In order to decide whether we should purchase the information we must be able to determine its worth. But first we might ask: what do we mean by the word information in this context? It is easy to see that if, for every lottery, that is for every value of i, there is a value of k such that $q_{ik} = 1$ and $q_{il} = 0$ for all $l \neq k$, then all lotteries would reduce to certainty outcomes, and the choice could be made in the absence of risk. Then, it is clear that the decision maker could choose the lottery that indeed maximizes his or her utility. The best choice is obvious, and no other choice will result in a better outcome. In general, some other choices will result in a lower utility. If information prevents the decision maker from taking choices that would result in lower utility, then that information has value, and its value is measured by the increase in expected utility resulting from the information.

The preceding case is for perfect information. But suppose we do not have perfect information. Still, we want to evaluate the value of improved information. Our approach is to construct a compound lottery as shown in Fig. 7.5. This lottery enables the choice to conduct an experiment or not. If no experiment is conducted, we enter a simple lottery. If an experiment is conducted, however, we then enter a compound lottery. The first stage in the compound lottery is the performance of an experiment. The experiment results in outcomes E_s, $s = 1, 2, \ldots, m$, each occurring with probability p_{es}. Then, given the outcome of the experiment, we enter the second lottery. Only now we have two interesting twists to our analysis.

First, the probabilities in the lotteries conducted after the experiment are all conditioned on the outcome of the experiment. Thus, instead of probabilities q_{ij}, we have the conditional probabilities $q_{ij}(A_i|E_s)$. Second, we can still choose which lottery to enter after the experiment, and so we must account for this decision.

In other words, although this is a compound lottery, it is not a compound lottery of the sort we discussed in the development of the von Neumann–Morgenstern axioms. The decision of which lottery to choose after the experiment is conditioned by the outcome of the experiment.

Finally, we can compare the utility of the case given the experiment with the simple lottery without the experiment to determine the worth of the experiment. To determine the

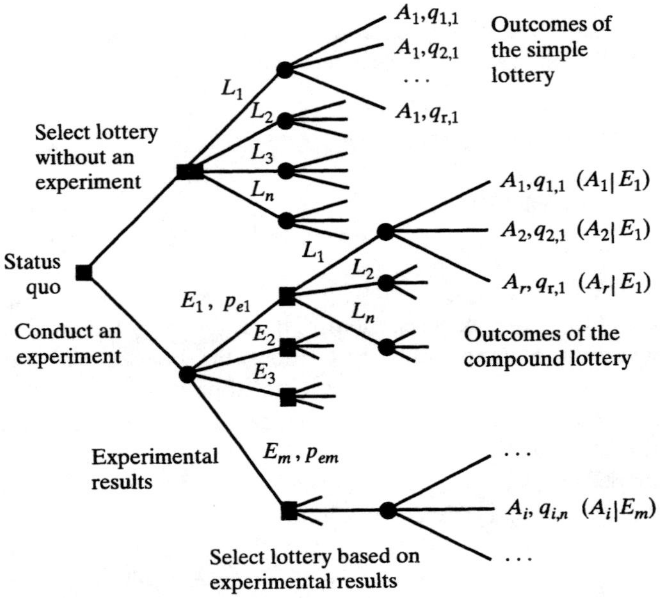

Figure 7.5 A compound lottery for the evaluation of improved information.

worth of the experiment, we include in our utilities a cost of the experiment, which we can vary. Given an experiment that we would decide to conduct if its cost were zero, we can then examine increases in the cost of the experiment. At some cost we become indifferent to the conduct of the test. This cost is the worth of the experiment. The net worth of the experiment is the indifference cost less the actual cost. Details of the solution for the compound lottery of Fig. 7.5 are given in Chap. 12.

7.9 THE VALUE OF IMPROVED SAFETY

Frequently, engineers design systems that impose a risk to human life. Generally, these systems cannot be designed to be perfectly safe at any reasonable cost, and risk-cost trade-offs must be made in order to proceed with the system design. Examples of such systems include automobiles, airplanes, the air route structure, and nuclear power plants. In cases where the system design choices are made to serve society as a whole, such as separation standards in air route systems, a determination of safety versus acceptable risk must be made. In other cases where personal and public safety is in question, engineers often choose to sidestep the safety issue by demanding that standards be imposed. They then design to these standards and, when fatalities occur, claim that they are not to blame. Yet, even in these cases, someone must decide what level of risk is acceptable.

Many approaches have been proposed and used. Some of the popular ones are reviewed briefly here.

- Gross expected earnings: This method is based on the premise that a human life is worth the total or gross future productivity of that life discounted to the present. Productivity is measured in terms of the potential future earnings of the individual, and the present value of this amount constitutes a loss to society.
- Net expected future earnings: This method is similar to the gross expected earnings, except that it recognizes that some portion of an individual's income is consumed by the individual and that the net economic loss to society is his or her gross expected earnings less the present value of his or her future consumption. This method is intended to estimate the net loss to others in society.
- Social decisions: This method recognizes that society takes risks and makes choices in investments to improve safety, and interprets these decisions as revealed preferences to derive an implicit value of a life.
- Insurance: This method examines the willingness of a person to buy life insurance and treats this as a measure of the value the person places on his or her own life.

None of these methods yields acceptable results, and arguments against them are easily made.

We will seek here to develop an approach to the valuation of improved safety by employing the von Neumann–Morgenstern lottery where one of the outcomes is death. For example, whenever we get into a car, we enter a lottery that has several possible outcomes, but we could simplify it to the two outcomes: (1) we get where we want to go safely, and (2) we die in an accident. Let us take P to be the probability of safely arriving at our destination. Then, the certainty equivalent utility, u_{ce}, is given by

$$u_{ce} = Pu_A + (1 - P)u_D$$

where u_A is the utility of arriving safely and u_D is the utility of death. It has been argued that this formulation will not work because u_D is negative infinity, and no non-zero value of P is therefore acceptable. But revealed preferences indicate quite clearly that people do choose to undertake activities that have a non-zero probability of death. This would imply that, in fact, u_D is not negative infinity. Since this equation is linear in P, we have

$$\Delta u_{ce} = (u_A - u_D)\Delta P$$

And since we expect u_D to be a large negative number, changes in u_{ce} with respect to P will be positive for increasing P. Also, we see that the more important to us a successfully completed activity is, that is, the larger u_A, the higher we will value improved safety. But how can we determine the quantity $u_A - u_D$?

One thing that we should keep in mind is that nothing in life is risk free. Life itself is a risk, and the average person lives only about 700,000 hours. Thus, even a young or middle-age person who is healthy has something like a one-in-ten million chance (10^{-7}) of dying in the next hour. Because of this background risk, risks that are much lower, say 10^{-9} per hour, appear negligible. Another thing we must watch out for is the units in which we measure changes in safety and certainty equivalent utility. To be sure, the von Neumann-Morgenstern lottery is for a single event. This would comprise a single ride in a car, or a single airplane flight, or a single bungee jump. It does not apply to events that are continuous in time. Nor is it clear how you might go about extending the theory to a continuous event.

We also must be careful to keep in mind that the von Neumann–Morgenstern lottery approach works only for individuals or groups that have transitive preferences. To be safe, we should consider applying the method only to individuals. If, however, we could determine a monetary worth of improved safety for each affected individual, we could then sum these amounts, interpreting them as the combined amount that the collection of individuals would be willing to pay to achieve the safety improvement. Note that this approach is very different than examining the actions of a society, which would generally act intransitively, and which would not only provide the added safety to those individuals who want it, but to everyone. In a case where society adopts a safety measure, the cost of the measure is often born by the entire society through taxation. In such a case of forced consumption, it is not appropriate to use the demand curve as an indication of value from which we could determine a consumer surplus.

One approach might be to determine preferences by questionnaire. But this approach is prone to distortion by the respondents, particularly when discussing subjects such as their own death. A better approach is to examine revealed preferences of individuals in cases where they had a choice, at some cost, to improve their own safety. Wearing a seatbelt might constitute a revealed preference. But here, the cost of putting on a seatbelt might be difficult to assess. Surely, except in special situations, it cannot be much more than a penny or two, representing the value of the time that it takes to put the belt on. We will leave this problem at this point. No satisfactory solution exists at this time, but the lottery approach seems to be a logical one if the problem of determining $u_A - u_D$ can be solved.

7.10 SUMMARY

In this chapter we develop and review the fundamental concepts of utility theory. Utility theory is a powerful method for measuring preferences. But it is limited to individuals or groups that express rational preferences, where rational is taken to mean transitive. We discussed the notions of both cardinal and ordinal utility measures. Ordinal utility provides for the ordering of preferences, but does not provide a measure of the strength of preferences. Although cardinal utility measures might be formulated in some instances, and we used them above extensively, ordinal utilities are adequate for the purpose of engineering design, which requires an ordering of preferences. The preference orderings provided by utility measures are insensitive to affine transformations.

After discussing the basics of utility theory, we extended the theory to the case where the utility function can be decomposed into a series of separate functions, each a utility for a given attribute of a system. We then examined four different multi-attribute utility function forms, the linearly additive form, the multiplicative form, the log-linear form, and the lexicographic form. The first three forms were shown to be equivalent and valid so long as the attribute utilities are independent.

Next, we examined one important way in which utility theory is limited. We looked at the preferences of a group that is comprised of rational individuals. Despite the fact that each individual in a group may have transitive preferences, Arrow showed that the group need not have transitive preferences. Therefore, only in special cases will a group possess a utility function.

If a group, in general, does not possess a utility function, the question arises, how can a group make decisions? We found that the method of group decision making that provides each member of the group the highest possible utility is to allow for side payments between members of the group. The side payments are conditioned upon the payee changing his or her stated preference as the result of the receipt of a payment to do so. Thus, the members of a group can negotiate between themselves to achieve a solution that makes no member worse off than he or she would be without the negotiation, and at least one member of the group is better off. The end point of this negotiation reaches a condition referred to a Pareto optimality.

Upon completion of our discussion of preferences under conditions of certainty, we turned our attention to decision making in the presence of risk. We introduced the von Neumann–Morgenstern lottery as a means of extending utility theory to include the case where the outcome of a decision can be expressed only in probabilistic terms. Finally, we applied von Neumann–Morgenstern utility to two problems frequently encountered in engineering design, the value-of-information problem and the value-of-a-human-life problem. We found that this theory provided interesting and unique insights.

REFERENCES

1. Atherton, W. N., *Theory of Union Bargaining Goals*, Princeton University Press, Princeton, NJ, 1973.
2. Chipman, J. S., L. Hurwicz, M. K. Richter, and H. F. Sonnenschein, *Preferences, Utility, and Demand,* Harcourt Brace Jovanovich, New York, 1971.
3. Fishburn, P. C., Independence in Utility Theory with Whole Product Sets, *Operations Research,* vol. 13, no. 2, March-April, 1965, pp. 28–45.
4. Greenberg, J. S. and G. A. Hazelrigg, Research on the Problem of Efficient R&T Program Formulation Under Conditions of Uncertainty and Risk, Annual Progress Report for the period May 1, 1975 to April 30, 1976, AMS Report No. 1268, The Aerospace Systems Laboratory, Princeton University, Princeton, NJ, April 30, 1976.
5. Heal, G. M., *The Theory of Economic Planning,* North-Holland/American Elsevier, NY, 1973.
6. Luce, R. D. and H. Raiffa, *Games and Decisions*, John Wiley, New York, 1957.
7. Raiffa, H., Preferences for Multi-Attributed Alternatives, The RAND Corporation, Memorandum RM-5868-DOT RC, April 1969.
8. *Readings in Decision Analysis*, R. A. Howard, J. E. Matheson, and K. L. Miller eds., Decision Analysis Group, Stanford Research Institute, Menlo Park, CA, 2nd ed., 1976.
9. Starr, C., Social Benefit versus Technological Risks, *Science,* 165, September 1969, pp. 1232–1238.

PROBLEMS

1. What constitutes a rational set of preferences for an individual? Provide an example.
2. What is the difference between ordinal and cardinal utilities?
3. Show that the preference orderings given by a cardinal utility function are insensitive to an affine transformation.
4. A person is confronted with the following gambles, and chooses to place the bets indicated.

Gamble	Entry price	Payoff for winning	Probability of winning	Decision
A	\$25	\$100	0.30	Reject
B	\$10	\$50	0.20	Accept
C	\$1	\$10,000	0.000001	Accept
D	\$20,000	\$100,000	0.90	Reject
E	\$1,000	\$2,000	0.5	Reject

   Are these demonstrated preferences rational?

5. Suppose you are designing a system whose utility is a function of five attributes. For example, the system might be an airplane whose utility depends on range, speed, payload, reliability, and cost. What can you say about the marginal utilities of each attribute for an optimal design?
6. Write a mathematical utility function for the airplane example in problem 5.
7. Suppose the airplane of problem 6 is a commercial airliner. Is a multiattribute utility function, such as that obtained for the solution to problem 6, appropriate? If not, what would be appropriate?
8. Suppose you are presented with the following gamble: You can bet \$10 on the toss of three fair coins. If all three tosses are heads, you will win \$50. Otherwise, you win nothing. What would be the value to you of knowing in advance of making the bet the outcome of the first two tosses? Hint: Would you accept this bet without knowing the outcome of the first two tosses?
9. You are reading your local newspaper and come across the headline, "City Council Optimizes Budget Allocations." What conclusions might you draw?
10. You are designing a weapons system that is intended to support the operations of 20 military units, each headed by a general. You have spoken to each general to determine their preferences regarding the attributes of the system. You want to design an optimal system. Do you need to define a utility function to do this? If so, whose utility function do you use or how do you generate one?
11. You question a friend and find that he does indeed have an irrational set of preferences. Can you say that your friend acts in his own best interest? Why?
12. We know that preferences expressed through utilities are independent of affine transformations. Does it therefore follow that we can write a valid linearly additive utility function such that all attribute utilities lie on the range $0 \le u_i \le 1$? What about a multiplicative utility function?

13. What does the following statement mean: "I like vanilla ice cream more than you do"?
14. Do we need von Neumann–Morgenstern utility theory in engineering? Why?
15. A group of people critique a family of products. They agree on a list of important attributes, and the attributes are such that, for every person, more of each attribute is better. But there is no agreement on the relative values of how much better are positive increments of each attribute or how much each attribute contributes to the overall value of the product. Despite this disagreement, the group agrees unanimously that one product is better than all the others. What preference ordering form would demonstrate this preference?
16. Does von Neumann–Morgenstern utility apply to many practical cases of decision making?
17. Comment on the following statement: "All systems are optimal according to some measure of worth."
18. You and two friends are arguing over what to do on Saturday evening. The only thing you agree on is that you all want to go out together. What can you do?
19. Show that, in general, it is not possible to optimize a product for a group of customers.
20. Nancy is chairing a meeting between herself, Bob, and Tom. She raises an issue that has three options, A, B, and C. On first vote, Nancy simply poles Bob and Tom for their preferred choice. Each option receives a single vote. Upon further questioning, Nancy discovers the following preferences:

    Nancy $A > B > C,\ A > C$
    Bob $C > B > A,\ C > A$
    Tom $B > A > C,\ B > C$

    At this point she says, "I know, we'll vote first to determine the best of two options, and then we'll vote for the winner of these options against the third option. Since there are three of us, there can't be a tie, and we will obtain our choice." They agree and Nancy proceeds, selecting two options for the first vote. Which option is selected?
21. Explain how revealed preferences can be used to determine a utility function.
22. Explain the linkages between personal preferences and demand. What is the difference between preference and demand? Why can we construct a demand function for a product but cannot construct a market preference function, both using the same product attributes?
23. What information of importance to the engineering design process is contained in a demand curve?

8

Forecasting

8.1 THE PURPOSE OF THIS CHAPTER

Most engineering systems are designed for the purpose of use over some extended period of time. Some systems are short-lived; they operate only a few minutes, hours, or days, and are then discarded. In some sense, an expendable launch vehicle for orbiting a satellite is an example. Its use lasts only for a few minutes, and then it is discarded. But most systems are intended for use over a much longer period. The lifetime of a stereo or television or an automobile may be several years. The lifetime of a refrigerator or other household appliance may be 10 to 20 years. The lifetime of a bridge or building may be 50 years to a century or longer. Any time that an engineering system is designed for extended use, it is important to forecast the conditions under which the systems will operate in the future. And, in some sense, this is even important in the case of the expendable launch vehicle. Although each individual rocket may be in use for only a few minutes, it is certainly hoped that the design will prove useful for several years.

Many aspects of the success or failure of an engineering design can depend on events or conditions of the future over which the engineer has little or no control. These factors can affect the economic demand for the system, the in-use economics of the system, the demand for other products, the need for regulation and the subsequent impact of regulation on the system, the future disposal of the used system, and the future economics of the manufacturer of the system. For example, the sudden upward price shifts in the cost of gasoline increased the demand for smaller automobiles at the expense of the demand for larger ones. Rising electric energy costs have made it economical for consumers to purchase energy-efficient appliances, and these appliances have reduced operating costs for users, thereby shifting their consumption patterns. A computer manufacturer that used small quantities of a now-defined toxic substance finds that it is required by regulation to buy back discarded computers to assure proper environmental handling and disposal of the toxic substance.

In these cases, the designs of the systems mentioned were influenced, explicitly or implicitly, by assumptions regarding the future. We will refer to these assuptions as *forecasts*. Sometimes, the effect of forecasts on engineering design is blatant and obvious. When an airplane manufacturer decides that it will design and produce a new wide-body jet in the 250-seat class for trip lengths of 2,500 nautical miles or less, it is using forecasts of future demand for air travel to estimate the market for an airplane in this particular market niche. The decision is made at high levels in the company to pursue this particular market niche, and the remainder of the engineering effort is directed toward fulfilling design objectives derived from the resulting design goal. However, many other design objectives relate to forecasts that are more subtle and less overtly communicated. What will be the future price of aviation fuels, and how should this price affect attitudes toward engineering tradeoffs that relate to fuel efficiency? For example, suppose marginally increased fuel efficiency could be achieved by forming a particular portion of the airplane's surface in a complex shape that significantly increases the manufacturing cost of the part. What is the best design choice? What is the best design choice between increased manufacturing cost and lowered maintenance cost or vise versa?

Tradeoffs in engineering systems design typically demand forecasts of many things: the future rate of inflation, future interest rates, future availability of fuels or raw materials, future policies with respect to the environment, the future legal environment, and so on. On another level, the use of a law such as $F = ma$ to compute the future position of an object is, in essence, a forecast. Thus, we can view the determination of system performance itself as a forecasting process. And to do this will give us new insights into the process of systems modeling.

Overall, forecasting is vital to successful systems design. This chapter is devoted to a discussion of the most fundamental aspects of forecasting. Again, this is a vast subject, and researchers have studied it from many different points of view. The view provided here is narrow and focused to the needs of the systems engineer.

8.2 TYPES OF FORECASTS

Predicting things is very difficult, especially the future.

When asked what the future has in store for us, there are many different approaches we could take to seek an answer. For one thing, we could just guess. For another thing, we could seek out a seer or a visionary—a fortune-teller. But generally, we seek an approach that is more rigorous, more traceable, more explainable, and more rational. Regardless of the approach, however, there are some principles that we can keep in mind. First, it is a rare event indeed that can be predicted with complete precision. Some things, such as the encounter of a spacecraft with a planet can be forecast with a very high degree of precision. Other things, such as future political events, cannot be accurately forecast. Second, however, it is also the rare event that we have absolutely no expectation on. This is to say that, generally, events behave according to certain expectations or relationships, but not precisely. Let us envision crudely that a future event, E, is dependent in part on conditions about which we have information available today, I, and in part on conditions about which we have no information or which are random, R. Conceptually, but certainly not with mathematical precision, we might write

$$E = f(I,R)$$

If we develop a rational forecasting model, that is, a model that seeks to make use of as much information as possible, the best we could do is to capture the effect of I on our ability to forecast E. There is no information available from the random process, R, other than the characteristics of the process itself, such as its standard deviation, which enables us to estimate the characteristics of the error in a forecast.

A forecast that makes effective use of all available information will have a pattern of error that is totally random. One way to view this is that it would not be possible to make money, on average, betting against the forecast. For example, consider the stock market. Stock prices are determined by the supply and demand for the individual stocks. These, in turn, are driven by the expectations of the aggregate set of investors. In essence, these investors provide, through stock prices, a set of forecasts on the future performance of each stock. If, on average, this forecast were not accurate, we could make money above and beyond the average rate of return by betting against the forecast. But in the stock market, normal market forces result in forecasts that prevent individual investors from gaining wealth at the expense of others. In essence, the natural market forecasts tend to contain all available information (with the exception of insider information, the use of which is illegal in the United States). Even if forecasts are made in secrecy, their use in stock trading divulges their results and in essence makes them known to all traders.

All forecasts make use of models. Some forecasts use models that rely purely on data from similar experiences in the past, and extrapolate these data to the future. For example, forecasts of population and electric power demand have been made by simply examining past trends in these quantities and extrapolating them into the future. In other cases, models have been used to unify data on past events, which has the effect of increasing our breadth of experience. The model $F = ma$ is one example. It unifies data from a wide variety of experiences in a predictive model for mechanical systems that has very strong forecasting ability. Indeed, if we can forecast F and m accurately over the period of the desired forecast, then we can use this model to very accurately determine the state of a mechanical system. Still, despite the accuracy of this model, it does not yield perfect precision in determination of the future states of a system. Our concept still applies of a forecast consisting of an information-based part as well as a random part. Between these extremes exist a wide range of forecasting models that mix unifying models and trend models.

We will refer to the use of unifying models in forecasting as the *problem-solving* approach. The problem-solving approach requires that we know the laws that apply to the situation and have the necessary data to exercise the model.

The success of trend models relies on the *law of large numbers*. When we forecast population of the earth, for example, we know that a future population equals today's population plus the number of births minus the number of deaths that occur between today and the date of the forecast. We can estimate births and deaths statistically, based on previous trends, and thus forecast population. But such estimates give no insight into specifically who will be alive and who will not be alive. Likewise, a forecast of electric power demand is an aggregate estimate of consumer behaviors. Although we cannot predict when or how each consumer will use a unit of electricity, we can predict the average usage. This is much like predicting how many heads will occur in 1,000 flips of a fair coin. Here, although we would expect about 500 heads in total, this forecast gives us no insight into the outcome of each flip. Indeed, the only reason we can obtain the forecast is because we are conducting a sufficiently large number of flips that each specific event makes an unimportant contribution to the overall outcome.

Some events occur only once. Still we would like to forecast the outcome. To the extent that these events include uncertainty, all we can do is to provide a *probabilistic* forecast. For example, perhaps we want to forecast the success or failure of a research program as a function of funding level, project mix, schedule, and other parameters. Probabilistic forecasts provide a range of possible events and their associated probabilities.

An important point relative to all forecasting is that forecasts of virtually anything include some amount of uncertainty. The forecast itself may be helpful in understanding the consequence of a policy or action, *but to provide information on the future, we must have an estimate of the accuracy of the forecast as well as the forecast itself.* Sometimes it is not too difficult to estimate the accuracy of a forecast; sometimes it is very difficult. For example, an application of $F = ma$ that is well within the established limits of this model is subject to error mainly from errors in the input data and the computational approach to its solution. Both can be estimated. However, in the case of a model that derives from a rather complex set of assumptions and simplifications, it may be impossible to estimate objectively the model error.

In the sections that follow, we will present some common approaches to forecasting, and then we will extend these models to provide information. We also will make use of Monte Carlo modeling once more to provide both forecasts and estimates of their accuracy.

8.3 REGRESSION ANALYSIS

Regression analysis is a set of techniques for finding and quantifying relationships between variables for which data exist. We generally express the relationships in terms of a curve or equation. For example, suppose we have electric energy demand data for a period of 25 years. It would be reasonable to plot these data in various forms, such as total annual consumption versus year and peak demand versus year. From these plots, we might see trends in the data that we would like to understand better. We would then turn to regression analysis to gain this understanding.

Normally, when we plot data, such as peak electric energy demand versus year, for which we plot the year on the abscissa and the peak demand on the ordinate, we recognize two different types of variables: those that we are free to choose and those that respond to our choice. The variables that we are free to choose we call *independent* variables, and the variables that respond to the choice of independent variables we call the *dependent* variables. We normally plot the independent variables on the abscissa and the dependent variables on the ordinate. In terms of measurements made to obtain data, we can think of choosing values of the independent variables and observing, at these values, the values of the dependent variables. In the case of peak electric energy demand, we could observe the peak demand during the period January 1 through December 31 for each year, and this observation would become the corresponding value of the dependent variable.

In a case such as peak electric energy demand, many factors influence the dependent variable: population, industrial activity, weather, consumer decisions, the price of electric energy, and so on. All of these factors taken together might produce trends that are of interest and that we seek. But variability in the influencing factors may also lead to significant year-to-year variability in the dependent variable. What we want to find is a trend line that summarizes in a simple way the data. A trend line is defined by an equation of the form

$$x(t) = \sum_{i=0}^{n} a_i f_i(t)$$

where $x(t)$ is the dependent variable and t is the independent variable. The a_i are constants and the $f_i(t)$ are linearly independent functions of the independent variable that we normally restrict to be single valued functions of the independent variable. Linear independence of the functions implies that

$$f_i(t) \neq \sum_{\substack{j=0 \\ j \neq i}}^{n} \alpha_j f_j(t)$$

That is, $f_i(t)$ cannot be expressed as a linear sum of the $f_j(t)$ for $j \neq i$. One form of the trend line equation is the polynomial

$$x(t) = \sum_{i=0}^{n} a_i t^i$$

Given that the $f_i(t)$ are linearly independent and single valued, the m data points can fit precisely if we take $n = m - 1$. If we represent the data points as the pairs (x_i, t_i), and denote by F the matrix of elements $f_{ij}(t_i)$ and define the vectors $\boldsymbol{x}^T = (x_1, x_2, \ldots, x_m)$ and $\boldsymbol{a}^T = (a_1, a_2, \ldots, a_n)$, then the solution for $\boldsymbol{a}$ is

$$\boldsymbol{a} = F^{-1}\boldsymbol{x}$$

and a solution exists as long as the determinant of F is non-zero. This formulation poses two problems. First, it does not smooth over the random elements of a process to show the genuine trends that underlie the process and, second, high-order equations generally cannot be used to forecast the events outside the range of the data because they have the ability to diverge wildly and rapidly as soon as they are no longer restricted by data. Thus, to use historical data as a predictor of the future, we must abstract trends from the data using low-order equations.

The approach we use in regression analysis is to formulate the problem as a problem in optimization. Define the vector $\boldsymbol{f}^T = (f_1, f_2, \ldots, f_n)$, and let $\boldsymbol{f}(t_i)$ be denoted by $\boldsymbol{f}_i$. Similarly, let $x(t_i)$ as computed by the equation

$$x(t_i) = \boldsymbol{f}_i^T \boldsymbol{a}$$

be denoted by x_i, and let x_i^* denote measured values of the dependent variable at corresponding values of the independent variable. We next define S to be the sum of the squares of the deviations between the measured and calculated values of x.

$$S = \sum_{i=1}^{m} (x_i - x_i^*)^2$$

Substituting for x_i,

$$S = \sum_{i=1}^{m} (\boldsymbol{f}_i^T \boldsymbol{a} - x_i^*)^2 = (\boldsymbol{F}\boldsymbol{a} - \boldsymbol{x}^*)^T(\boldsymbol{F}\boldsymbol{a} - \boldsymbol{x}^*)$$

and we now minimize S with respect to $\boldsymbol{a}$. Note that S is quadratic in $\boldsymbol{a}$. This fact leads to an easy solution for $\boldsymbol{a}$, which is the reason for the formulation of the problem in this form. The necessary condition for a minimum of S is

$$\left(\frac{\partial S}{\partial \boldsymbol{a}}\right)^T = 2F^T F\boldsymbol{a} - 2F^T \boldsymbol{x}^* = \boldsymbol{0}$$

The solution for $\boldsymbol{a}$ is

$$\boldsymbol{a} = [F^T F]^{-1} F^T \boldsymbol{x}^*$$

Notice that this solution holds for arbitrary functions f_j, with the condition only that they are linearly independent functions. An example for f_j was given in which $f_0 = 1, f_1 = t, f_2 = t^2$, and so on. This form is a polynomial in t. But other forms are equally valid. Another popular form for f in regression analysis is the logarithmic form:

$$f_0 = 0$$

$$f_1 = \log(bt)$$

$$f_j = 0 \qquad \text{for} \quad j \geq 2$$

The f_1 term can be written

$$f_1 = \log b + \log t$$

which is equivalent to the linear form. It models the wide variety of processes that grow or decay logarithmically. Another popular form is the exponential form

$$f_0 = 0$$

$$f_1 = e^{bt}$$

$$f_j = 0 \qquad \text{for} \quad j \geq 2$$

where b is another constant. This equation is not linear in b, but it can be made linear by taking the logarithm of both sides of the equation

$$\ln x = \ln a_1 - bt$$

and we see that this then takes the same form as the linear regression.

The choice of the f_j is critical to the success of a regression analysis. Information regarding the form of the f_j should be taken into account in their selection. As noted earlier, it is generally desirable to include as few terms as reasonable in a regression equation. This will give the equation greater stability as a predictor of the future. In fact, it is common to include only the f_0 and f_1 terms in a regression analysis. In the case of a polynomial, this provides a straight-line fit to the data. But it is possible to use more terms provided that you feel that the result will not go astray when used to predict the future.

The following figures show regression fits for two sets of data. Figure 8.1 shows a linear fit to data for which a linear fit appears appropriate. Figure 8.2 shows an exponential fit, and Fig. 8.3 shows a linear fit to the same data. Comparing these curve fits to the data that they fit, we see that we must worry about two potential problems. First, there is the question of how well the curve represents or fits the data. That is, do the data lie relatively close to the curve, or do they spread out widely around the curve? Second, we must be concerned with the nature of the difference between the curve and the data. As we see in Fig. 8.3, the data have a trend that is not represented by the curve. In other words, the difference between the data and the curve is not random. Since it is not random, there is information in the data that we have not picked up in our curve fit.

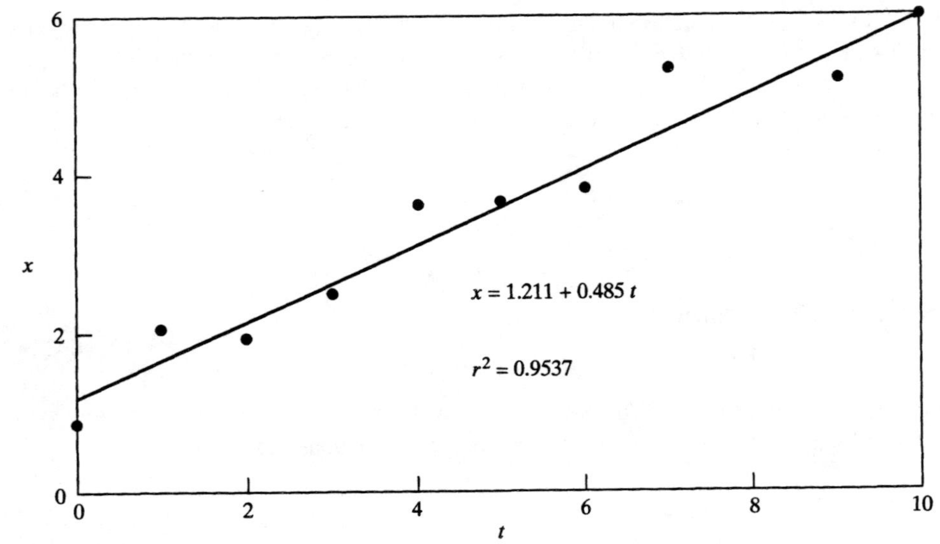

Figure 8.1 A linear regression line.

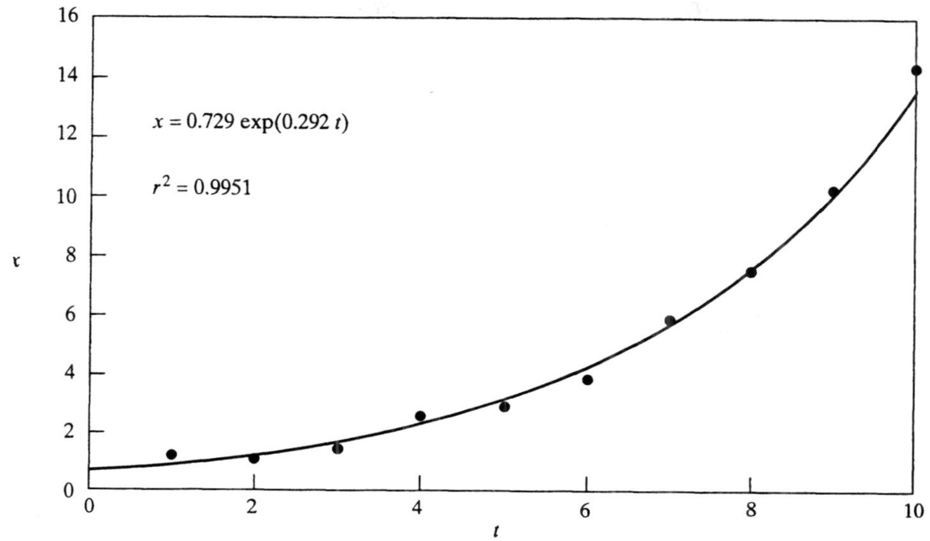

Figure 8.2 An exponential regression line.

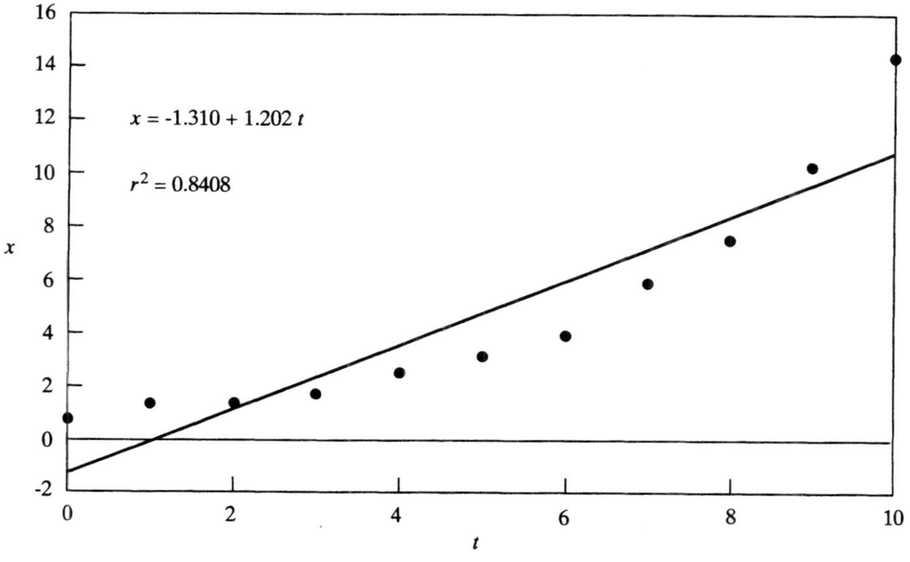

Figure 8.3 A linear regression line fit to the same data as in Figure 8.2.

As we might have suspected from the start, there is information in the vector of differences between the curve and the data. We define the vector $\boldsymbol{e}$ as follows:

$$\boldsymbol{e} = \boldsymbol{x}^* - \boldsymbol{x}$$

where $\boldsymbol{x}$ obtains from the result of the minimization of S. We call this the vector of residuals, and we can learn a lot about the goodness of the curve fit by examining $\boldsymbol{e}$. In matrix notation, it is not difficult to show that

$$\boldsymbol{e}^T\boldsymbol{x} = 0$$

This results because of the fact that the $\boldsymbol{a}$ are determined by an optimization process. Furthermore, if there is a non-zero f_0 term in the model for x, it can be shown that

$$\sum_{i=1}^{m} e_i = 0$$

One measure of the goodness of the regression analysis is the extent to which the regression line, as a predictor of $\boldsymbol{x}^*$, reduces the variance between the prediction of $\boldsymbol{x}^*$, namely $\boldsymbol{x}$, and $\boldsymbol{x}^*$ versus the variance between the mean of $\boldsymbol{x}^*$, as a predictor of $\boldsymbol{x}^*$, and $\boldsymbol{x}^*$. We would expect that the variance between $\boldsymbol{x}^*$ and $\boldsymbol{x}$ would be less than the variance between $\boldsymbol{x}^*$ and the mean of $\boldsymbol{x}^*$. Thus, we define the parameter

$$r^2 = \frac{\sum_{i=1}^{m} (\boldsymbol{x}^* - \bar{\boldsymbol{x}})^2 - \sum_{i=1}^{m} (\boldsymbol{x}^* - \boldsymbol{x})^2}{\sum_{i=1}^{m} (\boldsymbol{x}^* - \bar{\boldsymbol{x}})^2}$$

where the average value of the observed data is given by

$$\bar{\boldsymbol{x}} = \frac{\boldsymbol{x}^{*T}\boldsymbol{i}}{m} \qquad \text{where} \quad \boldsymbol{i}^T = [1, 1, \ldots, 1]$$

r is called the *coefficient of correlation*. This can be written in vector notation

$$r^2 = \frac{\delta^T\delta - \boldsymbol{e}^T\boldsymbol{e}}{\delta^T\delta}$$

where

$$\delta = \boldsymbol{x}^* - \bar{\boldsymbol{x}}\boldsymbol{i}$$

It must be true that

$$\delta^T\delta \geq \boldsymbol{e}^T\boldsymbol{e}$$

$$\text{where} \quad \delta^T\delta \geq 0 \quad \text{and} \quad \boldsymbol{e}^T\boldsymbol{e} \geq 0$$

Thus, $0 \leq r^2 \leq 1$. The value of r^2 is a measure of the extent to which the variable t "explains" the behavior of x^*. $r^2 = 0$ implies that there is no linear correlation between x^* and t, whereas $r^2 = 1$ implies that x^* is fully explained by t. Be careful, however.

Although we use the word "explains" to describe the relationship between the dependent and independent variables, all that we can conclude, even in the case that $r^2 = 1$, is that the variables are correlated. *In no way does correlation imply cause and effect, and the regression analysis itself provides little or no confidence that the future can be predicted in terms of the independent variable.* This point is so important that it bears emphasis.

Consider the demand for electric energy. For many years prior to 1973, it was observed that demand grew exponentially in time. Forecasts of future demand were based on the assumption that this demand growth would continue. Given the data and the regression analysis as performed, this assumption is equivalent to saying that the explanatory variable for electric energy demand is time. Demand for electric energy would grow simply by allowing time to pass, and nothing else would affect demand. Indeed, there was a very high value of r^2 associated with this conclusion. Few people thought of other variables influencing demand, such as population, industrial growth or decline, or price of electric power. The oil shortage of 1973 resulted in price increases of energy resources. This reduced the demand for electric energy and invalidated virtually all electric energy demand forecasts. *Regression analysis cannot be a substitute for thinking.* It is always important to think about cause and effect relationships, and to concern yourself with the nature of the independent variable as a predictor for the dependent variable. In the case of electric energy, a closer examination of potential independent variables would have shown that real electric energy prices (prices corrected for inflation) had fallen steadily from the turn of the century. The price shocks encountered in 1973 marked the end of that steady price decline.

Figure 8.4 shows summer peak demand for electric power over the years from 1960 through 1985, together with forecasts made over the years from 1974 through 1985. Looking at the actual summer peak demand, it is clear that something happened in 1975 to break the previous pattern of growth. As a result, each of the successive demand forecasts is lower than the previous forecast. Indeed, what happened is that prices ceased their steady downward trend. In fact, although it is not done here, the entire 25 years of data provide a sufficient basis for the determination of the demand elasticity as well as the constant-real-price rate of demand growth. Figure 8.5 shows these same data with an exponential regression line fitted to the first 15 data points on the curve. This regression line has a very high r^2 value, providing confidence in the curve fit, and it predicts an annual demand growth rate of 7.7 percent. It appears quite apparent from the figure that the exponential regression line formed the basis of the 1974 forecast.

The example forecasts discussed above failed to identify the important explanatory variables. They did so because these variables were masked in early data. In both cases, real prices fell with time in such a way that there was near perfect correlation between real price and time. Thus, time could be substituted for price as an explanatory variable in the regression analysis. Other factors were also important. Penetration of the potential market for electric energy as the rural electrification of the nation progressed and general population growth also spurred demand. These variables too were well correlated with time and hidden in the data. In cases such as these, the data may be insufficient for regression analysis alone to separate the contributions of one explanatory variable from others. Thus, you should always be concerned that important trends have been observed in the data used for the analysis.

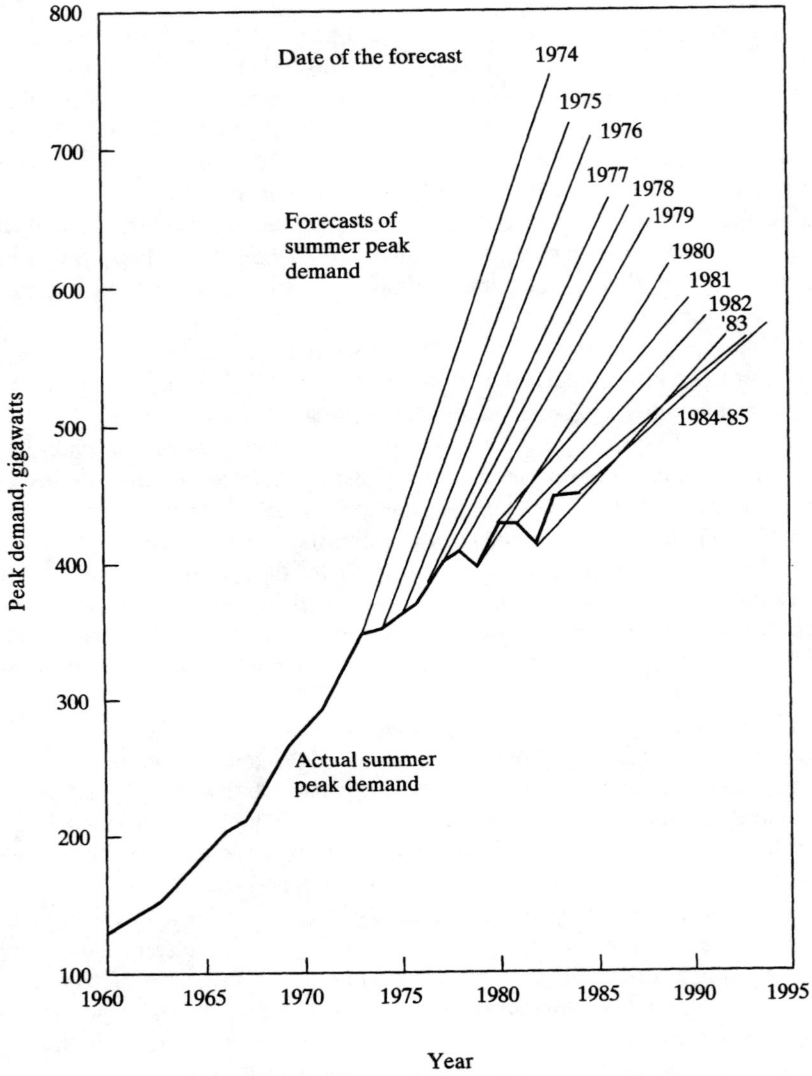

Figure 8.4 Summer peak demand projections made by the National Electricity Reliability Council.

In the event that we decide to consider more than a single explanatory variable in a regression analysis, we can proceed to use the same form of the analysis as we developed earlier. When we use more than a single explanatory or independent variable in a regression analysis, we refer to the analysis as a multivariate regression. It is easy to extend the equations and conditions developed earlier to the case of a *multivariate regression*. For example, consider the linear form

$$x(t,y) = a_0 + a_1t + a_2y$$

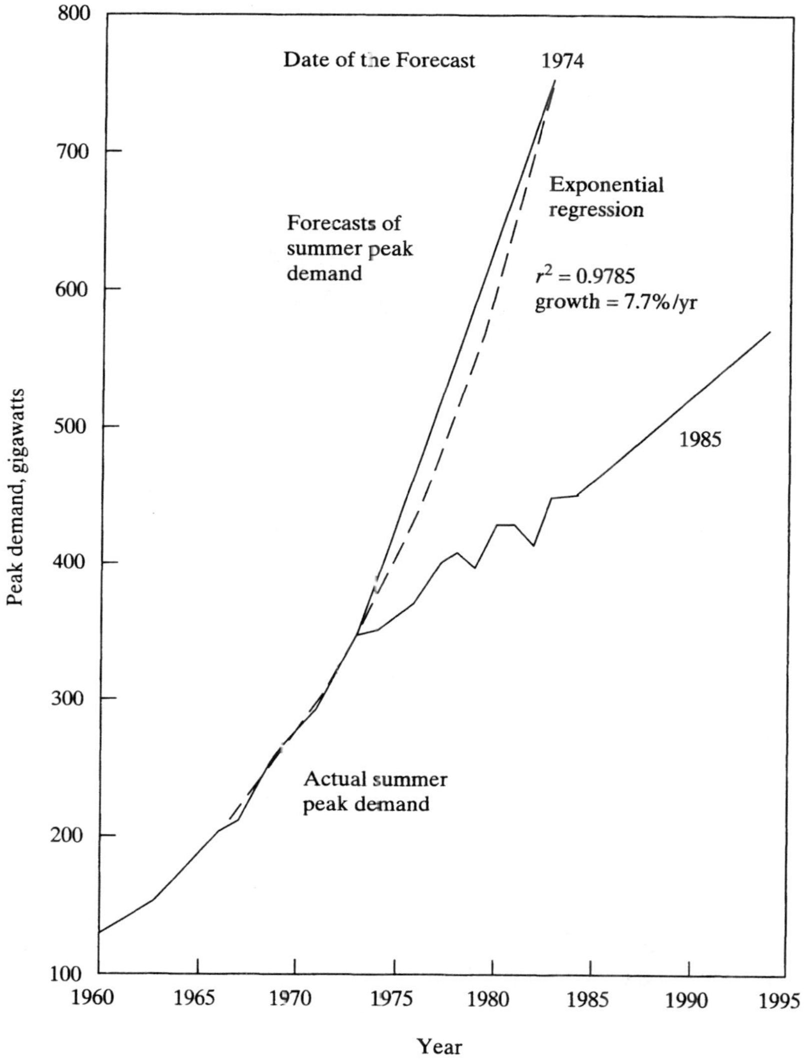

Figure 8.5 An exponential regression line fit to the data of Figure 8.4 covering the years from 1960 through 1974.

This form explains behavior of x in terms of two independent variables, t and y. Solution for $\boldsymbol{a}$ is obtained from the above.

This discussion by no means deals thoroughly with the subject of regression analysis. Entire and extensive books have been written on this subject alone. This brief review is intended merely to acquaint the systems engineer with regression analysis as one approach to forecasting or discovering relationships between variables in a system, and to hint at some of the powers and problems that the approach has.

8.4 LOGISTIC CURVES

Many phenomena exhibit exponential growth during some phase of their existence. Electric energy demand is a classic example. In these cases, there is a strong tendency to use regression analysis to fit the exponential form

$$x(t) = ae^{bt}$$

to the data. The resulting fit is then used to provide long-term forecasts. Indeed, exponential forecasting models may work well early in the lifetime of a phenomenon, but it is virtually certain that exponential growth does not continue unchecked for long. A good long-term forecast should, therefore, never rely blindly on an exponential regression line. For any phenomenon, you must ask, when will the exponential model become inappropriate? That is, when will the growth break off the exponential curve and follow a less enthusiastic growth pattern. And we will ask the additional question: what is the mathematical form that might be more appropriate for forecasting phenomena whose growth cannot continue indefinitely?

In the longer term, we find that most physical phenomena follow a pattern of population or intensity that is depicted by an S-shaped curve that we refer to as a logistic curve. A typical logistic curve is the Gompertz curve, given by the equation

$$x(t) = Ka^{b(t-1)}$$

Here, K, a, and b are constants, such that

$$K > 0, \quad 0 < a < 1, \quad \text{and} \quad 0 < b < 1$$

If K is taken to be 1, then $x(t)$ can be interpreted as a fraction that represents the level of saturation in the phenomenon achieved as a function of time. K itself can also be a function of time. Suppose $K(t)$ is a parameter that represents a population times a constant representing the ultimate electric energy demand per person. Then the Gompertz curve will yield total electric energy demand, but with a limit that is physically reasonable.

An example of the Gompertz curve is shown in Fig. 8.6 compared to an exponential curve. Both curves have a similar shape for low values of t, which might correspond to data taken early during a phenomenon. But as soon as the S shape of the Gompertz curve begins to become apparent, the two curves diverge, and their predictions of $x(t)$ are very different for larger values of t. Proper selection of $K(t)$ is the key to achieving success in the use of the Gompertz curve. A strength of the Gompertz curve is that $K(t)$ can be selected on the basis of physical considerations or limitations. For example, suppose we wish to predict the development of a relatively new technology as measured against some parameter. The density of transistors on a microelectronics chip might be such a parameter. If physical considerations lead us to a value for the maximum possible density, then we can use this as the value of K. Consider another case in which the ultimate demand for a product is determined by population. In this case, $K(t)$ can be a function of the population. In any case, $K(t)$ is an upper bound on $x(t)$, and it is a known function of t.

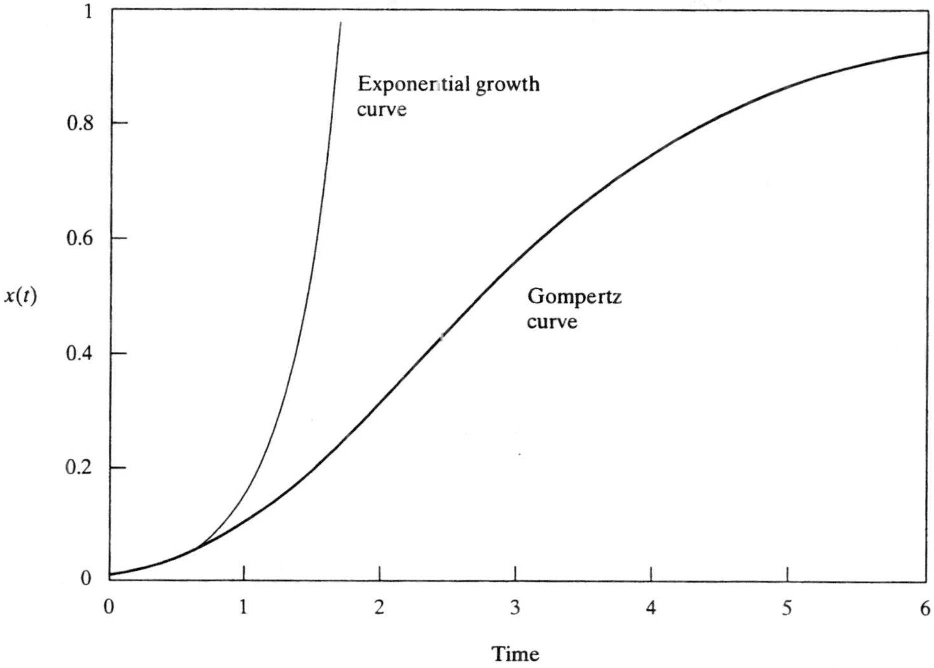

Figure 8.6 A typical logistic curve compared to an exponential curve.

We then rewrite the Gompertz equation in the form

$$\frac{x(t)}{K(t)} = a^{b^{(t-1)}}$$

Since $K(t)$ is a known function of time, we can rewrite this equation

$$y(t) = a^{b^{(t-1)}}$$

and define

$$y^*(t) = \frac{x^*(t)}{K(t)}$$

We notice that $y(t)$ is always less than 1. Now we can take the logarithm of both sides of the equation for $y(t)$

$$\ln y(t) = b^{(t-1)} \ln a$$

Both $y(t)$ and a are positive numbers less than 1. Therefore, we change signs and take the logarithm again

$$\begin{aligned} \ln[-\ln y(t)] &= (t-1)\ln b + \ln[-\ln a] \\ &= t\ln b - \ln b + \ln[-\ln a] \end{aligned}$$

which is a form that can be used in a regression analysis.

So we see that, despite the seeming complexity of the Gompertz curve, we can use it in a regression analysis, and that it is likely to provide much more realistic estimates of many parameters for which we have data early during a phenomenon or for which there exists a reasonable or physical limitation that we can express as a function of time.

8.5 FORECASTING UNCERTAINTY

One thing we can say with reasonable certainty regarding virtually every forecast we could make is that it is likely to be in error by some amount. In other words, forecasts contain uncertainty. If we want the forecast to give us information on the future, rather than simply a data point, then we must obtain an estimate of the uncertainty in the forecast as a part of the forecast itself. There are a few techniques for doing this, and we will review some of them here.

First, we will return to the regression analysis and look at the error associated with the prediction of a variable, x. Remember that the regression line is an attempt to explain values of a dependent variable x in terms of an independent variable, t. But, in general, t does not explain all of the variability in x. What we seek from the regression line is the expected value of x conditional upon a value of t, namely $E\{x|t\}$. Then our ability to predict x is limited by two factors: an inability to determine $E\{x|t\}$ precisely and the variability of x around $E\{x|t\}$. These factors are illustrated in Fig. 8.7. They are referred to as *lack of fit* and *pure error*. Note that the variance of a prediction of x will be higher than the variance in the prediction of $E\{x|t\}$. We will begin by analyzing our ability to predict $E\{x|t\}$. We have

$$x = \boldsymbol{f}^T(t)\boldsymbol{a}$$

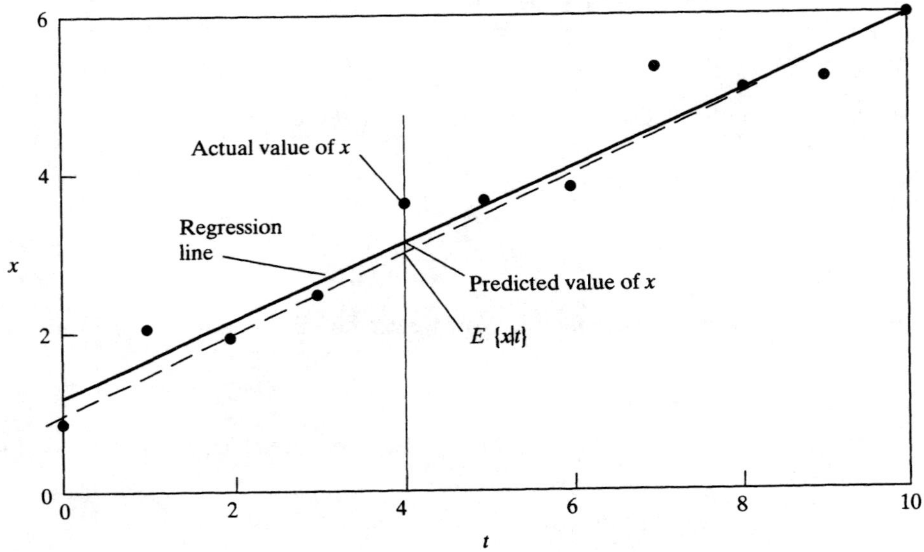

Figure 8.7 The nature of errors in a regression analysis.

Let us specifically consider the case where $f_0 = 1$, and $f_1 = t$, although we note that this form includes the exponential and Gompertz curves. We will also make some assumptions regarding the form of $\boldsymbol{e}$.

1. e_i is a normally distributed random variable with mean zero and variance σ^2.
2. e_i and e_j are uncorrelated for any $i \neq j$.

These assumptions are not always realized, but they are not unreasonable. Recall that the central limit theorem states that random variables that are the result of several random processes tend toward a normal distribution, regardless of the underlying phenomena. Given these assumptions, $E\{x|t\}$ is given by

$$E\{x^* \mid t\} = \boldsymbol{f}^T(t)\,\boldsymbol{a}$$

and

$$V\{x^*\} = \sigma^2$$

where V refers to the variance. Now we can compute the variance in our estimates of a_0 and a_1 as follows:

$$V\{a_0\} = \frac{\sigma^2 \sum_{i=1}^{m} t_i^2}{m \sum_{i=1}^{m} (t_i - \bar{t})^2}$$

$$V\{a_1\} = \frac{\sigma^2}{\sum_{i=1}^{m} (t_i - \bar{t})^2}$$

where

$$\bar{t} = \frac{1}{m} \sum_{i=1}^{m} t_i$$

and an estimate for σ is

$$\text{est}\ \sigma = V\{x^*\}$$

A forecast of x is given by the equation

$$x = a_0 + a_1 t$$

and the variance of an estimate x is given by

$$V\{x_k\} = V\{\bar{x}\} + (t_k - \bar{t})^2\, V\{a_1\}$$

$$= \frac{\sigma^2}{m} + \frac{(t_k - \bar{t})^2 \sigma^2}{\sum_{i=1}^{m} (t_i - \bar{t})^2}$$

But you must keep in mind that this is an optimistic estimate of the variance in an estimate for x_k. It assumes that the future will be the same as the past, and that there are no hidden factors that could lead to changes in x_k, such as the examples given earlier where prices were strongly correlated with time.

Next we will examine the case where we use the problem-solving approach to forecasting. Perhaps we wish to forecast the future demand for electric energy in an area where we do not believe that historical data provide a good indicator of future demands. For example, suppose it is proposed to develop a wilderness area, bringing industry to the area along with a significant population, and it will be necessary to supply the area with electric energy to meet the growing demand. For this case, we might create a forecasting model that makes use of both logic and historical data. We could write the demand for energy as a function of three sectors: industrial, commercial, and residential. It is reasonable to suppose in this example that we would know the nature of the industry that will occupy the area. Thus, we would know the energy consumption per unit output from the industry, although this would be subject to some uncertainty and to reduction through technological innovation over time. The industrial demand might be a function of the dollar value of product from the industry. We might also assume that population is a function of the dollar value of the industrial product, allowing for the fact that some initial population is needed to get the industry started, and the size of the commercial sector might be taken proportional to the population. We now obtain a forecast equation of the following form:

$$D_E = \alpha_I O_I(t) + \alpha_R P(O_I,t) + \alpha_C C[P(O_I,t)]$$

The α's are weighting factors that relate electric energy consumption to the activity level of each sector. They could be constants or functions of time, if technology innovation or other factors are thought to be important. O_I is the output of the industrial sector, measured for example in dollars, $P(O_I,t)$ is the population of the area as a function of industrial output and time, and C represents the activity level of the commercial sector, which includes support services such as grocery stores, medical care facilities, and so on.

In general, when an area undergoes a planned development, there will be a master plan that provides the planned industrial output as a function of time. This could be the starting point for the forecast on industrial output. But such a forecast involves assumptions regarding many external factors: cost of production, demand, cost of transportation and accessibility of markets, and so on. We cannot know that these assumptions will be satisfied, and so the forecast of O_I is uncertain. Likewise, all other variables in this equation also are uncertain. Still, there are constraints that relate to the physical realities of the system. These are of the following sort:

- The rate of growth of industrial output is limited by the output of the previous time period and the previous highest output rate. Growth beyond the previous highest output rate requires the installation of new equipment, and that takes time.
- Industrial output cannot grow unless it is accompanied with a growth in population, but output can decline without a reduction in population.

- Population will be supported only by industrial output. So if industrial output declines, population will decline, but the decline in population will lag the decline in industrial output.
- Development of the commercial sector will lag population growth.

These and other constraints can be written in mathematical form. For example, the first constraint described on the previous page could be written

$$O_I(t_{i+1}) \leq O_I(t_i)(1 + \beta_I)$$

$$O_I(t_{i+1}) \leq \max O_I(t \leq t_i)(1 + \beta_{I\,\max})$$

where the β's are time period-to-time period growth and maximum growth rates.

Having a model such as this, it is now subject to use in a Monte Carlo simulation, which could then yield a probabilistic forecast. The Monte Carlo model would step through time, from the current into the future, providing estimates of demand according to the above model and including uncertainty in the many parameters that drive the model. That is to say, all of the variables on the right-hand side of the equation for D_E would be treated in the Monte Carlo model as random variables. But there remains a question on how to treat them. We will address this question by looking at the variable O_I.

It is reasonable to take the master plan as the basis for a forecast of O_I. But we know that O_I will not develop exactly according to the master plan, and we want to include uncertainty in the value of O_I as a function of time. It is further reasonable to assume that there will be correlation in the values of O_I for successive time periods. In other words, it is unlikely that O_I would be well below the planned level in one time period and well above it in the next. Whatever factors caused the lack of growth in the first place are likely to persist, causing sluggish growth in successive time periods. Conversely, if growth in one period is above the planned level, it is likely to continue at an above-planned rate for successive time periods. For this reason, we cannot simply define a probability distribution on O_I and sample from the distribution to compute a sampled value for each time period in the simulation. Instead, we must concern ourselves with the rate of change of O_I.

Let us take as the reference level of industrial activity at any time $O_I^*(t)$, and as the actual level $O_I(t)$. Then define

$$\Delta O_I(t) = O_I(t) - O_I^*(t)$$

The derivative $d\Delta O_I(t)/dt$ represents the rate at which the actual level of activity diverges from the planned rate. What we want to simulate is the rate at which this rate changes. Thus, we write

$$\Delta O_I(t_{i+1}) = \Delta O_I(t)(1 + \gamma_I)$$

where γ_I is a random variable selected from a distribution that has zero mean. We could take γ_I to be a normally distributed random variable. There would be a good deal of rationality to this choice. But using a symmetrical triangular distribution is nearly as good. Now, at each time step in the simulation, we select a value for the random variable γ_I, and this value determines the rate of change of ΔO_I, which in turn determines O_I. This is also how we would treat the other similar parameters in the forecast equation.

A simplified flowchart for the overall Monte Carlo simulation is shown in Fig. 8.8. A typical forecast provided by such a simulation can be illustrated in the form shown in Fig. 8.9.

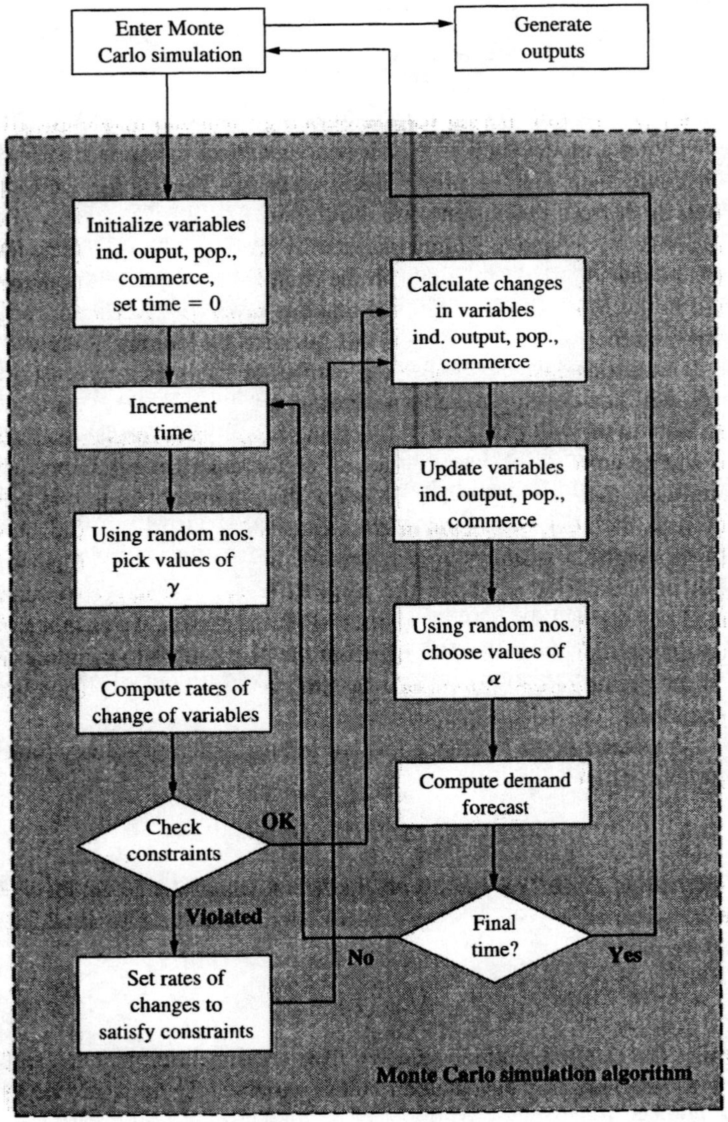

Figure 8.8 Flowchart for a Monte Carlo simulation forecast of electric energy demand.

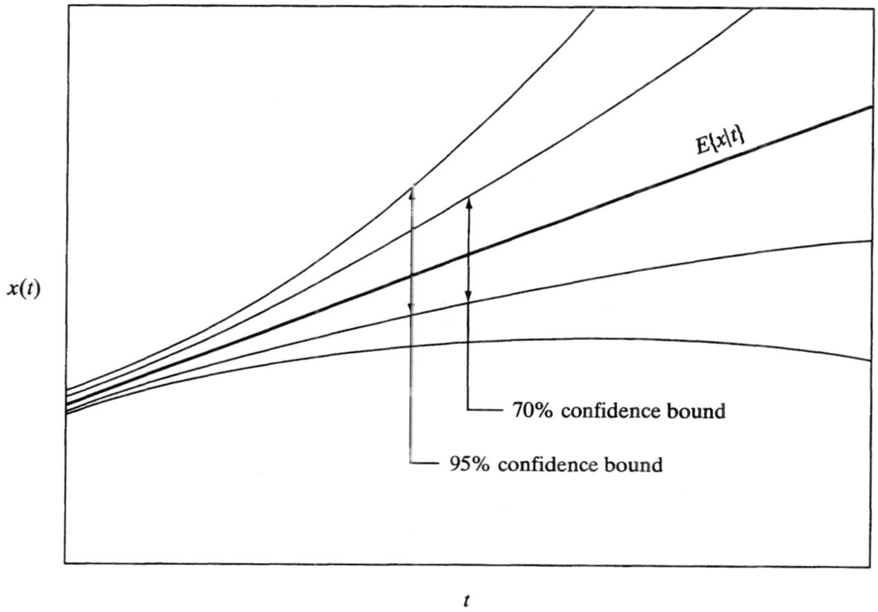

Figure 8.9 Presentation of output from a probabilistic forecast.

8.6 FORECASTING UNCERTAINTY WITH EMBEDDED DECISIONS

We can expand our ability to provide probabilistic forecasts to include embedded decisions. Consider the problem outlined above, where it is desired to forecast the electric energy demands of an area that is undergoing industrial development. It could be the case that a large company is planning to build a plant in the area. If it does, energy demand will increase rapidly. If it does not, demand will increase much more slowly. You, as the forecaster, cannot control the decision to build or not to build the plant. All you can do is assess the probability that the company will choose to build the plant. If the decision is made to build the plant, then at some future date, demand will begin to increase rapidly. This can be accommodated in the simulation by changing the function $O_I(t)$ and constraints appropriately, and two individual forecasts can be run, one with the decision to build and one with the decision not to build. The total, probabilistic forecast is then the weighted sum of the individual forecasts, weighted by the probabilities of each decision.

The problem becomes more interesting if the decision to build will be conditioned upon some future events. For example, the company planning to build the plant might divulge its plan to build contingent upon favorable tax laws and an adequate local infrastructure. This could include the availability of an adequate electrical energy supply. In other words, the decision to build might be conditional upon conditions that the forecast itself is intended to forecast. Such decisions can be added to the simulation if the simulation is thought of as a potentially real future rather than merely a forecast of the future.

The simulation process is shown in Fig. 8.10. Each simulation develops a possible future through to the point in time when the embedded decision is to be made. At that point in the simulation, all the conditions upon which the decision will be made are known, and the decision can be computed. The simulation then continues through the time period of interest, with the decision having been made.

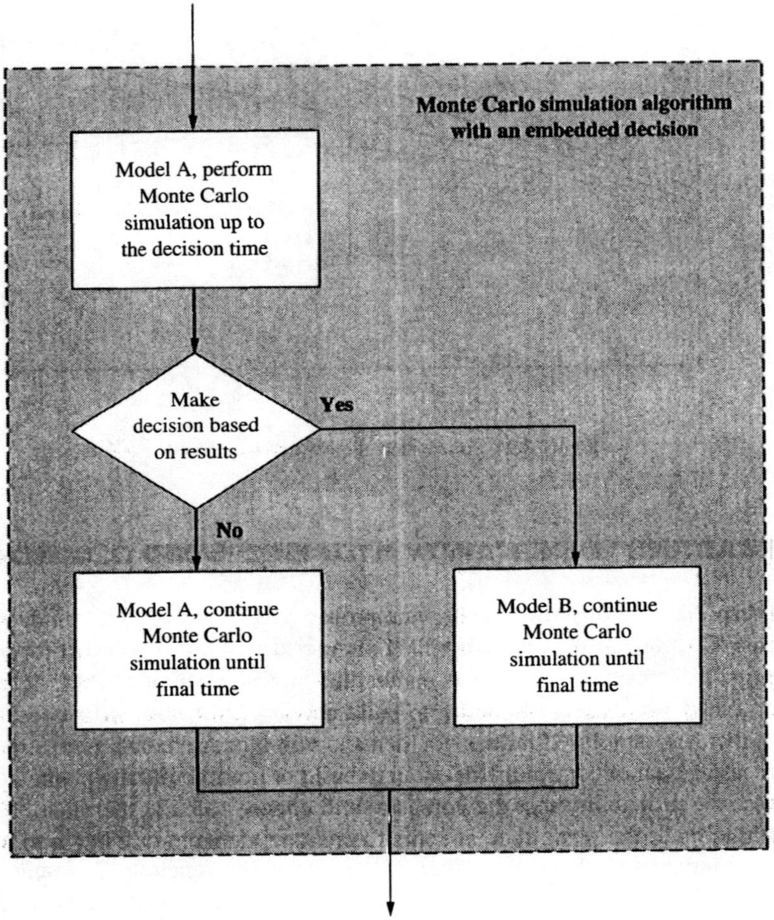

Figure 8.10 Simulation algorithm for a Monte Carlo forecast including an embedded decision.

We see here that the Monte Carlo approach can be applied to the generation of forecasts as well as simulations, although these are in essence the same thing. Forecast simulations such as this can incorporate embedded decisions that are purely random when based on elements of the forecast, purely determined by conditions of the future provided by the forecast, or a mixture of both.

8.7 SHORTCOMINGS OF FORECASTS

It is most important that we recognize fully our inability to predict the future. Virtually all forecasts contain a significant amount of uncertainty, and our tendency is to underestimate the uncertainty, sometimes by a large amount. Perhaps this tendency is because of the complexity of thinking in terms of uncertainty. It takes a lot of effort, and we have been trained to deal with certain futures, not with futures that include large amounts of uncertainty. Still, assuming that uncertainty does not exist does not reduce the uncertainty that does exist. We cannot eliminate it, we can only deal with it. Sometimes, careful modeling can reduce uncertainty. But just as often, modeling makes clear how large the uncertainty associated with some variable actually is.

The preceding sections developed a mathematical framework for forecasting that attempts to extract as much information as possible from the environment and apply it to the forecast. The methods include approaches for the quantification of uncertainty. Still, these methods model only effects on uncertainty that are known. They do not take into account factors that are hidden for some reason. As a result, the tendency of these methods is to underestimate the uncertainty present in our forecasts. A prudent practice would be to assume that the results of such analyses provide only a lower bound on the uncertainty present. Actual uncertainty could be much larger. We see this in the examples given for the forecasts of electric power (Figs. 8.4 and 8.5).

8.8 SUMMARY

Virtually all engineering systems or system designs are intended for use over an extended period of time. This period may be as short as weeks or months, but it typically ranges from years to decades. Indeed, buildings, bridges, and other major works may be designed for a useful life of up to a century, and sometimes longer. Any time that a system is designed for use over an extended period of time, it is important for the systems engineer to take into account conditions of the future—future costs, supplies, demands, environmental regulations, and so on. These factors will affect the net value derived by society from the system. For example, in the design of a building, it is important to project future energy costs in order that design decisions can be made relating to effective energy use. The purpose of rigorous forecasting approaches is to extract from the environment and use effectively all available information, in order that the forecast will be as complete and accurate as possible. Still, virtually all forecasts will contain some amount of uncertainty. Estimating the uncertainty in a forecast is an important part of the forecasting process.

We examined two basically different approaches to forecasting. The first made use strictly of historical data on the variable of interest in the forecast, such as a forecast of future population based on historical population trends. We recognized that forecasts of this sort work largely because of the law of large numbers. The forecasts provide estimates of averages; they do not give any information on specific members of the population. The second approach, which we referred to as the problem-solving approach, uses unifying principles to relate the forecast to a broader experiential base. With respect to the latter approach, we showed how Monte Carlo simulation can be used to provide both forecasts and estimates of the uncertainty in a forecast.

The final caution of this chapter is that you should always recognize the presence of uncertainty in forecasts and guard against underestimating how large the uncertainty actually is.

REFERENCES

1. Draper, N. R. and H. Smith, *Applied Regression Analysis*, John Wiley, New York, 1966.
2. Mendenhall, W. and R. J. Beaver, *Introduction to Probability and Statistics*, PWS-Kent Publishing Company, Boston, MA, 1991.
3. Wonnacott, R. J. and T. H. Wonnacott, *Introductory Statistics*, John Wiley, New York, 1985.

PROBLEMS

1. In the field of microelectronics, it has been observed that the number of transistors that can be placed on a 1 cm^2 chip doubles every 4 years. This relationship has held true for over 20 years, and today a transistor can be placed in an area that is 0.001 mm^2. How long do you think it will continue? Why?
2. Predict the price of gasoline over the next 20 years. Do you think that the oil companies believe that the world is running out of oil? Why?
3. Your company produces a product in a market that has grown by 10 percent per year for the past 15 years. According to your estimate, the market is currently at 30 percent of saturation. Predict sales for the next 20 years.
4. Your company plans to design a new commuter airplane in the size range of 30 to 50 seats. List things that should be forecast in order to properly assess alternative designs. Hint: Some of these things might not be quantitative.
5. Fit a trend line to the following data.

t	$x(t)$
1	0.928
2	2.712
3	6.065
4	10.009
5	10.250
6	12.883
7	9.413
8	31.460
9	14.931
10	42.358

How good is your fit?

6. For the data of problem 5, predict the value of x at $t = 20$. What variance do you expect in this forecast? Comment on the accuracy of the forecast.

7. You have an extremely accurate experimental system for testing Newton's law. You impose a force on an object of known mass and measure its acceleration. You get the following results:

Case	F, Newtons	m, kg	a, m/s^2
1	1.0967	0.8304	1.3207
2	0.00136	0.000204	6.6667
3	1587.8	256.83	6.18230
4	204.8×10^6	1389×10^3	147.44
5	2.107×10^6	7.865×10^{-3}	0.000267896
6	3708.954×10^{13}	694.80615	5.338113×10^{13}
7	1.0000000	1.000000	1.000000
8	6.5089241	3.4678952	1.8769091
9	375.689336	24.795630	15.1514333
10	52.002385	0.0012584	41324.2

Do these data prove that $F = ma$?

8. True or false: The current price of a stock reflects a forecast of its future price.

9. Your company is a major manufacturer of copy machines and printers. Are your sales data of any value to a company in the paper industry?

10. A regression model explains variability in a dependent variable x as a function of independent variables $\mathbf{t}$. How does this differ from the use of a physical model such as $V = IR$ or $F = ma$?

11. A bent coin is flipped. The probability that it will land heads is 0.65. Forecast the outcome of the next flip. Forecast the outcome of the next 1,000 flips.

12. You are designing an oil refinery to produce gasoline for automotive use. What forecasts would you like to have and what data might you use to obtain them?

13. You are designing a solar heating system for a private house. What forecasts would you like to have and where might you find them?

14. Over the past 15 years, fuel economy of automobiles has increased dramatically. How much will it increase over the next 15 years?

9

Engineering Systems Modeling

9.1 THE PURPOSE OF THIS CHAPTER

The essence of engineering design is decision making, and the ability to make good engineering decisions lies in having good information. Sometimes engineers have the luxury of building and testing real systems. But more often for information we must rely on models. Models, particularly computer models, give us a relatively easy and rather inexpensive way of trying many different designs or testing many different design decisions. Were it not for such models, we would have to build actual systems in order to gain design information. In such a case, it would be possible to test only a few designs at most. On the other hand, with a computer model, we might be able to test hundreds or even thousands of design options. This capability enables us to make much better design decisions.

Formal models provide, essentially, three things: *accuracy*, *resolution*, and *identification of non-intuitive cases*. The accuracy of a model relates to its ability to represent the real world on an absolute scale, whereas resolution relates to the ability of a model to distinguish between different cases on a relative scale. We can relate these notions to the measurement of temperature using a thermometer. We are limited in our ability to measure the temperature of an object precisely by the accuracy of the thermometer. Accuracy is limited by error present in the measurement. Suppose instead, however, we wish to know which of two objects is the warmer. Our ability to distinguish between the temperatures of the two objects is limited by the resolution of the thermometer. Merely to distinguish between the warmth of two objects, accuracy is less important, and sensitivity and noise become the limiting factors. We might suspect that a thermometer could distinguish the relative warmth of objects whose temperature differences are only a fraction of the error we would expect in a measurement of the absolute temperature. This is because we suspect that the absolute errors in two sequential measurements are positively correlated. So, for example, a given thermometer might have an absolute accuracy of ±1 °C, but it can distinguish between the warmth of two objects whose temperatures differ by only 0.2°C.

There are some cases that simply escape our intuition. By means of the modeling process, we sometimes *discover* potential physical phenomena that we would not have expected. Models sometimes discover new phenomena, such as the behavior of elementary particles or the existence of black holes. But models also find other, less dramatic system behaviors. For example, we might discover that a particular bridge design is prone to developing large oscillations in high winds.

In support of engineering design, we turn to models for help in answering three fundamentally different types of questions:

1. Will the system, as designed, work?
2. Which of two system alternatives is the better?
3. Do I properly understand the system?

To answer questions of the first type, a model must be accurate. To answer questions of the second type, a model must have good resolution, and for questions of the third type, a model must be sufficiently complete.

Sometimes physical systems impose on society risks or hazards. Models allow us to test approaches to the design and use of such systems without the imposition of risk on society. This concept even reaches into training. For example, airline pilots train on simulators that *model* actual flight, enabling them to train for hazardous situations with no actual risk.

The average person can manipulate concurrently perhaps five variables by means of intuition alone. But a typical engineering system consists of hundreds or thousands of variables. It is not possible for any individual to deal with the complexity of real systems without formal models.

Finally, models provide the backbone of systems engineering. They are the system engineer's tool for integrating inputs from disciplinary specialists to provide information for decision making. With models, a systems engineer can draw upon the expertise of physicists, chemists, materials scientists, engineers of different disciplines, economists, operations researchers, lawyers, and accountants, and integrate their knowledge to achieve an intelligent design. In this book, we emphasize that every aspect of an engineering system is subject to being modeled. This includes the design of the system, its use, its maintenance and repair, and its disposal. Indeed, it includes the research and development process that precedes the system and the system engineering process itself. The more capable of modeling you become, the more you will be able to draw on the expertise of others, and the better you will be as a systems engineer.

9.2 ENGINEERING SYSTEMS AND SYSTEM MODELS

We will begin our discussion of engineering systems modeling by providing a formal definition of a system. The following definition is provided by Rivett (1980).

A system is a set of entities with three properties:

1. *Every entity in the set affects at least one other entity in the set.*
2. *Every entity in the set is affected by at least one other entity in the set.*
3. *The set cannot be split into two subsets such that no member of one subset affects at least one member of the other subset or that no member of one subset is affected by at least one member of the other subset.*

The above definition provides a basis for much thought on what comprises a system. The requirements that each entity in a set of entities that comprise the system must affect at least one other entity in the set and be affected by at least one other entity distinguishes between elements that are part of a system and elements that merely affect or are affected by a system. The third property strengthens this distinction. We should keep in mind that interaction between entities in a system can be either physical or nonphysical, for example in the form of information. Figure 9.1 gives examples of some simple systems.

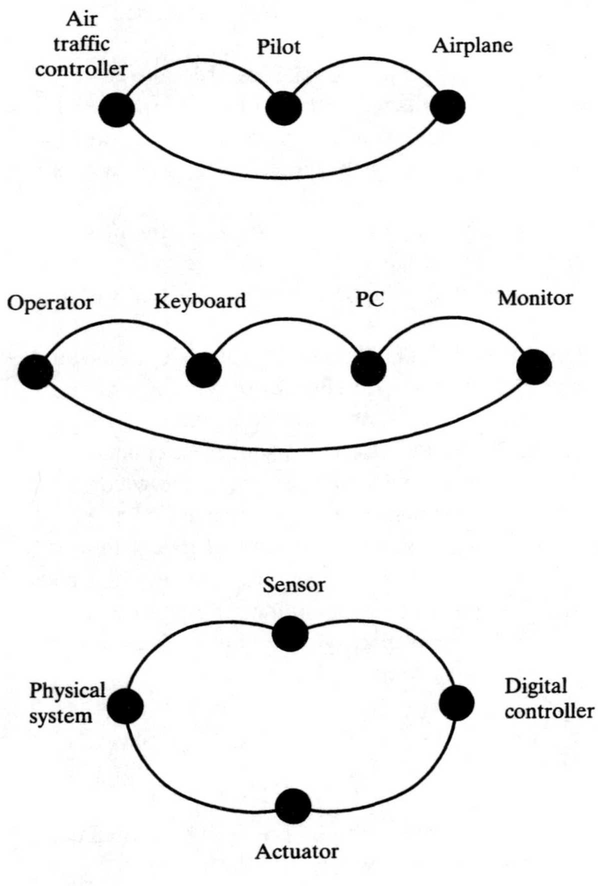

Figure 9.1 Some typical systems.

When we ask specifically which entities comprise a system, we see that we can draw a boundary around the entities of the system such that the entities inside the boundary comprise the system. Determination of the system boundary is an important step in the development of a system model.

We now turn to a discussion of models. We can define a model very simply as follows:

A model is an abstraction of reality.

The key is that a model is not reality. Still, a model attempts to capture some elements of reality. We could say that a model is a copy, an image, a representation, a system of postulates, data and inferences. The key, however, remains: a model is not reality. Therefore, a model will not behave precisely the same as the reality it is intended to model.

We will examine the three basic classes of models: *iconic* models, *analog* models, and *symbolic* models. Iconic models are scale representations, either large or small, of physical or real things. Examples include wind tunnel models of airplanes, rockets, buildings and bridges; pilot plants; model trains and automobiles; and ship models. The advantages of iconic models are that they tend to be smaller or at least simpler than the real system and thus less expensive to build, and they can be used to gain an understanding of the behavior of the real system, even in cases where the underlying physics that govern the system are not well known. Model airplanes can be used in wind tunnel tests to understand air flow, lift, and drag of a particular airplane design. Ship models can be used to understand hull-wave interactions and hull drag. By using iconic models, the engineer can evaluate a proposed design and make improvements without the expense of building the real system. Iconic models can be used when the engineer cannot write equations that adequately describe the behavior of the system. Still, some understanding of the physical phenomena may be beneficial in order to interpret or scale the test results.

Analog models are models that use one property to represent another. For example, colors on a relief map may be used to represent elevation. Other examples relevant to engineering include the use of models from one discipline to represent systems of another discipline. We might, for example, use an electrical circuit as an analog to heat transfer through an insulated wall. Voltage is the analog of temperature, current is the analog of heat flux, and resistance is the analog of thermal resistivity.

Symbolic models, or mathematical models, use symbols to designate properties of the real system. Symbolic models are more abstract than iconic or analog models in the sense that symbols do not resemble the real system or system elements in any way. We could argue the breadth of symbolic models. Indeed, we can define them to cover a vast range, including all of human thought. In at least some sense, a concept such as "the sky is blue on a sunny day" can be thought of as a symbolic model. Symbolic models are subject to manipulation by mathematics and, in many cases, they can be reduced to a form that is compatible with modern computers. When we can reduce a symbolic model to computer software, we often are able to use the model to examine system alternatives in a very inexpensive way.

The great benefits of symbolic models lie in the fact that they can be created at very low cost compared to other types of models; they are subject to analysis on computers;

they are easy to modify to make possible the consideration of alternatives; they are subject to human understanding because they are created by humans; and they can give consideration to physical conditions that may be difficult to achieve in nature or that are hazardous. Unlike nature, where it can be difficult at best to control the conditions of an experiment, through the use of symbolic models we can precisely control the conditions input to the model and precisely control changes in variables one by one. The main problem with symbolic models is that we rarely can be certain that they adequately represent the phenomena that they are intended to represent. In general, we must always seek to *validate* a model before we rely on it for engineering decision making. The language of symbolic models is mathematics.

Figure 9.2 gives a conceptual overview of models. Iconic models are closest to reality, and symbolic models furthest from reality. Yet, as we move toward symbolic models, the ease of development and use improves. The major thrust of systems engineering today is toward symbolic models. They alone are the subject of this book.

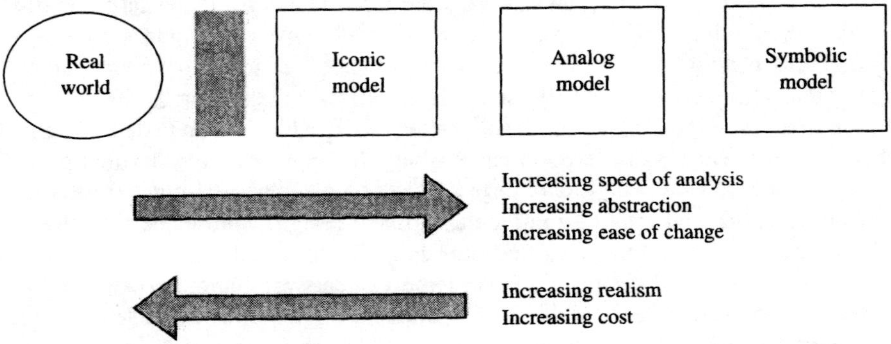

Figure 9.2 A conceptual overview of models.

To gain a further understanding of models, we will return once more to the example of guessing how many beans are in a jar. Recall that we could simply guess. In engineering systems design, there is rarely anything to prevent us from guessing. Certainly, we could guess how many beans are in a jar. The problem with a guess is that it often lacks the accuracy or resolution that we need for effective engineering design. We turn to modeling as a method of gaining greater accuracy and resolution than our guess. So, to begin with, we could simply make a guess and estimate the error in our guess. We could represent our ability to guess as a probability distribution such as that shown in Fig. 9.3 for the number of beans in a jar. This distribution provides all the information that we have regarding the number of beans in the jar at the time that we make the guess. Our feeling that our state of information lacks accuracy is indicated by the breadth of the probability distribution shown in the figure, and the lack of a well-defined peak.

In the case of guessing how many beans might be in a jar, we formulated a symbolic model:

$$n = \frac{V_j}{V_b} f$$

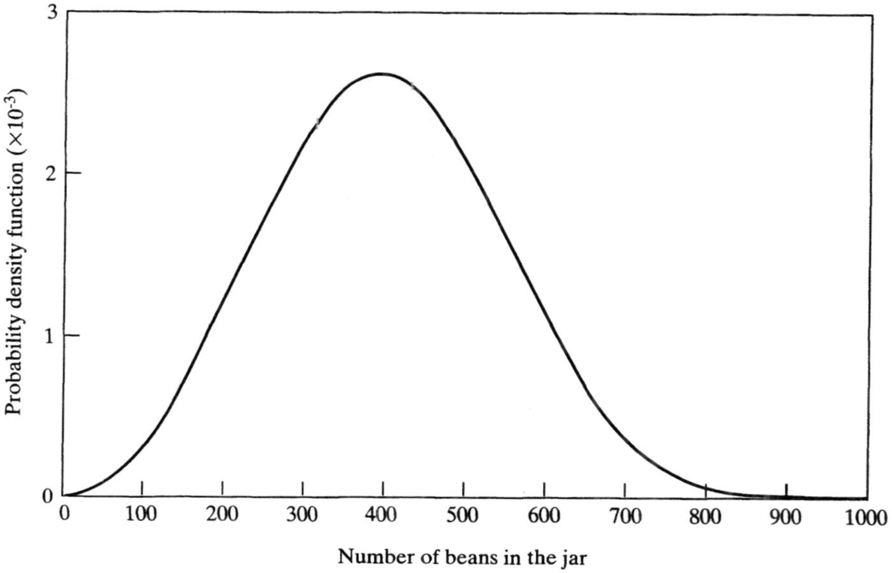

Figure 9.3 A probabilistic estimate of how many beans are in a jar.

where n is our estimate of how many beans are in the jar, V_j is the volume of the jar, V_b is the average volume of a bean, and f is the fraction of the jar that is filled with beans as opposed to air. What we have done in creating this model is to disaggregate our estimate from an estimate directly on the number of beans in the jar to estimates on each of three other variables, the volume of the jar, the average volume of a bean, and the packing factor f. Why would we want to do this? It might at first seem that all we have accomplished is to triple the work necessary to get a value for n. But underlying the development of this model is the notion that we can make better estimates for the values of the three variables of the model than we can for n directly. That is to say, we expect that the probability distributions, the pdf's, on V_j, V_b, and f will be much more peaked than our original distrib-=ution on n. Having the distributions on these variables, we next use the model to unify these data into the variable of interest, n. So we see that what models do in use is to unify data. In this case, the model unifies the variables V_j, V_b, and f into the variable n. If we use the model correctly, and we recognize that our estimates of the variables V_j, V_b, and f contain uncertainty, then we will compute not a single estimate for n but rather a probability distribution for n. We will say that the model is a *good* model if by using it we can obtain an estimate of n that is less uncertain than we started with. If using the model fails to reduce the uncertainty in our estimate of n, then we would say that it is a *bad* model. The concept of a good model is illustrated in Fig. 9.4.

- In developing a symbolic model, we disaggregate a variable into other variables, which we choose because we can estimate them with sufficient accuracy that we can improve our accuracy of the estimate of the original variable.

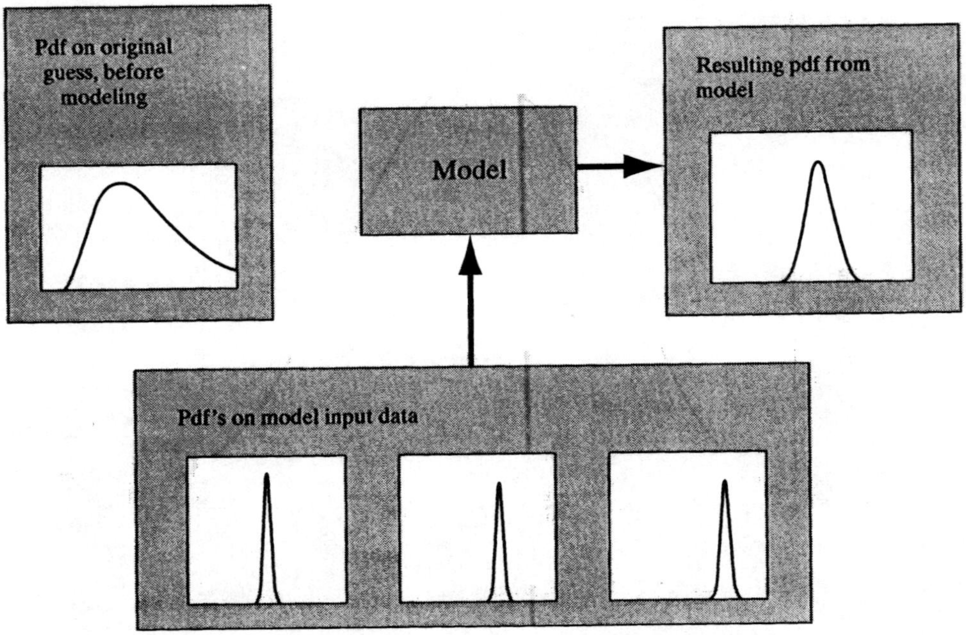

Figure 9.4 The concept of a good model as one that reduces uncertainty in an estimate of the variable or variables of interest.

- In using a symbolic model, we unify estimated variables that are the inputs to the model to obtain an estimate of the variable of interest.
- A good model is one that improves our ability to estimate a variable.
- A bad model is one that fails to improve our ability to estimate a variable.

Models can be broken into different types according to many different schemes. Here, we will consider two fundamentally different types of models: *predictive* models and *descriptive* models. (These correspond roughly to what Rivett (1980) refers to as inductive and deductive models.) A predictive model is a model that predicts the behavior of a system. $F = ma$ is a predictive model. It enables us to predict the behavior of a physical system subject to external and internal forces or other factors. Predictive models may come from the laws of nature, as we understand them. They can also derive by other means. For example, consider the model: new balance equals old balance plus deposits minus withdrawals plus interest. This model is a perfect predictor of a bank savings account balance. It does not derive from the laws of nature, but rather from our knowledge that that is how the bank computes the new balance. The method of computation of the new balance is determined by people, and so we can know it precisely and with complete confidence. Laws of nature are not determined by people and, in general, we cannot know them precisely or with complete confidence.

A descriptive model, as its name implies, describes a process. It is roughly equivalent to Rivett's deductive model, which results solely from statistical or mathematical analysis, "without any consideration of what ought to be happening." A typical descriptive model is the model commonly used for friction, $f = \mu f_N$, where f is the friction force, μ is the coefficient of friction, and f_N is the normal force causing the friction. This model can be derived by performing a linear curve fit to data correlating f to f_N. The result of the curve fit is a value for μ. To perform the curve fit, we must assume a form for the mathematical relationship between f and f_N, and the model will apply only over the valid range of the curve fit. Indeed, in this case, μ is defined to be the value of a constant that obtains the best possible linear relationship between f and f_N. Because in this strictly descriptive model there is no rationale for the model to apply beyond the range of the data from which the value of μ is estimated, it can be extremely dangerous to use the model outside this range. Further, the model cannot be exercised without measured or experimental data. Thus, it describes the friction that results from a normal force, but it does not predict it. This model will provide estimates of friction force whose error is comprised of the same error terms as exist in any regression model, and they may be analyzed by the theory presented in the previous chapter.

Our definition of a descriptive model will now depart from Rivett's definition of a deductive model. The model we used above to assist us in our guess of how many beans are in a jar derives from our intuitive notions regarding the way that beans fill a jar. In this sense, the model is inductive. Yet, we are well justified to believe that this model would adequately represent the physical world, provided that we could measure the number of beans, n, as a function of the inverse of the average volume of a bean, V_b, and the volume of the jar, V_j. We could then perform a regression analysis to determine μ. Further, we can be rather assured that, if we chose the *correct* values of these variables, we would obtain a precisely correct estimate of n. Thus, here, we define this model also to be a descriptive model.

There is a fundamental difference between predictive and descriptive models that a systems engineer must be aware of: *descriptive models can be and often are precisely correct; predictive models are rarely precisely correct.* Consider the examples given above. The model for friction depends on a measurement or estimate for the coefficient of friction at a specific value of the normal force f_N. Without a value for μ, which we would normally obtain from data, we cannot use the model. On the other hand, all of the error in this model can be subsumed in our inability to estimate the normal force and the coefficient of friction for that particular normal force. Given appropriate values for these variables, the model necessarily gives a precise answer. On the other hand, despite the awesome capability of predictive models to predict the future, they are virtually never precise. (Only in the case of a phenomenon that is determined by humans, such as the bank balance, can we be sure of the precision of a predictive model.) Error in estimation provided by a predictive model comes not only from the variables that must be estimated as inputs to the model, but also from the model itself. In one sense, this is the case because we want to use the model in a range of input variables for which we have never directly collected data. To be precise, for example, $F = ma$ is not a perfect model. We already know that this model neglects relativistic effects, which sometimes can be important, but in any event lead to some error. Further, we know that the inclusion of relativistic effects improves the predictive capabilities of this model, but we do not know that the resulting model is precise, and it is likely not to be. More commonly, however, our application of such laws is imprecise. For example, we may assume that solid bodies are rigid and we neglect many forces, such as radiation pressure, magnetic forces imposed by the Earth's magnetic field, and so on.

Because the error that results from the application of a descriptive model is subsumed largely in the variables that must be input to the model, it is relatively easy to estimate the error present in the model outputs. On the other hand, because the model itself contributes to modeling error in the case of predictive models, it can be very difficult or essentially impossible to predict the error present in the application of a predictive model. This can be very important when it comes to the use of predictive models to support the engineering design process.

9.3 THE MODEL-BUILDING PROCESS

Let us now turn our attention to the actual model-building process. The first step in model building is to decide what to model. This statement is so obvious that it seems even trivial. Still, many model builders fail to carefully identify precisely what their model seeks to emulate. A proper statement of the objectives of the model is key to the development of a successful model. Another way to look at this is that you need to decide what is the purpose of the model. Scientific models generally have a different purpose than engineering models, and their requirements are correspondingly different. The purpose of a scientific model is typically to predict a phenomena or to explain a phenomena. On the other hand, the purpose of an engineering model is typically to provide information for a decision-making process. In the case of a scientific model, for the purposes cited, error in the model result is of lesser importance. However, in the case of an engineering model, error in the model result is the primary issue. Therefore, virtually all engineering models must include measures of uncertainty to be of value. Also, many engineering models must deal with probabilistic phenomena. Thus, outputs from them often take the form of probabilities and probability distributions. It is important to recognize and specify this because, if you fail to do so, the model is likely not to compute the quantities that are necessary.

If we accept that a model is an abstraction, that is, it resembles reality but in an incomplete way, then we must accept modeling as a process of abstraction. Reality is simply too complex to model, and besides, in most cases, we cannot comprehend reality. To construct a model, therefore, we must decide what features of reality to include in the model; that is, what to keep and what to throw away. Too often, model builders decide only what to keep. We should be aware of both sides of the process. Knowing what we are throwing out when we build a model helps us to understand the assumptions that underlie the model, and thus it gives us an idea of the range of validity of the model and its accuracy.

Again we emphasize that if the model supports an engineering decision-making process, then it must provide information. We know from our previous discussions that models convey information only if they include some measure of the accuracy of the model results; that is, if they include probabilities associated with each numerical result provided.

We can now state a sequence of steps to take in the modeling process. No such list of steps will be complete or properly ordered for every case, but they do serve as a general guide.

1. Formulate a clear statement of purpose. What is the engineering objective to be undertaken? What decisions are the model to support? What is to be the output of the model?
2. Define the engineering design option space or set of alternatives that the systems engineer wishes to consider.
3. Select those options from the option space that you want to analyze using the model.
4. Determine what is to be modeled. Define the system to be modeled in terms of its boundaries and entities. List the phases of the system life cycle that are to be modeled. List the parameters that the model is to estimate.
5. For each system option that is to be modeled, identify the interactions between the system entities; that is, which entities are acted on by other entities and which entities act on other entities.
6. Identify available data to support the modeling process.
7. Develop a symbolic model of the system compatible with the available data.
8. Flowchart the logic of the system model.
9. Convert the flowchart to a computer code.
10. Validate the model.

9.4 SOURCES OF ERROR IN SYMBOLIC MODELS

We have already mentioned the difference between descriptive and predictive models and noted that error in the use of a descriptive model lies mainly in the inputs to the model (possibly also in the computational algorithm) and error in a predictive model includes model error as well. Figure 9.5 shows the sources of modeling error. To begin, no model is reality. All models abstract from reality. They bring some things along into the model and leave some things behind. Some aspects of reality we do not understand, and we do not model them. Consequently, in the conversion from reality to a symbolic model we accept that some aspects of reality are left out. We also accept approximations to some of those aspects of reality that we bring forward. For example, we might choose to model a mechanical process using the relationship $F = ma$. We know from the start that this is an approximation, but for many cases a rather good one.

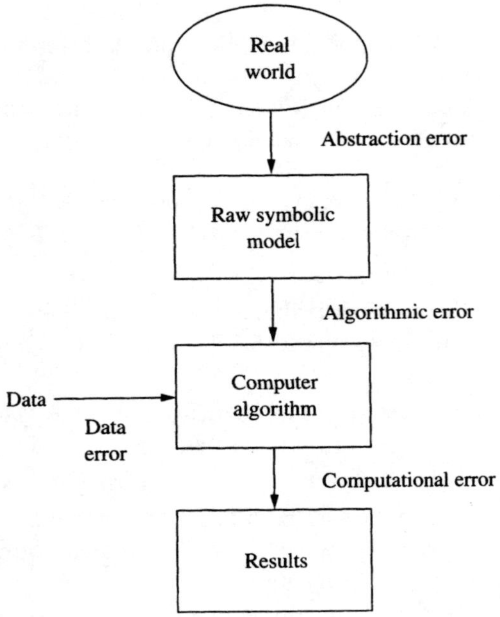

Figure 9.5 The sources of error in modeling.

We generally begin writing the actual model by developing a *raw* symbolic model. We use the word raw here to mean a general mathematical model expressed in conventional mathematical form. Such a model might include differential and integral equations. It might include maxima or minima of functions, and other such mathematical concepts. We also call this a raw model because it is in a form that often we cannot solve. Suppose the model contains a set of nonlinear differential equations. We cannot obtain a solution of these equations directly. We must go through intermediate steps. Thus, the next phase of model building consists of converting the raw mathematical model into an algorithm suitable for solution, typically using a digital computer.

Take the case where our model consists of a set of nonlinear differential equations that we want to solve. Digital computers cannot solve differential equations nor, for that matter, can they solve algebraic equations. What they can do is to perform numerical operations that resemble addition, subtraction, multiplication, and division over a set of rational numbers. And they can perform logic. When we implement a mathematical model on a digital computer, what we are doing is creating an *algorithm* that the computer can use to compute results that approximate the results that we would obtain from the raw model if we could obtain them. If we consider the process of division, say x/y, we begin by choosing approximations of x and y from a set of rational numbers whose range and resolution is limited by the bit size of the computer words that represent these variables, and we then approximate the result also with a rational number, typically of the same ilk. Only in special cases do the computed results equal precisely the algebraic answers. Algorithmic error builds according to many factors: how many variables the model contains, how many operations the model performs, and how the operations themselves stress the rational number set that we use for our approximations of the variables.

An example of how errors can build in algebraic operations follows. Consider the operation $(x - y)/z$, and take for x, y, and z the following values:

$$x = 457.83952$$

$$y = 457.83949$$

$$z = 0.00000217$$

For simplicity, we will work through to the result in a base-ten system. Computers use more efficient systems. We obtain first the difference between x and y:

$$x - y = 0.00003$$

Now, even though we had been carrying eight significant figures in x and y, their difference has only one significant figure, and we really need to question that. If x and y are themselves the result of earlier calculations, this figure might have no significance whatsoever. It could, in fact, have the wrong sign! Next, we divide this result by z.

$$(x - y)/z = 13.824885$$

This result appears perfectly plausible, and we could easily be lulled into thinking that it is actually accurate to all eight places. In reality, it contains no significant places. To be sure, there are ways around this problem. One approach is to perform all arithmetic operations using twice as large a word size as is maintained in the answers. This helps some, but it does not eliminate error altogether.

Many models contain differential equations that need to be integrated. To do this, we use various numerical integration algorithms. These are computer representations of mathematical forms that also are imprecise. Some numerical integration algorithms include sophisticated logic to reduce overall error. Typical of these are the predictor-corrector methods, which adjust integration step size to maintain a desired accuracy. These methods might be good at maintaining overall accuracy, but the fact that the step size adjusts during the integration process introduces noise into the integrand that could reduce resolution. For example, suppose successive integrations are to be used to obtain numerical derivatives. Now the noise introduced by the predictor-corrector algorithm could dominate a finite difference calculation of the derivative of the function.

Next, recognize that all models need data. The output from a model is only as good as the data input to the model. The commonly used phrase to describe this phenomena is GIGO: garbage in, garbage out. In designing a model, we should always keep in mind that, almost always, alternative models are available. The best model is not always the one that is the most mathematically sophisticated. Rather it is the one that produces the answer with the least uncertainty or, better still, with the least decision-making risk. The two can be different. The modeling process includes data gathering, and the form of the model should take into account the availability and accuracy of data.

On the one hand, it is important to understand modeling errors and to quantify them, but on the other hand, it can be very difficult to do this. But keep in mind that a model for which we have no estimate of accuracy at all provides absolutely no information for decision making. Results provided by a model provide information for decision making only after we assign a probability of their correctness or an estimate of their error. We have identified four sources of modeling error:

1. *Abstraction error*—error in the abstraction process that leads to a raw model, or lack of understanding of nature
2. *Algorithmic error*—error in the translation of the raw mathematical model into an algorithm suitable for computer application
3. *Data error*—error in input data
4. *Computational error*—error that results from the fact that computers deal only with a finite set of rational numbers rather than the full number set.

We know of no comprehensive method for the estimation of all these types of error. Generally, our hope is that we can maintain a negligible level of algorithmic and computational error. But this is neither easy nor always achieved. If we succeed at this, we have only to estimate the magnitude of the abstraction error and the data error. In the case of descriptive models, our task is less difficult. All the error in these models is the result of data error, and that error can be estimated by Monte Carlo techniques already discussed. The general procedure is shown in Fig. 4.5.

Estimation of the error in predictive models is very difficult. Most predictive models are developed from basic scientific principles. We can question these principles, but seldom are they presented to students complete with an analysis of the accuracy of the measurements that underlie them. And frequently we determine their limits of applicability only after we obtain a model to supersede them. Such was the case when relativistic mechanics superseded Newtonian mechanics. But today, we have essentially no idea by how much and when relativistic mechanics is in error. Much work remains to be done on the conversion of scientific theories and predictive models of science into informational models suitable for use in engineering decision making.

One type of model is inherently prone to error. This is the Monte Carlo model. Monte Carlo models *sample* the real world through a simulation process. Even in the absence of abstraction error, algorithmic error other than the Monte Carlo process itself, and computational error, Monte Carlo model results contain error. This obtains from the fact that no Monte Carlo model can conduct a perfectly random and complete sampling of the real world simulation model. All Monte Carlo models represent tradeoffs between model complexity, sample size, and computer run time. You have already been cautioned about the error in Monte Carlo models, and a procedure for analyzing and controlling this error is included in the discussions of Monte Carlo modeling in Chap. 4.

Finally, it needs to be pointed out that all of the errors cited and estimated above obtain under the assumption that the modeler makes no outright mistakes in any of the modeling steps, data collection, or data entry. That possibility always exits—it probably occurs more often than not. We are tempted to claim that all large computer simulations have errors or bugs of some sort. Sometimes these are subtle, and arise in situations that are difficult to notice. In the preparation of this manuscript, we made use of many of the latest versions of word processing, graphics, and spreadsheet software. We found errors or at least problems in all of the software. In one case, we found that a parameter was computed significantly in error for one case while in other cases the computations seemed to be correct.

Relative Error

In the preceding discussion, we are concerned with modeling error, which is the determinant of model accuracy. We need accurate models in order that they provide assurance that a design will work. But as soon as we obtain a feasible design—one that will work—our attention turns to improvement of that design. In order to improve a design, we must identify alternative feasible designs and distinguish between them. Let $\boldsymbol{x}_i$, a design point, be a set of inputs to the model $f(\boldsymbol{x})$, such that the predicted state of nature is $\boldsymbol{y}_i$

$$y_i = f(\boldsymbol{x}_i)$$

Let $\boldsymbol{Y}_i$ be the true state of nature corresponding to this design point. Then we can write

$$\boldsymbol{Y}_i = \boldsymbol{y}_i + \boldsymbol{e}_i = f(\boldsymbol{x}_i) + \boldsymbol{e}_i$$

Now consider a second design point given by $\boldsymbol{x}_j$.

$$\boldsymbol{Y}_j = \boldsymbol{y}_j + \boldsymbol{e}_j = f(\boldsymbol{x}_j) + \boldsymbol{e}_j$$

Model inaccuracy is determined by $\boldsymbol{e}$. The model is precise if, for all permissible $\boldsymbol{x}$, $\boldsymbol{e} = \boldsymbol{0}$. But suppose we wish instead to compare the two alternatives.

$$\begin{aligned}\Delta\boldsymbol{Y}_{ij} &= \boldsymbol{Y}_i - \boldsymbol{Y}_j = \boldsymbol{y}_i - \boldsymbol{y}_j + \boldsymbol{e}_i - \boldsymbol{e}_j \\ &= f(\boldsymbol{x}_i) - f(\boldsymbol{x}_j) + \boldsymbol{e}_i - \boldsymbol{e}_j \\ &= \Delta f(\boldsymbol{x}_{ij}) + \Delta\boldsymbol{e}_{ij}\end{aligned}$$

Here we see that the ability of a model to predict $\Delta\boldsymbol{Y}_{ij}$ depends not on the absolute model error $\boldsymbol{e}$, but rather on the differential model error $\Delta\boldsymbol{e}$. $\Delta\boldsymbol{e}$ determines the resolution of the model. We see that $\Delta\boldsymbol{e} \rightarrow 0$ as $\boldsymbol{e}_i \rightarrow \boldsymbol{e}_j$. That is to say, resolution depends on the extent to which the model error is correlated across the alternative design points. If the errors are positively correlated, the resolution of a model will exceed its accuracy.

An illustrative example might help to understand this concept. In the 1960s, a model was developed to estimate the risk of collision between two airplanes that are not under active radar control, where the collision risk is the result of the airplanes' inability to maintain precisely their assigned routes. This model made many assumptions, including that airplanes have the shape of a rectangular box whose dimensions are equal to the length, height, and wingspan of the largest airplane in the system, and second that pilots do not see and avoid other airplanes. These assumptions clearly bias the model to overestimate collision risk, perhaps by a great deal; that is, the model has poor accuracy. Nonetheless, the model accurately predicts that there would be less risk if all airplanes were smaller, if there were fewer airplanes using the system, or if airplanes are spaced further apart in the system. Using this model, we can study the effect on collision risk of system alternatives, even though those alternatives result in changes in collision risk that are a very small fraction of the error in the model's ability to estimate collision risk correctly.

It is obviously important to estimate model accuracy. The need to estimate model resolution is less clear. But model resolution limits our ability to optimize engineering designs. Thus, for the purpose of engineering design, it is also important that we pay attention to model resolution. It is especially important to account for correlation in model error between alternative system designs.

Resolution in the Case of Uncertainty

The theory of resolution presented above for the case of model error applies also for the case of uncertainty resulting from nature or other sources external to the system. A simple example will demonstrate this. Consider the electric energy output from a solar panel. We will use the very simple model

$$U = \phi T A g \eta$$

where U is the total energy produced; ϕ is the average normal solar flux; T is the exposure time; A is the overall solar array area; g is a geometrical factor, $0 < g < 1$, which gives the fraction of A that is an active energy-producing area; and η is the solar cell efficiency. We will assume for simplicity that the uncertainties in T, A, g, and η are negligible. But, because of variability in weather, the uncertainty in ϕ, and hence the uncertainty in U, are quite large. T, A, g, and η relate to the design of the solar array. ϕ is a factor of nature, not under the control of the design engineer. On the other hand, despite the uncertainty in ϕ, the value of ϕ that results for any given exposure does not depend on the design. Thus, we can compare alternatives using the differential model

$$\Delta U = \phi(T_1 A_1 g_1 \eta_1 - T_2 A_2 g_2 \eta_2)$$

and we can say with high confidence which design, 1 or 2, will produce the larger amount of electrical energy.

At this point you might be tempted to neglect the uncertainty in ϕ, and replace ϕ with its expected value. If we rewrite our original equation

$$E\{U\} = E\{\phi\} T A g \eta$$

such that

$$E\{\Delta U\} = E\{\phi\}(T_1 A_1 g_1 \eta_1 - T_2 A_2 g_2 \eta_2)$$

The problem with this is that it throws out information that is contained in the pdf of ϕ. We can see the impact of this information on design when we add economic factors. We could take as the value of the energy produced

$$V = \alpha E$$

where α is a constant in units of, say, \$/kW-hr. And a simple model for the cost of the system is

$$C = \beta(g, \eta) A$$

where β is a cost scaling factor, $\$/m^2$, which can be a function of the array geometry and materials reflected in the values of g and η. Now the net value of the system is

$$\begin{aligned} V_N &= V - C \\ &= \alpha U - \beta(g,\eta)A \\ &= A[\alpha\phi\, Tg\eta - \beta(g,\eta)] \end{aligned}$$

In this expression, it is clear that the uncertainty in ϕ imposes risk in V_N and thus can impact the selection of alternatives.

Nonintuitive Cases

In their barest essence, symbolic models can be thought of as a mathematical expression of our intuition. This leads to an interesting paradox: How can we obtain, from our intuition, a nonintuitive case? To be precise, we do not find nonintuitive cases using our intuition. What we discover is inconsistency in our beliefs. We begin our analysis of any system by applying our intuition. Indeed, our system models most often reflect our *a priori* assessment of which phenomena are important and which are unimportant. As we increase the range of phenomena and the degree of system complexity that we incorporate in our model, we also increase the degrees of freedom in the results that the model can produce. At the same time, the complexity of the system model grows beyond our ability to comprehend all of its facets intuitively. The preciseness and consistency of mathematics assures consistency and enables us to see and comprehend the fine structure in a system model.

Demanding complete consistency through the use of mathematical models often provides us with insights that we could not obtain from intuition alone. Indeed, the main purpose of system models is to enhance and fine tune our intuition, so that ultimately we can design the system by intuition, which is the final basis for all design.

9.5 DESIGN MODELS

There is a strong tendency to make engineering design into a top-down process. A general system configuration may be formulated at the system level, and at this level subsystem constraints are developed and applied. In this approach, a system is divided into subsystems and the function of each subsystem is defined. From these functions, a set of subsystem requirements, stated in the form of constraints, is defined. Subsystems are then designed compatible with these constraints. In the event that the subsystems themselves are complex, the process may be repeated at levels of increasing system detail, that is, closer to the actual parts or components. To some, this might appear to be a systems approach. But we must look at the design process in finer detail. What is the objective of the subsystem design decisions? Is it to obtain a subsystem design compatible with the constraints? And why must we have constraints in the first place?

Frequently we hear engineers speaking about the design of subsystems for which they are responsible in terms of the constraints that they must meet. Upon closer examination, however, we find that these are not constraints at all, but merely design decisions that have been made at higher levels in the organization or at higher levels of aggregation in the system design. When design is done by this approach, we remove many degrees of freedom from the individual design engineers, and force them to consider design options that might have significant penalties. The major problem is that we might never know what penalties we pay as the result of such a design process because we never ask the question. And we never ask the question because we start with the constraints and treat them as immovable. *It is almost always the case in engineering design that we use constraints to reduce the complexity of the design process.* Consider the following examples:

- In the computation of an optimal trajectory for a spacecraft on a flyby of a planet, we impose the mathematical constraint that it must not crash into the surface of the planet. This simplifies the mathematics of the trajectory computation, but what we really want is to maximize the value of scientific data returned from the mission and, if the spacecraft crashes into the surface of the planet, it will produce much less data.
- In the design of an aircraft autopilot for making automated landings, we impose constraints on the maximum acceleration or g loading that we want the autopilot to impose on the airframe and passengers. This simplifies the mathematics of the autopilot design, but the real objective is to develop a design that results in safe and comfortable landings. The problem is that safe and comfortable are very hard to measure and incorporate into an overall system objective function.
- In the design of a signal processing algorithm for TV image compression, we might impose a constraint on the compression algorithms that there be essentially zero probability of information loss in the reconstructed image. This obviates the need to define human preferences for image quality. But it may be possible to achieve higher overall image quality through compression if some loss of information is allowed.

In each of the above cases, we are seeking to simplify the problem by taking the larger system and dividing it into two or more smaller systems, each of which is, hopefully, easier to deal with alone. We separate these smaller systems by boundaries defined by the constraints, and no longer treat the full system for what it is. In the case of the autopilot, the airframe (the structural components of the airplane) and the passengers are parts of the larger system. We strip them away, however, with a simple constraint on g loading, and deal only with the autopilot itself and its interactions with the airplane dynamics. We will never know, unless we ask the question directly, whether allowing a 1 percent increase in g loading will significantly ease the autopilot design. Any time that we impose a constraint on an engineering design, we obtain a design that is either on the constraint boundary or inside the constraint boundary. If the design lies inside the constraint boundary, the constraint is *not active*, and the system is not penalized by the constraint. If the design lies on the constraint boundary, however, we say that the constraint is *active*, and it is generally the case that an active constraint imposes a cost on the system.

A purpose of system design models is to reduce the need for constraints to the absolute minimum. To the extent that system models can handle more complexity, constraints are needed less, and we can achieve more optimal overall designs. We can see how system design models work by means of an example case. Consider the design of a spacecraft consisting of ten subsystems as listed in Table 9.1. A very high level and simple model of the spacecraft is given in this table for some key physical system parameters: mass, volume, peak power consumption, and peak thermal load. In this table, we have taken m to represent mass, P to represent power, V for volume, T for thermal loading, and E to represent energy. Greek letters denote scaling constants in the appropriate units. For example, α_S is the specific mass of the structure in units of kilograms of structural mass per kilogram spacecraft mass.

TABLE 9.1 SPACECRAFT SYSTEMS MODEL

Subsystem	Mass	Volume	Peak Power	Peak Thermal Load
Structure	$\alpha_S m_{S/C}$	V_S	0	$\delta_S V_S^{2/3}$
Solar array	$\alpha_{SA} P_{SA}$	—	P_{SA}	—
Battery	$\alpha_B E_B$	$\beta_B E_B$	$\gamma_B E_B$	$(1-\eta_B)\gamma_B E_B$
Power conditioning	$\alpha_{PC} P_{PC}$	$\beta_{PC} P_{PC}$	$-\eta_{PC} P_{PC}$	$(1-\eta_{PC})P_{PC}$
Thermal control	$\alpha_S m_{S/C}$	$\beta_{TC}(P_{PC} + \text{const.})$	$\gamma_{TC} P_{S/C}$	$(1-\eta_{TC})\gamma_{TC} P_{S/C}$
Attitude control & station keeping	m_{AC}	V_{AC}	P_{AC}	$(1-\eta_{AC})P_{AC}$
Telemetry telecom & control	$\alpha_{TT\&C} B_{TT\&C}$	$\beta_{TT\&C} B_{TT\&C}$	$\gamma_{TT\&C} B_{TT\&C}$	$(1-\eta_{AC})\gamma_{TT\&C} B_{TT\&C}$
Propellant & tankage	$(1+\alpha_T)m_P$	$(1+\beta_T)m_P/\rho_P$	0	0
Thrusters	$\alpha_T T$	$\beta_T T$	$\gamma_T T$	—
Payload	m_L	P_L	P_L	P_L
TOTAL	$m_{S/C}$	$P_{S/C}$	$P_{S/C}$	$\delta_S V_S^{2/3} + \delta_{S/C} P_{PC}$

Symbols

α = Specific mass	β = Specific volume	ρ = Density
η = Efficiency	γ = Specific power	δ = Specific heat loading
m = Mass	P = Power	E = Energy

We could expand the table to include other parameters as necessary, such as average power consumption, average thermal load, propellant consumption, and so on. When the table provides models for a sufficient range of physical parameters, it can be used to model the subsystems as a function of a small number of design parameters such as payload mass, payload volume, and payload telecommunications requirements. This occurs when the model closes on itself, that is, when it predicts the values of all parameters contained in the model itself except for the design parameters themselves.

The objective in system modeling is to begin with a model, however simple, of the entire system. Then, rather than to select a set of constraints that isolate the individual subsystems from each other and proceed with the design of the individual system components, we seek to refine the system model by incorporating increasing levels of detail in each of the subsystem models. That is to say, for each of the boxes in this table that contains a model, we would seek to replace these models with models of increasing detail. At any point in this process, we can solve the system model to obtain a system design at the level of detail provided by the subsystem models.

A question that arises here is: how do we take design options into account in this process? The answer is that we develop separate subsystem design models for each design option. For example, we would have one subsystem model for an aluminum structure and a different model for a composite model. At the level of detail shown in Table 9.1, the different models are represented as different values of the scaling constants. At finer levels of detail, however, the models could take on entirely different forms.

The philosophical difference between the systems approach and the *conventional* approach to system design is this. In the conventional approach, the focus is on design, which is accomplished by a process of *divide and conquer*. That is, a system is divided into subsystems by means of the imposition of constraints, and design takes place essentially at the subsystem level. In the systems approach, the focus is on modeling at increasing levels of detail, and the system design emerges from the modeling process. Only at the final stages of system design do the two approaches merge, producing a final system description. Even at this point there are differences. As we shall soon see, part of the systems design as obtained from the systems engineering approach includes a design for system production or manufacture, deployment, operation, maintenance, repair, and disposal. *The systems approach does not end with the design of the physical elements of a system.*

9.6 SYSTEM LIFE CYCLE MODELING

In this section, we will consider modeling those aspects of the life cycle of a system that do not involve probabilistic events such as random failures. We will deal with the modeling of random failures in the next chapter.

Design Validation Costs

Virtually all system designs must be tested or validated after completion of design and before manufacture is begun. Different designs can impose significantly different validation costs. Validation costs may be considerably higher if human safety is a significant factor. The validation of an airplane design is one example. Many design choices can be strongly influenced by the need to qualify the design and prove it safe. For example, consider the choice of structural materials. Today, the major material for commercial airplanes is aluminum. Early airplanes were made of wood and fabric. The transition from wood and fabric to aluminum began not long after the invention of the airplane itself. But it took nearly 50 years to complete this transition. Much of this time was spent validating models for the use of aluminum in aviation applications. New materials, such as composites, are available. But their use is limited as a result of the design validation costs that occur when they are used in new applications.

We know that even toy designs must be validated to assure that no part of the toy can impose a significant hazard to a youngster. The cost of such validation depends on the complexity of the design, and the extent to which it uses new versus tried and proven approaches.

In the end analysis, design validation costs are major factors tending to limit design innovation, and we need to take them into account.

Manufacturing Costs

Although it might not seem so difficult, the modeling of manufacturing costs can be rather troublesome. Even when a system is designed using computer-aided design and manufactured using a computer-aided manufacturing system, there still remain undefined steps that complicate the modeling process. In particular, the sequence of manufacturing steps might not be defined. Also, human intervention in the manufacturing process can be difficult to model.

To the extent that the manufacturing process is capital intensive, cost may be subject to wide variability as the result of the method of depreciation chosen. This problem is discussed in Chap. 6.

One simple model used for manufacturing cost is to take this cost to be proportional to the weight or mass of the system. For example, for products made largely of steel, it might be possible to estimate manufacturing cost as, say, $10 per kilogram. For buildings, it is common to make a rough estimate of cost in terms of area or floor space. A building of a certain type might cost $2,000 per square meter to build. These models can provide adequate data for some analyses, but care should be taken to be sure they are not used to examine design tradeoffs that run counter to the model. For example, complexity of design might lower system weight and increase system cost simultaneously. This would not be reflected in a model that treats manufacturing cost as proportional to weight.

The systems engineer would really like to know how each little change in design affects manufacturing cost. How much money does it save to replace a bolt with a rivet? How much does it save to reduce by one the number of bolts holding a part together? If we had answers to these questions, we could also ask how the changes affect reliability, repair, and maintenance, and ultimately customer satisfaction.

We will discuss manufacturing costs further in Chap. 11.

Maintenance Costs

Sooner or later, everything breaks or wears out. Sometimes a relatively small or inexpensive part of a larger system wears out well before the larger system does. In this case, it is often desirable to *maintain* the system by replacing the worn out part. This is essentially what happens with oil in an automobile engine. The oil *wears out* after about 5,000 km and, rather than throwing out the entire automobile, we replace the oil. Other parts, such as brake pads, wear out less rapidly, but still long before the end of the useful life of the automobile. For parts such as these, we can predict their lifetimes with relatively good accuracy, and we can propose a maintenance schedule for the system.

The maintenance schedule provides us with much of the information that we need to model maintenance costs. Maintenance costs include many components: parts, labor, facilities and equipment, training, parts distribution, and parts stockpiling, for example. Many of these costs occur prior to the actual maintenance itself, and it is important to take the time value of money into account in the cost model. These costs include facilities and equipment, training, parts distribution, and stockpiling.

Distribution Costs

Most manufactured goods are manufactured in one location and sold or used in another. Thus, the goods must be transported or distributed for sale. The cost of distribution can be an important component in the overall system cost model, and it can be significantly dependent on product design. Indeed, some products are designed specifically to minimize distribution costs. Distribution costs can be dependent on the volume of a good. This is the case, for example, if the product is shipped in a vehicle that is volume limited, such as a truck shipping relatively light goods. In the shipping of soda bottles, bottles must be stacked in such a way that there is a lot of space between and around the bottles. Can designs that use flat tops make it possible to increase the volume of soda shipped in a single load, and thus reduce shipping costs? Some products, such as electronic equipment—computers, televisions, stereos, and so on—need special protection, and that protection demands volume. Consider design tradeoffs that reduce the shipping volume of a product but impose additional requirements on the product itself. By making the product itself stronger it may be possible to reduce the shipping cost to more than offset the added cost of the product.

Again, it might seem that distribution costs can be neglected, or that they do not have a significant effect on the design of the product. But this assumption should not be made without careful investigation. Also, you can be led to think that the modeling of distribution costs can be relatively simple. This also might not be the case. There is first the problem of packaging and the determination of packaged density, and second there is the extensive opportunity for the optimization of the distribution of the product. There can be some very complex problems embedded in these issues, and the mathematical models for their solution can become quite complex.

Sales Cost

There exist several different approaches to the distribution and sale of a product. We can see some of them in everyday use. Many products are sold by wholesale or retail distributors. These distributors stockpile the products and have them immediately available to the consumers. Stockpiles and displays cost. Also, in this case, other costs, such as shipping costs, are incurred prior to the end purchase and use of the product. Delays such as these impose costs that must be borne ultimately by the consumer.

A second approach to sales is the catalog approach. Here, customers order goods from a central facility. This approach can significantly reduce the cost of sales by reducing the need for storage, reduced inventory times, the potential for storage in locations that are less costly than major population centers, the need for fewer and less knowledgeable sales personnel, and so on.

Another sales cost is the cost of advertising. Although you might think that advertising and product design are separate functions, they are not. The linkage is clear in the case of automobile design, where style and other design features are closely linked with advertising approaches. Another technology where advertising and design are closely linked is the soda can, where it is imperative that it be possible to print labeling on the metallic can surface.

Disposal Costs

Disposal costs can be a particularly difficult design-related cost component to model. The major difficulty arises over the extended time period that must be considered for most systems. Extreme examples, where disposal costs can drive both design and economics of a system, include nuclear power plants and nerve gas for chemical warfare. In cases such as these, the toxicity of the end products can demand very expensive disposal procedures. For systems that lie at these extremes, it is obvious that disposal costs must be included as a component of overall system cost and taken into account in the system design.

Disposal costs of somewhat more benign systems are less obvious. What will be the disposal cost in 10 or 20 years of an automobile or a personal computer? How do these costs depend on design? The Corvette has a plastic body. How will its disposal cost compare to that of an automobile made with a metal body? In earlier times, we did not worry so much about disposal costs. But this policy has created toxic waste sites, and their cleanup is becoming very expensive. In the future, will manufacturers be asked to bear the burden of disposal costs for the systems that they design, manufacture, and sell? Some manufacturers have already been hit with disposal problems and costs.

Another form of disposal cost that can be important is the cost of disposing of chemicals or intermediate products used in the manufacture of a system. The use of these chemicals clearly relates to product design and manufacture, and their disposal will be increasingly important in the selection of design options. Design options for environmentally benign manufacturing should be given careful consideration, and the system life-cycle models used to evaluate design options should take these costs into account.

Liability Costs

Product liability has become a major design issue over the past 20 years or so. There are two basic approaches to dealing with product liability: to self-insure against liability or to purchase insurance. Either way, the cost of liability can be very high. The light aircraft industry of the United States is one industry that understands this issue all too well. For this industry, the cost of product liability insurance rose to as high as 30 percent of the cost of an airplane to the consumer. This appears to have been a significant factor in the choice of at least one manufacturer, Cessna, to stop making light aircraft altogether, even though their designs were among the most popular.

Product liability is not something that a company must simply accept as a consequence of being in a particular market. Products possess inherent risks, and the consumer is expected to accept those risks upon purchase. A chain saw is, for example, a dangerous product. In the hands of a careless or untrained operator, it can cause severe physical harm. But the risks of this product are obvious, and we do not expect large liability suits from people who blatantly misuse the product. On the other hand, an automobile is not as blatantly dangerous a product, and consumers have come to expect a level of safety built into the design. Large lawsuits arise when the manufacturer has failed to take responsible actions to prevent undue risk or hazard to the operators or users of a product.

Because system design and liability are closely linked, the cost of liability should be taken into account in overall system design. Thus, our system models should include liability cost, even though liability-cost models are likely to be rather subjective.

9.7 OBJECTIVE FUNCTIONS

In essence, system design is analogous to control engineering. In control engineering, we seek to control a system in such a way that it maximizes (or minimizes) some objective or measure of worth. We effect control of the system by choosing values of control variables as a function of time. The state of the system is monitored by sensors. In engineering design, we seek to manipulate nature in such a way that we maximize the utility of an engineered system. During the design process, we monitor the utility of alternative designs largely through the process of systems modeling, and we judge options in terms of the value or worth associated with their predicted outcomes. We refer to the mathematical formulation of the measure of system utility as the system *objective function*. It is important that a system model include modeling of the objective function. Sometimes the objective of a design is so far removed from the parameters of the design itself that, without such a model, it can be very difficult to assess the utility of a design. For example, the purpose of an automobile design could well be to provide a profit to the manufacturer. In light of this objective, we could ask how a specific design alternative, such as a smaller engine that would tradeoff acceleration for fuel economy, would affect the profitability of the company over, say, the next 10 years. The objective function in this case must include forecasts of the future price of gasoline, the market for automobiles in terms of their acceleration versus fuel economy attributes, future buying practices of potential consumers, and a myriad of other variables. The objective function must also account for the uncertainty inherent in the forecasts upon which it is based.

Engineers frequently attempt to escape the rigors of developing proper objective functions by making sweeping assumptions. "The consumers want the peppiest car they can get with a highway fuel economy of at least 10 km/liter." This sweeping statement uses a constraint, 10 km/liter, to obviate the need for developing a more complete objective function.

Another tradeoff that takes place during the design process is the tradeoff between reliability, lifetime of components, maintenance, and repair. Design engineers certainly consider these factors, but not generally as a part of the design process itself. Indeed, each component in a system might be designed to maximize reliability and/or lifetime subject to cost constraints, but rare is the design that considers all factors concurrently. We return to the example of the automobile design. The need for aftermarket parts supply can significantly affect both the profitability of a design and the consumers' image of the car. Thus, we see that modeling of the objective function can demand a system reliability model and a system repair and maintenance model. This requires special modeling approaches that we will discuss in the next chapter.

We noted earlier that system disposal can also pose a problem. Disposal is becoming an increasingly important aspect of system design. Thus, you should consider including in the system objective function a system disposal model that relates factors of system disposal back into the corporate profitability or other objectives.

The overall development of a system objective function is outlined in Fig. 9.6. This figure is only an example; it does not attempt to be complete, and many other categories of factors might be needed in the case of a specific system. We assume here that the corporate objective function for the product relates to profitability, but other objectives are possible. Profitability depends on costs and revenues generated by the system over its life cycle. Estimates of these require forecasts of parameters that relate to both costs and revenues. Estimates of consumers' preferences in terms of the physical attributes of the system are also needed. Given probabilistic estimates of costs and revenues, corporate financial parameters and risk preferences also are needed in order to convert probabilistic cost

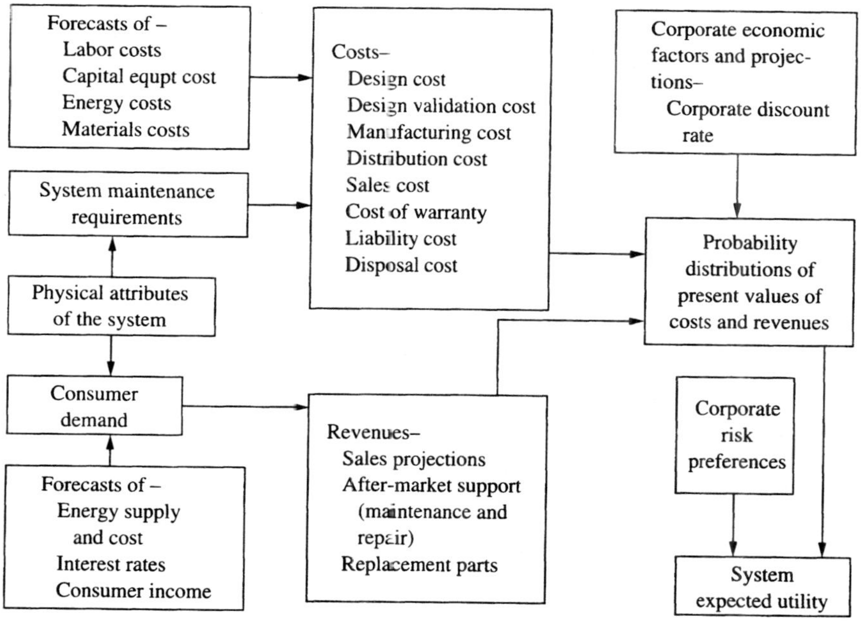

Figure 9.6 An outline for the computation of system utility.

and revenue streams into a utility measure. It does not take much study of this figure to see that objective function modeling can be as complicated, or even more complicated and time consuming, than modeling of the physical system. But it is the objective function that is the *raison d'être* for the company's involvement in the system in the first place. Failure to model the objective function is tantamount to overlooking the larger purpose of the system. The systems engineer cannot make good decisions without the aid of a well-thought-out and carefully developed objective function.

9.8 SUMMARY

In this chapter we begin to look at system models in the larger systems engineering context. We see that there is a philosophical difference between disciplinary design models and models for systems engineering. In the case of the conventional approach to engineering design, the focus of the modeling process is on subsystem design. Design is accomplished through a process of divide and conquer in which the system is divided into subsystems defined in terms of subsystem boundaries and facilitated by means of constraints imposed on each of the subsystems. In the systems approach, the focus is on the modeling process, and the design emerges from the system model at increasing levels of detail.

We found that a systems model goes beyond design of the physical system to include design validation, manufacture, distribution and sales, product support, liability, and disposal. And we will soon add to this list. Through the use of systems models, we hope to avoid the imposition of arbitrary constraints that limit design options and develop a better understanding of the consequences of each individual design option.

REFERENCE

1. Rivett, P., *Model Building for Decision Analysis,* John Wiley, New York, 1980.

PROBLEMS

1. Return to problem 4 in Chap. 1. You proposed three different models for the number of beans in a jar. Which of your models is most accurate?
2. What are the advantages of a symbolic model over other types of models? What are the disadvantages?
3. For the most accurate model in problem 1, what are the specific sources of error? Suggest ways to reduce the error.
4. Give three examples each of descriptive and predictive models.
5. Give an example of the difference between accuracy and resolution.

6. Choose a product. Identify all of the phases in the life cycle of the product. For each phase, identify the categories of costs associated with the product.
7. You need to know the voltage drop across a resistor. You can guess, 100 volts with a standard deviation of 12 volts, or you can measure the current flowing through the resistor and the value of its resistance. You choose to use the model

$$V = IR$$

You make measurements and obtain $I = 17$ amps with a standard deviation of 1 amp, and $R = 4$ ohms with a standard deviation of 0.5 ohms. Given these measurements and their accuracies, is $V = IR$ a good model?
8. Write a model to obtain a numerical derivative (finite difference) of

$$y = e^x$$

For this calculation, what is most important: accuracy or resolution? Given your model, how could you improve its resolution? Write a model with better resolution.
9. What are the purposes of a mathematical model?
10. True or false: Your best attempts at modeling the stresses in a mechanical part enable determination of the maximum stress to an accuracy of ±30 percent 2σ. Therefore you cannot confidently make design decisions that could reduce the maximum stress by only 3 percent.
11. Develop a simple life cycle cost model for a house. Identify sources of uncertainty and error in your model.
12. Develop a simple model to compare the life cycle costs of two houses located in the same neighborhood. Comment on the uncertainty in this model. How would the uncertainty change if the houses were located, say, 2,000 miles apart?
13. Under what conditions can you model components of an overall system for design decision making independent of the larger system?

10

Analysis of System Reliability

10.1 THE PURPOSE OF THIS CHAPTER

In the previous chapter we saw that the system objective function contained terms that involve system reliability. For example, system maintenance relates to reliability; the need for replacement parts relates to reliability; and consumer demand relates, at least indirectly, to reliability. In the case of some systems, such as nuclear power plants, public acceptance depends on reliability. In other systems, such as commercial aircraft, operational cost and system economics as well as safety depend on reliability. It follows that reliability should be a major consideration in system design.

Reliability is a function of several parameters. First, it is a function of system design. Often, but not always, the overall reliability of a system can be improved by including redundancy in components or system elements that are prone to failure. Second, reliability depends on how a system is used. In an automobile or airplane, system reliability depends greatly on the care with which the system is used and on the maintenance provided. Often, manufacturers have little control over how a system is used. On the other hand, the use of a system such as a spacecraft, which might cost over $100 million, is highly controlled and managed to get the maximum value from the system.

In any case, in order to assess the value or safety of a system, to assess the cost of unreliability, to assess the effectiveness of designs to improve reliability, or to assess and optimize system use-management options, it is necessary to develop and apply system reliability computer simulations. The purpose of this chapter is to develop the methodology for doing this.

10.2 NOTIONS OF SYSTEM RELIABILITY

The concept of reliability may at first seem deceptively simple. It is easy to think that a system is either working or broken, and that reliability is a measure of how often the system breaks. But, in reality, reliability is much more complex than this. A system might have redundancy in some components, so if one breaks, the system continues operation at the same level of performance that it had when all components worked perfectly. Has a failure occurred? Has the system exhibited unreliability? Suppose a relatively minor system feature fails, but the system operation continues relatively unaffected. An example might be that the cigarette lighter light bulb in an automobile fails. Is this a failure worth counting?

The importance of a failure depends on many factors. What might not be counted as a failure in one case can be critical in another. For example, an airliner that is used in transoceanic flight typically has triply-redundant navigation units. Suppose the airliner is departing from the United States bound to Korea and shortly after takeoff one of the three units fails. The failure is reason to turn the airliner around and land to have the unit fixed. The failure is key to the success of the flight, and the flight failed because one navigation unit failed. On the other hand, suppose the flight has progressed and the airliner is over the middle of the Pacific Ocean when the same failure occurs. There is no cause for alarm, the flight will continue to its destination, and the failure of the navigation unit will not cause failure of the flight.

What about the case of the cigarette lighter light bulb? To the driver who does not smoke, this failure may have no significance whatsoever. But to a smoker, it might be quite important.

What we see happening in the above examples is that the importance of the failure depends on how the failure affects the measures of system utility. In the case of the airliner, a fly/no-fly decision must be made based on the availability of certain critical system elements. The navigation unit is perceived as critical to the safety of the flight, and the decision rule is that the flight shall not proceed to cross the ocean if one or more of the navigational units has failed prior to reaching the oceanic segment of the flight. On the other hand, once the fly/no-fly decision has been made, and the airliner begins its ocean crossing, a failure in a navigational unit is no longer cause to terminate the flight. The utility of the third unit, on the margin, is to provide redundancy in case of failure. It is not to provide navigation information *per se*.

We see a similar issue with respect to the cigarette lighter light bulb. The light bulb is of no consequence to a person who does not use it, or who does not rely on it for any purpose. It has value only to someone who uses it for some purpose, in this case to provide information for decision making: the decision making necessary to locate the cigarette lighter and use it to light a cigarette.

Mean Time Between Failure

Often, we will see reliability of a system expressed in terms of a parameter such as mean time between failure (MTBF). But such a parameter has meaning only for systems that are either clearly operational or clearly broken. Most systems have many possible degraded modes of operation, and the MTBF is too simple a concept for them. Despite the limitations of MTBF as a measure of system reliability, it often suffices quite well at the component level. It often is possible to say that a component is either working or broken. Consider a component whose failure rate is given by the probability density function $f(t)$.

Then, the MTBF is the mean of this density function.

$$\text{MTFB} = \int_0^\infty tf(t)\,dt$$

Many different forms of the density function are possible. But a common form is the exponential, which applies to a memoryless system or a system for which the failure rate is a constant in time. For a memoryless system,

$$f(t) = \frac{1}{\mu}\,e^{-t/\mu}$$

and it follows that MTBF $= \mu$. Also, for such a system, the probability of a failure during any time interval Δt, given that the system is operational at the beginning of the time interval, is given by

$$P_F = \int_0^{\Delta t} f(t)\,dt$$

or, for the memoryless system,

$$P_F = 1 - e^{-\Delta t/\mu}$$

Other failure models might also be used. The Weibull distribution is often used to model failures in electric power systems. Another failure model obtains from the combination of several simpler distributions. An exponential distribution with a small value of μ may be used to model early failures, another exponential distribution with a much larger value of μ may be used to model random failures during the normal operating lifetime of the system, and a normal distribution may be used to model wearout failures. The total failure distribution is then the sum of the three distributions.

$$f(t) = \frac{\alpha_1}{\mu_1}\,e^{-t/\mu_1} + \frac{\alpha_2}{\mu_2}\,e^{-t/\mu_2} + \frac{1-\alpha_1-\alpha_2}{\sigma\sqrt{2\pi}}\,e^{-(t-m)^2/2\sigma^2}$$

where $0 \le t \le \infty$, and μ_1 is the time constant for early failures, μ_2 is the time constant for random failures during the normal lifetime, m is the mean wearout time, and σ is the standard deviation of the wearout time. Note that this equation is approximate since the normal distribution has the range $-\infty < t < \infty$, and it applies only when m/σ is large, say greater than 5. α_1 and α_2 are both positive numbers between 0 and 1 such that their sum is also less than 1. They represent the fraction of failures that occur as early failures and random failures, respectively.

The Impact of a Failure on System Utility

From the above discussions, we see that the impact of a component failure depends on how the failure affects the system utility function. Some failures might have no impact at all on the system utility function, others might have very great impact. Sometimes a failure occurring at one time has negligible impact on the system utility, while the same failure at another time has a major impact on system utility. The utilities of some systems depend only upon completion of some activity. Such might be the case with the delivery of a package by air freight. The utility of the delivery system is high if the package is successfully delivered, low if the package does not get delivered. For such a system, we are interested only in the probability of successful delivery.

For other systems, the system utility depends on the time during which the system operates. Take the example of a spacecraft designed to gather scientific data on, say, the surface of another planet. The spacecraft travels to the planet, lands on the planet surface, and then gathers data over time, making measurements of such things as temperature, wind, elements present in the atmosphere, and so on. The transmitted data might include several pictures taken over time with changes in season and weather. It is probably obvious that the most exciting and valuable information is that returned early after the spacecraft lands. These data are unique. As the mission continues, successive units of data have decreasing marginal utility. They validate earlier data and expand their range. But they are not unique in the sense that the early data are. In a case such as this, early system failures impact the system utility much more than do later system failures.

Frequently, engineers design systems to meet a target end-of-life system reliability. But merely specifying an end-of-life reliability does not fix reliability as a function of time across the entire lifetime of the system. Look at the reliability curves shown in Fig. 10.1.

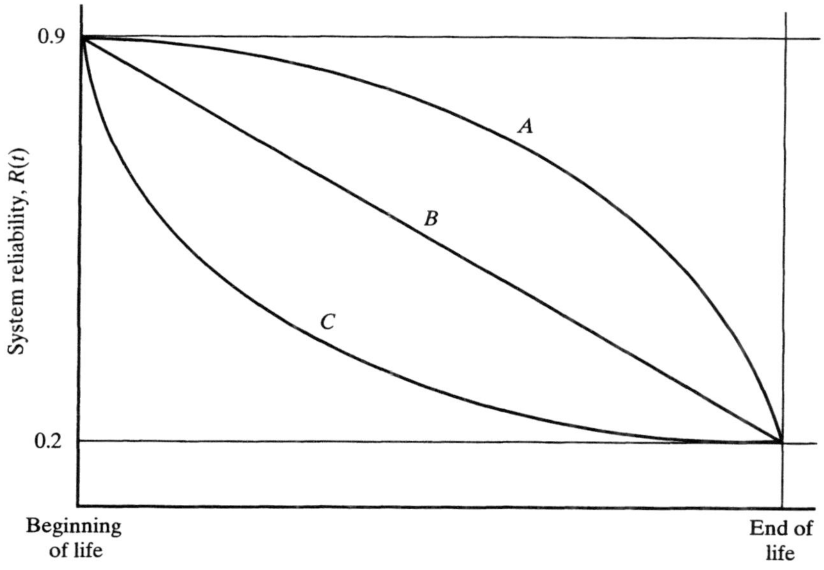

Figure 10.1 Different system reliabilities as a function of time, all achieving the same end-of-life reliability.

All of these curves have a beginning-of-life reliability of 0.9 and lead to an end-of-life reliability of 0.2, but the reliability-time paths, $R_i(t)$, taken from beginning of life to end of life for the three curves shown are quite different. Suppose now that the utility of the scientific data returned from the spacecraft is as shown in Fig. 10.2. For each of the system reliability curves, the probability of obtaining full mission utility, u^*, is 0.2. But the expected utility, given by

$$E\{u_i\} = -\int_0^{t_{eol}} u(t) \frac{dR_i(t)}{dt} \, dt$$

can be drastically different depending upon $R_i(t)$. From this example, we see the need to consider both system utility and system reliability as functions of time.

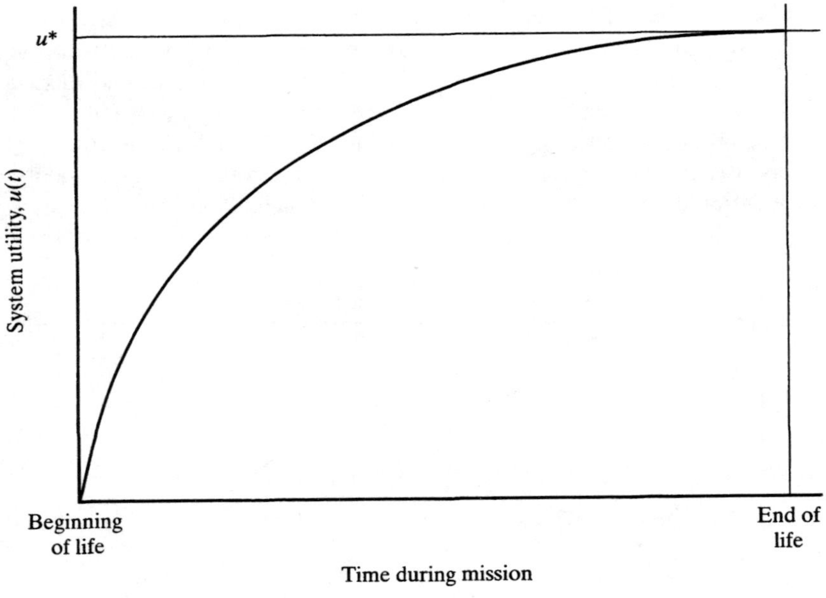

Figure 10.2 System utility generated as a function of time.

The matter becomes even more complicated when partial system failures are possible. In the above case, the utility function becomes a function of the level of operability of the spacecraft, and $R_i(t)$ is a vector of quantities representing the reliabilities associated with the different possible operational states of the spacecraft.

10.3 SYSTEM RELIABILITY DIAGRAMS

We will now develop methods for determination of the reliability of a system as a function of time. We will begin with systems whose only states are fully functional and broken, and then extend the analysis to include systems that can operate in degraded states.

We begin by considering a system that consists of n subsystems, each of which is necessary for successful operation of the overall system. A reliability diagram for this system is shown in Fig. 10.3. We say in this case that the subsystems are in *series* in the reliability diagram because every subsystem must be working in order for the system as a whole to be working. Let the probability that each subsystem is operational at time t be given by $P_i(t)$. Then the probability that the entire system is operational at time t is given as the product

$$P_S(t) = \prod_{i=1}^{n} P_i(t)$$

Conversely, the probability that the system has failed at time t is $(1 - P_S)$.

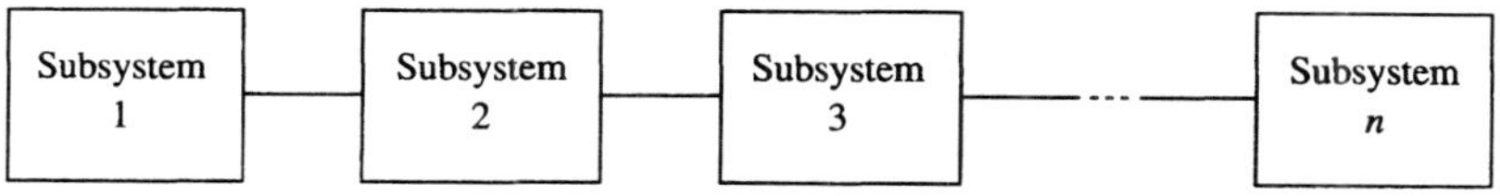

Figure 10.3 A system reliability diagram consisting of subsystems in series.

We can take the terms $P_i(t)$ in a very general way to cover more than simply the passage of time. For example, a particular subsystem might consist of a switch that fails when it is turned on or off. Each time that the switch changes position, there is a chance that it will fail. Let P_c denote the probability that the switch will operate properly on any particular change of position, and $m(t)$ denote the number of times the switch has changed position as a function of time. Then the probability that the switch is operational as a function of time is given by

$$P_{\text{switch}}(t) = P_c^{m(t)}$$

Likewise, $P_i(t)$ can also be a function of various state variables such as temperature, pressure, thermal load, power, and so on, and their histories. Indeed, you will note that the variable $m(t)$ in the above equation is, in essence, a state variable.

Consider another example where the failure rate of a system element depends on temperature. Let us arbitrarily take the model

$$\frac{dP}{dt} = -aPT^k$$

where P is the probability that the system element is operational at time τ, T is system temperature, and a and k are constants. Then

$$\frac{dP}{P} = -aT^k\,dt$$

$$\ln P = -\int_{t=0}^{t=\tau} aT^k \, dt$$

$$P = e^{-\int_{t=0}^{t=\tau} aT^k \, dt}$$

Note that we have taken $P(t = 0) = 1$. In this case, we can define a state variable X_T such that

$$\frac{dX_T}{dt} = -aT^k$$

with the boundary condition $X_T(t = 0) = 0$. We can then write P as a function of X_T

$$P = e^{X_T}$$

Next, we consider the case where there is redundancy. Suppose in Fig. 10.3 that we wish to improve the overall system reliability, and we feel that this could best be done by improving the reliability of Subsystem 2. But Subsystem 2 is of a design that does not lend itself easily to increased reliability. Another approach is to employ redundancy. To do this, we add copies of Subsystem 2 to the system in such a way that they substitute for one another if one or more fails. The resulting reliability diagram appears as shown in Fig. 10.4. For the redundancy shown, the reliability of the overall Subsystem 2 function,

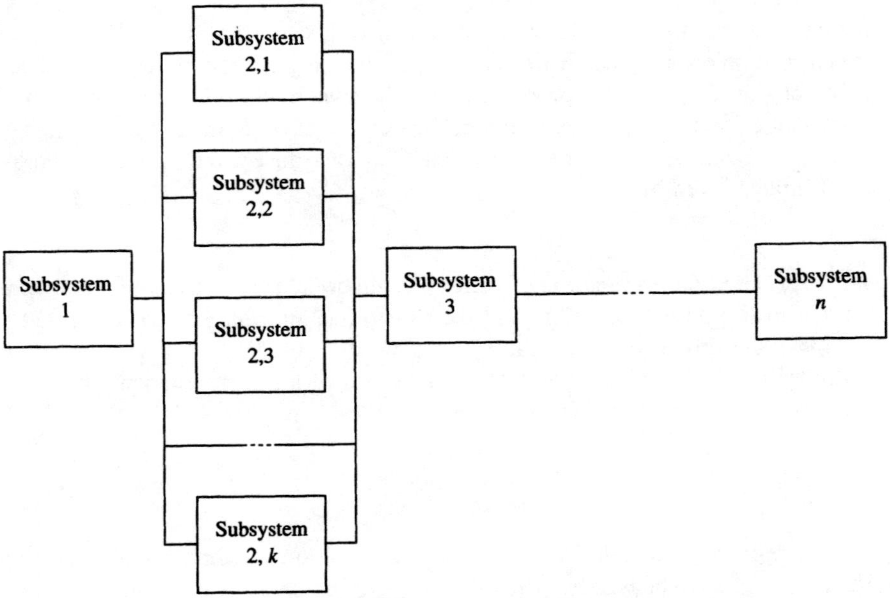

Figure 10.4 A reliability diagram for a system that has subsystem redundancy.

$P_2(t)$, is given by

$$[1 - P_2(t)] = \prod_{i=1}^{k} [1 - P_{2,i}(t)]$$

or

$$P_2(t) = 1 - \prod_{i=1}^{k} [1 - P_{2,i}(t)]$$

In the case that each of the redundant subsystems has the same reliability

$$P_{2,i}(t) = P_2^*(t)$$

the overall subsystem reliability reduces to

$$P_2(t) = 1 - [1 - P_2^*(t)]^k$$

If redundancy always took the form shown in Fig. 10.4, system reliability could always be improved by increasing k. But redundancy generally requires the addition of sensing and switching elements, the purpose of which are to detect faults in one or more of the redundant subsystem components, switch them out of the system, and replace them with functioning components. For example, suppose Subsystem 2 is a light bulb, and we want one bulb burning at all times during operation of the system. We may install redundant light bulbs, but then we also need a sensor to determine that a bulb is lit, and a switch

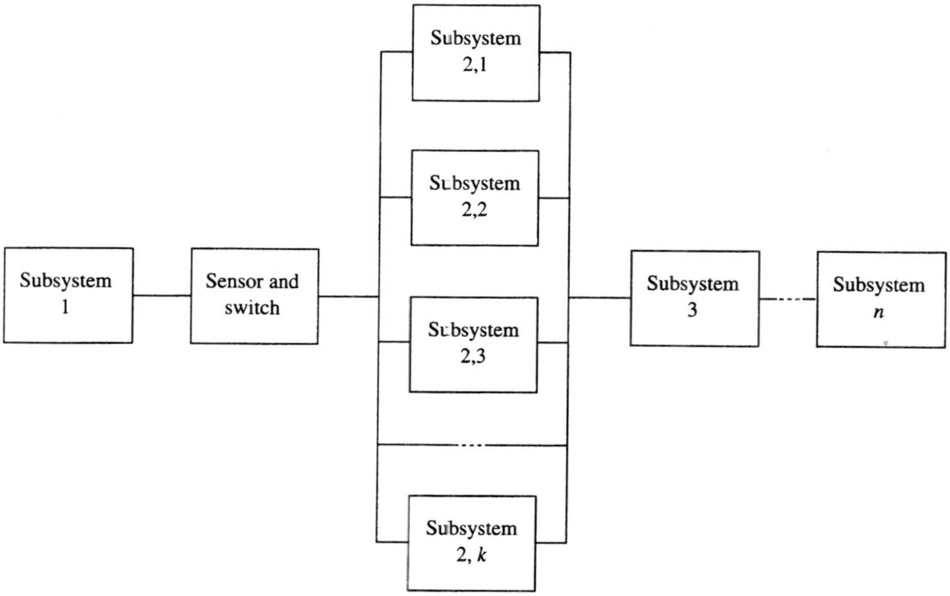

Figure 10.5 A reliability diagram for a system that includes redundancy and failure detection and switching components.

to turn on a new light bulb when the previous bulb burns out. The resulting reliability diagram now appears as shown in Fig. 10.5. The addition of the sensor and switch appear in the system as new subsystems that are in series with other elements of the system reliability diagram. Their presence degrades the overall system reliability. In general, the complexity of both the sensor and the switch will depend on the level of redundancy, k, and their reliabilities will decrease as k increases. Thus, a finite value of k, not $k = \infty$, will lead to a maximum overall system reliability, and we see that there is also a need for the sensor and switching systems to be of high reliability in order to improve overall system reliability.

Because redundancy generally requires that subsystems analogous to the sensor and switch of Fig. 10.5 be added as well as the subsystem redundancy itself, it is not always the case that adding redundancy improves overall system reliability. System safety is a form of reliability, and thus the arbitrary addition of new safety devices, for similar reasons, does not always improve system safety.

It is for this reason that it is so difficult to approach perfect reliability or safety in any system. As we approach perfect reliability, any attempt to further improve reliability by adding to system redundancy runs the risk of actually reducing reliability because of potential failures of the subsystems that we must concurrently add to monitor for failures and take corrective actions when they occur. This phenomena has been observed in commercial aircraft. Each new safety device installed on an airplane also has its own failure modes, and can itself present hazards that, under certain circumstances, lower overall safety. Such is possible with collision avoidance systems. Collisions involving commercial airliners are extremely rare events, on the order of one in 10^8 flights. Collisions may occur because pilots fail to see and avoid another airplane. But with the use of collision avoidance systems, pilots no longer feel as strong a need to watch for other airplanes, so collisions then occur because of the failure of the collision avoidance system. Now, unless the collision avoidance system itself has an extremely low failure rate, its use to prevent collisions could actually increase the collision rate. The thing to keep in mind is that, particularly in the case of systems that are already highly reliable or safe, the steps needed to incrementally improve reliability or safety might not be obvious, and careful analysis is needed to assure that the goal of improving reliability or safety is actually met.

Reliability diagrams such as that shown in Fig. 10.5 can be drawn for virtually any system. These diagrams can then be used in the computation of overall system reliability. Each of the subsystems in this figure may itself consist of many components. We can model each subsystem by drawing its reliability diagram. Thus we proceed hierarchically, by diagramming a system at increasing levels of detail, until we obtain a reliability analysis at the desired level of detail. More sophisticated models can be substituted directly into each of the boxes in a diagram such as that of Fig. 10.5.

Now we consider a system that has multiple functions, and where one or more functions can fail or degrade without the entire system failing. It is convenient to use a spacecraft as an example. A spacecraft contains several subsystems that are vital to the survival of the spacecraft itself. These include the structure, power supply, attitude control, thermal control, and telemetry and telecommunications. We call these subsystems the spacecraft *bus*. If any of these systems fail completely, the spacecraft will fail to operate, and all science or commercial function of the spacecraft will be lost. We will consider a spacecraft intended for scientific purposes that has on board, say, five scientific instruments. In the simplest case, we will take the spacecraft bus to be either functional or nonfunctional. A reliability diagram for such a spacecraft is shown in Fig. 10.6. Here, we depict the science instruments as being operationally independent. In this case, we can examine the reliability of each instrument as a function of time.

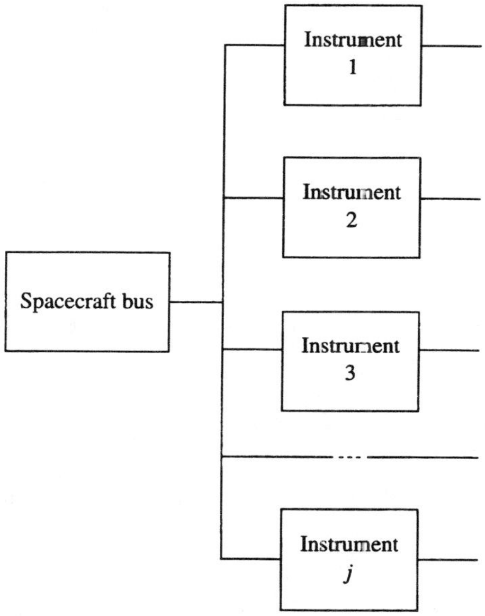

Figure 10.6 A reliability diagram for a spacecraft bus and independent instruments.

Our next job is to examine the utility gained from the data gathered by the instruments. Again, we encounter difficulties because a mission might be viewed as a set of science experiments, each of which relies on data from one or more instruments. But data gathered by an instrument for one experiment might not satisfy another experiment. Thus, each instrument might be programmed to supply data for experiments in some sequence. Let X_i denote the ith experiment, $i = 1,2,\ldots, n$, which requires data from instrument I_j. Let D_{ij} denote the data obtained from the jth instrument for the ith experiment. Then we write the utility obtained from the ith experiment as

$$u(X_i) = f[D_{ij}, j \in I(X_i)]$$

By this notation, we mean that the utility obtained from the ith experiment is a function of the data obtained from all instruments that are needed for this experiment. In general, the functional form of the utility will not be linear in terms of the D_{ij}. That is, linear utility forms, such as the multiplicative form, generally will not apply and more complex forms will be needed. Assuming that we can develop such utility forms, we will, however, take the total mission utility to be the sum of the utilities obtained from the individual experiments.

$$u_M = \sum_{i=1}^{n} u(X_i)$$

This too might be an oversimplification of the utility gained from a mission, but we will accept it for this example.

Next, we need a mission time line. This is a schedule of all events that take place during a mission, including which experiments are to be conducted at what times in the mission. The times at which many experiments are conducted might be dictated by constraints. For example, if the mission involves sampling the upper atmosphere of the earth, then data on the upper atmosphere must be collected while the spacecraft is in the upper atmosphere. The mission time line will determine the sequence of events that occur throughout the mission. From this time line it is possible to determine the subsystem reliabilities. For the case depicted in Fig. 10.6, we assume that, if a spacecraft bus subsystem fails, the mission is terminated. If a science instrument fails, the utility of all further science is limited by the unavailability of data from that instrument. Given the spacecraft

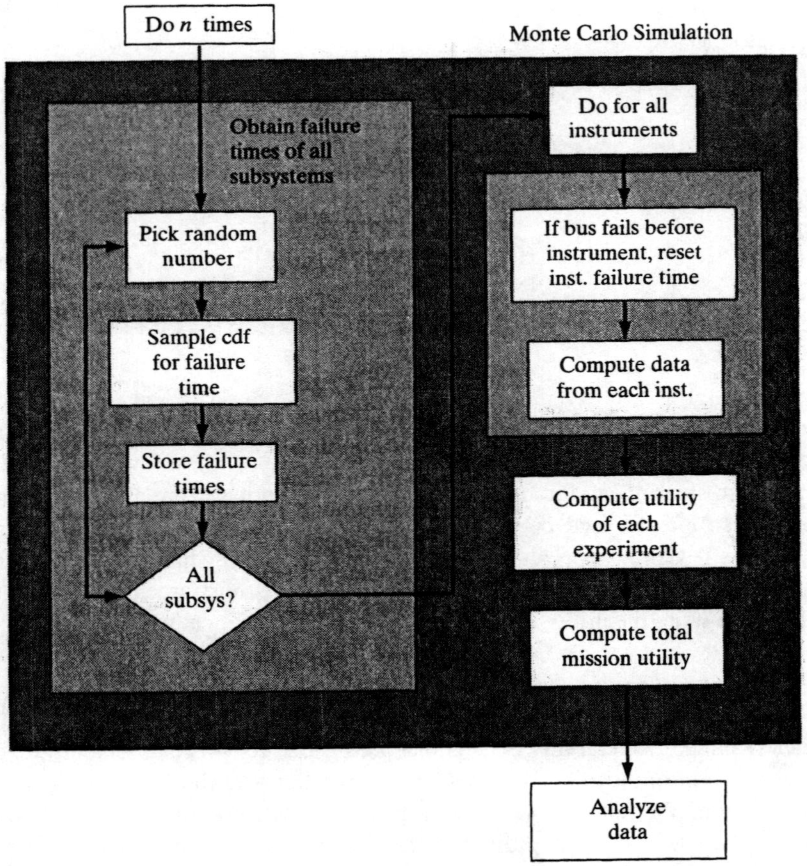

Figure 10.7 A Monte Carlo algorithm for computing expected utility of a mission using the spacecraft of Figure 10.6.

reliability diagram, the mission time line, and the science utility functions for each experiment, it is then possible to compute the probability distribution of mission utility. From this distribution, we can extract summary parameters such as expected utility. The computation of expected utility may be facilitated using the Monte Carlo method as outlined in Fig. 10.7, under the assumption that the mission time line is held to as much as possible given the failures that occur.

We can now complicate matters a bit more. Suppose the telemetry, telecommunications, and control (TT&C) subsystem of the spacecraft itself is comprised of three subsystems: a bus subsystem necessary for the operation of the overall subsystem, a narrowband datalink, and a wideband datalink. In normal operation, the narrowband datalink would be used for uplink of command data to the spacecraft and the wideband datalink would be used for downlink of scientific data. If one of these subsystems should fail, the spacecraft continues to operate, but in a degraded mode. In this case, the data provided by the spacecraft depends on its operational status. For the case at hand, there are four operational states of interest:

- S_1—Fully operational
- S_2—Wideband datalink operational, narrowband datalink nonfunctional
- S_3—Wideband datalink nonfunctional, narrowband datalink functional
- S_4—Spacecraft bus nonfunctional.

We note that the spacecraft can transition from state S_1 to state S_2, S_3, or S_4, or from state S_2 to state S_4, or from state S_3 to state S_4. But no transition from state S_2 to state S_3 or vice versa is permitted. We will denote the times of such transitions by the variables t_{kl}, where t_{13} is the time that a transition occurs from state S_1 to state S_3. Our simulation must now determine the D_{ij} as a function of the t_{kl}, that is $D_{ij}\ (t_{kl})$.

The complication that arises here is that the mission time line will, in general, change with a change in status of the spacecraft. If we had only one mission time line to consider, or even three, corresponding to the three operational states of the spacecraft, our problem would not be too difficult. We could still calculate the expected utility for the mission. Unfortunately, the spacecraft can transition from one state to another at any time during the mission, and to compute expected utility we need time lines for every possible transition. For this simple case where there is a nominal time line and two degraded modes of operation, we obtain the nominal time line and two additional sets of degraded-mode time lines, each containing an infinity of time lines. We cannot hope to solve for the expected utility of this case by enumeration of all possible time lines.

To get around this dilemma, we turn instead to the use of a set of rules for the generation of time lines given a transition from the fully operational state to a degraded state at any specified time. One approach would be to consider every feasible time line, with feasibility defined as meeting the mission and experiment constraints, and then to choose the time line that provides the maximum expected mission utility. This process could be significantly aided if we can formulate the problem for solution using an optimization algorithm such as integer programming.

We begin by simplifying the problem. Rather than to maximize the expected utility, we will assume that, after a failure resulting in a degraded mode, the spacecraft and all subsystems remaining functional will be completely reliable, and we will simply determine the mission time line that maximizes utility. Let I_i be a vector whose elements consist of zeroes and ones denoting which instruments are needed for the ith experiment, X_i. Let T_i be the time required to perform the ith experiment, and let $u(X_i)$ denote the utility of the ith experiment successfully completed. Also, let t_i be the time at which we plan to start the ith experiment, and define x_{ik} such that $x_{ik} = 1$ if the ith experiment is begun at time $t_i = k$, and $x_{ik} = 0$ otherwise. Finally, define I_{ijK} such that $I_{ijK} = 1$ if the jth instrument is required for the ith experiment at time K from the start of the experiment ($I_{ijK} = 1$ for $0 \leq K \leq T_i$ and all $j \in I_i$). We will also discretize time such that x_{ik} comprise a finite number of variables. Our problem then can be stated as follows: maximize u_M,

$$u_M = \sum_{k=t_f}^{k=t_F} \sum_{i=1}^{n} u(X_i) x_{ik}$$

where t_f is the time of the failure and t_F is the completion time of the mission. This function gives the total utility of the mission gained after a degrading failure takes place. This maximization is subject to the following constraints:

$$x_{ik} = 0,\ 1$$

$$\sum_{k=t_f}^{k=t_F} x_{ik} \leq 1 \qquad \text{for all } i$$

$$\sum_{j} \sum_{i} x_{ik} I_{ijK} \leq 1 \qquad \text{for all } K$$

$$\sum_{j} \sum_{i} x_{ik} I_{ijK} = 0 \qquad \text{for } \quad k + K > t_F$$

The first of these constraints assures that an experiment is either begun or not begun, that is, it is not fractionally begun. The second constraint assures that each experiment is begun only once. The third constraint assures that each instrument is used for only one experiment at any given time, and the fourth constraint assures that no experiments continue beyond the final time.

This formulation of the problem does not account for the time value of utility, which might reflect the fact that additional failures are possible. A possible improvement, which might move the solution closer to the expected utility solution, would be to add a discount factor to the equation for utility.

$$u_M = \sum_{k=t_f}^{k=t_F} \sum_{i=1}^{n} u(X_i) x_{ik} \alpha^k$$

where α is a positive number less than one. This formulation places higher value on experiments performed early in the mission, and would therefore emphasize the conduct of those experiments with higher utility earlier rather than later.

While the optimization approach may work to create mission time lines in some cases, there may be cases where it is too complicated or where the mission is such that it does not lend itself to formulation for optimization, or where such a formulation requires too much computing time. In these cases, we might simply state a set of priorities for the experiments, and conduct them as the priority list dictates, subject to constraints on availability of instruments, physical opportunities, and so on. For example, we might simply state that we want to conduct X_1 first, followed by X_2, followed by X_3, and so on. Whenever a degrading failure occurs, we simply continue to conduct experiments according to this logic, to the extent possible.

In the above analysis, we are seeing two important concepts. First, we see that when a failure occurs, we must adapt to the failure by replanning the remainder of the mission. This is a decision-making process that greatly complicates the overall mission-planning process. Not only must we plan for the case where everything works properly, but we must plan for the cases where there are failures as well. Second, we see that the very possibility of failures occurring in itself affects mission planning. In light of the possibility of failures, we tend to plan the more important activities of a mission to take place as early in the mission as possible.

The approach described above, while applied to the example of a space mission, is widely applicable to engineering systems.

10.4 CARTOONS AND SCENARIOS

We introduced the notion of cartoons back in Chap. 2, where they were used to assist in the creation of options. We will now expand on this notion to obtain a method for modeling activities involving an engineering system. We will use as an example for explaining the methodology of cartoons and scenarios a Space Shuttle mission in which it is desired to launch a communications satellite into geostationary orbit and retrieve another satellite from the same orbit. The cartoon shown in Fig. 10.8 depicts this mission under the condition that all activities proceed as planned. Our objective here is to analyze this mission allowing for the possibility of failures in each of the major activities of the mission. We are interested in the probability of success of the mission and its cost. Therefore, we must model these variables.

Our first task is to convert the cartoon of Fig. 10.8 into a scenario as shown in Fig. 10.9. A scenario considers all the events that take place in the cartoon, and considers their consequences whether they succeed or fail. The consequence of interest in this analysis is cost, so we will keep track of the cost consequence of each event. The launch of a Space Shuttle begins with lift-off. Both main engines and booster rockets are used to accomplish lift-off. When the boosters complete their burn, they are jettisoned, recovered, and refurbished for reuse. Recovery and refurbishment of the boosters is a probabilistic event. If the booster burn fails, it may cause loss of or damage to the Space Shuttle, and these are cost events that must be considered.

Next, we consider the Space Shuttle orbiter, and whether the orbiter successfully achieves orbit. Failure can result in cost penalties. Once in orbit, the Space Tug (ST), namely the vehicle that will take the satellite from low earth orbit to geostationary orbit and return the used satellite, and the satellite to be taken to geostationary orbit (S2) are checked out. If either are faulty, they are returned to earth for another launch attempt. If both check out, then a sequence of events is initiated that places S2 in geostationary orbit and returns the old satellite, S1, to the Space Shuttle. Finally, the Space Shuttle returns to earth for refurbishment and reuse. At this point, the mission is complete.

All of the probabilistic events that take place in the above scenario are depicted as the diamond-shaped boxes in Fig. 10.9, and all of the cost-incurring activities are shown in the rectangular boxes. For each of the probabilistic events, we can assign probabilities of success and failure. In general, it is possible for there to exist more than two possible outcomes, each with a separate probability, but we have, for simplicity, shown only success and failure outcomes here. For each cost-incurring activity, we can assign a cost or, if we choose, a probability distribution of cost.

The simulation model outlined above examines one single mission. It may be the case that this mission must be repeated over a period of many years in order to establish and maintain a communications satellite system. It is reasonable to expect that, over a period of time, there will be learning that will change both the probabilities of success and failure for each event, and the costs of each cost-incurring activity. Table 10.1 shows typical success data that might be used in running this model, and Table 10.2 shows typical cost data. If these data seem out of line with current space mission costs, it is because they were based on data relevant to the 1970s. The data given in Table 10.2 are also subject to learning, and may thus be a function of time.

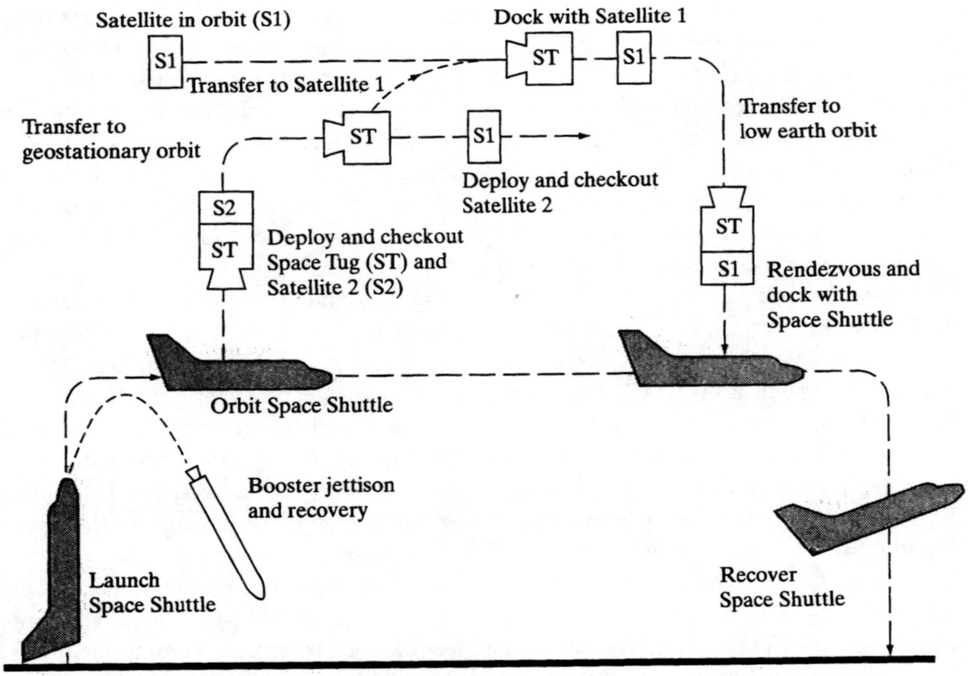

Figure 10.8 A cartoon showing the placement of a communications satellite into geostationary orbit and the retrieval of a failed satellite.

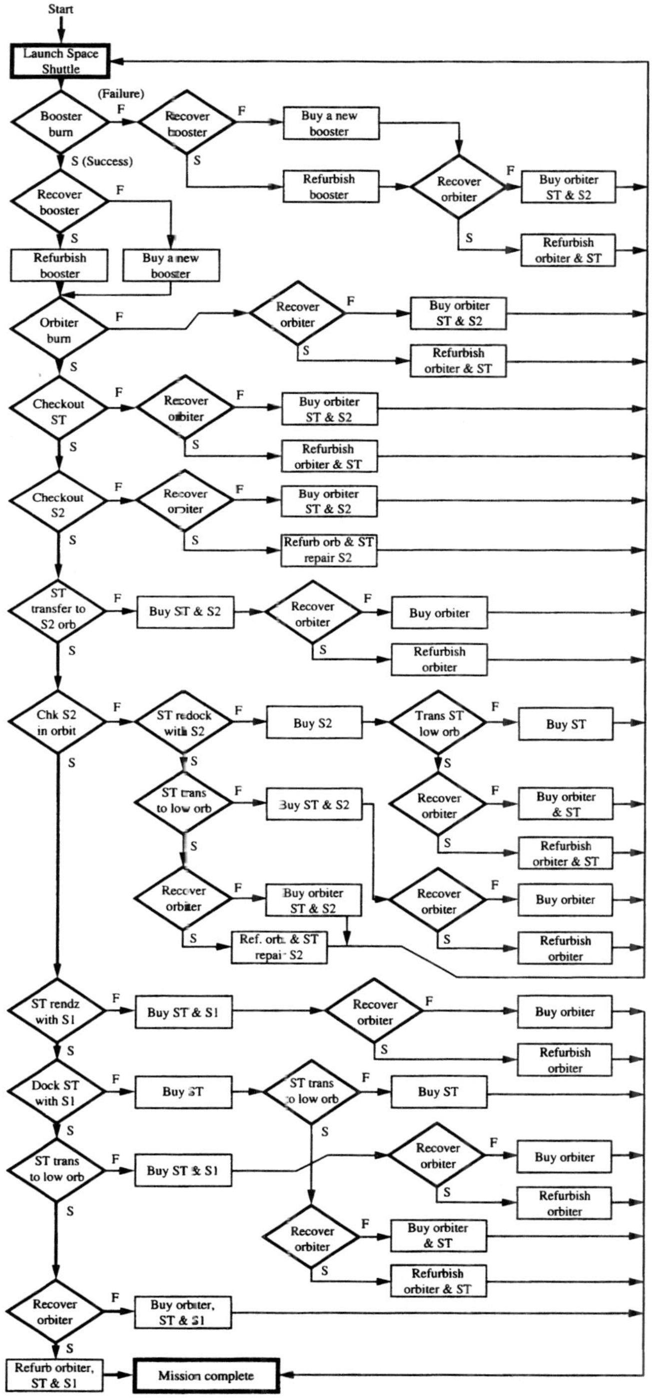

Figure 10.9 A scenario diagram for the mission depicted in Figure 10.8.

TABLE 10.1 SUCCESS DATA FOR THE MODEL OF FIG. 10.9

Probability of success for	Event probability of success as a function of time from start of analysis, years			
	1	2	3	4
Booster burn	0.99	0.992	0.994	0.994
Booster recovery given booster success	0.93	0.94	0.95	0.95
Booster recovery given booster failure	0.40	0.45	0.50	0.50
Orbiter recovery given booster failure	0.97	0.98	0.99	0.99
Orbiter achieves low earth orbit	0.99	0.994	0.998	0.998
Orbiter recovery given abort to orbit	0.97	0.975	0.98	0.98
Orbiter recovery from low earth orbit	0.99	0.995	0.999	0.999
Space Tug checks out successfully in low earth orbit	0.97	0.975	0.98	0.98
Space Tug transfers to satellite placement orbit	0.98	0.984	0.988	0.988
Space Tug transfers from terminal orbit to low earth orb and rendezvous with Shuttle	0.98	0.984	0.988	0.988
Space Tug docks with satellite S1	0.97	0.975	0.98	0.98
Satellite S2 checkout in terminal orbit	0.97	0.975	0.98	0.98
Space Tug reacquires S2 if S2 does not check out interminal orbit	0.98	0.985	0.99	0.99
Probabilities not given above	1.00	1.00	1.00	1.00

Data source: Economic Analysis of Advanced Propulsion Systems, *Mathematica, Inc.*, Princeton, NJ, April 30, 1973.

Let us assume that the overall purpose in this modeling process is to aid in the design of the satellite. There are many tradeoffs that we might make that affect satellite lifetime, reliability, and cost. Our objective is to maintain an active satellite system with a high probability and, at the same time, minimize the cost of the system over time. One factor affecting satellite lifetime is station-keeping propellant. When the station-keeping propellant runs out, the satellite can no longer be used. Loading more propellant on the satellite increases its lifetime. But the total satellite mass must not exceed the launch mass capability of the launch vehicles, the Space Shuttle and the Space Tug in this case. So, if we add propellant, we must reduce weight elsewhere, typically by removing redundant components. This increases the probability of failure during the normal operating life of

TABLE 10.2 COST DATA FOR THE MODEL SHOWN IN FIG. 10.9

Cost component	Cost estimate, $ millions			Pdf form
	Most likely	Min	Max	
Booster expendable cost	3.5	3.1	5.25	3
Booster reusable cost	5.5	5.0	6.0	4
Booster refurbishment cost	1.0	0.8	1.8	3
Orbiter expendable cost	2.2	2.0	3.0	3
Orbiter reusable cost	300	250	350	5
Orbiter refurbishment cost	3.7	3.0	4.5	3
Space Tug expendable cost	0.01	0.008	0.011	3
Space Tug reusable cost	18.9	15.2	28.5	3
Space Tug refurbishment cost	0.38	0.3	.57	1
Satellite cost	20.0	15.0	28.0	3
Satellite refurb. cost (% of satellite cost)	15.0	10.0	20.0	3
Operations cost per flight	0.15	0.12	0.17	2

Uncertainty profiles (pdf's) for cost data:

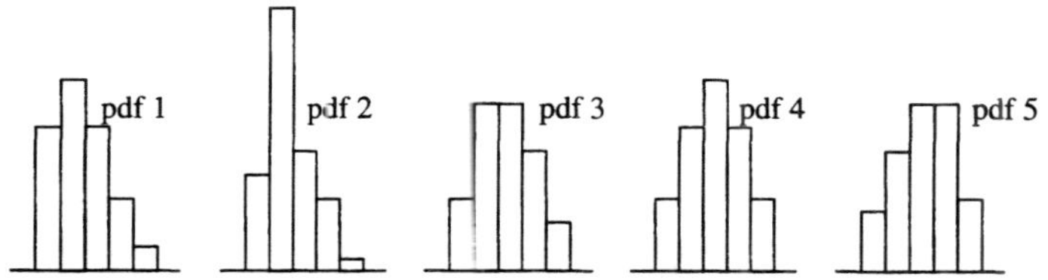

the satellite. What we need to do to examine these tradeoffs is to link together the launch model described with a satellite reliability model in order to obtain a model for the establishment and maintenance of a satellite system. Fig. 10.10 gives an overview of a Monte Carlo model that does this.

Using the model described, we can compute the probability distribution of mission costs. This distribution is shown in Fig. 10.11. The important thing to observe is that explicit consideration of uncertainty leads to estimates of mission cost that range up to two times the estimate obtained when uncertainty is not considered, even using the very optimistic data given in Tables 10.1 and 10.2. In addition, using the Monte Carlo analysis, we can make the computations necessary to compare system reliability and redundancy

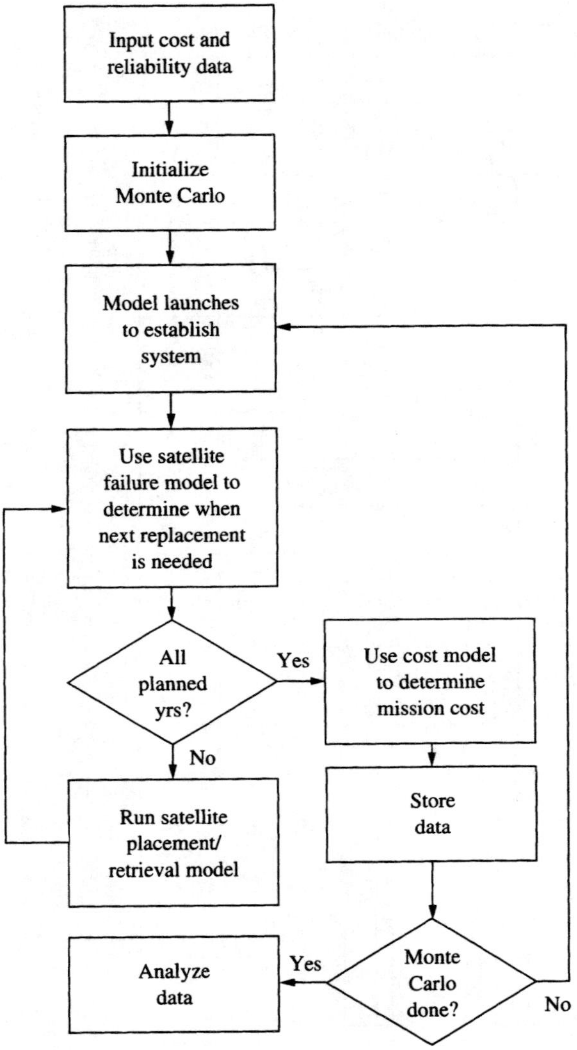

Figure 10.10 A Monte Carlo model for computing the probability distribution of mission costs.

options. Thus, this modeling procedure enables us to examine the impact of system alternatives on total mission cost, enabling carefully considered choices rather than guesses.

10.5 DECISION AND EVENT TREES

The cartoon and scenario approach to system modeling provides a powerful tool for modeling systems that undergo a series of discrete events, such as the launch of a satellite. In the example provided, each probabilistic event results in outcomes for which the subse-

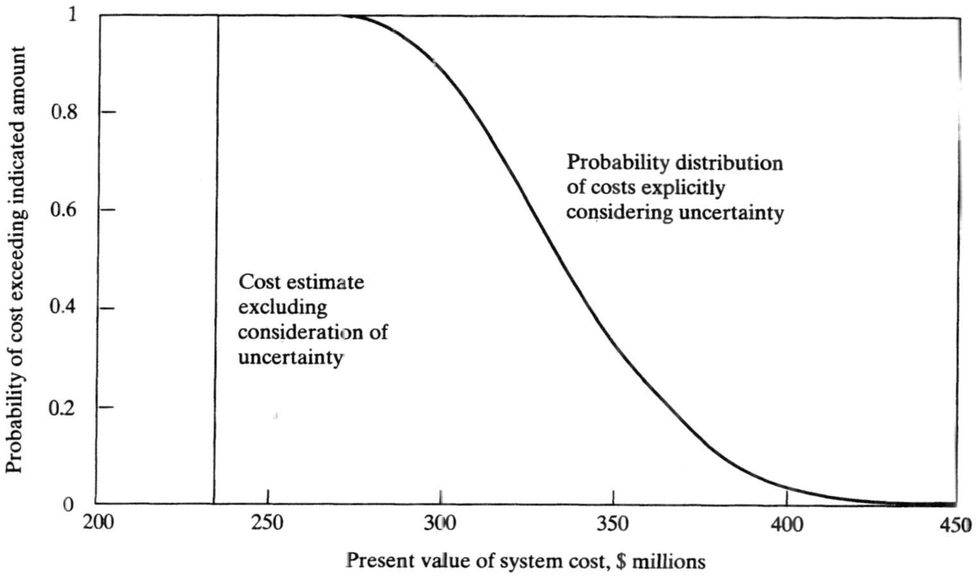

Figure 10.11 Probability distribution of costs for the mission shown in Figure 10.9 compared to costs assuming deterministic events.

quent actions are defined. For example, if booster recovery fails, the cost-incurring action is to buy a new booster. The weakness of this method is that it does not provide a mechanism for decision making within the simulation itself. In fact, the example shown in Fig. 10.9 is simple enough that we can enumerate all outcomes and prescribe all decisions in advance of the simulation. Unfortunately, not all systems can be simulated in such a simple manner. Sometimes, a simulation results in so many options that arise at so many different decision points that it is not possible to enumerate them all in advance of a simulation. In this case, we have to develop a method for simulating decisions as well as random events.

It is helpful to understand that the scenario depicted in Fig. 10.9 could be drawn in the form of a decision/event tree where the diamond-shaped boxes are represented as random-event nodes (dots), and cost-incurring activities are represented as the branches of the decision/event tree. Fig. 10.12 shows Fig. 10.9 partially translated into a decision/event tree. Notice the difficulty in drawing the decision/event tree posed by the potential for looping back to the beginning box in Fig. 10.9 because of failure to deploy satellite S2. Comparison of these two figures shows the advantage of the scenario figure. When a scenario figure can be drawn, it provides a clearer and easier-to-understand view of the proposed activity. It clearly shows a success path and the failure paths with their respective failure recovery actions. This is somewhat harder to see and understand in the decision/event tree format. Indeed, some activities can be so complex to plan that we cannot draw either a scenario or a decision tree. Rather, we are forced into developing shorthand notations from which we can develop computer simulation codes.

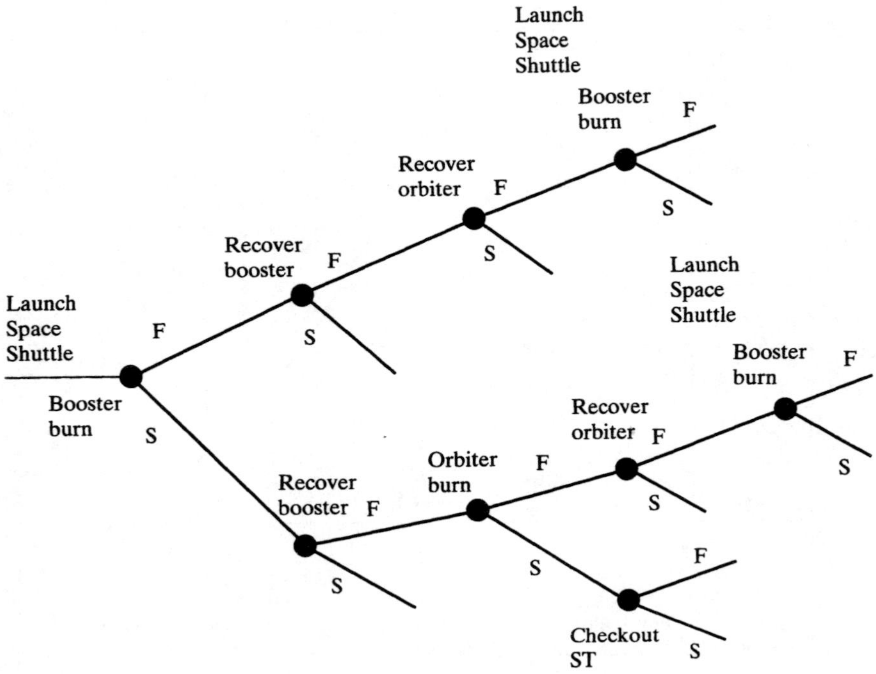

Figure 10.12 The scenario of Figure 10.9 partially converted into a decision/event tree.

We can illustrate the need for and use of short-hand notations in the use of decision and event trees by considering an example that involves the system design process itself. Suppose that we want to construct a process plant to produce a new material. We know of two processes by which the material can be produced, A and B. Process A is well defined, and we could design a plant given the knowledge we currently possess relative to this process. Process B is much less well defined, but it promises to be a substantially more economical process than A. The success of process B depends on two reaction rates which are not well known, but which could be determined by a test. However, the test is rather expensive. The design question is: should we proceed to design the plant based on process A, or should we conduct the test in hopes that we will be able to design a more economical plant based on process B? For simplicity, we will assume that we know precisely the cost of the material per unit for the case where it is produced by process A. For process B, however, we can only write the cost as a function of the two reaction rates, R_1 and R_2,which as yet we do not know. For simplicity, we will say that C_A is the cost per unit by process A, and the function $C_B(R_1,R_2,r)$ provides the cost per unit by process B. r is a random component in the functional form for C_B, which emphasizes that, even after determination of R_1 and R_2, it is not possible to predict C_B precisely.

We can draw the decision/event tree for this design activity using shorthand notation as shown in Fig. 10.13. Notice here that both R_1 and R_2 comprise infinite sets of possible outcomes. The test results in a two-dimensional infinity of outcomes, followed by a

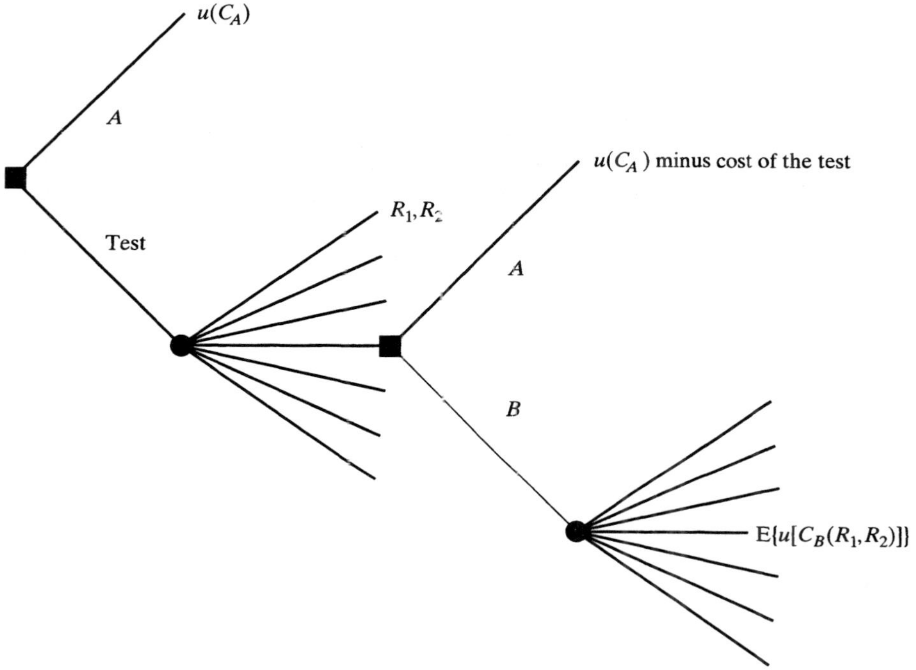

Figure 10.13 A decision/event tree showing a shorthand notation for events with an infinite number of possible outcomes.

decision to choose A or B. It is impossible to physically draw even a reasonable sampling of the actual decision tree. Also, it will not be possible to examine every branch of the decision tree, as there are an infinite number of branches. As a result, we must evaluate this decision tree using a Monte Carlo simulation to sample the branches of the decision tree. For simplicity, we will assume here that the test determines R_1 and R_2 precisely. In the case where the test merely reduces the uncertainty in R_1 and R_2, we can still use this methodology, but the mathematics becomes much more complex. In the case where uncertainty remains in R_1 and R_2, the post-test decision to choose option A or B must be made subject to conditional probabilities on the distributions of R_1 and R_2. It is relatively straightforward, using the Monte Carlo procedures outlined previously, to model this decision and then select the initial option, A or test, that provides the higher expected utility.

10.6 TWO DIFFERENT MONTE CARLO APPROACHES

There are two distinctly different approaches to the formulation of Monte Carlo models for the problems outlined in this section. They are the time-driven approach and the event-driven approach. We will discuss them here.

Time-Driven Monte Carlo Models

A time-driven Monte Carlo works with discretized time. It steps through time, moving from time t to time $t + \Delta t$. At each step, the status of all systems and activities is checked, decisions are made in the simulation, and the important variables are tracked. We can see how this works by means of an example. Consider the simple system reliability diagram shown in Fig. 10.14, which depicts a spacecraft consisting of a bus and a single instrument. Let us assume that we are interested in computing the expected utility of data collected from the instrument, and that utility accumulates in time at a marginally decreasing rate, such that we can write

$$u_M = a \ln t$$

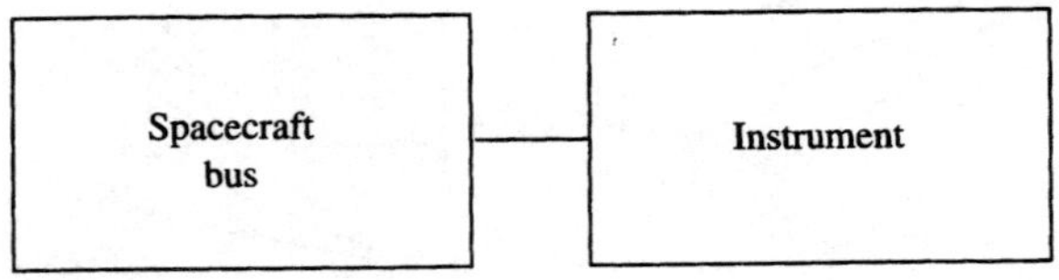

Figure 10.14 A simple reliability diagram to illustrate alternative Monte Carlo approaches.

A time-driven Monte Carlo model that computes $E\{u_M\}$ is shown in Fig. 10.15. The model begins at time zero and steps through time in increments of Δt. At each time step, the model determines whether either subsystem has failed. It does this by generating a random number for each subsystem and comparing that random number to the computed probability of failure during the time interval. If the random number is less than the probability of failure, the subsystem is taken to have failed. The probability of failure of subsystem i during time interval j, which occurs at time $t = j\Delta t$, is given approximately (for small Δt) by

$$P_{ij} = \frac{f_i(t_j)\Delta t}{\int_{t_j}^{\infty} f_i(t)\, dt}$$

where $f_i(t_f)$ is the probability density of the failure rate for subsystem i. Note that this probability is conditioned on the subsystem working at the beginning of the time interval. Any decisions that are made during the operation of a system can be simulated in the Monte Carlo simulation, and we see that the time-driven approach is easy to follow and straightforward to code.

The major complication in the application of the time-driven Monte Carlo approach is in computing the probabilities of failure. You must be careful to compute the proper

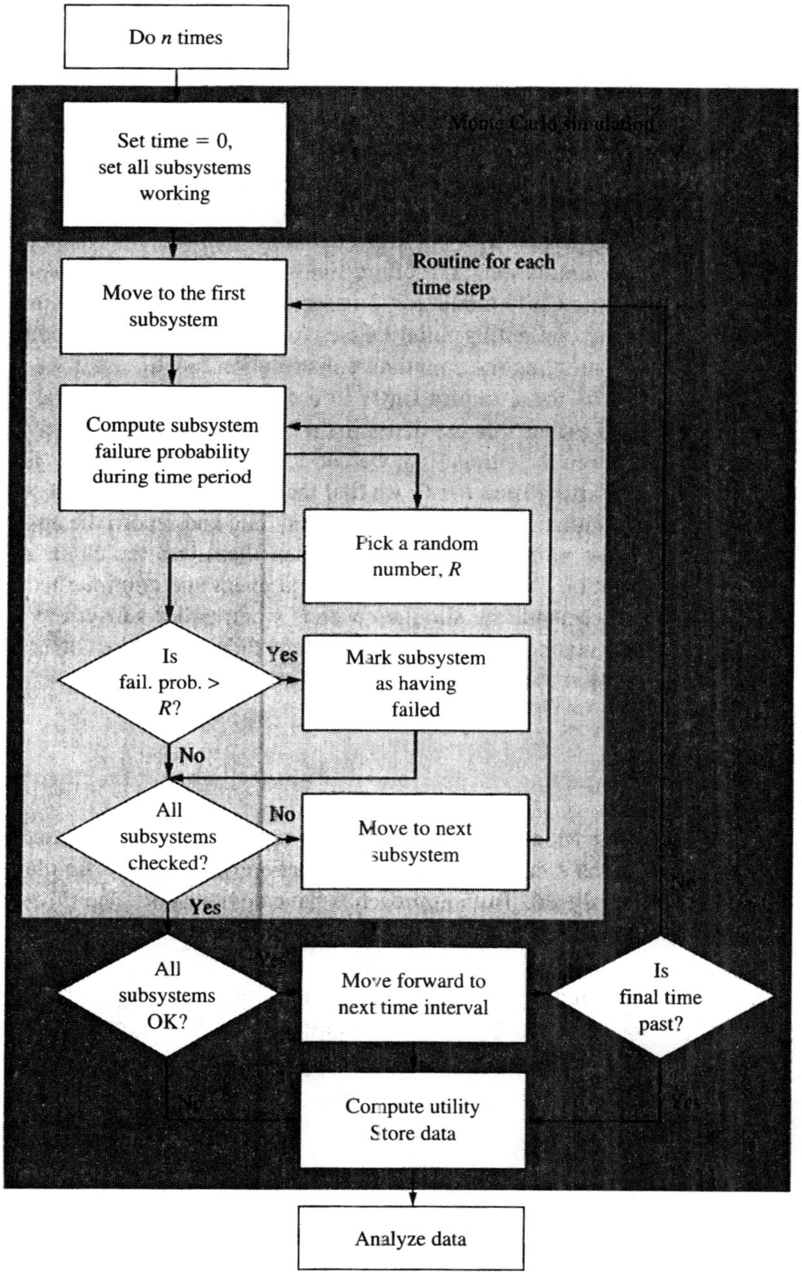

Figure 10.15 A flowchart for a time-driven Monte Carlo simulation.

values of P_{ij}. Generally, these are conditional probabilities, and you must be careful to take the conditions into account.

Event-Driven Monte Carlo Models

The event-driven Monte Carlo approach is structurally different from the time-driven approach. This approach does not discretize time. Instead, all systems are checked at the beginning of the simulation, and failure times or *event* times are computed for each. An event-driven Monte Carlo model for the case of Fig. 10.14 is shown in Fig. 10.16. In this case, we begin by computing failure times for each of the subsystems in the spacecraft. This is done by sampling the cumulative distribution function, not the probability density function. We store these failure times in a future event clock. The future event clock keeps track of all events that are determined but that have not occurred yet in the simulation. At any time in the simulation, we can look at the *future event clock* to find the next event in time. Starting from $t = 0$, we find the time at which the first subsystem fails. We then move the simulation clock forward to that time and record the appropriate quantities. Next, we generate new future events and enter them into the clock as appropriate, then move the simulation clock forward to the next event and continue the simulation. In the example case presented, the simulation ends when either subsystem fails or if the final mission time is reached without a subsystem failure. Having determined the mission time, the simulation then computes the mission utility.

Comparison

In either case, the Monte Carlo simulation is repeated many times, and data stored on the particulars of each simulation. Upon completion of the set of simulations, the data collected can be analyzed. Both approaches have advantages. The time-driven approach is conceptually simple, it uses the pdf directly to compute the probabilities of system failures, and it can be visualized as a process in time that equates to the actual physical process to which the system in question will be subjected. On the other hand, the event-driven approach requires far fewer computations, since it seeks out the failures immediately and steps directly to them. Also, the event-driven approach does not discretize time and, therefore, provides better resolution. In general, the event-driven simulation will be more efficient in terms of computational time. But both approaches should yield comparable results.

10.7 PROBABILISTIC RISK ASSESSMENT

Probabilistic risk assessment (PRA) is a tool that was developed about 20 years ago, principally for the assessment of the threat to society posed by nuclear power plants. It is similar to expected utility analysis, except that outcome states are not converted into utilities

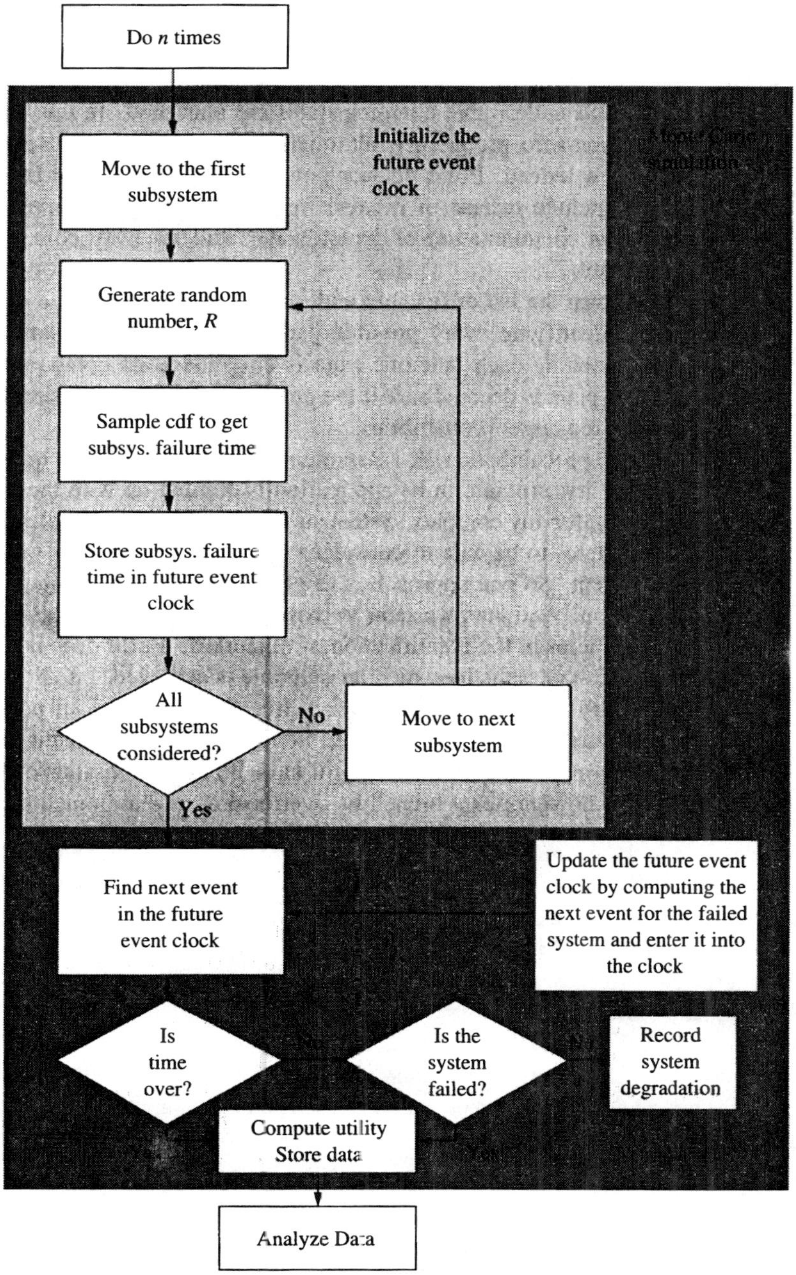

Figure 10.16 A flowchart for an event-driven Monte Carlo simulation.

and the analysis focuses on the potential outcomes that are significant threats to human health, economic operation of the plant, a hazard to the ecology surrounding the plant, and so on.

First, feasible undesirable outcome states are identified. In this context, *feasible* is taken to mean non-zero probability, although the probabilities of some feasible states might be very low indeed. For a nuclear power plant, examples of feasible undesirable outcome states include release of radioactive material into the atmosphere, core meltdown, radioactive contamination of groundwater, and runaway core and breach of the containment vessel.

Second, given the list of feasible undesirable outcome states, a decision/event tree is constructed, identifying every possible path to each given outcome state. Then, the probability of reaching each outcome state is computed and compared against a target probability. The plant is deemed safe if the probabilities of all undesirable outcome states is below the desired target probabilities.

In concept, probabilistic risk assessment is very simple. It is quite something else in practice. The first problem in its application is coming up with the list of undesirable events. Certainly, for any complex system such as a nuclear power plant, there are far too many outcome states to be able to consider them all, particularly if we distinguish carefully between them. So one approach is to group outcomes into broad classes: Class I, Class II, and so on. And then we refer to these as a Class I accident, a Class II accident, and so on. This helps in the identification of outcomes. But it does not particularly help in the identification of paths by which an outcome is achieved.

Next, we try to draw a decision/event tree that identifies all possible paths to all important outcomes. This is perhaps even more difficult than identifying the classes of outcomes of importance. Such a tree might have literally thousands of branches. And it is important that no significant branch be overlooked, as that branch might be the dominant one in determining the probability of a specific outcome.

Having drawn the event tree, probabilities must be assigned to the outcomes of each event. One thought here is that, by separating a major accident into a sequence of events, the accident itself might have a near zero probability, but each independent event can have a much higher probability. An event might be a valve failure or pipe breakage. We have enough experience with these events to assess their probability of occurrence. Then it is system redundancy and safety measures that cause these not-so-rare events to combine into the extremely rare catastrophic accidents that are the subject of our assessment. If we have all the possible paths to an accident and all the probabilities for every event on every path, then it is a straightforward, albeit perhaps large, job to compute the probabilities of each class of accident. But now we encounter the next major problem. In general, we assume that the probabilities of all events in the decision/event tree are unconditional probabilities.

For example, the probability of a valve failure is simply the probability of a valve failure. It does not change if a pipe fails. Opponents of the PRA methodology often cite this as its Achilles heel. Suppose a valve is in a system to shut off a fluid flow in the event that a pipe breaks. It can be argued that the conditions to which the valve is subjected are different depending upon whether the pipe is broken or not. Therefore, it does not seem unreasonable that the probability of a valve failure should be conditional upon whether the pipe has failed or not. The problem is, we rarely have the data necessary to estimate such conditional probabilities and, even if we could, using them could vastly complicate the analysis. The problem of conditional probabilities leads to thoughts that a system can have systemic problems that are overlooked by the PRA methodology. Thus, in general, a PRA is thought to provide a lower bound of the risk posed by a system.

Another problem with PRA is the determination of the target probabilities that might be acceptable. One chain of thought is that the system is unsafe if there is any non-zero probability of catastrophic accident. Then, even the existence of a plausible accident path in the decision/event tree is enough to disqualify the system. But this approach does not make sense, because it invalidates many well-accepted systems from use. Another approach is to examine alternative systems. For example, if coal power plants are an accepted form of electric power supply, then a nuclear power plant should be at least as safe. And typically, since the nuclear plant is a new technology, we might ask that it be safer by a margin of, say, two. Again, this sounds simpler than it really is. The problem is that the classes of accidents for the two plants can be inconsonant. How can we compare, say, the threat of global warming to the threat of a release of radioactive material that contaminates a large area of land? Still another approach is to compare the risks of the activity at hand to general background risks. This form of comparison sometimes leads to a conclusion like: the risk of cancer from a nuclear power plant should not exceed one-tenth of the risk of cancer from cosmic radiation. These kinds of comparisons are also difficult to deal with in the PRA format. Still another method might be comparison with another similar activity. A new nuclear power plant should be at least as safe as those already in operation, or preferably safer by, say, a factor of two. Despite all the logic that people try to employ in their safety arguments, there is no right method, and the PRA methodology, while useful in providing guidelines, falls short of being totally authoritative.

10.8 AN ALTERNATIVE TO PROBABILISTIC RISK ASSESSMENT

One thing that we as systems engineers need to understand is that we can seek to improve the safety of a system to the point where it results in the imposition of higher risks on society. Take air travel as an example. Today it is far safer to travel across the United States by commercial air carrier than it is to drive. Still, we would like to see air travel even safer. But safety comes only at some cost. If we demand high enough safety, we will drive the cost of air travel up to the point where people will choose to drive. Then, they will be exposed to a greater risk as a result of our attempt at improving the safety of air travel.

Likewise, if we compare nuclear power to coal or solar energy (taking the broad definition of solar energy here to include wind energy, hydropower, and other sun-driven forms of energy) we see that nuclear energy has enjoyed a far better safety record than its alternatives. Still, we would all like to see it safer yet. But, if we insist on something approaching absolute safety, we will force the price of nuclear power up, resulting in the choice of a less safe alternative.

This notion of overall safety leads to a method for the safety assessment of major new systems. Rather than demanding that they be unduly safe, we can take the approach that we are seeking to better our life on the planet and improve safety overall on the margin. Now we can see that there is an optimal level of safety. That level trades off absolute safety of the system at hand with the resulting cost of the system and, hence, its use in the marketplace. If we have a safe way of doing something, rather than to force it to be even safer yet, the systems approach is to choose the level of system safety that provides a level of use in the marketplace that maximizes overall safety. This optimization can be accomplished by comparing level-of-safety alternatives in the larger context of a total system analysis. For example, in the case of nuclear power, we might compare the total safety for all society of different levels of nuclear power plant safety, each of which results in different plant costs and, hence, different levels of demand for nuclear power versus alternatives.

10.9 SUMMARY

In the previous chapter, we found that we cannot properly design a system if we cannot take into account its operation. The operation of virtually all systems involves the chance of system failure. Thus, system design requires that we take into account system reliability. This chapter is devoted to developing necessary methods of analysis. To begin, we found that system reliability is easy to define only for very simple systems that tend to be either fully operational or totally failed. Most systems, however, have degraded modes of operation that reduce the system utility, but do not destroy all system functionality. Based on this notion, we looked at the computation of system expected utility, taking into account the possibility of failures. We developed two important methodologies. The first involves drawing a system reliability diagram, followed by Monte Carlo computation of system expected utility. We saw how this method works in the case of a system that functions over an extended period of time. The second method involved cartoons and scenarios. This method applies to systems that operate through a set of discrete events. We showed how to use Monte Carlo analysis to evaluate these systems. Then we expanded on these methods to include the case where decisions are embedded in the operation of a system. Finally, we covered the method of probabilistic risk assessment, and described some of the shortcomings of the method.

The underlying thesis of this chapter is that it is necessary to model the operation of a system in order to properly design the system in the first place. Secondly, it is the thesis of this chapter that system operation is amenable to modeling using standard Monte Carlo techniques applied in innovative ways.

REFERENCES

1. Greenberg, J. S. and G. A. Hazelrigg, A Reliability, Cost and Risk Analysis of Establishing and Maintaining a Space Communications Satellite Network, AIAA Paper No. 73-582,

AIAA/ASME/SAE Joint Space Mission Planning and Execution Meeting, Denver, CO, July 10–12, 1973.

2. Greenberg, J. S. and G. A. Hazelrigg, Methodology for Reliability-Cost- Risk Analysis of Satellite Networks, *Journal of Spacecraft and Rockets*, vol. 11, no. 9, September 1974, pp. 650–657.

PROBLEMS

1. What is the MTBF of a system element whose failure rate pdf is given by the following?

$$f(t) = 2\,\frac{t - t_a}{(t_b - t_a)^2} \qquad \text{for} \quad t_a \le t \le t_b$$

2. Suppose the utility of a system is given as a function of the time that the system remains operational as follows:

$$u(t) = \log t$$

Further, suppose that the system has a failure pdf that is memoryless with an MTBF of 4. What is the expected utility of the system? Now consider redesigning the system to provide redundancy. The system is redesigned to include two parallel elements, each with MTBF 3.5, and a sensor/switch combination that has a reliability of 0.9. That is, the sensor/switch will detect a failure and switch to the redundant element with probability 0.9. What is the incremental utility of this redundancy design?

3. Determine the MTBF for a system that consists of two elements, A_1 and A_2 in series. Both A_1 and A_2 have memoryless failure pdf's with $\mu_1 = \mu_2 = 1{,}000$ hours.

4. Determine the MTBF for a system that consists of two redundant elements, A_1 and A_2. In series with A_2 is a sensor and switch. Both A_1 and A_2 have memoryless failure pdf's with $\mu_1 = \mu_2 = 1{,}000$ hours when on. Assume that $\mu_2 = \infty$ when A_2 is off, that the probability of the switch working is 0.98, and for the sensor, $\mu_{sensor} = 2{,}000$ hours.

5. Refer to the notation of problem 4. If the utility of a system is given by

$$u - \ln\left(\frac{t}{1{,}000}\right)$$

what is the expected utility of each of the following systems, where $\mu_1 = \mu_2 = 1{,}000$ hours:

a. A system consisting only of element A_1.
b. A system consisting of A_1 and A_2 as redundant elements.
c. The system of problem 4.
d. If the utility for money, C, is

$$u = \ln\left(\frac{C}{1{,}000}\right)$$

what is the redundancy of cases (b) and (c) worth?

6. Solve problem 4, but take $\mu_2 = 2{,}000$ hours when A_2 is off.

7. Consider a system in which A_1 and A_2 are redundant elements each with memoryless failure probability distributions, and with $\mu_1 = 500$ hours and $\mu_2 = 1{,}000$ hours. A switch and sensor

is used to turn on the redundant element when the primary element fails. The switch works with probability 0.98, and the sensor has a memoryless failure probability with $\mu_{\text{sensor}} = 2{,}000$ hours. If system utility is given by

$$u = \alpha t$$

which element, A_1 or A_2 should be the primary, that is, which element should be on first? What is the difference in expected utility between the preferred choice of system operation and the nonpreferred choice?

8. Consider a system in which two elements, A and B, are in series. Assume each element to have a memoryless failure distribution with $\mu_A = \mu_B = 1{,}000$ hours and with

$$u = \alpha e^t$$

What is the incremental expected utility obtained if element A is redesigned so that element B can be eliminated? Assume that the redesign of A causes its MTBF to be reduced to 750 hours.

9. You are given a system with a memoryless failure distribution and an MTBF of 100 h. The system is to perform 10 different jobs, each of which takes 10 h to perform. The utility of the jobs is as follows: $u_1 = 1$, $u_2 = 2$, $u_3 = 3$, $u_4 = 4$, $u_5 = 5$, $u_6 = 6$, $u_7 = 7$, $u_8 = 8$, $u_9 = 9$, $u_{10} = 10$. What order of performing the jobs maximizes the expected utility? What is the difference in expected utility between the best and worst orders for performing the jobs?

10. A system consists of three elements, A, B_1, and B_2. Both systems B_1 and B_2 require that A be functional in order to produce results. But B_1 and B_2 are independent of each other. All elements have memoryless failure distributions with MTBFs of 500 hours. The system utility is given by

$$u = 5u_1 + u_2$$

where u_1 refers to the utility produced by element B_1 and u_2 refers to the utility produced by element B_2. Take

$$u_1 = t$$

$$u_2 = t^2$$

What is the expected utility for the system?

11. A communications satellite has a 0.90 probability of successful launch and deployment in orbit. After deployment, it has a memoryless failure distribution with an MTBF of 15 years. The design life of the satellite is 10 years, and its wearout distribution is normal with mean 10 years and standard deviation 1 year.

a. Draw a curve that gives the probability that the satellite is operational as a function of time from launch.

b. If the satellite generates revenue at the rate of \$20,000 per hour, what is the expected present value of revenues that the satellite will generate over its lifetime?

c. What would be the expected value of a design change that would reduce the standard deviation of wearout time to zero?

d. What would be the expected value of a design change that would increase the mean of the wearout distribution to 11 years?

12. Consider a system that has two possible designs. In the first, elements A and B are arranged in series. A has a memoryless failure distribution with an MTBF of 10 hours, and B works with certainty for 6 hours, then fails. In the second alternative, the system consists of only one element with a memoryless failure distribution and an MTBF of 5 hours.

a. If system utility is given as ln (t), which system has the higher expected utility?
b. If system utility is given as t, which system has the higher expected utility?

13. You have been asked to design a system to deliver supplies to a remote base in Antarctica. This operation requires that supplies be delivered by ship to McMurdo base, and then by helicopter to the remote site. Weather is frequently a problem, and both the ship and the helicopter flights can be delayed. Draw a cartoon for this process.
14. For problem 13, draw the scenario diagram.
15. An airplane crash occurs because the pilot forgot to extend the flaps for takeoff. It is recommended that an interlock be installed so that, in the future, a pilot cannot advance the throttles for takeoff unless the flaps are extended. How might this interlock reduce safety?
16. An auto manufacturer adds an antilock braking system to its cars. Is their safety improved, worsened, or unchanged?
17. Do single engine or twin engine light planes have more fatal accidents as a result of engine failure?
18. You want to improve your safety while driving. What can you do?
19. Some airliners use a two-person crew, others use a three-person crew. Which is safer? Why?

11

Cost and Benefit Analysis

11.1 THE PURPOSE OF THIS CHAPTER

It is probably safe to say that no engineering project is ever undertaken without some prior assessment of both its cost and benefit, either formal or informal. Certainly, these two factors, perhaps more than any others, drive system design. Therefore, an understanding of them is key to understanding the objectives of system design and to the success of the system.

Frequently, we wish to minimize system costs and maximize system benefits. But ultimately these goals are contradictory, and benefits can be increased only if costs are allowed to increase, or costs can be decreased only if benefits are allowed to decrease. The question then is: how should we trade off costs for benefits in a design?

Another problem that we have encountered is that engineering system option spaces tend to be very large, and we are always interested in any potentially viable approach to paring the option space without extensive analysis. Sometimes, consideration of costs and benefits can help us pare the option space. For this reason alone they are worthy of study.

11.2 COSTS

Cost is a measure of the value of the inputs to a process. It is the sum of the economic prices paid for those inputs. *It is important to recognize that cost is not a physical parameter of a system.* Rather it is a measure of the worth that people place on the things that comprise a system or that go into its production. The inputs to any engineering system can be classified as labor, capital, and resources. Labor is the sum of human effort required to design, test, manufacture, operate, repair and maintain, inspect and regulate, and dispose of a system. Capital has been defined in Chap. 6 to consist of the goods produced by the economic system itself to be used as inputs for further production or services. And resources are what is provided by nature for use by man. They include land, air, water, minerals, and so on. Price, on the other hand, is the amount charged for something. Price includes earnings or profit (or loss), and it is not necessarily reflective of the cost of something. Price is usually higher than cost, but it can be lower.

We will distinguish two fundamentally different types of system costs: nonrecurring and recurring costs. *Nonrecurring costs* comprise the investment that is made in a system, to create and own the system. Nonrecurring costs are generally taken to be the costs that are incurred to reach operational status for a system. In concept, nonrecurring costs occur only once in the lifetime of a system, generally, but not always, at the beginning of the life cycle of the system.

Recurring costs, as their name implies, continue to occur or recur throughout the lifetime of a system. For example, for a building, the nonrecurring costs include the cost of land, architectural and engineering costs, and the cost of construction. Recurring costs include the cost of energy (heat, air conditioning, and lighting), the cost of repair and maintenance for the building, and the cost of operating the building (security, janitorial costs, etc.). For an automobile, the nonrecurring cost is the cost of purchasing the vehicle, perhaps including interest on money borrowed. The recurring costs include license fees, insurance, taxes, fuel, maintenance, and repairs. The nonrecurring cost may be spread over time if the vehicle is purchased with a loan that is repaid over, say, four or five years. Nonetheless, it is a cost that is incurred only once, and the decision to incur it is made the day the vehicle is purchased. This cost does not depend on the future use of the vehicle. On the other hand, the recurring costs are determined by use. License, taxes, and insurance costs are typically a function of time, and they are incurred each year that the vehicle is kept in use. Fuel, maintenance, and repair costs tend to be proportional to the distance that the vehicle is driven. Recurring costs can be, and generally are, influenced by decisions made after the acquisition of a system. The nonrecurring costs are largely determined at the moment of the acquisition decision.

Now let's return to the example of a building and consider the tradeoffs between nonrecurring and recurring costs. Many design options affect both nonrecurring and recurring costs. In the building, we could add insulation or better windows to reduce heating and air-conditioning costs as well as providing better temperature control. We also could use materials that require less maintenance or last longer. Each option will have a particular nonrecurring and recurring cost. For simplicity of comparison, we will reduce both the nonrecurring and recurring costs to their present values at a particular date, such as the system acquisition date. Then, for each option, we can plot the present value of recurring costs against the present value of nonrecurring costs. A typical plot of these variables is shown in Fig. 11.1. A natural goal is to reduce both nonrecurring and recurring costs simultaneously, but we are limited by the available technology. So the plot of Fig. 11.1 shows a technology frontier that is comprised of the system options that simultaneously minimize both nonrecurring and recurring cost. Because this plot shows present values of costs, the optimum system design is the design that minimizes the sum of the two costs, namely it is the design that lies on the minimum total cost line as shown.

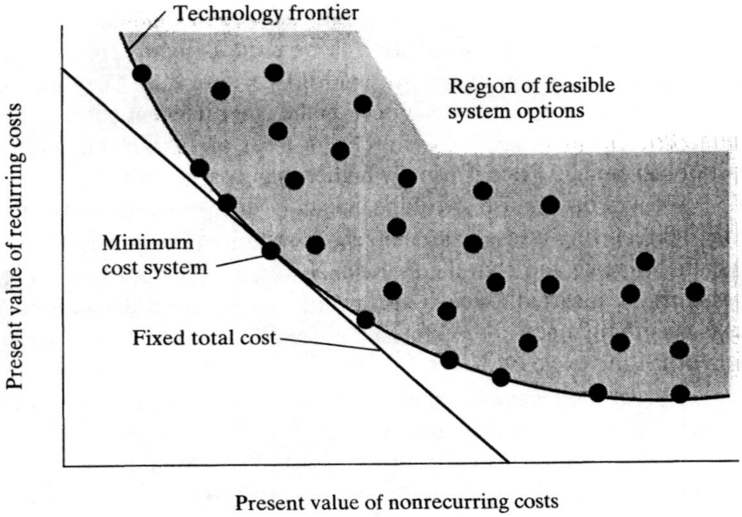

Figure 11.1 The concept of a technology frontier in the space of nonrecurring and recurring costs.

The equation of this line is

$$C_{\text{total}} = C_{\text{nonrecurring}} + C_{\text{recurring}}$$

It is important to recognize that the optimum system designs, namely those designs that comprise the technology frontier, depend on the discount rate that is used in the determination of present values. If we change the discount rate, in general, different system options will emerge as the optimal options. Furthermore, the determination of recurring costs frequently depends on the prediction of future costs for various commodities, such as heating fuel, electricity, and labor. Therefore, the optimum system options might depend rather strongly on the assumptions made with respect to the future.

One additional category of cost is important in system design. This is *sunk cost.* Sunk costs are costs to which we are already committed. They are not costs that will occur as a result of future decisions. An example shows how sunk costs can influence system design. Consider once more the building. Suppose that we had designed the building and then ordered some long lead-time items. Let's assume, for example, that windows are such an item, and that we ordered a relatively expensive, energy-conserving window in conjunction with the use of a particular heating system. Then, some time after we have committed to paying for the windows, we find that a new heating system has been developed that will reduce heating costs substantially, and we change the design to use this system. No longer would the expensive windows be the cost-minimizing choice. But they have already been ordered, and we have to pay for them whether we use them or not. We say that their cost is sunk because we have made the commitment to pay for them. Therefore, their use in the design is, essentially, free from the point of view of future decisions. The concept of sunk cost is illustrated by the decision tree of Fig. 11.2.

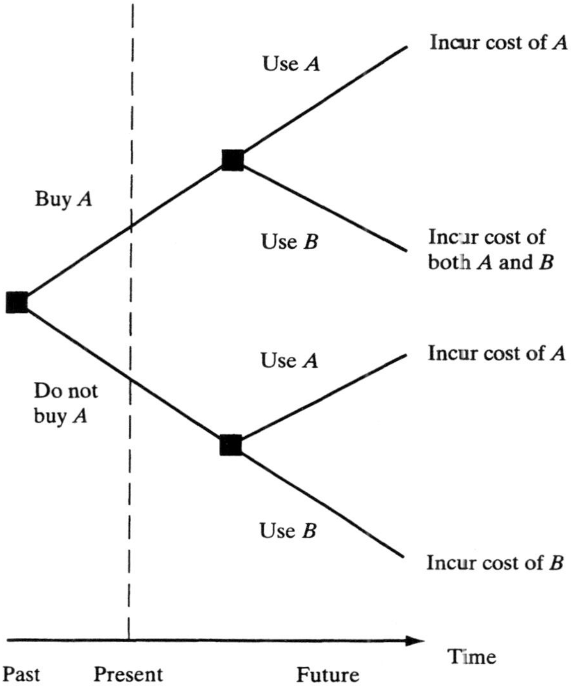

Figure 11.2 The concept of sunk costs illustrated by a decision/event tree.

Sunk costs can have profound effects on system design. Consider an automobile manufacturer who has a very substantial investment in the production of a particular vehicle design. The vehicle design uses dies for stamping body parts and specialized tooling for the manufacture of each of the thousands of parts that comprise the entire automobile. Design changes might be strongly desired, but inhibited by the cost of retooling the entire plant. The fact that designs that strongly resemble the current design can be produced without the extensive retooling costs that would be required of more radical design changes make the more radical design changes uneconomical. This can lead to a failure to update a design, and eventual failure of the product line as more modern competing designs emerge. It also becomes apparent that, sometimes, we must consider not only design changes in the present design, but future design changes as well when evaluating design options. Choices we make today will result in sunk costs that will affect future design choices.

11.3 MODELING SYSTEM COST

There are many different approaches to the estimation of system costs. Regardless of the method of cost estimation used, however, you must always keep in mind that cost estimating is a form of forecasting, and it is susceptible to all of the vagaries and uncertainties of predicting the future. With this in mind, we will discuss two different approaches to cost estimation. We call these the top-down approach and the bottom-up approach.

Top-Down Cost Estimating

Top-down cost estimating begins at the system level and adds detail by sequentially dividing the system into smaller and smaller parts. For example, suppose we desire to estimate the cost of a new commercial airplane. One approach is to examine the cost of already-developed airliners and characterize their costs according to some set of parameters. The simplest approach might be simply to note that cost depends on gross takeoff weight. We might then estimate the weight of the airliner and, from that, determine its cost. The relationship between system weight and system cost is expressed as an equation and it is called a cost estimating relationship (CER). The CER approach to cost estimating has been widely developed in the United States, particularly for aerospace systems. It is sometimes called parametric costing.

One significant advantage of the CER approach is that it separates the functions of engineering and cost estimating. Engineers tend to be capable of estimating the physical parameters of a system, such as its mass, volume, power consumption, etc. They are less capable of estimating cost at such an aggregate level as an entire system. The CER converts the estimates of the physical parameters of a system into a cost estimate. To begin, data are gathered for similar systems that have been built in the past. These data include physical parameters of the systems and their cost. They are then plotted, and a regression analysis is performed on the data. CERs have been developed at varying levels of sophistication. Done properly, the data must first be corrected by conversion to constant-year monetary units. They must also be corrected for any obvious vagaries, such as a major undercharge or overcharge for capital items resulting from accounting practices that do not correct for inflation. Then the curve fit represents an estimate of cost given a value of a physical parameter, such as weight. The procedure is shown in Fig. 11.3.

The CER method is often extended to finer levels of detail. At the next level, the system may be divided into subsystems and CERs provided for each subsystem. Sometimes, additional factors are added to the CER to account for the complexity or "newness" of the system being costed. Always, however, the CER proceeds from physical parameters to cost.

The CER method relies on several assumptions. It is important to understand them and to guard against using the CER methodology in ways that violate these assumptions.

1. In gathering data for the estimation of CERs, only data from successfully completed programs are included. This is reasonable because the cost to completion of a failed program cannot be measured—it simply does not exist. Because of this, however, an assumption inherent in the CER method is that it must be an absolute certainty that the program can be completed successfully. This means that CERs should be applied only to

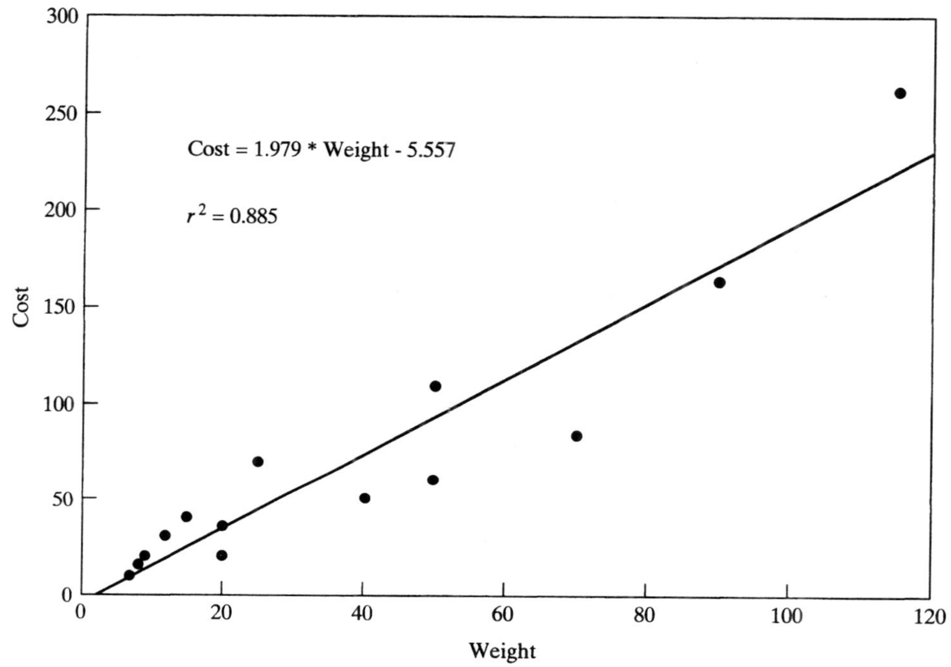

Figure 11.3 An illustration of the Cost Estimating Relationship (CER) methodology.

estimate the cost of systems that use established technology, that is, within the set of things that have already been built.

2. In gathering data for the estimation of CERs, actual costs and physical system parameters are taken, not estimates of these parameters made early in the program. Thus, it is assumed that physical parameters can be estimated with precision and certainty, even as early in a program as the time when the first cost estimates are needed. This again reinforces the notion that CERs apply only to systems that are comprised entirely of tried and proven technologies.

3. In using cost and physical parameter data as the basis for estimation of a CER, standard regression analysis techniques are employed wherein the independent variables are assumed to be determined precisely, and all estimation uncertainty lies in the dependent variable. Thus, in order to properly use CERs it must be assumed that the system physical parameters are known precisely.

These conditions are rarely met in practice, and therefore, strictly speaking, the CER method of cost estimating has limited application. Still, it is commonly used well beyond the bounds of these assumptions, and extraordinary efforts are made to correct the problems that arise from such misapplications.

In developing and using CERs, you should ask the following questions: Is the CER useful? Does the correlation coefficient for the CER suggest strong correlation? Is the correlation transferable to the universe from which the sample points came? What are the effects of including more variables to explain cost? Does the existence of correlation between variables in a CER imply that a cause-effect relationship exists?

One of the major shortcomings of the CER methodology is that it is a descriptive approach to costing, not a predictive approach. Therefore, it cannot be used to predict the cost advantages of new technologies nor can it resolve cost differences between two different designs that result in similar gross physical parameters. For example, it cannot predict the cost difference between two systems designed with subtle dimensional differences, but which have similar weight, volume, etc. In fact, CERs often give counterintuitive results. For example, if a weight constraint on an aerospace system is relaxed, that is, if the system weight is allowed to grow, we would expect that the relaxation of the constraint would enable the cost of the system to be decreased. But a CER that relates cost to weight would predict that relaxation of the constraint would increase cost.

Bottom-Up Cost Estimating

Whereas top-down cost estimating begins at the system level and proceeds to form cost estimates at finer and finer levels of detail, bottom-up cost estimating begins at the finest level of detail in a system and aggregates cost upward from that point. Basically, bottom-up cost estimating seeks to estimate the costs of all the components, elements, or subsystems that comprise a system; the cost of integrating or assembling these into a whole system; and to develop the system cost as the sum of component costs. The steps one follows to do bottom-up cost estimating are the following:

1. Break the system into its elements, including assembly, test, etc.
2. Estimate the labor, capital, and resources required for each element in the system.
3. Build up the element costs from the requirements imposed on each element. System models can be used to assist in cost estimating, and these models might include models of the manufacturing process, testing and quality assurance, and so on.
4. Build up the system costs by aggregating the element costs, including costs for system integration, test, etc.

The bottom-up cost estimating process is depicted in Fig. 11.4.

CERs comprise a descriptive model for system costs. To develop a CER, we make use of historical data to measure the cost of similar systems previously built and operated. We then disaggregate these data into *explanatory* variables, which involve physical parameters of the system such as weight, volume, power consumption, etc. Bottom-up cost estimating is typically more of a predictive modeling process, using physical models to predict cost incurring activities. Each activity is then costed separately and summed to obtain the system cost.

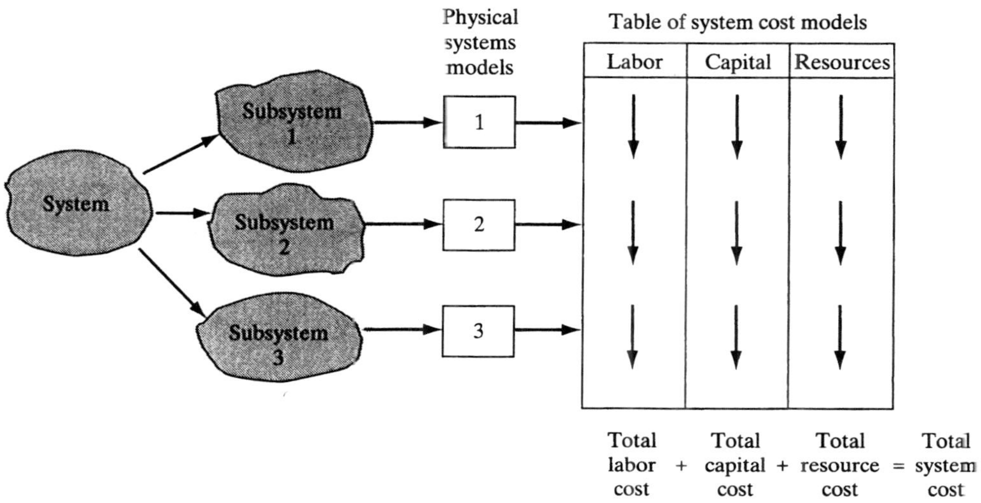

Figure 11.4 A bottom-up cost estimating procedure.

Cost Learning

Cost learning is defined as the reduction in system cost that derives from the experience of building and operating systems of the same type. If you have ever bought something that required assembly or installation, you know that it can take quite a bit of time to do the assembly or installation properly. Much of the time you spend learning about the system and how it should be assembled rather than actually doing the assembly. If you buy two such systems, you know that the assembly of the second system proceeds much more quickly than the first. This is because you have *learned* how to perform the assembly or installation. If you continued to assemble similar systems, you would continue to learn methods or tricks to accomplish the assembly even more quickly. And if you needed to assemble a large number of similar systems, you might invest in equipment, such as an electric screwdriver or a jig, to further speed your work. The process of continually improving production with experience results in cost reductions that are referred to as *cost learning*.

The commonly accepted model for cost learning is based on the notion that each successive doubling of production leads to an equal reduction in cost. Thus, the cost of the *i*th unit is given as a function of the cost of the first unit, C_1, by the following relationship:

$$C_i = C_1 i^{\alpha}$$

where α is called the learning exponent given by

$$\alpha = -1 + \frac{\log_{10} r - 1.699}{0.301}$$

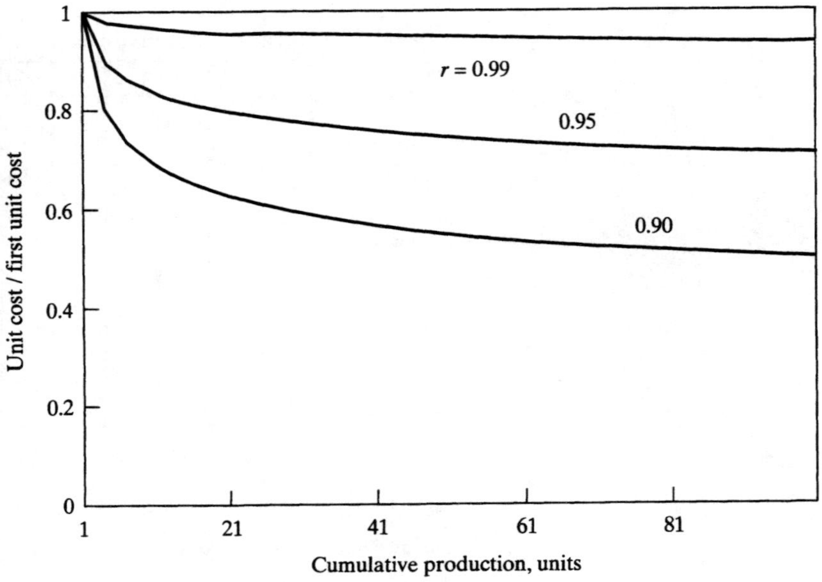

Figure 11.5 Illustrative cost-learning curves.

where r is the cumulative average learning rate in percent. Some typical learning curves are shown in Fig. 11.5.

Cost Spreading

Many costing methodologies provide only estimates of total cost for a system, a subsystem, or an activity when, in fact, the cost is incurred over an extended period of time. For example, consider the construction of a power plant that might take place over, say, five to seven years. An estimate only of the total cost is not sufficient for many purposes. The present value of this cost may be substantially affected by the fact that it is incurred over an extended time. Therefore, we need to find an approach to estimating how costs *spread* over time.

Suppose that the total cost for a system is given by C_T, and that the expenditure of this sum occurs over a period of T years. Now let the rate of expenditure be given by $c(t)$, such that

$$c(t) = C_T g(t)$$

where

$$\int_0^T g(t)\, dt = 1$$

So we see that $g(t)$ has the properties of a probability density function. We can change the limits of integration by making the transformation

$$\zeta = \frac{t}{T}$$

and we obtain

$$\int_0^1 f(\zeta)\, d\zeta = 1$$

where

$$f(\zeta) = Tg(T\zeta)$$

There is a class of probability density functions that have this property. It is called the beta distribution, and it has the form

$$f(\zeta) = \frac{\Gamma(\alpha + \beta)}{\Gamma(\alpha)\Gamma(\beta)}\, \zeta^{(\alpha-1)}\, (1 - \zeta)^{(\beta-1)} \qquad \text{for} \quad 0 \le \zeta \le 1$$

Γ refers to the gamma function, where, as shown in Chap. 3

$$\Gamma(i) = (i - 1)! \qquad \text{for integer } i$$

Example beta distributions are shown in Fig. 11.6.

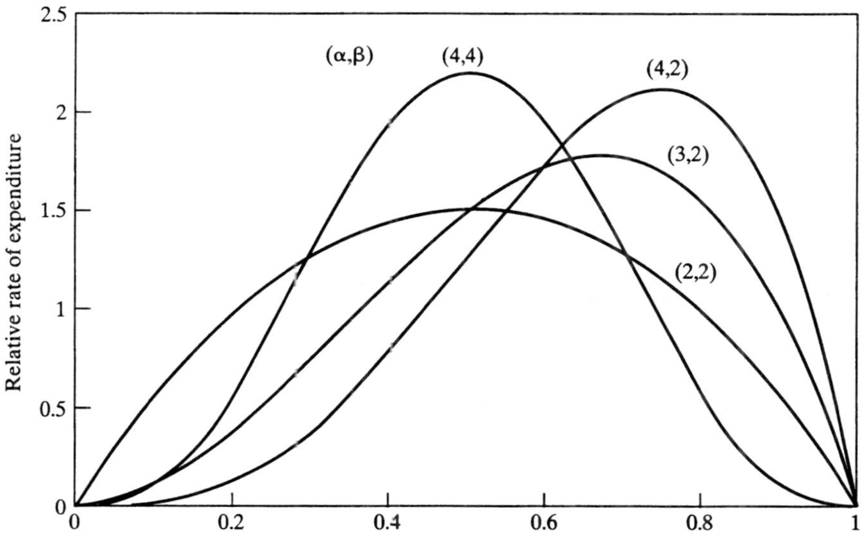

Figure 11.6 Illustrative beta distributions.

Life Cycle Cost Estimating

The life cycle of a system is comprised of several phases, which may differ from system to system. Typical phases in a system life cycle are the following:

1. Research and development to lay an adequate technology base
2. System design and development
3. System prototype and test
4. Manufacturing
5. Distribution
6. Sale
7. Use, including inspection, maintenance, and repair
8. Disposal
9. Installation of a replacement system

All of these phases are amenable to modeling and cost estimation using methods outlined in this book. The process of life cycle cost estimating is to recognize these and any other phases in the life cycle of a system, to cost every phase, and to sum costs across all phases to obtain the total life cycle cost.

Opportunity Costs

The *opportunity costs* of a particular project are defined as the net benefits that are foregone by undertaking that project rather than the next best alternative project.

11.4 PROBABILISTIC COST ANALYSIS

Before developing a methodology for probabilistic cost analysis, we should take note of the fact that cost estimates are frequently in error, and almost invariably they err on the side of being low. There are several reasons why this is the case. First, almost all system cost estimates are obtained by adding together component or subsystem costs. So a total system cost might be comprised of the sum of as many as 100 or even 1,000 subsystem costs. Subsystem cost is always a positive number—we never see negative subsystem costs. And there is almost always some uncertainty in the cost of each subsystem. For these reasons taken together, our knowledge of subsystem costs can be represented as a probability distribution that is virtually guaranteed of being a skewed distribution with a mean value higher than its most likely value, as shown in Fig. 11.7. Failure to treat costs using the mathematics of probability theory as described in Chap. 3 is mathematically incorrect, and it leads to underestimation of system cost.

Second, cost estimating is a form of forecasting and can, in itself, be a self-fulfilling prophecy. If a system is divided into subsystems, and each is allocated a budget, then

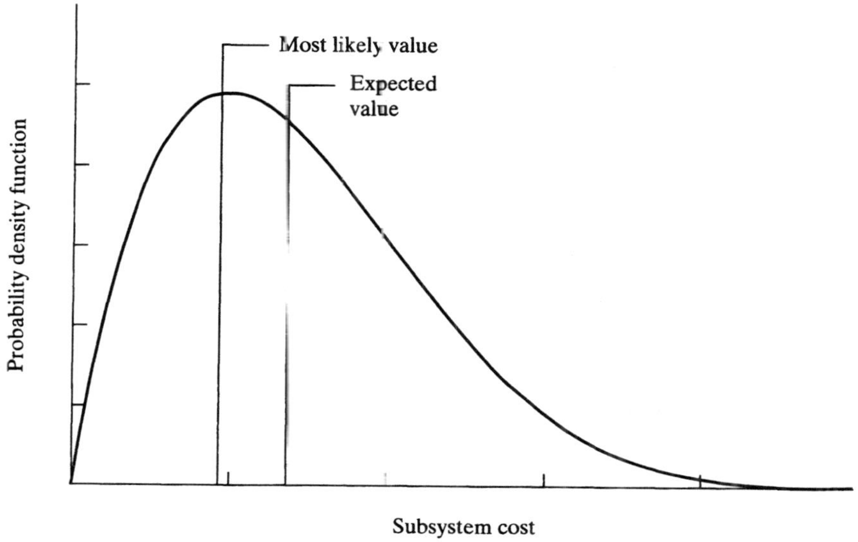

Figure 11.7 A typical subsystem cost probability density function.

it is generally safe to say that the total system will cost more than the sum of the estimated subsystem costs. The reason is this: each subsystem design manager will plan to spend his or her budget. So, if cost underruns the estimate, the manager will add to the design effort—this can always be done, and it frequently results in an improved design—in order to spend the entire budget. But if the manager overruns the budget, the overrun must be absorbed by the overall system budget. Only strong incentives for subsystem managers and a lack of information on the total available budget tend to prevent such cost escalation.

Third, once the sunk costs of a project begin to mount, there is growing pressure to finish the project regardless of the cost. Unless a reward system is in place to provide incentives to keep costs down, the sunk cost arguments developed above provide ample reason for cost overruns.

These arguments in themselves spell trouble for the cost estimating process. Indeed, it is a process that can, with proper incentive, feed on itself to increase system cost. We see that cost estimation in itself is not enough, but rather that the cost estimation process must be accompanied by an incentive scheme that encourages management to meet or better cost targets. With this caveat, we will now develop a probabilistic cost estimating methodology.

We begin by assuming that we do not know, in advance of building a system, the physical parameters of the system precisely. Therefore, we will need a system model to define the physical parameters for use in a probabilistic model. We can illustrate the procedure for doing this by means of a simple example. Suppose we want to build a solar energy system based on a new type of solar cell, and we want the system to produce a fixed amount of power, say 1 megawatt average power output. The question we face is: what will it cost to build and operate such a system? We will assume a system in which the solar arrays are fixed to the ground, rather than rotated to point toward the sun. We take the average solar flux to be given by Φ, which in a relatively clear area would be on the order of 1 kW/m^2. Because the solar arrays are fixed, however, they see on average a fraction of this flux that is only $\mu/2\pi$. The factor μ accounts for the fact that the local noon elevation of the sun changes from day to day throughout the year, and the factor of 2π accounts for the fact that the earth rotates on its axis so that the sun is not always directly overhead. We must further decrease the flux to account for weather. We denote this decrease by the fraction ν. Thus, the average solar flux shining on the solar arrays is $\Phi\mu\nu/2\pi$. The solar cells do not cover the entire area of the solar arrays as they must be separated from each other. We denote the fraction of the total array area that is an active solar cell by ε. Finally, we denote the conversion efficiency of the solar cells by η. Then the average power output per unit area, p, is given by

$$p = \frac{\Phi\mu\nu\varepsilon\eta}{2\pi}$$

Now, to get the desired average power output, P, we require a total array area given by

$$A = \frac{P}{p} = \frac{2\pi P}{\Phi\mu\nu\varepsilon\eta}$$

In our simple example, this equation describes the system in terms of its area, which in turn depends on 6 parameters that we must estimate: P, Φ, μ, ν, ε, and η. Φ is uncertain because of our inability to measure solar flux precisely, and because it depends on factors such as pollution, which could vary over the lifetime of the system. μ is a geometric constant, which we can compute rather precisely, and we will not consider further uncertainty in this variable. ν is a variable that depends on future weather. Over a long period of time, we can estimate ν rather accurately, although it is possible that there are long-term trends that we might fail to properly account for. Finally, ε and η relate to the technology of the solar cells. Most of our inability to predict A during the early phases of the design process will probably be the result of our inability to predict ε and η. The final physical parameter of the system that we will deal with in this model is the land area, A_L, that the system will occupy. Because the solar arrays must be oriented toward the sun, they will not lie flat upon the ground. Therefore, the area of the solar arrays will be somewhat less than the area of the ground that they cover.

$$A_L = \frac{A}{\omega}$$

where ω is a geometric factor computed with precision.

Next we turn to the cost model. The major elements of the system include land, structure, solar arrays, and wires. We will take the cost of land to be proportional to the area of land needed for the system, and the cost of structure, solar arrays, and wires to be proportional to the area of the solar arrays. We will also take the operating cost to be similarly related to area. The cost model is given in Table 11.1. The cost components listed in this table relate costs to the physical parameters, which are in turn dependent on the

TABLE 11.1 A SOLAR POWER SYSTEM COST MODEL

Subsystem	Capital cost	Operating cost
Land	$\alpha_L A_L$	$\beta_L A_L$
Structure	$\alpha_S A$	$\beta_S A$
Solar arrays	$\alpha_A A$	$\beta_A A$
Wires	$\alpha_W A$	$\beta_W A$

technology of the solar cells. The scaling parameters denoted by the α's and β's can be estimated for use in this model, or they can derive from more sophisticated models as they are developed. To use this model, we would make probabilistic estimates of the α's and β's, typically in the form of triangular probability density functions. Then we would embed this model in a Monte Carlo simulation to obtain probabilistic output data.

The parameter of most interest in this analysis is the present value of total system cost. This can be computed as follows:

$$C_{PV} = C_{cap} + C_{op} \sum_{t=0}^{t_{life}} (1 + r)^{-t}$$

where r is the discount rate and C_{cap} and C_{op} are the total capital cost (nonrecurring cost) and total operating cost (recurring cost), respectively; namely, they are the sums of the costs in the columns of Table 11.1.

Modeling costs, by means of a physical system model coupled to a cost model embedded in a Monte Carlo model, enable us to account for the uncertainty in our estimates of the cost parameters, here given as the α's and β's, and also to account for uncertainty in the physical parameters of the system as they also affect system cost.

11.5 BENEFITS

We turn our attention now from system costs to system benefits. We define a benefit as something that promotes well being or value received. There are many types of benefits. Among them are monetary; personal and national security; environmental improvements such as pollution control and noise control; aesthetic improvements such as parks, forests, and buildings; and health care and safety. Benefits of different types are not always easily compared to each other. The benefit of a new or improved engineering system is the sum of the benefits, of all types, resulting from the system as they accrue to all individuals in the society.

In attempting to quantify benefits, it sometimes helps first to identify the types of benefits resulting from a system according to a taxonomy of benefits. Such a taxonomy is given in Table 11.2, but this taxonomy is certainly not complete.

Sometimes a change, such as a new engineering system, benefits some people and disbenefits others. Now the question arises: how should different people's attitudes be

TABLE 11.2 A TAXONOMY OF BENEFITS

Intangible	Tangible	
	Direct	Indirect
Political Prestige Satisfaction of man's curiosity Social Promote future options	Cost savings Added use New capability Risk reduction Accelerate adoption of a new technology Reduced capital expenditures Increased productivity	Fallout benefits that result from direct benefits (for example, increased safety resulting from a system that is installed to decrease costs)

"weighted" together to form an aggregate benefit? In concept, a change is beneficial only if the benefits that result from the change can be distributed in such a way that at least one person is benefitted, and no one is disbenefitted. This is the concept of Pareto optimality discussed in Chap. 7.

Monetary Benefits

The estimation of monetary benefits are of paramount importance to us here, as they, more than any other kind of benefit, affect engineering decision making. There are three major categories of monetary benefits: cost saving, added use, and new capability. The theory of monetary benefits builds on the fundamentals of engineering microeconomics presented in Chap. 6.

A cost-savings benefit is illustrated in Fig. 11.8 for the case of an inelastic demand curve. A cost savings benefit occurs if a price is reduced. Price can be reduced as the result of a shift in the supply function or due to changes in taxation. In the case of an inelastic demand curve as shown, the consumer benefits, whereas the producer may or may not benefit depending on the nature of the supply-curve shift. The magnitude of the benefit to the consumer is given by

$$B_{cs} = q(P_1 - P_2)$$

Although it sounds obvious that cost savings is a benefit to a consumer, we can ask: how do cost savings benefit society? The answer lies in the fact that the cost savings allows society to make choices that are otherwise inaccessible and that can result in higher welfare for at least some segment of society without making any individual of the society worse off.

Added-use benefit is the benefit that accrues along with a cost savings benefit in the case that the demand curve has a non-zero elasticity. In this case, as the price of a good is reduced, the demand for the good increases with the increased consumer benefit as shown in Fig. 11.9. The added use benefit to the consumer may be estimated by

$$B_{au} = \frac{eq_1}{2P_1}(P_1 - P_2)^2$$

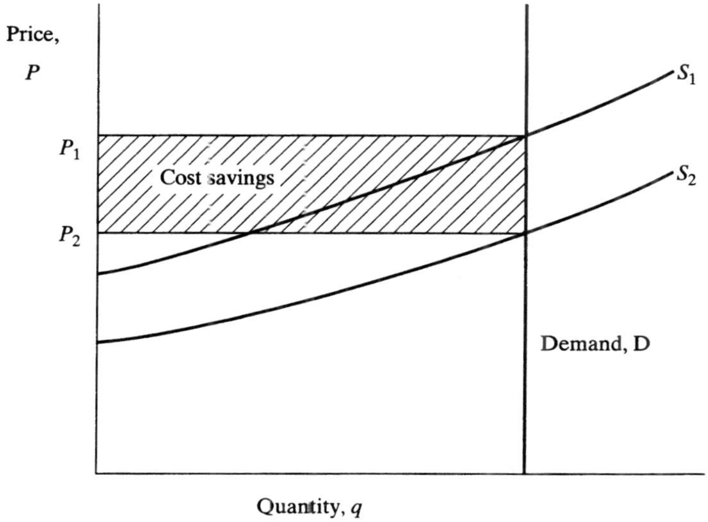

Figure 11.8 Cost savings benefit in the case of inelastic demand.

Fig. 11.9 shows both cost-savings and added-use benefits. The added-use benefits are distributed to new consumers and to previous consumers through increased use. The cost-savings benefits are distributed to previous consumers only.

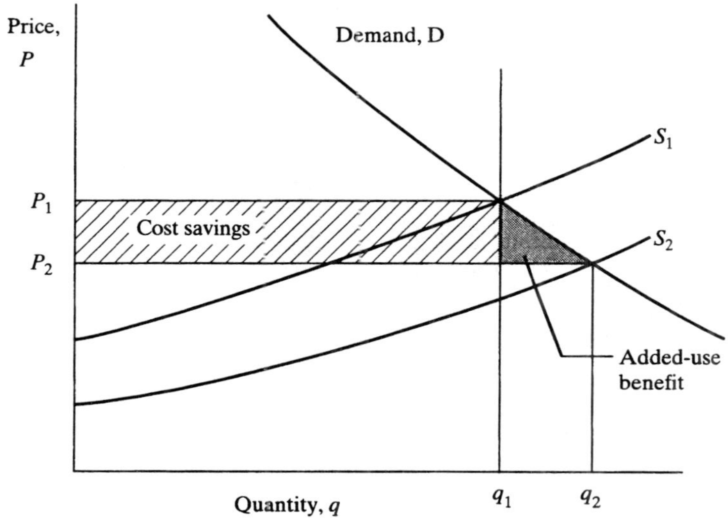

Figure 11.9 Cost-savings and added-use benefit in the case of non-zero elasticity.

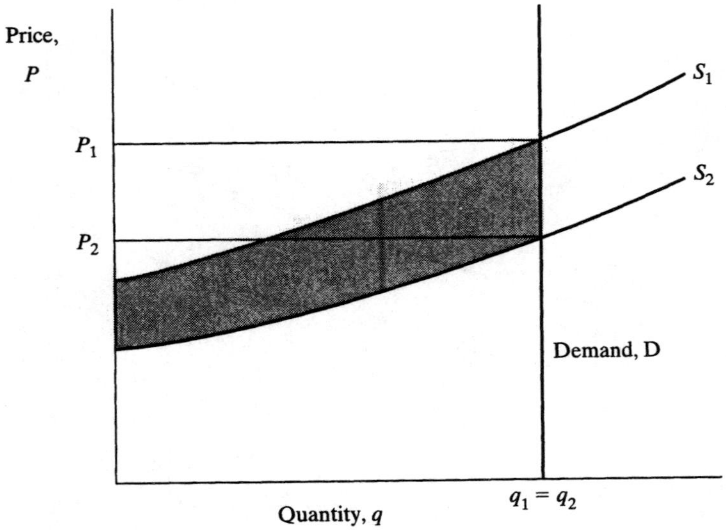

Figure 11.10 Net incremental benefits in the case of inelastic demand.

For the cases of cost-savings benefits and added-use benefits, the net incremental benefits consist of the sum of consumer and producer benefits. These are shown in Figs. 11.10 and 11.11 for the cases of inelastic and elastic demand, respectively. The net incre-

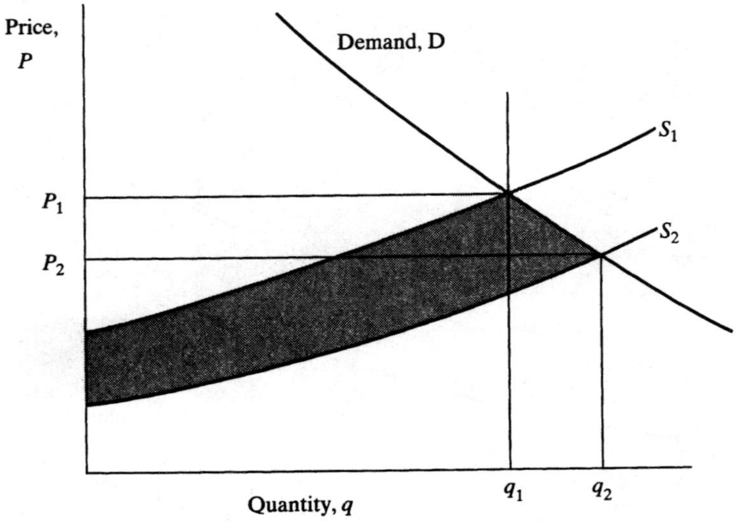

Figure 11.11 Net incremental benefits in the case of non-zero demand elasticity.

mental benefit is given by

$$B_{net} = \int_0^{q_1} [S_1(q) - S_2(q)]\, dq + \frac{eq_1}{2P_1}\left(1 + \frac{e}{E}\right)(P_1 - P_2)^2$$

where E is the supply elasticity, defined similar to the demand elasticity. We see from this equation that, in order to estimate the net incremental benefit, we must know the entire supply curve, not merely the supply elasticity.

Added use benefit may also derive as the result of a demand shift as shown in Fig. 11.12. This figure illustrates that consumers may obtain added use benefits even though price increases. Also, in this case, the suppliers clearly benefit. The added-use benefit resulting from a demand shift is estimated by

$$B_{aulds} = \int_0^{q_1} [D_2(q) - D_1(q)]\, dq + \frac{Eq_1}{2P_1}\left(1 + \frac{E}{e}\right)(P_2 - P_1)^2$$

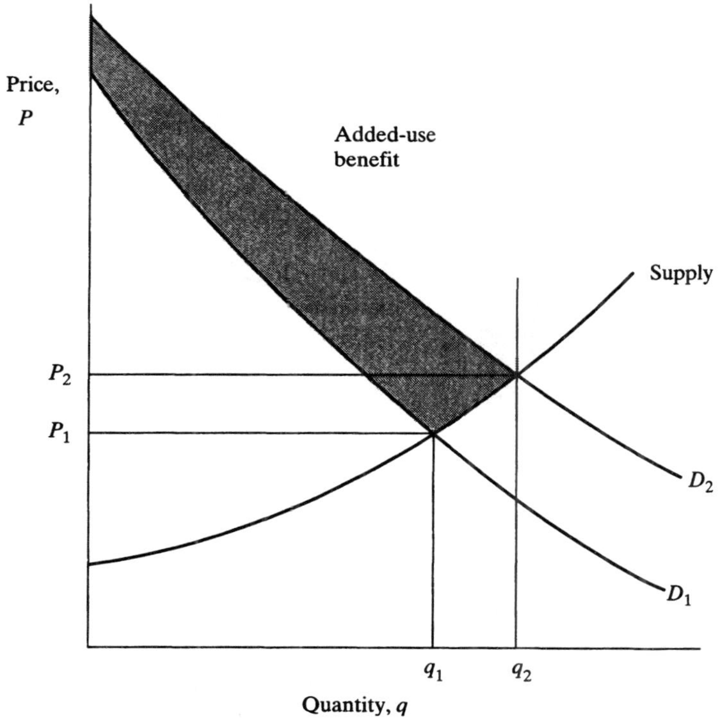

Figure 11.12 Added-use benefit resulting from a demand shift.

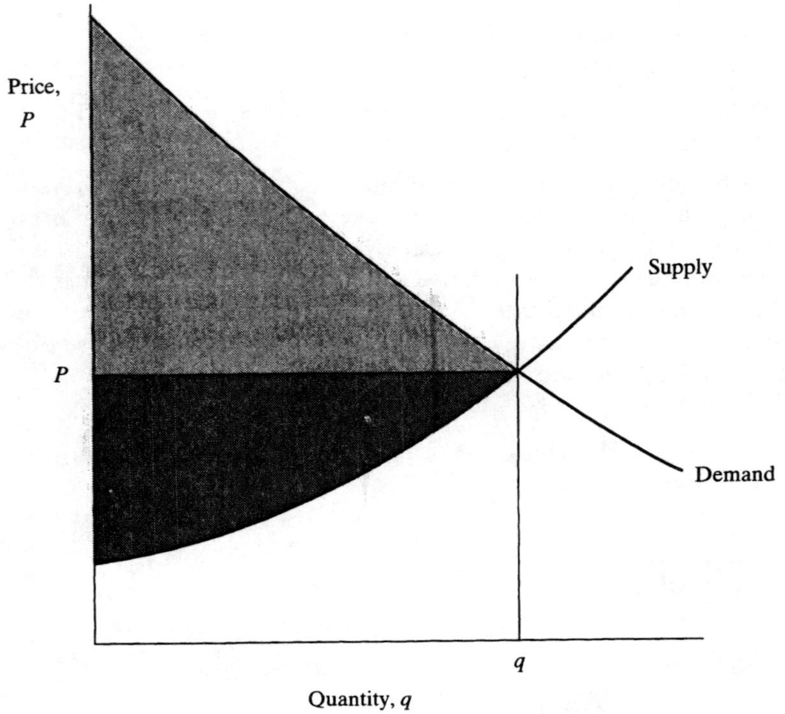

Figure 11.13 New capability benefits.

We see that, to estimate the added-use benefit resulting from a demand shift, we must know the entire demand curve.

The final type of monetary benefits are new capability benefits. The new capability benefits result from the production and consumption of a new product or service. They include both producer and consumer surplus as shown in Fig. 11.13, and they are estimated by

$$B_{nc} = \int_{0}^{q_1} [D(q) - S(q)]\, dq$$

Benefits of Government Funded Projects

The estimations of monetary benefits given above apply in the case of a free market. Projects that are undertaken by a government agency generally do not involve free-market goods. To estimate direct monetary benefits of these projects, we need to establish an appropriate context. We will examine two possible contexts for benefits, equal capability, and equal budget.

The *equal capability* context assumes that a decision has been made to conduct a project, and the question is: what are the benefits of new or different ways of doing it? In this case, the project objectives or goals are assumed given, and the value of the project is neither questioned nor measured. The measure of project benefits is the present value of cost savings that accrue from the project as conducted versus the project as conducted by means of a baseline alternative that also achieves the project goals. Equal capability does not address the worth or value of pursuing the project objective or goal in the first place, nor does it address possible modifications in the project objectives or goals that could make the project more attractive.

The *equal budget* context assumes that a decision has been made on how much to spend on a project, and the question is: what are the benefits of new or different ways of doing the project at a fixed budget? The equal budget estimator of benefits is the sum of the equal capability benefits and the consumer surplus that accrues from spending on additional capability at the same budget level as required of the baseline system.

Externalities

Externalities are defined as benefits or costs that accrue to bodies other than the one sponsoring a project. The possibility of externalities raises the question: how far should the sponsoring body go in taking them into account? In examining the benefits of a project, the sponsoring body should take into account external effects of actions insofar as they alter the physical production possibilities of other producers or the satisfactions that consumers can get from given resources. For example, suppose a project consists of the construction of a reservoir by the upstream authority of a river basin. And suppose this necessitates more dredging by the downstream authority. The added costs to the downstream authority should be included in the costs of the project. However, you should not take into account side effects if the sole impact of the side effect is through prices, products, or factors.

Miscellaneous Comments on Benefit Estimation

When is a benefit not a benefit? Answer: when it is not received. Consider the following, reported by the Wall Street Journal, March 19, 1974.

> *To hum or not to hum. An ultra-soundproof FAA [Federal Aviation Administration] facility near a Seattle airport is so quiet that designers built in a background 40-decibel hum. Tomorrow, officials will turn off the "white noise" for 30 minutes so workers can decide "whether it is nicer with or without" the sound, a spokesman says.*

On quantifying and placing monetary values on intangible benefits: do not bother. Instead, list and surface these benefits as a part of the overall analysis.

On private costs and benefits versus social costs and benefits: *Private costs and benefits* are defined as those costs and benefits that are directly experienced by the private decision maker, such as a business firm evaluating the project under consideration. *Social costs and benefits* are the costs and benefits experienced by society at large. They include all private costs and benefits, since the private decision maker is part of society. For example, consider the costs of a trip by automobile. The private costs include fuel, wear and tear on the automobile, the possibility of accidents, and the value of travel time. The social costs include the private costs plus the discomfort caused to others by added air pollution, congestion, noise, and so on.

The Benefits of Research and Development

To a business sponsoring research and development, the benefits of this activity include mainly shifts in producer surplus as described above. However, the benefits of research to society are much broader. They include the following:

> *Production*—These are the benefits to society of introducing new technologies into the marketplace. They include both producer and consumer surpluses.
>
> *Alternative uses*—These are the benefits to society of having new technologies available for uses other than that originally intended.
>
> *Information*—Research and development provides information for decision making on many issues of current importance. These can include decisions affecting current consumption patterns. For example, information on the effect of CFCs on the ozone layer surrounding the earth affects the decision of consumers to use CFCs versus alternative refrigerants.
>
> *Insurance*—Research and development can provide a storehouse of alternatives that can be called upon should the need arise.
>
> *Expectations*—Research and development creates expectations that, at some time in the future, there will be alternative technologies to supplant those we rely upon today. These expectations can change current consumption patterns, benefitting society. A good example is the case of a limited resource, which might be priced higher today because there are no substitutes. Research and development can, however, create the expectation of future substitutes, which lowers the current resource price.

Timing of Benefits

In addition to benefits themselves, the timing of benefits is also important. It is a benefit to accelerate the rate at which benefits are obtained from a project or activity, even if the total (undiscounted) benefit remains constant. The benefit of accelerating a benefit stream is given as the difference in the present value of the accelerated benefit stream and the baseline benefit stream. This concept is illustrated in Fig. 11.14. The benefit is the shaded area.

11.6 PROBABILISTIC BENEFIT ANALYSIS

As is true with costs, benefits also cannot be estimated with certainty. It is therefore important to incorporate uncertainty into their estimation. The present value of benefits

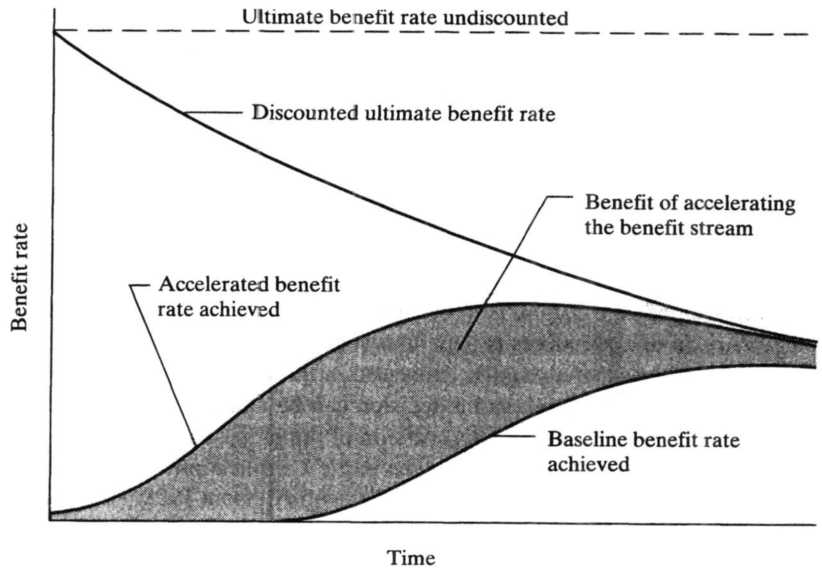

Figure 11.14 The benefit that results from accelerating a benefit stream.

of a project is given by an equation of the form

$$PV\{B\} = \sum_{t=t_i}^{t=t_f} B_t(1 + r)^{-t}$$

where t_i is the time at which the benefit stream begins, t_f is the time at which the benefit stream ends (the project horizon), B_t is the undiscounted benefit derived during time period t, and r is the discount rate. These parameters should be modeled with uncertainty. Note, however, that under conditions of risk, it is necessary to use the utility of the present value of profit to properly represent corporate decision preferences.

In general, the B_t are functions of a number of parameters that must be forecast, often well into the future. If you are designing a system for a company that plans to manufacture and sell the system, then the objective of your design is likely to be statable in economic terms, such as to maximize the present value of the profit derived from the system. A simple model for profit is revenue less cost. The revenue generated by a system in the ith time period, R_i, might be given by

$$R_i = P_i q_i$$

where P_i is the selling price and q_i is the quantity sold. (We discussed cost previously, so we will not include it here.) P_i is a control variable; the manufacturer selects the selling price. q_i is determined by the marketplace as a function of variables such as personal income, availability of capital, price of alternatives, and so on. There is no general form for revenue that we can state here. Rather, you must examine each case and develop an appropriate form. Then, you should model R_i and, using a Monte Carlo model, estimate the probability density function of R_i for each year within the project horizon. The process for doing this is very similar to the process used for the probabilistic estimation of costs.

11.7 BENEFIT-COST ANALYSIS

Frequently, the option space for an engineering system is so complex that we seek methods to simplify the comparison of alternatives for decision making. One approach to the comparison of alternatives is benefit-cost analysis. The notion behind benefit-cost analysis is that the cost of a project, expressed in monetary units, can be compared to the benefits, similarly expressed, and a decision can be made based on this comparison to fund, not to fund, or to fund an altered version of the project. A project is a candidate for funding if its benefits exceed its cost. The earliest applications of benefit-cost analysis in the United States date back to the turn of the century, about 1900, and dealt with the selection of harbor and river navigation projects. Later, in the Flood Control Act of 1936, federal participation in flood control schemes could take place only "if the benefits to whomsoever they may accrue are in excess of the estimated costs."

A typical project generates a cost/benefit flow as shown in Fig. 11.15. During the early stages of the project, costs exceed benefits. Most projects have a research phase, a design phase, and several successive phases prior to the time that benefits begin to accrue. Later in the course of the project, benefits begin to accrue, typically slowly at first and then more rapidly with time. At some time into the operational phase of the project, benefits may begin to exceed costs, and ultimately offset the initial investment. For a typical

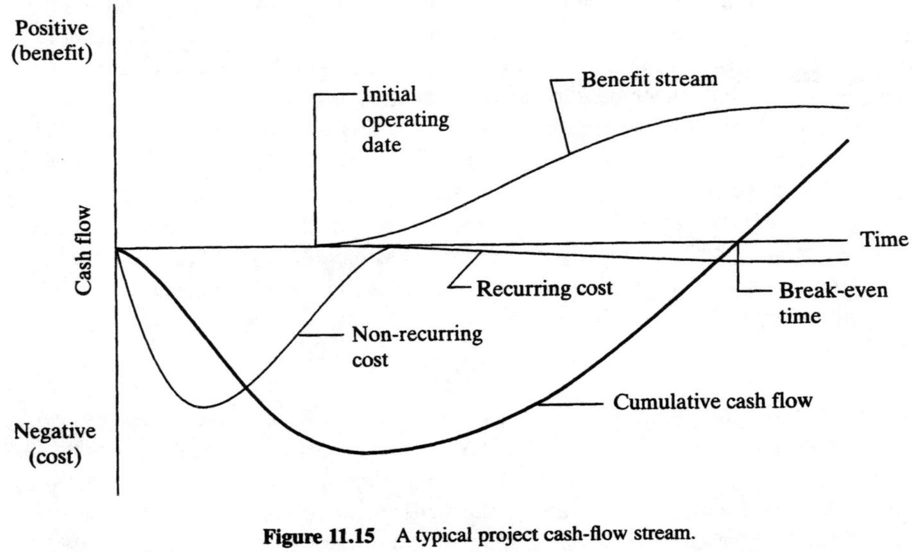

Figure 11.15 A typical project cash-flow stream.

project, nonrecurring costs might be expressed as a function of time using a beta distribution. Recurring costs might have a constant component and a component that is proportional to benefits, and benefits might grow in time according to a Gompertz curve. The cumulative cash-flow curve shows the cumulative present value of benefits less costs. The project break-even time occurs when the present value of benefits received equals the present value of costs incurred. The cumulative cash-flow curve in Fig. 11.15, which is an amount rather than a rate, uses a different vertical scale.

If we separate costs and benefits, we see that we can create a number of different benefit-cost measures:

Net benefit—The net benefit of a project is measured as the present value of project benefits less the present value of project costs, taken from the start of the project through to the project horizon.

Zero-base net benefit—This measure is the same as the net benefit measure, except that the time period of the measurement begins at the present time rather than the time of the start of the project. In this way, costs and benefits accrued to date are treated as sunk, and the resulting measure is more useful for decision making in the present.

Benefit-cost ratio—This measure is given by the ratio of the present value of the project benefits to the present value of project costs, measured from the start of the project to the project horizon.

Zero-base benefit-cost ratio—This measure is the same as the benefit-cost ratio, except that it uses benefits and costs measured from the present time.

Return on investment (ROI)—This measure applies to a project whose undiscounted benefits exceed its undiscounted costs. The present value of benefits, B, less costs, C, is given by

$$NPV = \sum_{t=t_0}^{t=t_F} (B_t - C_t)(1 + r)^{-t}$$

where r is the discount rate or the rate of return. ROI is the value of r for which the net present value of benefits less cost is zero. As you can see from this equation, this is a transcendental function in r, and so solution for the ROI is not easy.

Break-even time—The break-even time for a project is the time from the start of the project that is needed for the present value of benefits to equal the present value of costs. This measure is of interest because it is the time period over which an investor is exposed to risk of loss.

11.8 PROJECT SELECTION

Investment managers, research and development managers, and other corporate managers are often confronted with options for selection or rejection. The investment manager seeks to select a set of investments that simultaneously maximize return on investment and minimize risk. Investment decisions must be made under conditions of uncertainty. The research and development manager typically is given a research budget, and must allocate that budget across a set of potential projects to improve the overall performance of the corporation, possibly measured in terms of profits. This is done in the presence of considerable uncertainty on the outcome of each individual research project. The engineer may be faced with a choice between alternative system designs as well as alternative systems, and choices must be made in the presence of uncertainty. In each case, the decision maker must determine a utility for each choice, and rate choices by their utilities. Benefit-cost measures, strictly speaking, can be used in the absence of uncertainty for this purpose. They also can be used in the presence of uncertainty, provided that they are first converted to utilities. In practice, however, they are frequently used in the presence of uncertainty without conversion to utilities. In essence, this practice assumes that the utility of money is equal to money, but this point is frequently missed.

The benefit-cost measure that you would choose for comparison of alternatives depends on the nature of the investment situation. The use of the measures identified in the previous section is as follows:

Net benefit—The net benefit measure can be used in the situation where the manager seeks to maximize the net present value of benefit received from some past date through a time horizon, typically subject to a budget constraint. Net benefit project selection can be formulated as a problem in integer programming. Let x_i denote the choice to fund or decline the ith project, such that if $x_i = 1$ the project is selected, and if $x_i = 0$ the project is declined. Let NPV_i represent the net present value of the ith project. Then we want to maximize

$$J = \sum_i x_i \, NPV_i$$

subject to the constraints

$$x_i = 0 \quad \text{or} \quad 1$$

$$\sum_i X_i C_i \leq \text{Budget}$$

Budget constraints could be imposed on a year-by-year basis. In addition, we could impose further constraints on the x_i of the form

$$\sum_{i=i_a}^{i=i_b} x_i = 1$$

These constraints would enable us to include in the selection optimization several different alternatives of the same project, such as different designs of the same system.

Thus we see that the use of the net benefit measure can be to select for pursuit a set of systems, each chosen from a set of alternatives, that is designed to maximize the long-term profitability of the corporation or the long-term welfare of the organization, such as a society. In applying this measure, the decision maker can choose to assume that, once started, a project runs through to completion, or the decision to continue the project can be made each time that a project selection is made. The advantage of the latter approach is that it leads to a higher total net present value for the program of projects as a whole. The disadvantage is that projects can be canceled, and so it is possible to find yourself in a situation where technology is improving so rapidly that at each successive decision you decide to cancel all previous projects and start new ones. In this case, you will never finish any project and thus never get any benefit from the money spent, even though the projected present value of total benefits continues to increase.

Zero-base net benefit—This measure is used for much the same purpose and in much the same way as the net benefit measure. The difference lies in the costs and benefits that are used in the measure, in this case restricted to costs that will be accrued in the future, and benefits that will result from the incurrence of future costs. The zero-base net benefit measure enables a decision maker to consider committed costs to be sunk and look only at costs that can be controlled by future decisions.

Benefit-cost ratio—Both the net benefit method and the zero-base net benefit method generally require an optimization to achieve an optimal project selection. This process can be rather difficult, and it requires some degree of mathematical and computational sophistication. The benefit-cost ratio measure is intended to circumvent these difficulties. Projects eligible for selection are first ranked according to their benefit-cost ratio, from highest to lowest. Next, each subset of projects is examined, and only the project with the highest benefit-cost ratio in each subset is retained for eligibility in the final selection. Then, projects are selected in order beginning with the project that has the highest benefit-cost ratio, and projects are selected until the budget constraint is reached.

A problem with this approach arises if the costs of projects are large with respect to the budget constraint. In this case, it is possible that, in order to keep from exceeding the allowable budget, projects will be declined leaving a large portion of the budget unspent. One remedy for this is, when the budget constraint is exceeded by choosing the next project, proceed to the next best project and the next and so on until you find the next best project that fits within the budget constraint. The problem with this procedure is that it can lead to rather sub-optimal project selections.

Zero-base benefit-cost ratio—This measure is similar to the benefit-cost ratio, and it has all the benefits and problems of this approach. However, it examines project selections in terms of unsunk costs and benefits attributable to unsunk costs only.

Return on investment—This method is similar to the benefit-cost method except that the objective is the maximization of the rate of return rather than the maximization of the present value of net benefit. The return-on-investment measure is more appropriate when there is no budget constraint but when there is a cost of money. In this approach, the decision is to fund projects as long as they provide a net return on investment, that is, a return above the cost of money. Again, a simple, top-down selection process is used, and sophisticated mathematical optimizations are not necessary, although it may be difficult to compute the rate of return for each individual project.

Break-even time—This measure may be used when the decision maker perceives that there is considerable risk. The notion is that the sooner one's investment is returned the less the risk associated with a project. Break-even time also relates to the return on investment, albeit only indirectly, so the notion behind the break-even-time method is that it simultaneously minimizes risk and maximizes rate of return. However, it only approximates this.

As we note above, it is not correct to use these measures directly in the presence of risk. Rather, in this case, the decision maker should express measures of project worth in expected utility terms, accounting for the risk in project outcomes.

11.9 SUMMARY

The attributes of a system that are of primary importance to the system engineer are frequently its cost and its benefits. This chapter examined these in detail. We defined cost as the value of the inputs to a process and, for engineering systems, noted that two types of cost are important: nonrecurring costs and recurring costs. We observed that trades can generally be made between these two types of costs, and we examined the notion of a technology frontier defined in terms of costs. Next we considered the concept of sunk costs, and found that sunk costs should not affect the evaluation of future decisions. Then we examined two approaches to cost estimating: a top-down approach and a bottom-up approach. We saw the value of physical-system and cost models, and found that they can be used in a Monte Carlo analysis of system cost and cost uncertainty. We also developed methods for the estimation of cost learning and cost spreading.

In the second part of the chapter, we considered project benefits. We defined the benefits of a project as value received from the project. Then we developed a taxonomy of benefits that included both monetary and nonmonetary benefits. The remainder of the chapter considered the analysis of monetary benefits, however. Benefits were examined in the contexts of both a free market and the absence of a free market. In the absence of a free market, we introduced two approaches to benefit estimation: the equal capability approach and the equal budget approach. We completed our analysis of benefits by touching on the probabilistic analysis of benefits.

Finally, we looked at benefit-cost measures and their use in project selection. We identified several different measures and discussed their advantages and disadvantages.

REFERENCES

1. Ackoff, R. L., *Scientific Method: Optimizing Applied Research Decisions,* John Wiley, New York, 1962.
2. Anderson, L. G. and R. F. Settle, *Benefit-Cost Analysis: A Practical Guide,* Lexington Books, 1977.
3. Ansoff, H. I., Evaluation of Applied Research in a Business Firm, *Technological Planning on the Corporate Level,* J. R. Bright ed., Proc. Conference at Harvard Business School, Harvard University, Cambridge, MA, 1962, pp. 209–224.
4. Baker, N. R. and W. H. Pound, R&D Project Selection: Where We Stand, *IEEE Trans. Eng. Manag.,* vol. EM-11, Dec. 1964, pp. 124–134.
5. Devons, E., *Essays in Economics,* Allen and Unwin, London, 1961.
6. Fromm, G., Civil Aviation Expenditures, *Measuring Benefits of Government Investment,* R. Dorfman ed., Brookings Institution, Washington, D.C., 1965, Allen and Unwin, London, 1968.
7. Gee, R. E., A Survey of Current Project Selection Practices, *Research Management,* vol. 4, no. 5, Sept. 1971.

8. Greenberg, J. S. and G. A. Hazelrigg. A Methodology for Reliability-Cost-Risk Analysis of Satellite Networks, *Journal of Spacecraft and Rockets,* vol. 11, no. 9, Sept. 1974.

9. Greenberg, J. S., The Economic Implications of Unreliability, *Proceedings 1976 Annual Reliability and Maintainability Symposium,* 1976.

10. Hertz, D. B. and P. G. Carlson, Selection, Evaluation and Control of Research and Development Projects, *Operations Research in Research and Development,* B. V. Dean ed., John Wiley, New York, 1963, pp. 170–188.

11. Hirsch, J. H. and E. K. Fisher, A New Procedure Provides a Basis for Determining What Projects to Start, Stop, Commercialize, *Research Management,* vol. 11, no. 1, Jan. 1968, p. 21.

12. Hitchcock, L. B., Selection and Evaluation of R&D Projects, *Research Management,* vol. 6, May 1963, pp. 231–244.

13. Layard, R., *Cost-Benefit Analysis,* Penguin Modern Economics Readings, Middlesex, England, 1974.

14. Mishan, E. J., *Economics for Social Decisions, Elements of Cost-Benefit Analysis,* Praeger, New York, 1973.

15. Ridker, R. G., *The Economic Costs of Air Pollution,* Praeger, New York, 1967.

16. Rubenstein, A. H., Studies of Project Selection Behavior in Industry, *Operations Research in Research and Development,* B. V. Dean ed., John Wiley, New York, 1963, pp. 189–205.

17. Silk, L. S., An Optimal Method for Selection of Product Development Projects, Presented at the 15th National Meeting of the Operations Research Society of America, Washington, D.C., May 1959.

PROBLEMS

1. The manufacture of a product requires 2 h labor, at $35 per h with an overhead of 70 percent added to this cost, and 4 h of time on a machine that cost $200,000 and has an expected life of 5 years (assume 4,000 h of operation per year) and an hourly cost of operation and maintenance of $50. The cost of engineering the product was $5 million. The company produces 4,000 such products per year and expects a production run of 5 years. The company sells the product for $1,500. The company's cost of money is 10 percent.

a. What are the sunk costs?

b. What is the marginal cost of producing one unit?

c. Is the company going to make money on the product?

d. Should the company discontinue production of the product?

2. A company has designed for production and sale a home appliance. Costs are estimated as follows:

Research and development (over 2 years)	$1,500,000
Product engineering (over 2 years following R&D)	3,000,000
Tooling for production (invested 6 months before prod.)	4,500,000
Amortization of production tools per appliance	150
Production labor and overhead (per h)	75
Operation and maintenance of production tools (per h)	250
Other direct costs (per h)	500

Assume that it takes 6 minutes of labor and 10 minutes of production machine time to complete manufacture and assembly of the appliance. Take the company's discount rate to be 10 percent, and assume a ten-year production run at a production rate of 500 appliances per day, 200 days per year.

a. What are the nonrecurring costs?
b. What are the recurring costs?
c. What is the cost per appliance?

3. Your company constructs large, multistory office buildings. You have the following data on buildings recently completed:

Height, stories	Floor space, m^2	Cost, \$ millions
15	27,000	79.5
12	12,000	48.6
22	38,500	167.5
9	18,900	68.1
18	25,740	98.8
14	30,100	119.2
17	33,150	138.2

Your company is bidding on a 16-story building with a floor space area of 28,000 m^2. What your estimate of the cost of this building? What is your estimate of the error in your estimate?

4. A company makes commercial airplanes. The first airplane of a 10-year production run of 500 total airplanes cost \$40 million to produce. The company expects a 7 percent learning rate. What will the 500th airplane cost? What will be the average real cost per airplane at a discount rate of 5 percent?

5. In problem 4, assume that the airplane design required 5 years and cost \$700 million. Assume a beta distribution of engineering and test effort with $\alpha = 3$ and $\beta = 2$. What is the present value of engineering cost? If the company charges a fixed price for the airplane over its entire production run, how much will it have to charge in order to earn a 10 percent profit?

6. You are called upon to check a cost estimate for a system. The cost estimate has been made by dividing the system into 10 subsystems, and best estimates of the cost of each subsystem have been provided as follows: (dollars in thousands) \$95, \$15, \$37, \$153, \$74, \$58, \$29, \$86, \$42, and \$108. You check that the total of these is correctly given as \$697,000. But you know that there is considerable uncertainty in many of these estimates. So you solicit the data provided in the following table (again, dollars in thousands):

Subsystem	Minimum cost	Most likely cost	Maximum
1	80	95	250
2	10	15	50
3	30	37	70
4	140	153	350
5	70	74	170
6	50	58	140
7	25	29	75
8	75	86	200
9	35	42	100
10	90	108	300

You reestimate cost using the above data and assume that they apply to triangular distributions. What is your new expected cost? What is the standard deviation of your estimate? What is the shape of the resulting probability distribution of cost?

7. Five potential projects have the same cost, but each produces a different benefit. Assume that the benefits of the projects are as follows: project A, $1500; project B, $1375; project C, $1925; project D, $1225; and project E, $750. You choose to fund project C, as it has the highest benefit. What is the opportunity cost of this project?

8. You ask a friend if you can borrow her car for a trip. She agrees to loan you her car, but only if you pay all expenses. And she informs you that there is a problem with the car that might require major repair. If the problem does not arise, the car will cost $0.16 cents per km to drive. If the problem does arise, you will have to pay an additional amount of $400 to get it fixed. You want to drive the car 1,000 km. You assess the probability of the problem arising at 0.25. How much will the trip cost?

9. Your company produces circuit boards, which cost $3.50 each to manufacture and sell for $5 each, at the rate of 100,000 per year. A three-year production run is expected. An employee suggests a change in the manufacturing procedure that could save $0.20 per board. What is the benefit of this suggestion if implemented? What kind of benefit is it? Assume 10 percent discount rate.

10. In the case of problem 9, suppose your company implements the suggestion and then reduces its price by $0.10 per board. As a result, sales increase to 120,000 per year. Compute and discuss the benefits in this case

11. You invent a new product. It costs $10 to manufacture, and you can sell it for $12 at a production rate of 2,000 per month. You have a market research firm determine your consumers' preferences with respect to the product and deduce that their demand curve is liner with a demand elasticity of 0.9. What are the benefits of your invention to you? To society? Are you charg-

ing the best price for the product? What would happen to the total benefits if you changed your price?

12. Your company is tooling up to produce a product. It is estimated that production can begin in one year, and that the company will obtain a net present value of $150 million from the product. You discover that a design change can simplify production without altering the attributes of the product and enable the product to get into production in 11 months. At a discount rate of 10 percent, what is the value of your discovery?

13. You are examining four ongoing projects. Each of the projects incurs costs during its first 3 years, and from the fourth year on, produces a continuing benefit stream, assume forever. Data on the project costs and benefits are the following:

Project	Cost in year 1, $	Cost in year 2, $	Cost in year 3, $	Annual benefit, $
A	1,000	1,000	1,000	500
B	2,000	700	300	300
C	300	700	2,000	600
D	500	2,000	500	500

Assume a 10 percent discount rate. You are now beginning the third year of funding these projects. For each project, compute the net benefit, the zero-base net benefit, the benefit/cost ratio, the zero-base benefit/cost ratio, ROI, and break-even time. By each of the above measures, rank the projects. If your budget for these projects in year 3 is $1,500, which projects will you fund? Why?

14. Under what conditions might cost minimization be a valid goal of design and manufacture? When might cost minimization be invalid?

15. An engineer optimizes the timing of traffic lights on a stretch of a highway. What are the benefits of the action? To whom do the benefits accrue? How would you measure or estimate the benefits? What is it worth to the local government? What is the maximum allowable cost of analyzing and making the changes? How could the local government raise the money to pay for the cost of the engineering work?

16. Discuss the costs and benefits of a dam. To whom does each cost and benefit accrue? How would each be estimated?

17. An auto manufacturer develops a more efficient engine. Discuss costs and benefits. Who benefits and how? Who disbenefits and how?

18. An electric power company expects to experience power shortages during the hot months of summer. In response, it proposes imposing a rate hike. If the rate hike is approved, who will benefit? Who will disbenefit? Are there alternatives to a simple rate hike by the company that could provide higher overall benefit?

12

Methods of Decision Analysis

12.1 THE PURPOSE OF THIS CHAPTER

The earlier chapters of this book focus on the placing of engineering systems design into a decision-making context. We have examined approaches to the formulation of system options, the computation of expectations on each option, and the formulation in mathematical terms of values. In this chapter, we will examine decision-making approaches for the taking of choices from among options that arise in the context of engineering systems design. We start here from the position that we have already defined the system options, expectations, and values, and that our goal is to choose one option from the set of options.

The fact is, even given options, expectations, and values, it is not always a simple matter to make choices. Engineering systems design, as with many decision-making situations, is most often a matter of compromise, where we have many conflicting objectives for the system, and we have to give a little with respect to one objective to gain a little with respect to another. In some situations, the best choice is clouded by uncertainty, and there may even be the risk of catastrophe should the design be faulty. We seek, therefore, decision-making approaches that provide rational choice—choice that can be documented and defended, and choice that satisfies the goals of the system.

There exist many mathematically rigorous decision-making approaches. We examine a few of them here.

12.2 DECISIONS AND DECISION ANALYSIS

We began our discussion of systems engineering based on the premise that engineering design is a process of decision making. In this context, we defined a decision as an irrevocable allocation of resources. In the engineering systems context, the choice of design parameters constitutes an allocation of resources to the design. Design cannot be wishy-washy. We cannot specify a design in terms of "about's" or "sort-of's." Design requires that specific choices be made. Recall again our example of guessing how many beans are in a jar. You are not permitted to make a guess that is like, "oh, something between 100 and 2,000." To enter the game, you had to choose a specific number. Similarly, to enter the game of engineering design, we must choose specific design parameters. When we make our choice, we allocate resources to that choice. In the case of the bean jar, we paid an entrance fee to get into the game. In the case of an engineering system, we pay the cost of system design, test, manufacture, and all other costs associated with the business of that system.

There are two important characteristics of a decision:

A *decision* is made at an instant in time.

A *decision* must be made based on the information available at the time that it is made.

Decision analysis is the analysis of choices in a particular situation. Decision analysis can be defined in many ways:

Decision analysis is the systematic evaluation of specific alternative choices.

Decision analysis is a term used to describe a body of knowledge and professional practice for the logical illumination of decision-making situations.

Decision analysis is a normative, rather than descriptive, approach. It does not attempt to describe a decision process. It shows how a decision maker subscribing to a particular set of logical rules would make decisions in order to maximize attainment of his or her objectives.

Decision analysis provides philosophical foundations and logical quantitative procedures for decision making.

Decision analysis provides a basis for documentation of the state of information at the time a decision is made and a basis for communication with others.

As we develop the science of decision analysis, one distinction becomes very important. This is the difference between a decision and an outcome. A *decision* is, as we defined above, an allocation of resources. An *outcome* is the result of a decision. A decision is good if it is the best choice under the circumstances; an outcome is good if the decision maker gets what he or she wants. This concept is illustrated by the following coin-toss situation. The decision maker places a \$1 bet on a coin toss, calling for either heads or tails. In the case of this example, a head pays \$1.50 and a tail pays \$2.50, and we assume that the coin is fair. The decision/event tree for the bet is shown in Fig. 12.1. Again, we use the convention that decisions are denoted by squares, and probabilistic events are denoted by dots or circles. It is clear that the expected value of a bet on tails is higher than the status quo or a bet on heads. If the decision maker bets on expected val-

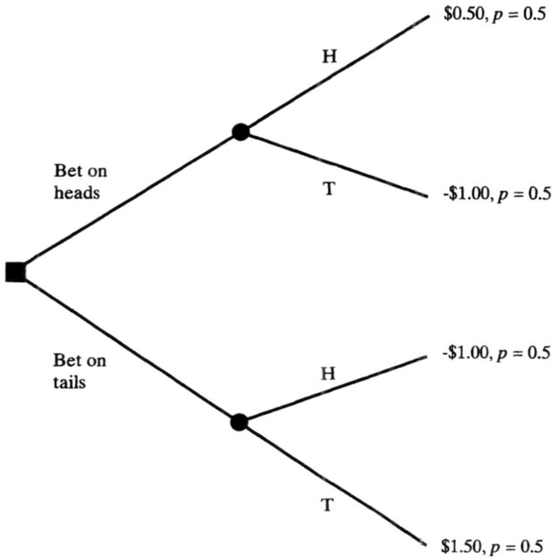

Figure 12.1 Decision/event tree for the coin toss example.

ues, the bet on tails is a good decision; the bet on heads is a bad decision, given the alternatives.

The outcome of this decision is given in Table 12.1. A good outcome is winning the bet, whether it is heads or tails.

TABLE 12.1 ILLUSTRATION OF DECISIONS VERSUS OUTCOMES

	Outcome	
Decision	Heads	Tails
Heads	Bad decision Good outcome	Bad decision Bad outcome
Tails	Good decision Bad outcome	Good decision Good outcome

12.3 BACKWARD INDUCTION

The first approach to decision analysis that we will examine is referred to as the method of backward induction. This method is, perhaps, the most rigorous from a mathematical point of view and, although it is conceptually simple, it can be computationally overwhelming.

The method of backward induction begins examining a decision by drawing its consequences in the form of a decision/event tree, as we have done already several times in the previous chapters. We will begin with an example. Suppose we are offered the chance to enter a game in which a coin is tossed three times and a prize is awarded as follows:

Number of heads	Prize
0	\$0
1	\$10
2	\$30
3	\$100

The cost of entering the game is \$30. The decision/event tree for this game is shown in Fig. 12.2. For the time, we will analyze this decision in terms of expected values. The game has four possible outcomes: lose \$30 with probability 1/8, lose \$20 with probability 3/8, break even with probability 3/8, and win \$70 with probability 1/8. The expected value of the game is thus

$$\begin{aligned} E\{\text{Game}\} &= \sum_{i=1}^{4} \text{Prob}_i \, \text{Prize}_i \\ &= \frac{1}{8}(-\$30) + \frac{3}{8}(-\$20) + \frac{3}{8}(\$0) + \frac{1}{8}(\$70) \\ &= -\$2.50 \end{aligned}$$

We next redraw the decision/event tree, replacing the probabilistic event, play, with its expected value. This gives us the decision tree shown in Fig. 12.3. The decision/event tree of Fig. 12.3 enables an easy choice between the options. Compared to the status quo

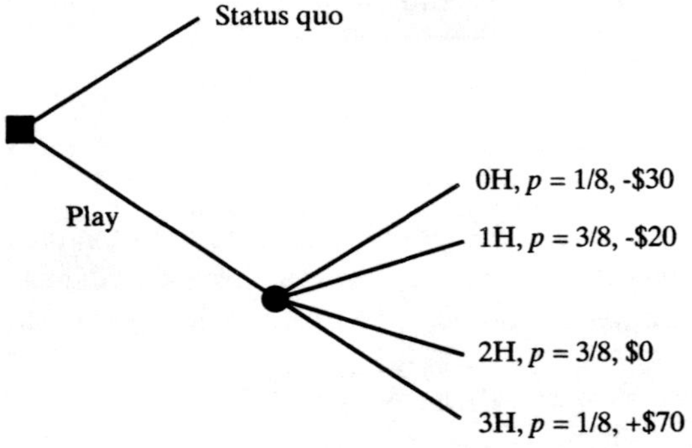

Figure 12.2 Decision/event tree for a multiple coin flip game.

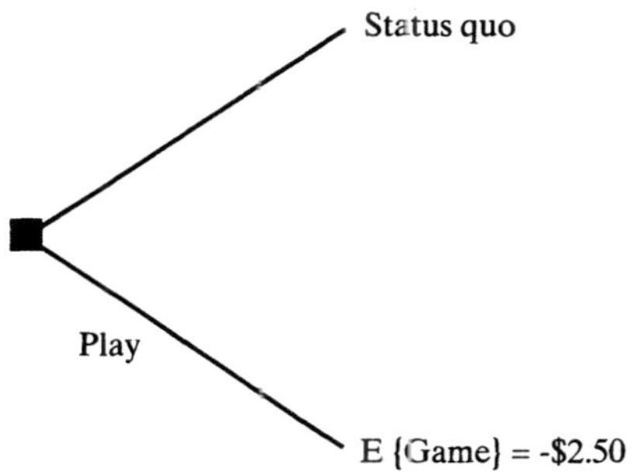

Figure 12.3 The certainty equivalent decision/event tree for Figure 12.2.

(that is, choosing not to enter the game), which has an expected value of $0, an expected-value decision maker would not play this game.

To apply the method of backward induction, we first drew the decision/event tree. Next, we went to the right-hand side of the decision/event tree and reduced the terminal branches of the event following the decision "play" to their expected value. We refer to this as the *certainty equivalent value.* Then we replaced the random event with its certainty equivalent. If there were more than one such random event on the right-hand side of the decision/event tree, we would do this with each event. We would then have worked backward through the tree by one event, simplifying it for further analysis.

To illustrate this procedure, we will add an additional level of complexity by changing the nature of the game. Instead of requiring that the entire $30 entry fee be paid at the beginning of the game, we will allow for sequential decision making. While the game and the prize remain essentially the same, we will allow the player to pay $10 per toss and to back out of the game at any time before all three tosses are made. There is no prize, however, regardless of the number of heads, if the player backs out before all three tosses are made. We refer to this game as involving *sequential decisions,* because the player can decide after each flip whether to pay for the next flip or not. The decision/event tree for the sequential decision game is shown in Fig. 12.4. It is considerably more complex than the decision/event tree of the game shown in Fig. 12.2. At the right-hand side of this tree there are four sets of probabilistic outcomes. Our first task is to reduce each of these to its certainty equivalent value, and we then replace the terminal events by their certainty equivalent values. The resulting decision/event tree is shown in Fig. 12.5.

Now we are faced with a decision/event tree that has on its right-hand side four decisions. A decision is something that is under the control of the decision maker, and we assume that each decision will be made rationally. The decision maker can choose to take the option that has the highest certainty equivalent value. Therefore, we can replace each decision node with the certainty equivalent value of its best option. The simplified decision/event tree is shown in Fig. 12.6.

Again, we are faced with probabilistic events at the right-hand side of the tree. Once again, we reduce these to their certainty equivalent values, and the tree simplifies to that shown in Fig. 12.7. Next, we replace the decision nodes of Fig. 12.7 with their maximum certainty equivalent values, obtaining the decision/event tree of Fig. 12.8. We are left with a tree that contains one probabilistic event, and it reduces to the equivalent decision shown

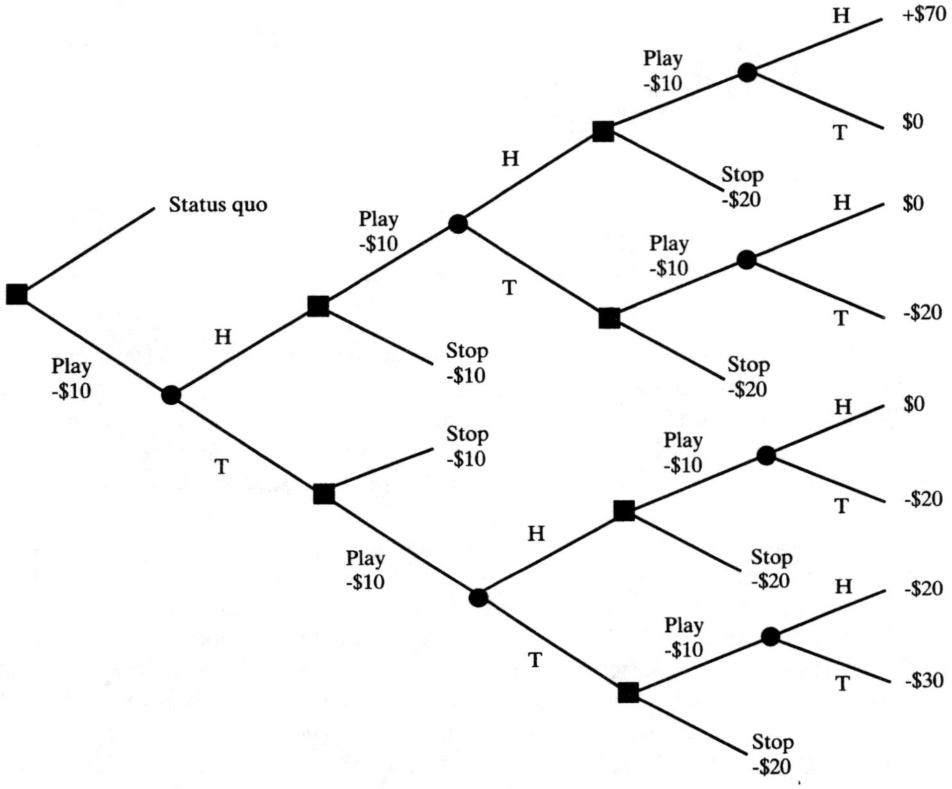

Figure 12.4 A decision/event tree for a sequential decision game.

in Fig. 12.9. This completes the analysis of the initial decision, to play the game or not, by the method of backward induction. We see that the method gets its name from the fact that it progresses by analyzing the decision/event tree from the right-hand side backwards in time to the decision in question.

Now we see something that might be rather surprising. Although the game remained essentially the same in the sequential decision case as it was in the all-or-nothing case, the existence of sequential decisions leads to a positive expected value for the decision to play the game. We can make the following generalizations:

> *In otherwise similar decision-making situations, the ability to make decisions sequentially always is advantageous.*
>
> *Some "games" or activities that have negative expected value or utility when taken as all-or-nothing choices have positive value or utility when taken with sequential decisions.*
>
> *The value of future decisions is to control risk.*

These generalizations weigh heavily on the organization of engineering projects. It is very important to structure programs so that they allow for sequential decision making.

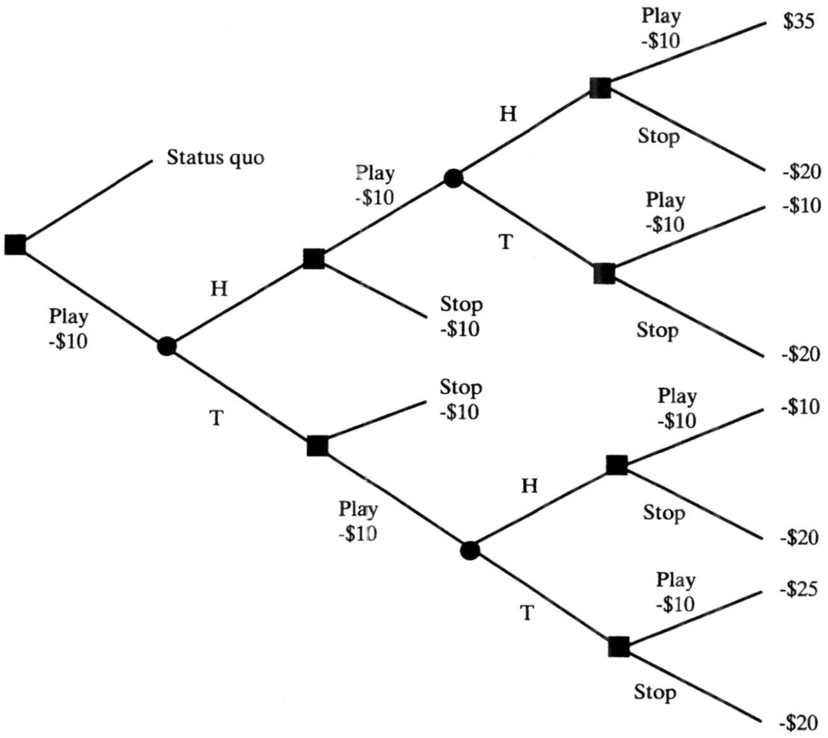

Figure 12.5 The first step in the solution of the decision/event tree of Figure 12.4.

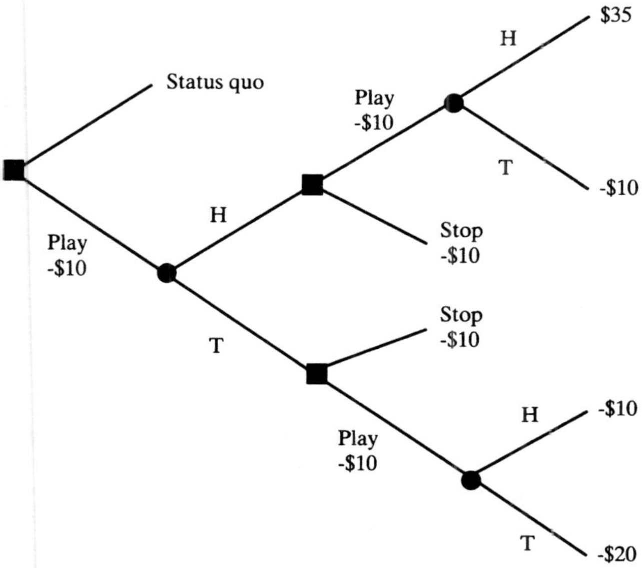

Figure 12.6 The second step in the solution of the decision/event tree of Figure 12.4.

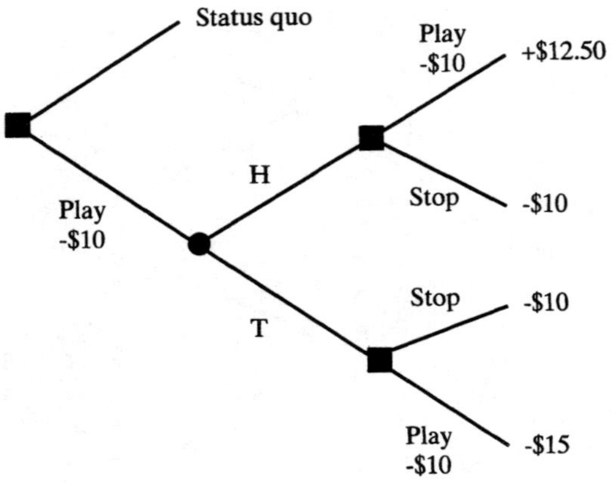

Figure 12.7 The third step in the solution of the decision/event tree of Figure 12.4.

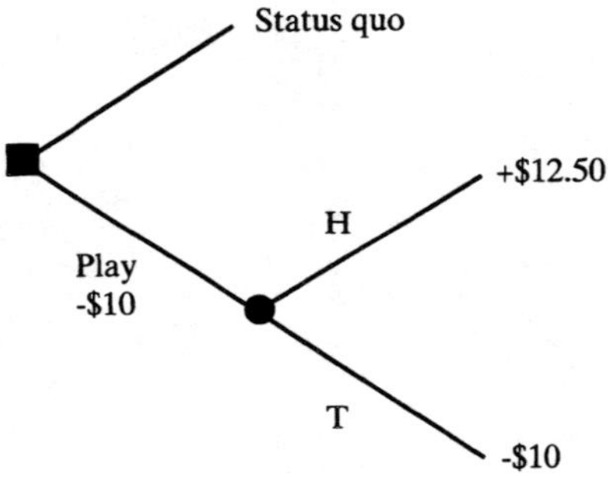

Figure 12.8 The fourth step in the solution of the decision/event tree of Figure 12.4.

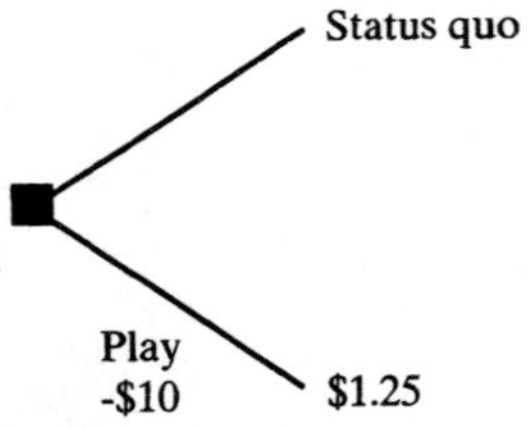

Figure 12.9 The solution to the decision/event tree of Figure 12.4.

We can see immediately from the backward induction method of project evaluation that periodic review and decision making always improves project management, and it always increases the expected value or utility of a project.

12.4 EXPECTED UTILITY ANALYSIS

We observed in Chap. 7 that the use of money directly as a measure of worth is generally incorrect. While many analyses, including the expected value analysis of the above example, do make use of money as a measure of worth, the mathematically correct approach is to use utility as the measure of worth at any time that there is risk in the outcome. Implicit in the use of money as the measure of worth in the presence of risk is that money and utility must be similarly valued, and we know that this is frequently not the case. Conversion of a decision analysis from expected value of money to expected utility is conceptually trivial. We need only replace monetary amounts in the outcomes with their utility equivalents. The analysis itself remains the same. The only difficulty lies in the conversion of monetary outcomes to utility outcomes.

Determination of Utility

One approach to the construct of a utility function for yourself is given below. First, recall that utility is arbitrary within an affine transformation.

$$u^* = au + b$$

Second, recall that if three future states are valued such that

$$W_0 > W_1 > W_2$$

then there exists a probability, p, where $0 \le p \le 1$, such that

$$u(W_1) = pu(W_0) + (1 - p)u(W_2)$$

The steps to defining your utility curve are as follows:

1. Choose two futures states, W_0 and W_2, that represent practical extremes of your future wealth. W_0 might represent a state of considerable wealth, such as having won a multimillion-dollar lottery, and W_2 might represent a state of extreme poverty.
2. Assign numerical utilities to these two states, such that $u(W_0) > u(W_2)$. Other than this condition, the values assigned are arbitrary because utility is arbitrary within an affine transformation.
3. Choose W_1 to be your current wealth. Then determine the value of p for which you are indifferent between your current wealth and a lottery with the outcomes W_0 with probability p and W_2 with probability $(1 - p)$. The utility of the state W_1 is then computed from the above equation.
4. Repeat step 3 with other values of W until you have mapped your utility function.

Sometimes it can be difficult to determine probabilities for sets of outcomes that are too broadly spaced, as in the example above. It is then possible to alter the procedure and consider values of W_0 and W_2 that are much closer to your current wealth. You could then work outward to values of W_0 and W_2 that are further and further from W_1.

The procedure outlined above works because you are the decision maker, and you can assess your preferences. How could you go about developing a utility function for a company, such as the company you work for, or for a government? In the case that there is a single decision maker, if possible, you might want to interview the decision maker.

But sometimes this will not be possible. In this case, the best approach is to look at previous decisions that have been made in the company and try to assess their outcomes and probabilities of success from which you can then construct a utility function. The mathematics of this process is essentially the same as the preceding process. Be prepared to experience some inconsistency in corporate decision making.

In the case of a government, Arrow's impossibility theorem tells us that it is unlikely that a utility function exists. But most projects, at least in terms of money, tend to be very small in comparison to the nation's budget. Therefore, taking the utility of money to be money itself is generally a good approximation. Still, managers of specific programs do exhibit risk aversion that would alter this situation.

An Engineering Example

You are an engineer working for an architect-engineer firm involved in the design of a building. Your job is to design the heating system for the building. You design a conventional system and determine that its installation cost will be $100,000. You also estimate that, at current energy prices, the cost of fuel and maintenance for the conventional system will be $20,000 per year. Then you find that an advanced heating system has just come on the market. Its installation, as a major modification to the conventional system, will cost $195,000. But at current energy prices its cost of fuel and maintenance will be only $15,000 per year. A quick calculation leads you to the conclusion that, at current energy prices, the advanced system is not a cost-effective choice. But you also feel that there is a chance that real energy prices will rise. As an example case, you select an energy price scenario in which real energy prices escalate at a rate of 4 percent per year, and in this case the advanced system does appear to be a cost-effective choice. You estimate that there is equal probability that real energy prices will stay constant or escalate at 4 percent per year.

Before completing your design, you discover that another alternative is available. It is possible to install the conventional system, but with the inclusion of an adaptor that will make possible later installation of the advanced system. The adaptor costs an additional $4,000, bringing the total initial cost of installation of the conventional system with adaptor to $104,000. Then, when the system is modified, the cost of the upgrade will be an additional $95,000.

To facilitate an analysis, you consider three options. In the first you install just the conventional system; the second is the direct installation of the advanced system; and in the third case, you assume a scenario in which you install the conventional system with an adaptor. Then you wait five years to determine if the price of energy escalates and base a decision to upgrade the system on the rate of escalation of real energy prices. Your assumption for this analysis is that, if energy price escalates in the first five-year period, it will continue to do so for the life of the building. You take the life of the building to be 25 years and the discount rate of the owner to be 5 percent in real or constant dollars. To facilitate the analysis you assume that utility is equivalent to money, and you seek to minimize the present value of total system costs.

The decision/event tree for the analysis of this decision is shown in Fig. 12.10. The

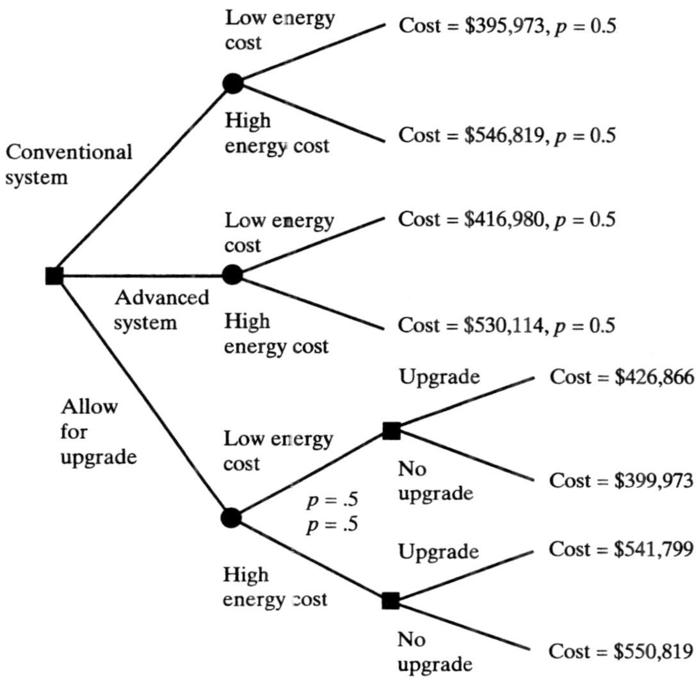

Figure 12.10 A decision/event tree for election of a heating system alternative.

outcome for the conventional system alternative includes the cost of installation, assumed payable on the day of first occupancy, plus the present value of the cost of energy and maintenance. For the low-energy-cost scenario, this is

$$PV(\text{Cost}) = \$100{,}000 + \sum_{t=0}^{t=24} \$20{,}000(1 + 0.05)^{-t}$$

And, for the high-energy-cost scenario,

$$PV(\text{Cost}) = \$100{,}000 + \sum_{t=0}^{t=24} \$20{,}000(1 + 0.04)^{t}(1 + 0.05)^{-t}$$

The present value of costs for the advanced system are similarly determined, with a higher installation cost and a lower energy and maintenance cost. The case of the upgrade is somewhat more complex. Now the installation cost involves the initial installation cost, \$104,000, and the subsequent upgrade cost, \$95,000, discounted to the present, giving a total present value of installation cost of \$182,156. Also, the cost of energy and maintenance drops at the time of the upgrade. The results of these calculations are given in Fig. 12.10.

The evaluation of the decision/event tree by backward induction is straightforward, and the results are summarized.

Alternative	Expected present value of total system costs
Conventional system	$471,396
Advanced system	$473,547
Conventional system with upgrade	$470,886

The first thing you notice here is that all of the alternatives have about the same expected present value of total system costs. Therefore, any alternative is about as good as any other. But there is an advantage to the conventional system with the adaptor to provide for later upgrade should energy prices escalate, and this alternative feels good as well, as it gives the building owner, at modest present cost, the opportunity to respond to future changes in energy prices. This, then, is your recommended design.

12.5 THE USE OF DECISION/EVENT TREES WITH AN INFINITE NUMBER OF POSSIBLE OUTCOMES

Occasionally we will encounter a decision-making situation for which the decision/event tree has an infinite number of possible outcomes. This is possible in two different ways. The first is the example shown in Fig. 10.13. A second is the decision/event tree for the game of Bernoulli's paradox discussed in Chap. 7. In this game, we flip a fair coin until it lands heads. The decision/event tree for this game is shown in Fig. 12.11. Note that we

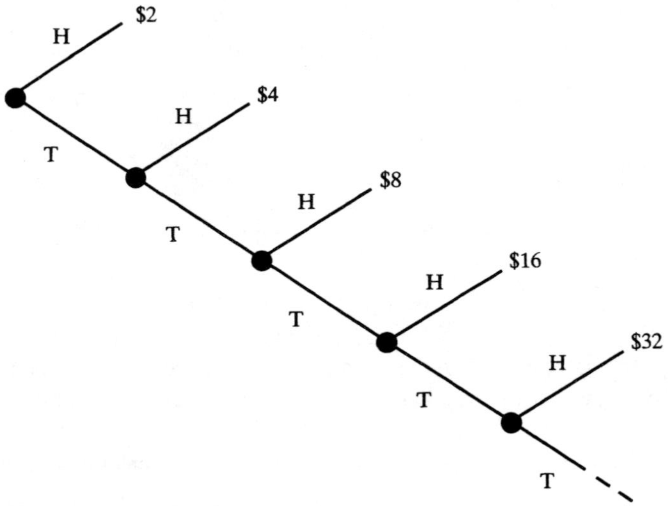

Figure 12.11 A decision/event tree for the game of Bernoulli's paradox.

cannot draw the entire tree because it has an infinite number of branches. And, since the backward induction approach requires that we begin with the entire decision/event tree drawn, we cannot use this method for an analysis of the decision to enter the game. Can we salvage the use of decision/event trees in this case? The answer is yes, but by means of a new method that we shall call the method of forward simulation. This method was first developed by Hazelrigg in 1978 for the analysis of alternatives in fusion research programs.

First, however, we must recognize that any simulation or computation we perform to evaluate the decision/event tree must be finite. That is, we can never simulate every branch of the tree because there is an infinite number of branches. Thus, any simulation of this tree will truncate the tree somewhere. The problem with this is that the truncated part of the decision/event tree will always have an infinite expected value—it will always be greater than the part of the tree that our simulation includes. Thus, we must first convert from monetary outcomes for each branch to utility outcomes. Using a logarithmic utility, such as we did in Chap. 7, this problem goes away, and an approximation to the decision/event tree that effectively truncates the tree after the tossing of several successive tails is acceptable. The conversion from monetary value to utility makes it possible to apply the forward simulation method, as it is no longer mandatory that we sample every branch of the decision/event tree.

The method of forward simulation applies specifically to the analysis of decision/event trees that, because of the complexities of the decision-making situation, cannot be fully drawn. The decision/event tree of Fig. 12.11 is such an example. Rather than to analyze the tree in all its detail as does the method of backward induction, the method of forward simulation uses a Monte Carlo approach to sample the tree. It does this by working from left to right through the tree, forward in time rather than backward. By determining probabilistic outcomes at each point in the tree as we move forward in time, we need contend only with a single scenario of the future for each simulation, rather than with all possible scenarios of the future. Again, there are two different cases in which the method of forward simulation applies. In the case of the example shown in Fig. 10.13, we can schematically draw the decision/event tree using the shorthand notation of the figure. We simply cannot draw all of the branches corresponding to the probabilistic events because they are infinite in number. In the case of Fig. 12.11, on the other hand, we cannot draw the tree, because its length is infinite. In the first case, the length of the tree is finite, but the numbers of branches at some nodes are infinite. In the second case, the numbers of branches at each node are finite, but the length of the tree is infinite.

We will consider the tree of Fig. 12.11 first. In this tree, all decisions are well prescribed in advance: if the outcome of a flip is heads, stop; if tails, flip again. In fact, we do not even need to show these implicit decisions in the decision/event tree. In the forward simulation method, we use this prescription to construct the decision/event tree as part of each simulation. In the Monte Carlo simulation, we *sample* the decision/event tree many times, recording the outcomes of each sample for later analysis. A flowchart for simulation of this case is shown in Fig. 12.12.

The case shown in Fig. 10.13 is somewhat more difficult because the future decisions are not clearly prescribed. So, for example, when we get to the second decision

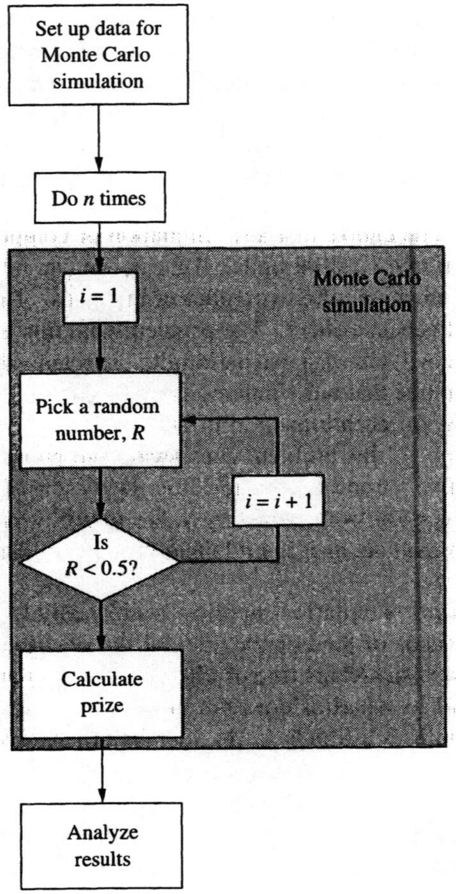

Figure 12.12 A flowchart for Monte Carlo evaluation of the decision to enter a Bernoulli's paradox game.

point in this figure, we do not know what to do. There are two approaches to continuing the analysis at this point. In the first, we do a nested Monte Carlo analysis. That is to say, we embed the part of the decision/event tree beyond the second decision point into a second Monte Carlo analysis that provides the information for making the second decision. With this information, we take the higher of the expected utility of decision A or decision B, and treat this as the expected utility of the outcome from each outcome of the first probabilistic event. We are now capable of sampling the tree in this manner to determine the expected utilities of the initial options, A or B, and we can prescribe a choice based on expected utility. The problem with this approach, of course, is that it can require vast amounts of computation. If we demand 1,000 samples from each Monte Carlo run, because the simulations are nested, we must perform 1,000,000 overall simulations. And this is with only one future decision. Additional future decisions would make this method completely intractable. We could, of course, perform fewer simulations for the future decision, but then we sacrifice accuracy.

In the second approach, we acknowledge that programs are not run perfectly anyway, largely because they pose decision-making problems that are too complex to deal with. In this approach, we accept a simulation of the real decision-making process. So therefore, we substitute a simpler algorithm for the rigorous expected-utility decision-making solution. In the example shown, we could simply prescribe a decision based on combinations of values for R_1 and R_2. Other simplified decision algorithms might also be used. In this case, the simulation proceeds forward in time, that is, from left to right across the decision/event tree without the need for nested Monte Carlo analysis. The flowchart for this approach is shown in Fig. 12.13.

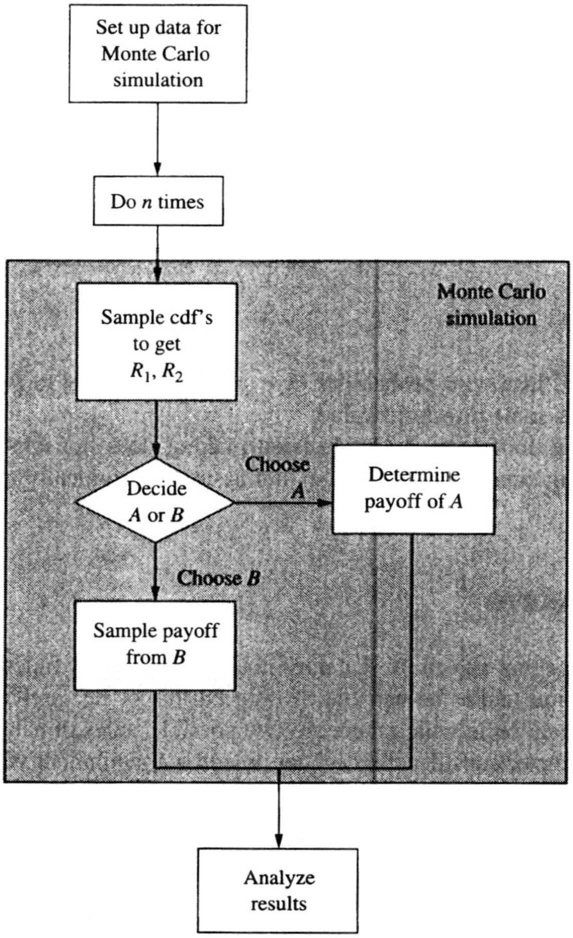

Figure 12.13 A flowchart for the Monte Carlo evaluation of a decision/event tree with an infinite number of branches.

12.6 CONFIDENCE LEVEL DECISION MAKING

Sometimes a decision maker feels uncomfortable with utility theory. Perhaps it seems a bit too impersonal, and the decision maker is not willing to allow that his or her preferences can be expressed in the form of an equation. A number of alternative decision-making algorithms have been developed that do not require explicit determination of a utility function. Confidence level decision making is one of these.

Confidence level decision making applies to win/lose situations in which the decision maker is averse to losing. Therefore, the decision maker's decision is to enter the game or activity only if the chance of winning is sufficiently high. A decision/event tree for the case of confidence level decision making is shown in Fig. 12.14. The outcomes

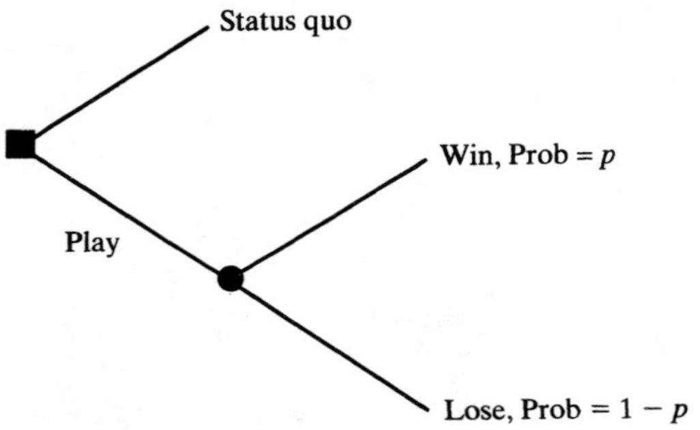

Figure 12.14 A decision/event tree for confidence level decision making.

are win with probability p, and lose with probability $(1 - p)$. The decision to play the game is made only if p exceeds some threshold value.

While simple in concept, confidence level decision making does not apply to a broad range of decision-making situations, and it does not assure economically optimal decisions.

12.7 MINIMAX DECISION MAKING

Minimax is another decision-making algorithm that does not require a utility function. It merely requires that the decision maker be capable of ordering his or her preferences. Consider a decision-making situation in which there are two possible states of nature and two possible actions. For example, consider the decision to add a containment vessel to a nuclear reactor. The two possible states of nature are (1) that the reactor remains stable throughout its life and never requires the containment vessel, and (2) a major accident occurs that would result in a large release of radioactive material unless the reactor is enclosed by a containment vessel. Under scenario 1, there is no need for a containment vessel. If we build one, we incur a considerable cost and receive no benefit. Under scenario 2, the containment vessel is vital to human safety. These states of nature, possible actions, and outcomes are reviewed below.

Possible states of nature	Possible actions	
	Build containment vessel	Do not build containment vessel
Reactor remains stable	(*A*) Added cost of containment vessel	(*B*) Status quo
Major accident occurs	(*C*) Human safety maintained	(*D*) Loss of life
Maximum loss	*C*	*D*

The preference ordering for the four possible outcomes is $B > A > C > D$. For each possible action, we examine the maximum possible "loss" (loss here is defined as lost utility from the failure to achieve the most desired outcome), which as indicated above are cases C and D. In this example, since $C > D$, outcome C is the minimum maximum possible loss. Thus, we choose to build a containment vessel to avoid the possibility of outcome D. This is the action that minimizes the maximum loss: the *minimax* action.

The important thing to observe regarding the minimax solution is that it does not require assessment of detailed utility functions and it does not require assessment of probabilities. It requires only a preference ordering for the possible outcomes.

12.8 REGRET

A minimax risk decision-making criterion was suggested by Savage (1951) as an alternative to the minimax criterion. The method requires that we assign utilities to all possible outcomes, but we are not required to determine probabilities. Consider a case in which there are two possible states of nature, S_1 and S_2, and two possible actions, A_1 and A_2. For each state of nature and possible action we assign a utility. Note that the utilities for each state of nature are comprised of the combination of the state of nature and the action taken, as actions typically have costs that affect the final overall state. The following data comprise a typical example.

Actions	Utilities	
	S_1	S_2
A_1	5	8
A_2	3	22

Now suppose the true state of nature is S_1. If we choose action A_1 we obtain utility 5, and if we choose action A_2 we obtain utility 3. The *regret* associated with the choice of A_2 is then obtained as 5-3 or 2. It is the utility or opportunity that is lost by choosing a nonoptimal action as opposed to the optimal action. The regret associated with the optimal action is 0. Using this approach, we translate the preceding utility table into a regret table.

	Regrets	
Actions	S_1	S_2
A_1	0	14
A_2	2	0

Note that every column of the regret table has at least one zero in it. The decision rule is to choose the action that has the smallest maximum regret, in this case A_2. The difference between the minimax criterion and the minimax risk criterion is this: the minimax criterion selects the action that minimizes the maximum possible loss; the minimax risk criterion selects the action that minimizes the maximum possible risk.

An example shows the difference between the minimax criterion and the minimax risk criterion. Consider a case with five available actions and two states of nature and with the loss function shown below.

	Losses	
Actions	S_1	S_2
A_1	2	3
A_2	4	0
A_3	3	3
A_4	5	2
A_5	3	5
Minimum loss	2	0

The loss table converts into a regret table much the same as does the utility table.

Actions	Regrets S_1	S_2
A_1	0	3
A_2	2	0
A_3	1	3
A_4	3	2
A_5	1	5

We now compare choices by the minimax criterion and the minimax risk criterion. By the minimax criterion we obtain the following.

Actions	Losses S_1	S_2	Maximum loss
A_1	2	3	3
A_2	4	0	4
A_3	3	3	3
A_4	5	2	5
A_5	3	5	5

In this case, the smallest maximum loss is 3, and we choose A_1 or A_3. Next we look at the choice by the minimax risk criterion.

Actions	Regrets S_1	S_2	Maximum regret
A_1	0	3	3
A_2	2	0	2
A_3	1	3	3
A_4	3	2	3
A_5	1	5	5

In this case, the smallest maximum regret is 2 for choice A_2.

We should note that the minimax risk decision-making approach has several faults. First, there is no demonstrated mathematical correspondence between utility and regret. Thus, there is no reason to believe that regret is an appropriate measure for decision making. Second, it is possible to construct pathological cases in which an arbitrarily small advantage in one state of nature outweighs a large advantage in another. Finally, it is possible for the choice of action to be influenced by the introduction of an *irrelevant alternative*. For example, suppose that there are three possible actions, A_1, A_2, and A_3, and that from the actions by the minimax risk criterion we select A_1. Now, it is sometimes possible to introduce a fourth possible action, A_4, which is not selected, but which changes the action selected to, say, A_2. This is disturbing indeed.

12.9 BAYES SOLUTIONS

The final approach to decision making that we will consider here is referred to as Bayesian decision making. This approach constitutes a modification to the minimax or minimax risk approaches to make use of probability information. Recall that neither the minimax nor the minimax risk approaches require probability information. Rather they deal only with outcomes or states of nature themselves allowing decisions to be swayed by highly unlikely outcomes. Bayes solutions are obtained by weighting loss in the case of minimax or regret in the case of minimax risk by the probabilities of occurrence of each outcome state. To illustrate, let us extend the example given in the previous section. For the case of minimax decision making, we get the following result.

Actions	Losses		Expected loss
	$S_1, p = 0.2$	$S_2, p = 0.8$	
A_1	2	3	2.8
A_2	4	0	0.8
A_3	3	3	3.0
A_4	5	2	2.6
A_5	3	5	4.6

The minimum expected loss is 0.8 for action A_2. Going through the same computations for minimax risk decision making, we get the following result.

Actions	Regrets		Expected regret
	$S_1, p = 0.2$	$S_2, p = 0.8$	
A_1	0	3	2.4
A_2	2	0	0.4
A_3	1	3	2.6
A_4	3	2	2.2
A_5	1	5	4.2

The minimum expected regret is 0.4 also for action A_2.

The actions chosen above are called Bayes actions. At this point, the method approaches closely the method of expected utility decision making. Both utilities and probabilities are necessary, and so it is recommended that, if you have these quantities, you simply use the expected utility method.

12.10 SUMMARY

In this chapter we took a more detailed look at formal approaches to decision making. We began by reiterating the definition of a decision as an irrevocable allocation of resources, which is an action taken at a point in time based upon the information available at that time. This led to a definition of decision analysis as the systematic evaluation of alternative choices. We carefully distinguished between decisions and outcomes, and noted that a good decision is one that is made rationally, in the context of the available information and across the range of available alternatives, to pursue the desired system goals. We saw that a good decision can lead to a bad outcome, and a bad decision can lead to a good outcome.

Next, we developed the method of backward induction. This method has an established mathematical basis, and it is a powerful tool for the analysis of a broad class of decision-making situations. It fails to work, however, in the cases where the situation is such that it is impossible to draw a decision/event tree. We saw two types of decision-making situations in which it is impossible to draw a decision/event tree: those that have an infinite number of branches at some nodes and those that have an infinite number of nodes. For these cases, we presented the method of forward simulation. The method of forward simulation is an expected utility approach that seeks to yield results similar to backward induction by sampling the decision/event tree using Monte Carlo simulation rather than by enumeration of all branches on the tree.

Then we considered some decision-making approaches that do not require the development of utility functions or probabilities. The method of confidence level decision making requires the assessment of probabilities, but not utility. The minimax approach requires neither probabilities nor utilities, and the minimax risk approach requires utilities but not probabilities.

The methods developed here can help us to evaluate alternative choices in cases where we are aware of the options, where we have expectations on each option, and where we have assessed at least in some sense our values. They provide both a better understanding of the advantages and disadvantages of each option, and they provide better resolution between options than we could obtain by intuition alone.

REFERENCES

1. Clemen, R. T., *Making Hard Decisions, an Introduction to Decision Analysis,* PWS-Kent Publishing Company, Boston, MA, 1990.
2. Cyert, R. M., M. H. DeGroot, and C. A. Holt, Sequential Investment Decisions with Bayesian Learning, *Management Science,* vol. 24, no. 7, March 1978.
3. Dyer, J. S. and R. F. Miles, Trajectory Selection for the Mariner Jupiter/Saturn 1977 Project, Proposed Technical Memorandum, Jet Propulsion Laboratory, Pasadena, CA, July 1, 1974.
4. Fishburn, P. C., *Utility Theory for Decision Making,* Lively, New York, 1970.
5. Fudenberg, D. and J. Tirole, *Game Theory,* The MIT Press, Cambridge, MA, 1992.
6. Hammond, J. A., Better Decisions with Preference Theory, *Harvard Business Review,* Nov./Dec. 1967, pp. 123–136.
7. Hausner, M., Multidimensional Utility, *Decision Processes,* R. M. Thrall, C. H. Coombs, and R. L. Davis eds., John Wiley, New York, 1954.
8. *IEEE Transactions on Engineering Management,* Special Issue, vol. EM-21, no. 4, Nov. 1974.
9. Keeny, R. L., Multiplicative Utility Functions, *Operations Research,* vol. 1, 1974, pp. 23–25.
10. Keeny, R. L., A Decision Analysis with Multiplicative Objectives: The Mexico City Airport, *Bell Journal of Economics and Management Science,* vol. 4, 1973, pp. 101–117.
11. Lindgren, B. W., *Elements of Decision Theory,* Macmillan Company, New York, 1971.
12. Luce, D. R. and H. Raiffa, *Games and Decisions,* John Wiley, New York, 1958.
13. Magee, J. F., How to use Decision Trees in Capital Investment, *Harvard Business Review,* July/Aug., 1964, pp. 126–138.
14. Magee, J. F., Decision Trees for Decision Making, *Harvard Business Review*, Sept./Oct., 1964, p. 79.
15. Matheson, J. E. and W. Roths, Decision Analysis of Space Projects: Voyager Mars, American Astronautical Society National Symposium, Saturn/Apollo and Beyond, June 11, 1967.
16. *Readings in Decision Analysis,* Stanford Research Institute, Menlo Park, CA, 1976.
17. Savage, L. J., The Theory of Statistical Decision, *Journal of the American Statistical Association,* 46, 1951, pp. 55–67.

PROBLEMS

1. Give five examples of decisions. For each, explain when the decision is made. What resources are being committed by the decision?

2. In the case of guessing how many beans are in a jar, where the closest guess wins a prize, what are the possible outcomes?
3. Draw a decision/event tree for the game of guessing how many beans are in a jar.
4. You are assigned to a relatively small project in your company. As you are conducting the project, a somewhat expensive piece of needed equipment fails. To replace the equipment will cost \$100,000 and take 2 months. This will delay the benefit of the project, estimated at \$150,000, also by this time. On the other hand, it is possible that the equipment might be fixed. You estimate that, for an investment of \$25,000, you have a 0.3 probability of fixing it. If the piece of equipment had not failed, the cost to complete the project would have been \$50,000, and this cost will be incurred when the equipment is fixed or replaced. What should you do: try to fix it, replace it, or quit the project?
5. You are designing a subsystem on a piece of heavy equipment. This subsystem is prone to breaking, and there appears to be no way to design it better so as to reduce the failure rate. However, it seems feasible to design sensors into the subsystem to warn of incipient failure and to pinpoint failures so that the subsystem can be repaired in considerably less time. What do you need to know in order to evaluate this design option?
6. You are managing a project that potentially could be accomplished by either of two approaches, A and B. But neither approach has a 100 percent chance of success. A has a 0.75 probability of success and costs \$10,000. B has a 0.5 probability of success and costs \$8,000. Each approach will take 6 months to check out. The benefit of the project completed is \$250,000. Your options are:
   a. Try approach A. If that does not work, quit.
   b. Try approach B. If that does not work, quit.
   c. Try approach A first. If that does not work, then try B.
   d. Try approach B first. If that does not work, then try A.
   e. Try both approaches simultaneously.
   f. Quit the project, and do not try either approach.

   Which option is best? What is its probability of success? What is its expected cost? What is the expected time to complete the project? Assume a 10 percent discount rate.
7. Determine your utility function for money. Using your utility function for money, determine how much you would pay to enter a game of Bernoulli's paradox.
8. You are given a chance to play a coin-flip game. A fair coin will be flipped ten times. If heads occurs three or more times, you will win. You decide to play if your chance of winning is more than 0.8. Will you play this game?
9. A device has been invented to rescue airplanes in an emergency. It consists of a parachute that can be deployed to lower an airplane to safety in the event of a major malfunction. The device is expensive, and occasions for its use are rare. Thus, possible outcomes are the following:
   a. The device is not installed, and all airplanes land safely.
   b. The device is not installed, and some airplanes crash with loss of life.
   c. The device is installed, and all airplanes land safely (that is, the device is not needed).
   d. The device is installed, and some airplanes use it to save lives.

   The preference order of these outcomes is given as $A > C > D > B$. Using minimax decision making, decide what to do.
10. In problem 9, assume that the utilities of the outcomes are as follows: $u_A = 100$, $u_B = 8$, $u_C = 10$, and $u_D = 20$. Use the method of regret to decide what to do. Suppose that the probability of an accident that could be prevented by the device is 0.07. Use Bayesian decision making to obtain a choice.

11. Design a decision-making strategy for buying a new car. What data would you like to have?
12. You drive to a local store to buy some exercise equipment. You believe that you can buy the equipment you want on sale for $140. But, when you get to the store, the equipment you want is sold out and the store has raised the price to $160. What do you do?
13. Do you want to get a Ph.D.?
14. What time should you go to bed tonight?
15. Assume that you have a car that is 10 years old. It has been a good car, but now it needs a new transmission. What do you do?
16. You want to fly a light airplane from City A to City B, a distance of 600 km. The airplane flies at an airspeed of 250 km/h while using 38 liters/h of fuel. What is the minimum amount of fuel you should have on board at takeoff?

13

State Transition Matrix Models

13.1 THE PURPOSE OF THIS CHAPTER

Many systems that appear to have complex behavior are subject to analysis by state transition matrix (STM) models. The general form of such systems is that of a linear set of differential equations, approximated as a set of difference equations. STM models describe the behavior of many systems that occur in nature and that are important in systems engineering. They are also useful for simulation of various policy issues and social problems, such as the spread of a disease, antidrug use policies, and anticrime policies. In systems engineering, STM models are useful for determining the impacts of various product liability or warranty policies. They also help us to determine impacts of product unreliability.

The purpose of this chapter is to introduce the STM model and its applications. STM models comprise a subset of the larger topic referred to as Markov models.

13.2 THE STATE TRANSITION MATRIX MODEL

Consider the system of differential equations

$$\frac{d\boldsymbol{x}}{dt} = B\boldsymbol{x} + \boldsymbol{b}(t)$$

where B is a matrix of coefficients, which might be a function of time, and $\boldsymbol{b}(t)$ is a vector forcing function. This system can be written in the approximate form

$$\frac{\Delta \boldsymbol{x}}{\Delta t} \cong B\boldsymbol{x} + \boldsymbol{b}(t)$$

If we take

$$\Delta \boldsymbol{x} = \boldsymbol{x}_{i+1} - \boldsymbol{x}_i = \Delta t B_i \boldsymbol{x}_i + \Delta t \boldsymbol{b}(t)$$

we can rearrange this system of equations into the form

$$\boldsymbol{x}_{t+1} = A\boldsymbol{x}_t + \boldsymbol{b}_t$$

This is the system of equations that is addressed by the STM method. In the above form, the elements of A may be constants or functions of time, but linearity demands that they do not depend on the state $\boldsymbol{x}$. Computationally, however, it is possible to use the STM method even if the elements of A include state variables.

We refer to the values of the state variables, $\boldsymbol{x}_t$, as populations, and to the matrix A as the state transition matrix. The elements of A, a_{ij}, can be interpreted as the probabilities that a member of state j will transition to state i over the next unit of time, or deterministically the fraction of members of state j that transition to state i. The vector $\boldsymbol{b}_t$ is a forcing function that enables externally caused changes in the populations of the states.

13.3 EXAMPLE PROBLEMS

We can demonstrate the STM model and its applications best through the use of some examples. To begin, we will consider a very simple case consisting only of two states, X and Y. We will take as our initial population of X, $x = 1{,}000$, and as the initial population of Y, $y = 0$. We will take the probability of transition from X to Y to be 0.05, and the probability of transition from Y to X to be 0.025. First, we illustrate this problem as shown in Fig. 13.1. Note that the transition probabilities are shown in the figure. Implicit in the figure is that members of a population that do not transition from one state to another remain in their original state.

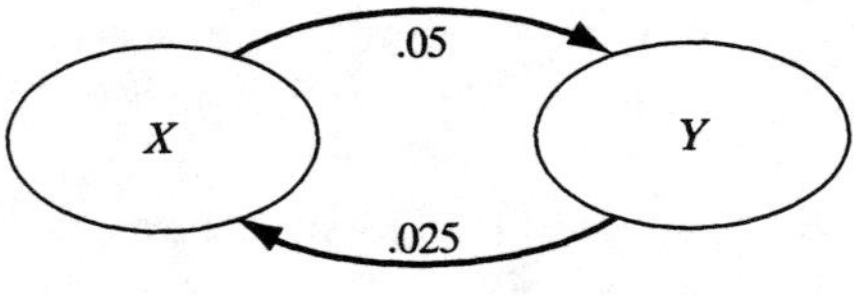

Figure 13.1 Illustration of a simple STM model.

The equations that describe this system are

$$\begin{pmatrix} x_{t+1} \\ y_{t+1} \end{pmatrix} = \begin{bmatrix} 0.950 & 0.025 \\ 0.050 & 0.975 \end{bmatrix} \begin{pmatrix} x_t \\ y_t \end{pmatrix}$$

Initial values are put in for x and y, and the system is solved by repetitive calculation. The result of this calculation is shown in Fig. 13.2. This figure shows a simple exchange of state from X to Y, and a new equilibrium condition with the population of Y eventually exceeding that of X.

In concept, the STM method is very simple. It is merely an accounting tool that, through the use of computers, enables examination of rather complex situations. The next

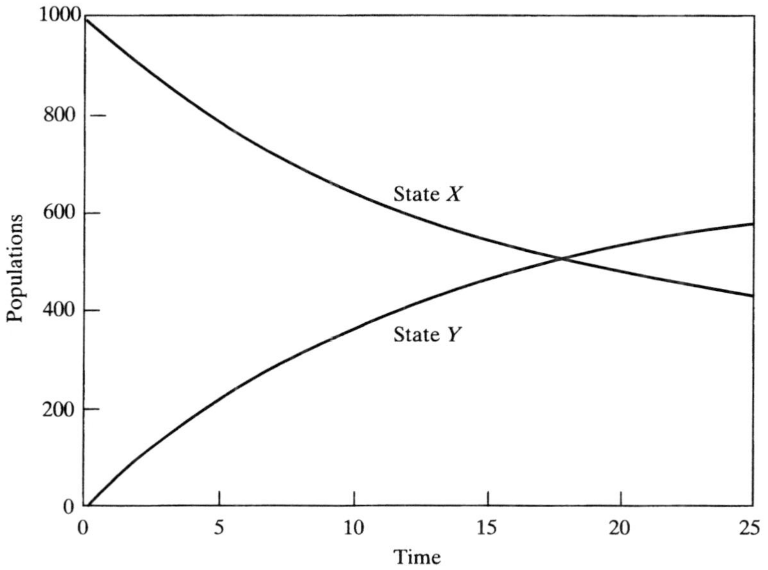

Figure 13.2 Results of the simple STM model given in Figure 13.1.

example is somewhat more complex and realistic. It is a typical model of the spread through a population of an infectious disease. The STM model, shown in Fig. 13.3, involves six populations:

1. People who have not been exposed to the disease,
2. People who have been exposed to the disease,
3. People who, after exposure to the disease, remain well or who recover from the disease,
4. People who are sick,
5. People who are sick and under treatment,
6. People who have died from the disease.

Again, we show the state transition probabilities in the figure. In this case, however, we have chosen a nonlinear transition from the unexposed state to the exposed state. We take this probability as the ratio of the population of states 2 and 4 above to the population of states 1, 2, 3, and 4. Here, we are assuming that people who are unexposed are protected from exposure to those people who are undergoing treatment, and that unexposed people are not at risk of exposure from dead people.

The output of this model is shown in Fig. 13.4. It shows several rather interesting features typical of infectious diseases. First, we see that a major portion of the total population is rapidly infected by people who are sick or who have been exposed and have yet to become sick. Interestingly, however, we see that not all members of the unexposed population become exposed. In fact, the disease runs its course over a short period of time, leaving some individuals unexposed. This model can offer information for policy

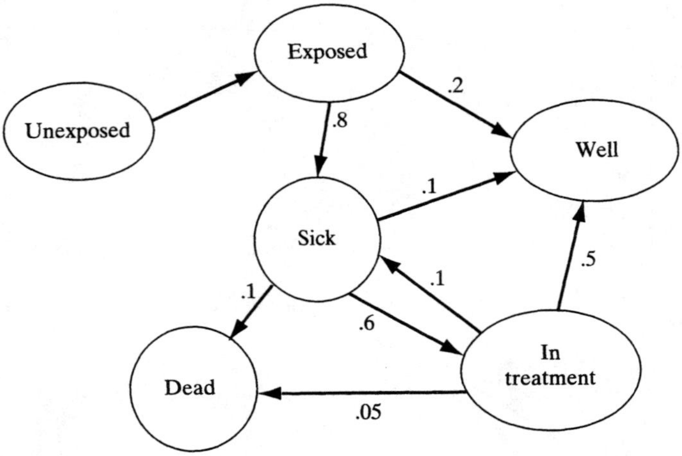

Figure 13.3 An STM model of the spread of an infectious disease.

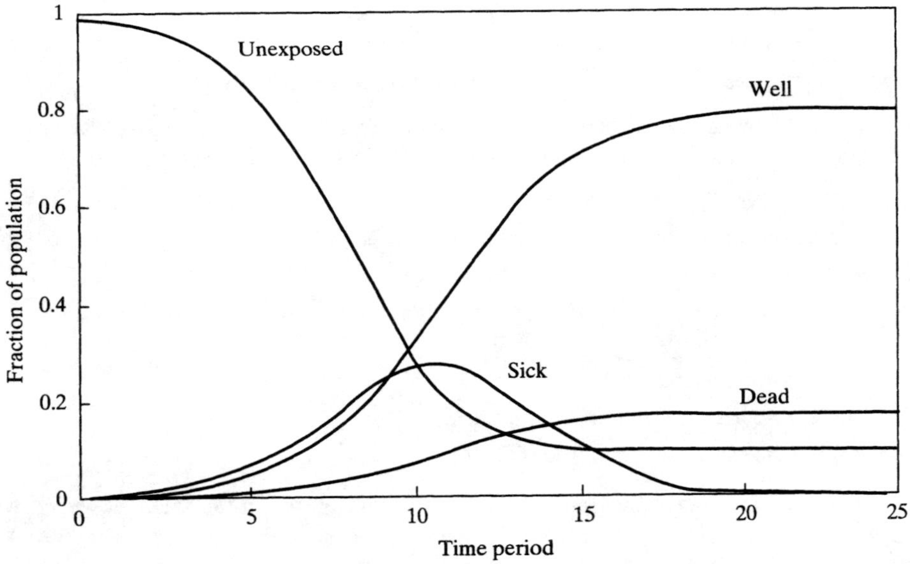

Figure 13.4 The output from the STM infectious disease model.

making in two important areas. First, it can tell us how much care will be required to take care of sick persons, and second, it can tell us how effective different policies to limit exposure might be in prevention of the disease and reducing the need for treatment.

Our third example relates to systems engineering. We will consider a product that lasts up to five time periods, after which it is discarded. Sometimes it is discarded earlier. At various times during the lifetime of the product, it can break and require repair. We have several potential questions regarding the product.

1. If we warranty the product, what cost will we incur to fix it as a function of the duration of the warranty?
2. If we make the product using a hazardous material, such that we are required to repurchase the product for disposal, what will the disposal of the product cost?
3. If we improve the reliability of the product, what will we save in warranty costs?

We can illustrate this problem as shown in Fig. 13.5. The rate of production of products is given by a Gompertz curve of the form

$$\text{Production rate} = 5{,}000 \times 0.01^{0.9^t}$$

The five time periods that comprise the life of the product are denoted by the ages one through five. Repair actions are denoted by the states denoted Rep. We assume that all failures are fixed by repair.

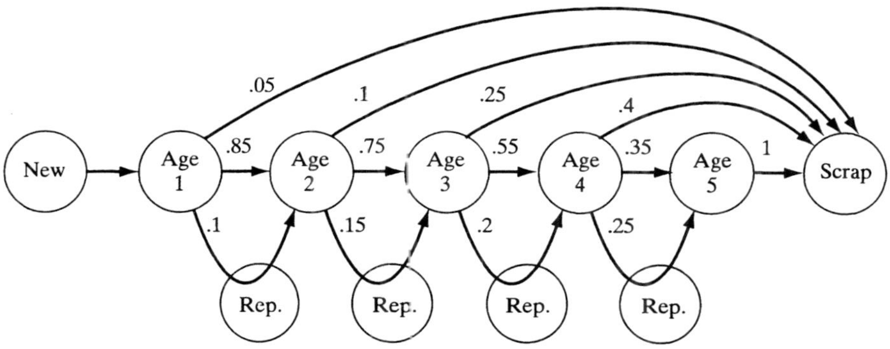

Figure 13.5 A product breakage, repair, and disposal model.

The outputs from this model are several, and worthy of discussion. Sample outputs are shown in Figs. 13.6 through 13.8. Figure 13.6 shows, as a function of time, how many products are in service and the rate at which products are being scrapped. These data could be useful in determining the cost of product disposal, if that cost will fall on the company producing the product.

Figure 13.7 shows the rate at which products are being scrapped in comparison to the rate at which they are being produced. This could be important information if the cost of scrapping products must be borne by the products that are being produced. Future prices and the competitiveness of the product could be determined by policies set as a result of such information.

Finally, Fig. 13.8 shows the impact of various warranty policies. The cost of a warranty policy will relate to the duration of the warranty period, and therefore the cost of repair that will be borne by the company. These data are typical of the data that can be obtained using the STM method. We conclude that it is a powerful method, useful for the analysis of a broad class of issues.

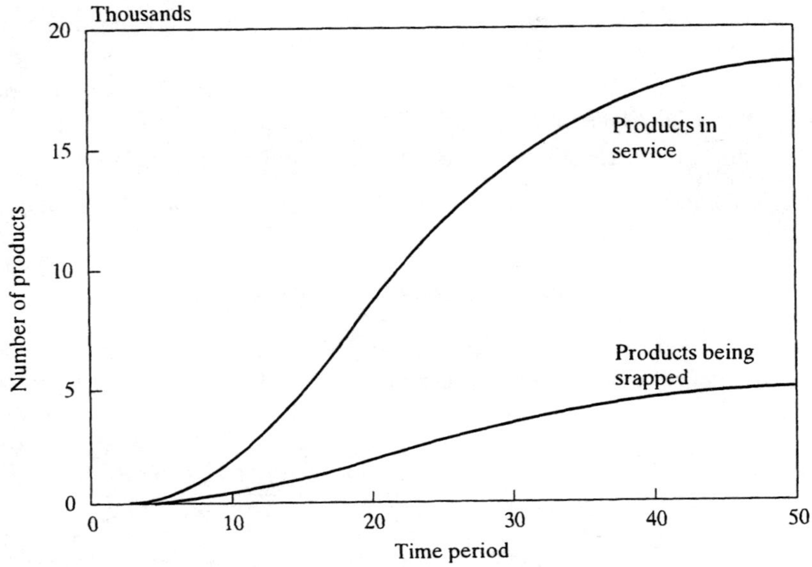

Figure 13.6 Products in service and the rate at which products are being scrapped.

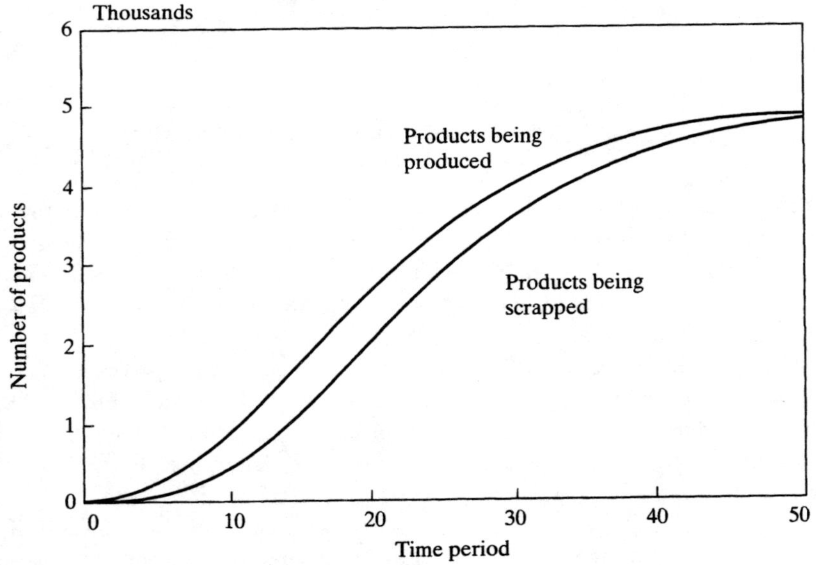

Figure 13.7 Comparison between the rates at which products are being produced and scrapped.

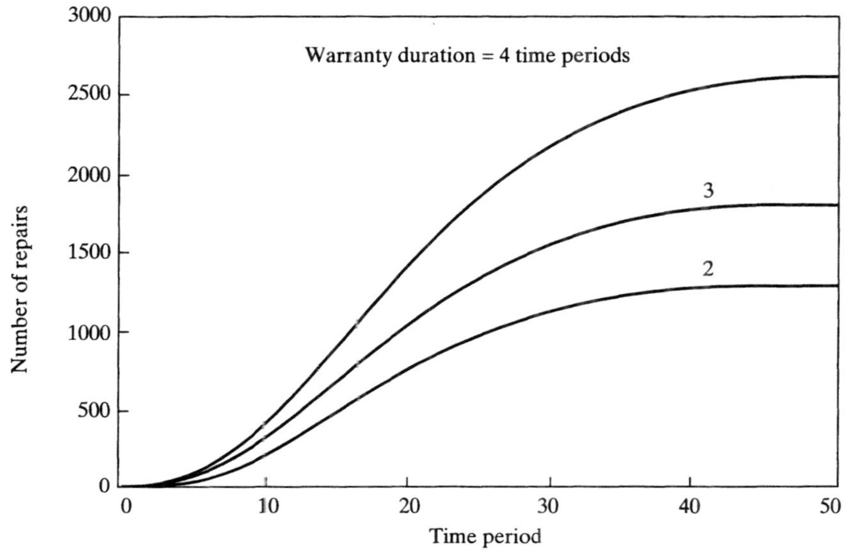

Figure 13.8 STM modeled rates of repair for different product warranty policies.

13.4 INCLUDING UNCERTAINTY IN STATE TRANSITION MATRIX MODELS

The STM method, as described previously, seems to be a probabilistic method. Indeed, we even describe the state transition matrix as a matrix of probabilities, and for members of a population that is essentially what this matrix is. But, despite its appearance, the STM method is a deterministic method. Yet, as with most other variables, we do not know the transition probabilities with precision. Perhaps we do not know even the initial states with precision. To be precise, therefore, all of these variables must be described as probability distributions. And, just as we have converted other deterministic analyses to probabilistic analyses using the method of Monte Carlo simulation, we can convert the STM to a probabilistic analysis. To do this, we describe all the input variables for an STM analysis as random variables, each described by a probability distribution. We then sample each variable and run the corresponding deterministic STM model, following which we repeat the procedure many times, say 1,000 to 100,000 times, to gather the desired probabilistic results. A conceptual flowchart for this procedure is shown in Fig. 13.9.

13.5 SUMMARY

We present, in this brief chapter, a method for the analysis of systems that are comprised of populations, each in a unique state. The method is referred to as the state transition matrix method. Its basic form closely resembles that of a set of linear differential equa-

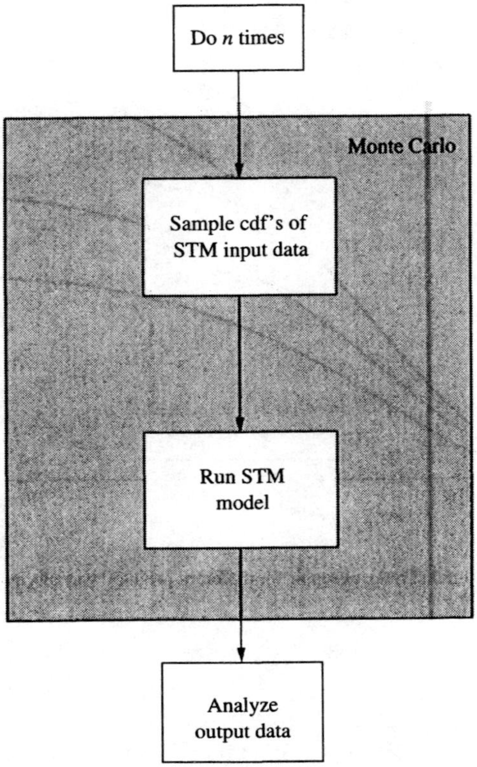

Figure 13.9 A flowchart for incorporating uncertainty into an STM model.

tions, although the method can handle some nonlinearities. It is an important method of analysis for the systems engineer.

PROBLEMS

1. Your company plans to introduce a product that has a failure rate of 2 per 100 operating products per year. The plan is to provide a 3-year warranty for all defects or failures. The cost of repairing the product, on average, is \$50. A 5-year production run is planned, with 1 million products produced in the first year, 2 million in the second year, 3 million in the third year, 2 million in the fourth year, and 1 million in the fifth year. Purchasers also dispose of the product at the rate of 7 per 100 per year. What is the present value of the cost of the warranty? Assume a 10 percent discount rate.
2. In problem 1, suppose an engineering option would reduce the failure rate to one per 100 operating systems per year. Further, suppose that this option costs an additional \$1.50 per product. Now what is the present value of the cost of the warranty? Would you advise making the change?

3. In problem 2, suppose that the added reliability gains customer satisfaction such that sales would increase in years 2 through 5 by 20 percent. Now what is the present value of the cost of the warranty? If profit is $10 per product, does the change make economic sense?

4. Your company produces a product that sometimes requires repair. Part A breaks with frequency five per 100 operational products per year, and part B breaks with frequency eight per 100 operational products per year. The product itself is discarded at the rate of 10 percent per year. Part A costs $5 and part B costs $3. The direct cost of storage of each part is $0.50 per year. A two-year production run is anticipated, with 1 million products produced each year. The policy of your company is to maintain a stock of spare parts for 10 years after the end of the production run. How many of each part must your company stock? If all spare parts are stocked at the end of the production run, what will this spare parts policy cost your company? Assuming that the price for each part will be fixed over all time, how much would your company have to charge for each part in order to be 90 percent certain of recovering its spare parts costs?

5. Consider two options to the policy in problem 4: (1) reduce the spare parts coverage period from 10 years to 8 years, but then, after the 8-year coverage period, buy back any broken products at 20 percent of their purchase price; and (2) reduce the failure rates from five and eight per 100 operational products per year to two and four per 100 operational products per year. What will be the impacts of these two alternatives? Which would you recommend?

14

Modeling the Research and Development Process

14.1 THE PURPOSE OF THIS CHAPTER

Throughout this book, it is emphasized that all phases of the life cycle of an engineering system are subject to being modeled. The modeling emphasis provided through most engineering curricula, however, is on modeling of the physical system. This is difficult enough. But there are many other aspects of engineering systems and life cycle phases of engineering systems that need to be modeled if we are to obtain all the information that we need to make good engineering decisions. And these nonphysical aspects of engineering systems can be, at least conceptually, even more difficult to model. Research and development (R&D) is an activity that comprises a phase in the life cycle of many engineering systems. It is important to the success of the system that this activity be well planned and managed, and that requires modeling. The purpose of this chapter is to present a method for the modeling of the research and development process, and the evaluation of options in this activity.

14.2 ATTRIBUTES OF RESEARCH AND DEVELOPMENT ACTIVITIES

Research and development activities possess many attributes that differentiate them from activities whose output is a physical system:

- The product of research and development is knowledge and information. These lead to the ability to make good engineering decisions and to carry them through to their desired conclusion. The output of a research and development activity cannot be measured in terms of physical parameters.
- Unlike a program to build a physical system, which seeks a single design for implementation, research and development activities may pursue goals through several

parallel approaches. Each approach may itself have parallel and overlapping activities.

- When there exist parallel and overlapping activities, there may be correlation in the probabilities of success or failure between these activities.
- There might be a variety of ways of achieving success in any particular research and development activity. Each successful outcome would provide a technology capable of satisfying the goals of the program.
- There might be a significant probability of failure.
- There can be considerable uncertainty in the schedule by which the research and development activity reaches success or failure. Along with uncertainty in the schedule, there can be a significant uncertainty in the cost of a research and development activity.
- A research and development activity can have several phases, across which programmatic decision making can be very important. For example, in the early phases of a research activity, the objective might be to prove the feasibility of certain approaches. Then, based on proven feasibility and other factors, a decision might be made to eliminate certain approaches from further development, carrying forward to completion only one technology capable of satisfying the requirements of the program.

Any research program can be measured in terms of three types of parameters: technical accomplishments, costs, and schedules. And, at least in theory, it is possible to hold two of these three constant, allowing the third to vary. Clearly, it is possible to hold cost (or cost and schedule) constant if technical accomplishments are allowed to vary. Less clearly, technical accomplishments and cost could be held constant and schedule allowed to vary. The latter may not always be possible, but the point is that, if the schedule can be varied, some variation in technical accomplishments is possible. A methodology for modeling research and development activities must take into account the management approach; what will be held constant and what will be allowed to vary.

14.3 A SIMULATION APPROACH TO EVALUATING RESEARCH AND DEVELOPMENT ACTIVITIES

An approach to the modeling of research and development activities that incorporates these considerations was developed by Hazelrigg (1985) and tested for the case of magnetic confinement fusion research. The approach is called RADSIM, the Research and Development SImulation Model. The RADSIM approach is a probabilistic forecasting method that accounts for uncertainty in schedules, degree of technological accomplishment, and correlation between related activities, and also allows for major decisions to continue or discontinue research and development activities at selected decision dates. The approach, which is based on the method of forward simulation as described in Chap. 12, uses the Monte Carlo technique to simulate a research and development activity. Each element and each decision of the research and development activity is simulated forward in time from the present through either successful completion of the activity, termination of the activity because of technical failure, or termination because of time or budget constraints. Each simulation represents one possible outcome of the overall research and development activity. By conducting several (perhaps 1,000) such simulations, a statistical sampling is obtained of the possible outcomes, including the probability of success and activity cost.

We begin our formulation by defining the general structure of a research and development activity as shown in Fig. 14.1. A total program may be comprised of research activities on a number of *alternative approaches*. Each approach has associated with it a number of *generic problem areas* (*A*, *B*, *C*,...) within which significant technology advancement is required in order to achieve program success. Within each alternative approach, technological advancements are structured into a sequence of *research and development phases* (*I*, *II*, *III*,...), each referred to as a *project*, ending in the development of a useful technology. The overall research and development activity, which we shall refer to as a *research program*, can thus be viewed as a set of alternative approaches, each of which consists of a set of projects that together could satisfy the objectives of the program.

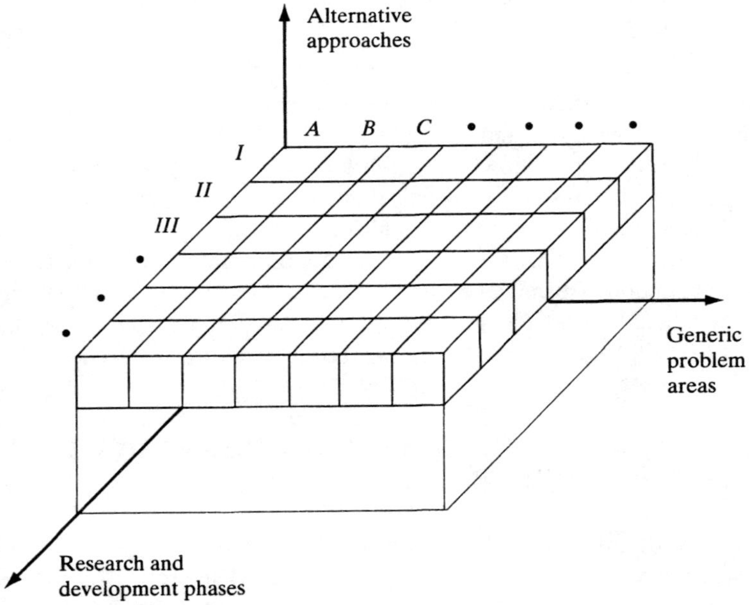

Figure 14.1 Dimensions of an R&D program.

Finally, the activities may progress through research and development *phases* such as basic research, applied research, subsystem development, and prototype and test, from concept to technology readiness. Research program cost estimating and a forecast of program technical results must account for these three dimensions of a research program, and it must take into account the uncertainties, overlaps, and correlations between disparate activities and across the alternative approaches. Each project is comprised of a set of *sub-*

projects, of which each is a discrete activity with a definable set of potential outcomes. Subprojects are illustrated as the small blocks in Fig. 14.1. Projects are illustrated as a row of blocks in a particular horizontal plane of this figure, including all subprojects in a given research and development phase. An alternative approach is represented as all blocks in a given horizontal plan, and the program consists of all blocks in all planes of the figure.

There are five elements of research and development programs that we must model: (1) the research process, (2) technological breakthroughs, (3) relationships among subprojects, (4) the performance of alternative approaches, and (5) the decision or program management process.

Modeling the Research Process

We model the research process at the level of the research and development subproject. First, we must identify each research and development activity that might take place in the overall research program, that is, each plausible subproject that could be conducted relating to research and development in a particular generic problem area, research and development phase, and approach. We can accomplish this by listing the alternative approaches that appear viable, and then listing generic problem areas associated with each approach. Using these lists, we then associate each problem area with each approach using an incidence matrix as shown in Fig. 14.2.

Generic problem areas	Alternative approach — nuclear propulsion example			
	Solid core	Gas core	Electric propulsion	Dual mode
Cryogenic tanks	✓	✓		✓
Electric propellant tanks			✓	✓
Nozzle	✓	✓		✓
Electric thrusters			✓	✓
Thermal energy conversion cycle			✓	✓
Shielding	✓	✓	✓	✓

Figure 14.2 An incidence matrix.

Based on the incidence matrix, in each generic problem area for each alternative approach and for each research and development phase, we set minimum technology requirements that represent a successful level of technology development. It is this set of requirements against which both subproject costs and probability of success are gauged. Within each subproject, the technology requirements are given as a lexicographic set: failure to achieve the value stated for any element of the requirements is interpreted as failure of the subproject.

Having identified the minimum technology requirements, our next job is to assess the probability of success and budgetary requirements as a function of time, given a specific set of research and development activities. In general, the planning and evaluation process proceeds from the set of technology requirements to a specifically planned activity designed to achieve the technology requirement. Research and development activities can be planned at a variety of different activity levels, as well as with different approaches. Each approach or activity level will have a unique relationship between success and time, and it will have a unique set of associated costs. Thus, for example, a very vigorous program of technology development may result in success in a very short period of time, whereas a less ambitious level of activity and cost might have both a lower probability of success and longer duration.

Figure 14.3 shows a format that you could use to input subproject probability of success and cost data. Shown is the probability that a particular subproject activity has been successfully completed as a function of time from the start of activity on the subproject. You could develop these curves in any of several ways. One relatively easy approach is to obtain from appropriate technical specialists answers to the four questions:

- What is the probability of ultimately achieving the stated technology requirement?
- What is the minimum subproject activity time required to achieve the stated technology requirements?
- What is the most likely subproject activity time required to achieve the stated technology requirements?
- What is the earliest subproject activity time at which the subproject would be deemed a failure if the stated technology requirements have not yet been met?

Based on the four numbers obtained from these questions, we can determine the probability distribution of the successful completion time as a triangular function, with the four numbers representing, respectively, the area under the triangle, the left corner of the triangle, the peak of the triangle, and the right corner of the triangle, The integral of this triangular probability density function is the cumulative distribution function that we require.

We must also specify subproject costs. These can be estimated after the subproject activities are clearly defined. Subproject activities, in general, call for the procurement and installation of capital equipment and subsequent support and operation of the equipment. Accordingly, as shown in Fig. 14.3, subproject costs are conveniently divided into two parts, a fixed or nonrecurring part and a variable or recurring part, with the fixed costs being committed upon initiation of the subproject and the variable costs comprising an ongoing component of cost that terminates upon completion of the subproject.

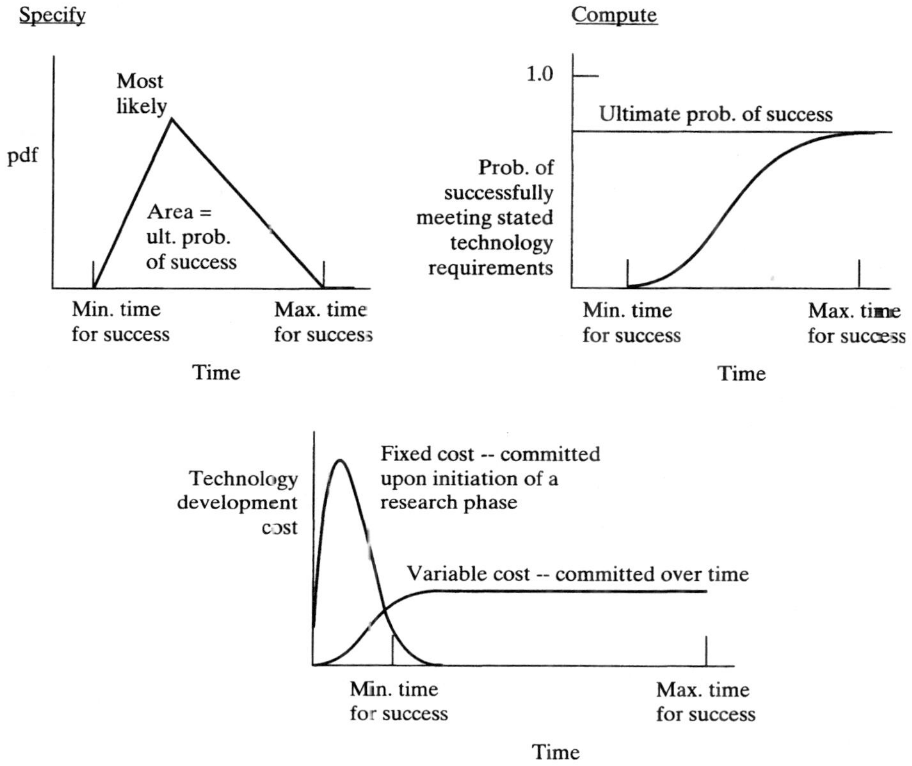

Figure 14.3 Subproject input data.

The subproject costs outlined here are the marginal or incremental costs of the subproject above and beyond the basic costs of the research and development phase. The basic cost of performing each research and development phase must also be accounted for. Research and development "phase" costs include major capital investments in large test facilities, and the basic cost of operating these facilities over the duration of the research and development phase. These costs may be input in a manner similar to the subproject costs.

The *conduct* of research and development requires a set of activities such as construction of facilities, the fabrication of equipment, purchase of instrumentation and other capital items, and labor. All of these things can be estimated with the aid of cost estimating relationships as described in Chap. 11.

Modeling Technological Breakthroughs

If the nature of research and development was such that research subprojects were planned only to accomplish the stated technology requirements for each research and development phase, the process of program evaluation would be considerably simpler. Often, however, subproject research and development activities are planned to provide an opportunity for technological breakthrough while, at the same time, assuring that the minimum technology requirements are met. Expressed differently, research and development is planned as a stepwise process in which each of the steps is sufficiently small that success is reasonably well assured, and yet the activities undertaken are such that there is some non-zero probability of technology advancement well beyond the minimum required level for success in each research and development phase. When such breakthroughs occur, it might not be necessary to undertake certain future research and development subprojects that had originally been designed to provide for the technology development accomplished by the breakthrough. The process of skipping future research and development activities as the result of breakthroughs is referred to here as *leapfrogging*. Here, we account for leapfrogging through the use of a parameter referred to as the *technology advancement factor*. This factor specifies the extent to which technologies in each generic problem area are developed beyond the minimum stated technology requirement.

A technology advancement factor of unity (1.0) indicates that the stated technology requirements are precisely met and not exceeded. A technology advancement factor of 2 indicates that the minimum stated technology requirements were exceeded by an amount such that the minimum technology requirements of the succedent research and development subproject in this generic problem area have been accomplished, and therefore it is not necessary to perform the subsequent subproject. A technology advancement factor of 3 indicates that the technology requirements of two subsequent subprojects in this generic problem area have been met.

Technology advancement is not a deterministic process. Thus we must specify a probability distribution on the technology advancement factor. A typical such distribution is shown in Fig. 14.4. As this figure shows, some subprojects may be planned in such a way that the minimum technology requirements are likely to be met but might not be exceeded.

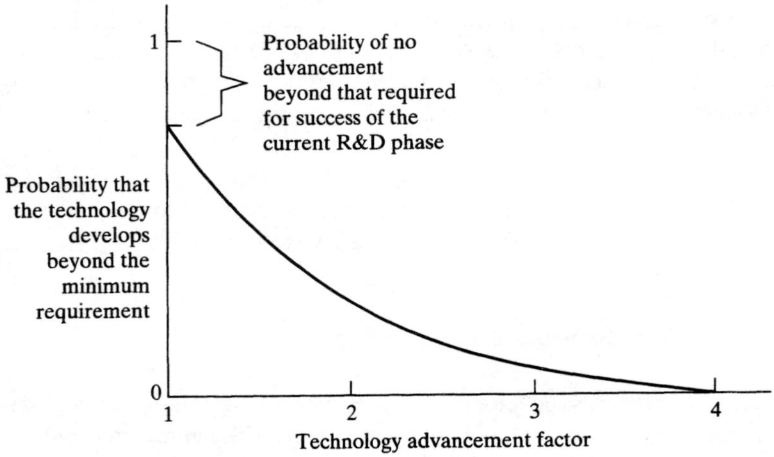

Figure 14.4 The technology advancement factor.

Noninteger values of the technology advancement factor are also permitted. A technology advancement factor of 1.5, for example, indicates that, while the technology has not totally leapfrogged the next subproject within a particular generic problem area, the probability that the next subproject will succeed is increased. Whenever we obtain a noninteger technology advancement factor, we must calculate an updated estimate, $P^*(t)$, of the probability of success for the next subproject in each phase and each approach, using the formulae:

$$P^*(t_f) = P(t_f) + r[1 - P(t_f)]$$

and

$$P^*(t) = P(t)\left[\frac{P^*(t_f)}{P(t_f)}\right]$$

with $P(t)$ being our original estimate of the probability that the subproject would succeed by time t, t_f the final possible time for success, and r the fractional part of the technology advancement factor. Figure 14.5 shows the result of applying the technology advancement factor to the subsequent subproject within a particular generic problem area.

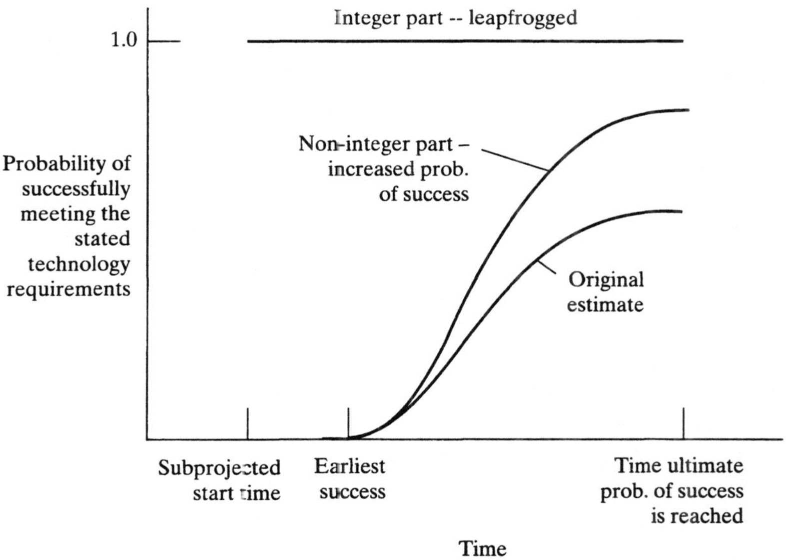

Figure 14.5 The effect of leapfrogging on the probability of success in a subsequent subproject.

The technology advancement factor described here is arbitrary; other descriptions could be used. This model was chosen because of its relative simplicity and its ability to capture the essence of the phenomena that it models. It is based on the approximation that, if the goals of a particular research activity have not been accomplished by earlier research, the time required to accomplish them will be about the same as originally planned. The probability of ultimate failure is, however, decreased by a factor of one minus the fractional part of the technology advancement factor. You could, if you rather, specify alternative relationships for technologies that are not well approximated in this way.

In every simulation, a specific technology advancement factor is computed, based on the estimated probability distribution, each time that a subproject is successfully completed. This technology advancement factor is then used to modify future subprojects according to the rules set forth earlier.

Modeling Interrelationships among Subprojects

Another aspect of research and development activities that we must account for is the correlation between activities that occur in projects associated with alternative approaches. For example, if a particular material is under development as part of two different approaches, and if it is successfully developed under the research of one approach, then it is also successfully developed for the other approach as well. We can account for these correlations by allowing subprojects for different approaches to be identified as *equivalent* subprojects. Where subprojects are so identified, we equate successful completion of one subproject to successful completion of the equivalent subproject or subprojects. This holds not only for ongoing subprojects but also, through the technology advancement factor, in the case of leapfrogging future subprojects.

There are two additional interrelationships among subprojects that must be accounted for. These are referred to as *prerequisite* subprojects and *excluded* subprojects. Prerequisite subprojects are subprojects that must be completed before another subproject may begin. We must allow certain subprojects to be identified as prerequisites of other subprojects. All prerequisite subprojects must be successfully completed before the subproject to which they are prerequisites may begin.

Excluded subprojects allow the formulation of parallel research activities, but prevent precise duplication of research activities in the parallel projects. For example, suppose a research strategy to develop a nuclear rocket engine provides for concurrent development of two concepts. Both may use the same nozzle, and therefore each program taken separately must have a nozzle development program. But if both configurations are developed concurrently, it is reasonable to have only one nozzle development program. By specifying nozzle research subprojects as exclusory, it is possible to configure a research program so as to have only one subprogram on nozzle research common to the research on both rocket concepts.

The subproject property of equivalence is bidirectional. That is, if subproject A is equivalent to subproject B, then subproject B is equivalent to subproject A. Bidirectionality does not apply, however, in the case of prerequisite or excluded subprojects. Clearly, the prerequisite property is not bidirectional. The exclusory property may sometimes be bidirectional. In those cases where it is, the property can be made bidirectional by specifying each of two subprojects as exclusory of the other.

Modeling Performance of Alternative Approaches

We would like our model to normatively model the decision or selection process relative to the program manager's preferences. To do this requires: (1) a method for identifying options, (2) a method for ranking all options presented by the model at every decision point, and (3) a decision-making strategy. Normative decision modeling is needed to account for the variability in technical success, cost, and schedule that is always present in research and development programs. Options at each decision point are identified through the specified structure of the program as shown in Fig. 14.1 and the specified program logic (prerequisite subprojects, excluded subprojects, etc.). We now discuss the generic problem of ranking options.

The ranking of options must take into account their significant differences in each relevant attribute. New technologies typically must pass three tests to win acceptance: they must be technologically successful, they must have an acceptable cost, and their implementation must have an overall beneficial effect. One means of ranking options is in terms of their expectations for success at these three tests. To do this for all possible options at all possible decision points, as is necessary for an automated simulation, requires some form of mathematical model. It is appropriate for this model to reflect the preferences of the manager. Although any such model is likely to be neither totally accurate nor relatively simple, so long as the differences between options are substantial, even approximate models may well provide similar rankings. If, on the other hand, the differences between options are small, errors in ranking may not be highly consequential.

Here, our simulation could take many different approaches; however, our discussion for application deals with two different measures of performance: the one taking into account only the likelihood of a technological success and the other combining cost and implementation success with technological success. Other measures of success could be established if necessary, but the logic here is that, in the early stages of a program, it may be difficult enough to estimate the probability of technological success. The estimation of cost and implementation success must await demonstration of technological feasibility. In many research and development programs, the first measure of success would be used to assess decisions for continuing or terminating research and development phases whose purpose is to demonstrate technological feasibility of a new idea or concept. Measures of cost and implementation become important after technological feasibility has been demonstrated. Each of these measures is used in different ways: (1) to indicate when an alternative approach should be terminated (if the measure drops below a specified threshold, the approach will be deemed to have failed and will thus be terminated), and (2) to prioritize alternatives as candidates for new starts.

For the first measure of performance, the likelihood of technological success of a specific alternative approach, we will use the estimated probability of successful completion of a specified research phase, which in turn is dependent on successful completion of specific subprojects. For a research and development project that is yet to begin, the probability that one of its subprojects will succeed is the *a priori* probability of success of that subproject within the allowable time. Subprojects that have begun in the simulation, but have not yet succeeded or been terminated, require different analysis as illustrated in Fig. 14.6. For example, consider a subproject having an *a priori* probability of success of 0.9

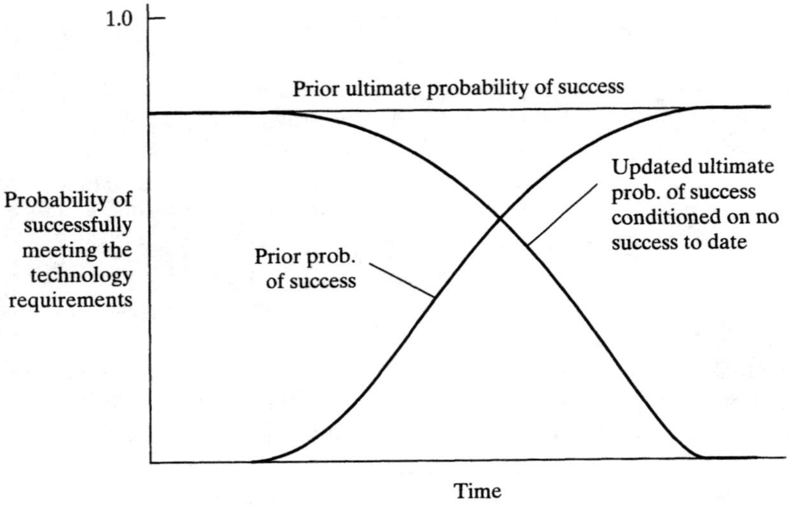

Figure 14.6 Probability of subproject success versus time in a subproject.

in its allotted time of 5 years. If the subproject has not succeeded as the fifth year draws to a close, one's expectations of success are likely to diminish.

According to Bayes' law, the probability $P_t(t_f)$ that ultimately a subproject will succeed, as estimated at time t, given that success has not occurred prior to time t, is

$$P_t(t_f) = \frac{P_0(t_f) - P_0(t)}{1 - P_0(t)}$$

where $P_0(t_f)$ is the probability that the subproject will succeed by time t as estimated at the beginning of the subproject, and t_f is the final possible time for success. (Note that $P_t(t_f)$ may depend on the status of other subprojects, and thus may change during the course of the subproject.) Thus, as the subproject proceeds without successful completion, the probability of that subproject ultimately succeeding decreases to 1 (unless $P_0(t_f) = 1$). At some point in time, the probability of failure may become so close to 1 that the manager will conclude that it is not desirable to continue the subproject. If the subproject were a crucial link in achieving success for a particular approach, then it would no longer be desirable to continue to pursue that approach or any of the research and development activities associated solely with it.

The second ranking measure that we will present here combines the ultimate usefulness of the technology with technological success. It is based on the benefit of successfully completing a research and development program. This benefit may depend on the date when technology development is completed and on the particular technological alternative that is developed. These dependencies may be expressed, for example, in terms of a total value that varies with time, a measure of the extent to which the new technology is adopted and used, and some measure of benefit to each application in which the technology is used. The benefit to each application can be expressed in terms of cost savings, improved reliability or probability of success, enhanced capability, or some combination of the above.

Here, we will use the expectation of a value-cost ratio as a second measure of performance, based on the estimated benefits of successful program completion. This measure is an extension and adaptation of the benefit-cost ratio to multiphase programs and is based on the notion that research is a process of buying information for future decisions. It makes use of an independent valuation of each alternative approach in a program as of a given date, computed by the following procedure, as shown in Fig. 14.7: (1) the benefit B of overall success, discounted to the present, is estimated; (2) the costs C_1, C_2,... for each phase of the research and development process are calculated and discounted to the present; (3) the current estimates of the probabilities P_1, P_2,... of go decisions for each research and development phase are calculated; and (4) the expected net value V of the alternative approach at that time is calculated. Note that this is not a trivial process because research schedules are not known precisely. We have two alternatives. First, we can simply use the research plan as a master schedule, and treat it as deterministic. Or, second, we could compute the expected values of costs and benefits by means of a Monte Carlo simulation on only the research activities in question here. At the beginning of a three-phase program, for example, V is equal to:

$$V = P_1P_2P_3B - C_1 - P_1C_2 - P_1P_2C_3$$

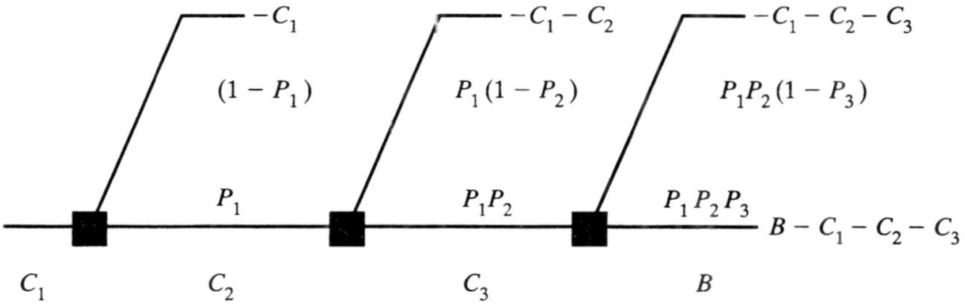

Figure 14.7 A project evaluation decision tree.

To obtain the desired measure of performance, V must be compared to a cost. A standard benefit-cost ratio would compare a project's benefit to its entire cost. However, V is the expected benefit of pursuing an approach net of the expected cost of all pursued future research and development phases, taking into account that future phases might not be pursued. Since the only commitment under consideration is to the next research and development phase, the nonsunk cost of that next phase is thus the cost to which V is compared, and the economic measure of performance we propose here for the nth phase is V/C_n.

Modeling the Decision Process

We can assume that decisions take place on a periodic basis (such as yearly) with respect to every ongoing and potential research and development effort on every alternative approach. In particular, every ongoing research activity is checked against the value measures outlined above and is terminated if the value of continuing the research drops below the specified threshold. Additionally, after each year that a research and development phase is successful or meets a particular milestone, the opportunity arises to begin new research and development phases, and the decision either to begin or not to begin a new phase must be made. We can allow a great deal of flexibility in considering this decision.

Figure 14.8 shows some of the process involved in making such a decision with respect to a research and development program that includes two alternative approaches. Alternative 1 is in Phase *I* with research proceeding in four generic problem areas. The shaded bars indicate the time over which the research and development continues in each generic problem area prior to successful achievement of the required technology developments. When the subprojects in each generic problem area of the project succeed, Phase *I* is completed and the decision to begin Phase *II* can be made. In the example shown, Phase *II* begins immediately upon completion of Phase *I*. In the meantime, research is continuing on Phase *II* for Alternative 2. As shown, Phase *II* for Alternative 2 is completed before Phase *II* for Alternative 1. Upon completion of Phase *II* for Alternative 2, the decision could be made to proceed with Phase *III* directly or, as shown, to wait until Phase *II* for Alternative 1 is complete, at which time the decision is made to continue with only one of the two alternatives. Thus, the decision to undertake a new research and development phase can occur upon completion of the previous research and development phase, upon completion of certain milestones in the previous research

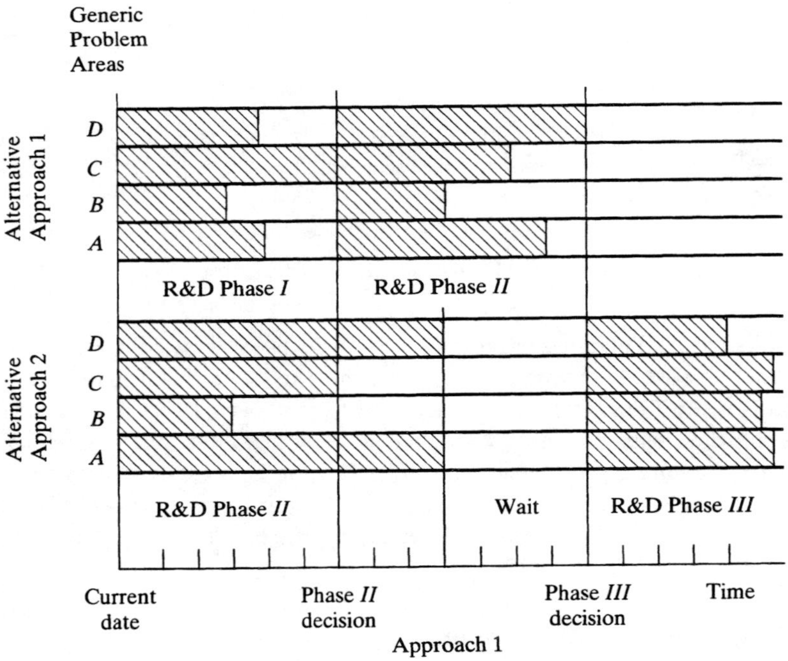

Figure 14.8 Simulation status and decisions.

and development phase, or upon successful completion or failure of specific phases of other approaches.

At the time that it becomes possible to begin a new research and development phase, the decision to initiate that phase is made consistent with a set of specified constraints. This approach allows establishment of any number of constraints of the class that mandate that the number of subprojects in any specified subset of the full set of possible subprojects at a specified state of completion must be no greater than a specified number. This class of constraints permits a broad range of practical program management constraints to be imposed on the selection of new research and development phase starts. For example, we could consider such constraints as:

- Allow research and development on only one alternative after Phase *III*.
- Allow research and development on only one alternative in each of two classes of alternatives after Phase *II*.
- Do not allow research and development on more than one alternative in each research and development phase beyond Phase *II*.
- Do not allow any new research and development phase starts after one alternative reaches a specified phase.

These decision rules are merely examples that show the range of possibilities available within the framework of a research and development simulation. Decision rules appropriate to the circumstances of a variety of particular research programs can be constructed.

New research and development starts can be selected from the set of candidate new starts according to any of many possible logics. For example: (1) the candidate new starts are divided into two groups depending on whether or not technological feasibility has been successfully demonstrated; (2) the group representing those approaches that have not yet demonstrated technological feasibility is ranked using the probability of ultimate technological success measure, and the group representing approaches beyond this phase is ranked on the expected value-cost ratio measure; and (3) new starts are chosen from among the candidates according to a list of priorities and within the specified set of constraints. Priority could be given, for example, to the potential new start that is farthest advanced in the overall research and development program. Thus, candidates for the final research and development phase would be selected first, those for the next-to-last phase would be selected next, and so on. Once decisions have been made on new research and development phase starts, the appropriate initiation of relevant subprojects begins. Subprojects may be started upon initiation of the research and development phase of which they are a part or at some specified time after the initiation of the research and development phase, or they may be keyed to the accomplishment of specified prerequisites.

Typical Management Results

We have outlined a simulation approach to the evaluation and costing of large-scale research and development programs. As such, it obtains data on events at the level of detail of the input data. All events occurring in the simulations can be stored for retrieval and analysis. Thus, we can produce a vast array of data on virtually any aspect of a program. But we must keep in mind that neither this nor any other mathematical approach can generate scientific data, nor can we predict specific events. But we can display the consequences of our input data, that is, of the estimates of costs and probabilities provided at a highly disaggregate level. Typical outputs include estimates of the following:

- The probability of successfully completing the research and development objectives of the program as a function of time for various management strategies.
- Program cost versus time for various management strategies. Cost is often a random variable. Probability distributions of cost may be obtained for each time period (typically, a year) in the simulation. There are a variety of ways of displaying such distributions. One way is to provide cost-probability distributions for each year. Another is to plot expected value of cost as a function of time together with 10 and 90 percent confidence cost curves. Other formats are also possible.
- Probabilities of each technological approach succeeding or failing. If a particular approach has a very high failure rate, you might want to see if program cost can be reduced without significant effect on the probability of success by removing that approach from the program. You could also investigate the effects of increasing the level of activity on those approaches that are most successful.
- Sometimes research and development programs hang up on needed technologies that are not ready in time. This can be discovered in the output, and variations in the program strategy can be examined in which added emphasis is given to such technologies to assure their readiness in a timely manner.
- Sometimes research and development programs fail because, in the manager's quest to conserve resources, program options are eliminated too soon. Using a simulation model, you often can uncover such problems and examine the merits of broader strategies.

We also can produce many other outputs, including counts and probability distributions on virtually all programmatic variables. The preceding list is only a sample of the capabilities of a simulation approach to the evaluation of research and development activities.

14.4 SUMMARY OF THE SIMULATION APPROACH

We have presented above a simulation approach to the costing and evaluation of large-scale research and development programs that takes into account the uncertainties inherent in the research process. In this approach, the overall costing and evaluation problem is dealt with by disaggregating the entire program into a set of research subprojects, each of which can be evaluated by experts in the particular field of the subproject. Our underlying goal in using this approach is to make use of the expertise that resides within the individual disciplines of which the program is comprised. First, we obtain detailed technical information regarding the various research and development activities that make up an overall program. Then, we use software to facilitate the aggregation of these data, obtained at the subproject level, to provide cost and evaluation information for decision making at the aggregate level of the entire program. Thus, by using an approach such as that shown previously, we can make use of the full range of expertise available across the scientists, engineers, and others who will staff the research program. And we can frame that expertise in a format that can be useful for both cost estimating and program planning.

REFERENCE

1. Hazelrigg, G. A. and F. L. Huband, RADSIM—A Methodology for Large-Scale R&D Program Assessment, *IEEE Transactions on Engineering Management*, vol. EM-32, no. 3, August 1985.

PROBLEMS

1. You seek to develop a product that requires the development of two new technologies, A and B. Each technology requires a research program that will cost $500,000 per year to support. Your best estimate is that each project has a probability of success characterized by a memoryless distribution with mean time to success of 2 years. The product, when developed, will produce a net present value of $5 million referenced to the date of completion of the development of both technologies A and B. Structure a good research program. What is the probability of success of your program? What will your program cost? How long will it take? Under what conditions would you consider that your research program has failed? When would you terminate your research program if it has not succeeded?
2. Consider problem 1, except that now you assess that there is only a 0.7 chance of developing technology A and a 0.8 chance of developing technology B. How does your research plan differ from your plan of problem 1?
3. Consider problem 2. Suppose that there is a 0.3 chance that technology B, if it is successfully developed, will be sufficiently better than planned so that the net present value of the product will be increased from $5 million to $10 million. How does this change your research program plan?
4. Your company manufactures a product that is currently successful in the marketplace, and it generates $200 million per year in gross revenues of which about $50 million is net profit. Your boss asks you how much you think the company should be spending in research and development to support this product. What factors would you want to consider in giving your response, and what data would you need to perform a rational analysis?
5. Show how the RADSIM approach can be used to do probabilistic cost estimation.

15

Information

15.1 THE PURPOSE OF THIS CHAPTER

The underlying thesis of the approach to systems engineering outlined in this book is that systems engineering is the treatment of engineering design as a decision-making process. In the first chapter, we acknowledged that good decision making is facilitated by good information. We then noted that the study of systems engineering reduces to the examination of three questions:

1. What information do we need to facilitate a good decision?
2. How do we get that information?
3. How do we use the information when we get it?

Indeed, good information is the essence of good systems engineering. The question that we will address in this chapter is: how do we know how good our information is? Can we actually measure information? The obvious answer, since this is a chapter titled *Information*, is yes. The measure that is presented below provides a measure of information that the astute reader will recognize as essentially a measure of the informational entropy of a system.

15.2 A QUANTITATIVE MEASURE OF INFORMATION

In the very beginning of this book, we defined information to be different from knowledge. Knowledge we defined as an agreed upon set of facts. Knowledge does not relate to decision making directly, and we make no pretense here of measuring or valuating knowledge. Information, on the other hand, we defined as relating to a specific decision, and we noted that it is quantitatively measured as the probability that the preferred choice in a decision will lead to the most desired outcome. Three things are of importance in this definition: the notion of choice, the notion of outcomes, and the notion of preferences. In the absence of a decision-making situation, by this definition, there is no such thing as information, although there may be knowledge. Thus, in order to talk about information, we must be presented with a decision-making situation, and this infers that we must have options among which we are to make a selection.

The notion of outcomes is also key to the valuation of information. Much of the conventional engineering curriculum deals with subjects that provide the student with an ability to predict physical outcomes of design choices. But here we go beyond physical outcomes. We recognize that the purpose of most engineering systems is to obtain benefit. The benefit we obtain from a system can take many forms. For a company it can be a revenue stream generated by the sale of the company's products. For a government it can take the form of national security, a cleaner environment, more full employment, better health care, a better educated population, and so on. For individuals, benefit can take the form of a better standard of living. It is important in examining the outcome of a decision to look at the real outcome, not a surrogate parameter that is a measure of some engineering property of a system.

And finally, there is the notion of preferences or values. We cannot valuate information in the absence of values or preferences on the outcomes of the available choices.

Let us recall the game of guessing how many beans are in a jar. The physical parameter of importance in this game is the actual number of beans in the jar. This parameter can be estimated through the use of physical models, albeit perhaps with considerable uncertainty. The number of beans in the jar is not, however, an outcome of the game. To the person playing the game, there are only two outcomes: win and lose, and these are monetary outcomes. If we want to play the game just for the fun of playing a game, and we really do not care whether we win or lose, then any guess produces an outcome that is as good as any other guess, and we have perfect information. But if we want to win the game, then the numerical value of our guess becomes important: we have a preferred outcome. It may come as a surprise to engineers and scientists, but the valuation of information depends as much on the totally nonphysical preferences of the decision maker as it might on the physical parameters of a system itself.

With these notions in mind, we will reiterate the definition of information:

> *In a specific decision-making situation, the measure of information is the probability that a decision maker's preferred choice will lead to the most desired outcome achievable.*

To implement this definition, we will examine a typical decision using the method of backward induction. The decision is represented by the decision/event tree shown in Fig. 15.1. For simplicity, this figure shows a decision with three options. Each option leads to a probabilistic outcome. Note here that, if the outcomes were deterministic and if the decision maker's preferences were known precisely, the preferred choice would be the choice that would lead to the most desired outcome with probability one. In this case we would say that the decision maker has perfect information, and we would give that infor-

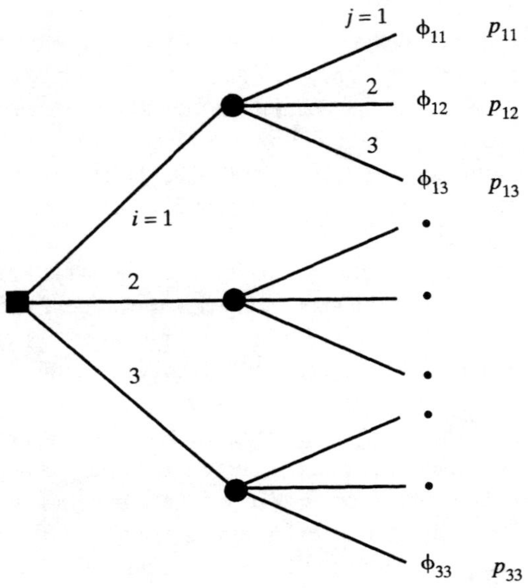

Figure 15.1 A decision/event tree for deriving a measure of the state of information.

mation the numerical value of 1.0. Now suppose that the outcomes, as shown, are probabilistic. No longer can we be sure that the preferred choice will lead to the most desired outcome.

As shown in the figure, we will denote the outcomes as ϕ_{ij}. We assume that we know that each outcome occurs with probability p_{ij} and has utility u_{ij}. Comparing the outcomes, the utilities are such that if

$$\phi_{ij} \succ \phi_{IJ}$$

which, as we know from Chapter 7, is read ϕ_{ij} is preferred to ϕ_{IJ}, then

$$u_{ij} > u_{IJ}$$

Further, if

$$\phi_{ij} \sim \phi_{IJ}$$

which is read, neither outcome is preferred to the other, then

$$u_{ij} = u_{IJ}$$

Now define the variable

$$\delta_{ij}^{IJ} = \begin{cases} 0 \text{ if } u_{ij} < u_{IJ} \\ 1 \text{ if } u_{ij} \geq u_{IJ} \end{cases}$$

Given these definitions, the probability that decision i will result in an outcome that is preferred to or at least as good as decision I is given by

$$P_i^I = \sum_j p_{ij} \sum_J \delta_{ij}^{IJ} p_{IJ}$$

The state of information H is now determined as the probability that the following statement is true:

Choice i will produce an outcome that is preferred to or at least as good as any other choice.

This probability is

$$H_i = \prod_{I \neq i} P_i^I$$

This definition of the state of information is such that H_i is unity if you can be sure that i is the choice that will lead to the most preferred outcome. The value of H_i decreases as the state of information is degraded. Thus, if H_i is near unity, you can expect that you are making a decision under a relatively good state of information. If H_i is substantially less than unity, there is reason to believe that a choice may lead to an undesirable outcome, and this is good reason to seek better information.

Note that the equation for H_i can be written in a logarithmic form:

$$\ln H_i = \sum_{I \neq i} \ln P_i^I$$

In this form, we see the similarity between the state of information and the entropy of a system. We define the informational entropy of a decision as

$$S_i = -\ln H_i$$

where $S_i = 0$ in the case of perfect information and S_i tends to infinity as the quality of information degrades.

15.3 SUMMARY

In this chapter we conclude our discussion on information by deriving a numerical measure of the state of information. The measure that we derive provides that the state of information is the probability that the preferred choice in a decision-making situation will lead to the most desired outcome. Thus, a state of perfect information is denoted by 1.0. We show that this measure is similar to the concept of entropy in a physical system. It is of interest to note that this definition of the state of information applies only to the case of a specific decision. The measure of the state of information includes the available choices, the expectations on the outcomes of each choice, and the values or utility placed on each outcome. Thus, the measure goes beyond the physical system alone to include the use of the system and the value derived from that use.

PROBLEMS

1. You bet $1 on the flip of a fair coin. What is your state of information?
2. You bet $10 on the roll of a single die. What is your state of information?
3. You are in charge of the design of an engineering system. You are confronted with three options. Option A has probability of success of 0.95 and a payoff given success of $100 million. Option B has probability of success of 0.99 and a payoff given success of $20 million, and option C has probability of success of 0.999 and a payoff given success of $10 million. In all cases a failure will cost $1 billion. Assuming that your company has a logarithmic utility for money, that is, its utility for money is the log of the amount of money, what is the best choice, and what is your state of information on this choice?
4. Consider the bean jar game of Chap. 1. What data might you collect, and what models might you use to improve your state of information?
5. You are forced to play the bean jar game of Chap. 1. The winner will receive an award and will have to make an acceptance speech to the class. But you really do not want to make a speech, so you would prefer to lose the game. What is your state of information?

16

System Life-Cycle Modeling and Optimization

16.1 THE PURPOSE OF THIS CHAPTER

Much of this book is devoted to techniques for defining and modeling system alternatives and the selection of a system design from the alternatives. The objective is to provide a basis for good decision making, and our thesis is that information is that basis. In the rare cases where we can actually define and analyze the entire option space, as we did in the case of guessing how many beans are in a jar, we can prescribe a decision by purely mathematical analysis. But few decision-making situations regarding engineering systems are so simple. First, it is rare that we can fully define the option space. Most of the time, we can only define the major alternatives in the option space, and sometimes we have trouble doing only that. The problem is that there are just too many options to deal with them all. Formulation of alternatives in an optimization context can help in deciding between many alternatives. But still, there may be many alternatives that cannot be contained within the optimization context.

Second, decision making depends on expectations, and nondeterministic expectations, that is, probabilistic outcomes, are at least to some extent subjective. It can well be the case that different engineers hold different expectations on the outcome space and, therefore, they would make different decisions, even given the same alternatives and values. We emphasize here that we cannot say that one is right and the other wrong. Unlike problems, whose answers are right or wrong, decisions can only be classified as good or bad based on expectations and values.

Third, decision making depends on values. But now we get into complications. On the one hand, we know that good decision making should be rational. It should be consistent with a utility function. Yet, on the other hand, we know that groups, in general, do not have a utility function. So, while a design should please the chief engineer as measured by his or her utility, if it is for a product that is intended to sell in a diverse marketplace, we cannot expect to represent the preferences of the marketplace by a single utility function. Instead, we will use demand curves to represent the preferences of the marketplace. Demand curves, which provide demand as a function of product attributes and price, typically show that there are niche markets for particular attribute bundles and price combinations. Niche markets are not unique, and several different niches may exist for a particular product.

For the reasons given above, it is our opinion that there is no correct design in the sense that other designs are incorrect. And because of this, we do not ascribe to the notion that design, in general, can be reduced to a mathematical process devoid of human input. Also, for the above reasons, virtually all system designs are suboptimal. Suboptimal means that they can be improved upon, and that immediately begs the question, how? In the face of all the complexity described earlier, it remains necessary to seek design improvements. In this chapter, we will look at approaches to comparing systems with the goal of obtaining incremental improvements.

16.2 OBJECTIVE FUNCTIONS OF A FIRM

We begin our discussion by looking at the objective functions of a firm because the worth of a particular system design is ultimately measured in terms of the value that it brings to the company or companies or government that is responsible for it. We have discussed objectives of a firm above, and intimated that one reasonable objective of a firm might be to maximize profit. In some cases this might be the case. But there can be other objectives, and they are worthy of note. One reasonable objective that applies to many companies is that they wish to stay in business. This objective should not be confused with making a profit. Profit and loss are *not* the economic determinants of whether a company stays in business or fails. A company is economically viable as long as it can pay its bills. It ceases to exist as soon as it fails to pay its bills. Some bills can be deferred by creditors who are willing to wait for payment. But one bill that a company has that can rarely be deferred is its payroll. In many states, if a company fails to meet its payroll, it is essentially defined by law to be out of business. Thus, *cash flow* is often more important to a company than profits.

Particularly for a start-up company or for a company that does not have ready access to large amounts of cash, a reasonable system design objective may be to minimize the probability or running out of cash. Most engineering systems projects have a cash flow that resembles that shown in Fig. 11.15. An initial investment must be made to get the project started. This usually involves a research and development phase, an engineering phase, and a test and evaluation phase before manufacturing and sales begin. During these early phases of a project, cash flow is mainly negative, that is, out of the company. Only after sales begin does a project generate revenue, or after the initial operation of the system do benefits begin to accrue. And it may be some time after this yet before revenues equal expenses. During this entire period, the present value of the cumulative cash outlays continues to grow. The maximum present value of cumulative cash outlay occurs at about the time when revenues finally equal expenses. We say about, because there can be fluctuations, and net cash flow can go positive and negative during successive accounting periods, allowing maximum cumulative outlays to increase even after the first occurrence of break-even cash flow. We will refer to engineering systems designed according to such a cash flow objective as *cumulative-cash-flow-optimized systems*.

There are many examples of cumulative-cash-flow-optimized systems in existence. And many engineers have been in a position where design choices were dictated by a need to get a product to market quickly. Usually, such choices are made in an ad hoc manner. The unique position that we take here is to state that it is possible to model a cash-flow objective function directly, and to optimize the system design with respect to that objective function. In essence, we are saying that it is best to state explicitly the objective of the design, model that objective, and use the model as the basis for decision making. There are two ways to include cumulative cash flow criteria in the objective function of a system design. The first is to compute the present value of net cumulative cash flow, and then constrain this amount to remain less than an allowable maximum. This tends to lead the evaluation in the direction of a deterministic estimate of net cumulative outlays, which can then be constrained. This approach has several problems, and it should be avoided. First, it leads to modeling as a deterministic process, an activity that may contain a great deal of uncertainty. Second, in reality, there may be no way to satisfy a net cumulative cash outlay condition with certainty, as it must be if it is stated as a constraint. And third, the use of constraints should be avoided wherever possible as they distort the decision-making process.

The preferred approach to cash-flow-optimized objective functions is to compute for each alternative the probability distribution of the maximum present value of the net cumulative cash flow, and to seek designs that minimize the probability of exceeding a target net cumulative cash flow amount. This approach acknowledges the true design objective, and provides information in accordance with this objective.

Many other monetary objective functions might also be appropriate. For a small project that is conducted in a large firm, a valid objective could be the return on investment (ROI). The ROI model, which is given in Chap. 11, clearly differs from the profit maximization model, which uses a fixed discount rate. The ROI objective function will also seek to keep net cumulative cash flow low but as a means of increasing the ROI rather than to assure that the company does not run out of cash. The ROI model also places emphasis on obtaining revenues early in the lifetime of a project.

Other objectives can be much more difficult to model mathematically. For example, a company might value its credibility in the marketplace. In this case, the company might place a high value on product reliability. An objective function involving product reliability must first define reliability, and that must be done, as we have seen in Chap. 10, in terms of the utility of the system to its customers. We will not attempt to develop a credibility model, as each case would have to be examined on its own. We will emphasize, however, that generally models can be developed. The following list reviews briefly some objectives that might apply to a project that you are working on:

- *Profit*—Maximize revenues less costs, subject to discounting and correction for inflation. Emphasis on short-term versus long-term profit depends on the discount rate used.

- *Cumulative cash flow*—Minimize the probability of exceeding a specified present value of negative cumulative cash flow, or minimize a parameter of the probability distribution of negative cumulative cash flow, such as the expected value or 90 percent confidence level.
- *Return on investment*—Relates to minimizing investment and to obtaining the earliest possible time to the marketplace for a product.
- *Credibility*—Seek to obtain or maintain a good name for the manufacturer by assuring quality in a product. Emphasis is placed on reliable design and quality control.
- *Customer loyalty*—Seek to maintain close touch with the consumer, and place a high emphasis on giving the consumer what he or she wants in the product. Place high emphasis on the utilities of the consumers.
- *Product availability*—Particularly for products that are difficult or impossible to store, such as electricity, availability upon demand can be very important.
- *Benefit to society*—This objective applies particularly to government projects, where there is no profit motive. Here the objective is to provide the maximum benefit to society by the conduct of a project.

Most of the above objectives can be formulated directly in the form of a utility function without great difficulty. In those cases where it is difficult to formulate direct measures of utility, we recommend that you resort to the use of multiattribute utility functions, such as those developed in Chap. 7.

16.3 OBJECTIVE FUNCTIONS OF CONSUMERS

It is not possible, in general, to write a utility function for a group of consumers. However, this does not preclude each consumer within the group from having his or her own individual utility function. It follows that we can use utility theory to describe the purchasing behavior of consumers in the aggregate but not by means of a single utility function. Instead, we must examine the purchasing decisions of each individual consumer, and sum across all consumers to obtain aggregate consumption patterns.

Suppose you are an engineer with the XYZ Corporation, which makes product A_1. Competitors to your firm make products A_i, for $i = 2, 3, 4, \ldots$ Each consumer will purchase according to his or her own utility such that, if

$$u(A_j) > u(A_k) \qquad \text{for all } k \neq j$$

and if

$$u(A_j) > u(\mathit{Status\ quo})$$

then the consumer will choose to purchase product A_j. We will denote by $x_m = 1$ that this condition is met for consumer m, and by $x_m = 0$ that it is not met. The total demand for product A_j is then given by

$$D_{A_j} = \sum_m x_m$$

This would be an acceptable approach to predicting market behavior if we could obtain the utility functions u_m for each and every potential consumer. Alas, however, if there are a million or ten million or even more potential consumers, enumeration of the utility functions of each and every one is an impossible task. So we turn to approximations. We can define classes of consumers, such that, within each class we assume uniformity of preferences, and we then write utility functions for the consumers of each class. We now have n classes of consumers, each containing m_n consumers, and we can write total demand for product A_j as follows.

$$D_{A_j} = \sum_n m_n x_n$$

Here, $x_n = 1$ if the nth class of consumer purchases the product. This form is still rather complex, and it will take a considerable amount of effort to exercise. Furthermore, the results will contain a certain amount of granularity, as the number of consumer classes generally will be considerably smaller than the number of consumers. But, if done with care, it can provide a great deal of insight relevant to design choices.

Consider the effect of price on consumer utility. As price is lowered, consumer utility for a product increases. Thus, for the mth consumer, $u_m(A_j)$ increases as price is reduced, and demand for product A_j increases, with some of the increase coming from increased overall consumption and some from increasing market share.

16.4 PHASES OF A PRODUCT LIFE CYCLE

In earlier chapters of this book, we describe typical phases in the life cycle of a product. We also describe approaches to the modeling of each phase. Here we give a checklist of actions that could improve the economics of a project.

Research to Proof of Concept

Before establishing a research program related to a particular product, model the product through its entire life cycle. This model will serve as a basis for research management. Next, using the model, determine the possible impacts that various technology improvements could produce. To measure the benefit of each improvement, be sure to reoptimize the product given each technology improvement. Then compare the benefit of the product given the technology improvement to the benefit of the product without the technology improvement. The difference is the gross benefit of the technology improvement. Next compute the expected net benefit of the technology improvement as the product of the gross benefit less the cost of research and probability of success of the research activity to produce the improved technology.

$$\text{Expected net benefit} = (\text{Gross benefit} - \text{Cost}) \times \text{Prob. of success}$$

This process enables you to examine alternative technology improvement activities and to select a portfolio of research activities aimed at improving the product benefit, generally as measured for the producer of the product.

We see that the first issue to face in managing product support research is the selection of research activities. A second issue is the management of the research itself. Research management is a process of establishing reasonable but productive technology improvement targets and then setting milestones or decision points at which times the research activities will be reviewed, and consideration given to their continued funding or the termination of some or all of the activities underway.

Always keep in mind that there are two ways to solve a design problem. The first is to find a solution to the problem, and the second is to find a way to avoid the problem altogether. The story of the development of computers provides several lessons here. Early computers were based on vacuum tube technology. Many people discussed the ultimate limitations of computers based on early designs. One often cited limitation was the need for cooling as the result of all the heat generated by the vacuum tubes, and one projection noted that a computer that approached the capacity of the human brain would require all the water flowing over Niagara Falls just to cool it. Another limitation dealt with the unreliability of vacuum tubes. Then transistors were invented, and computers turned to magnetic storage. This provided a quantum leap in computational capacity without encountering or solving the problems inherent in vacuum tubes. Still, computers were limited by the need for manual assembly and wiring of the memory. Surely this represented an ultimate limit on practical memory sizes. Then along came the integrated circuit and the paradigm of design limitations changed again. Today we think of memory as nearly free, and demand it in units of megs where, only a few years ago, K's of memory were prohibitively expensive for all but central mainframe computers.

Design paradigms are always changing, and it is one purpose of research and development to force this change. Thus, when you evaluate the potential benefits of research, do not limit your thinking to the current paradigms. Sometimes paradigm shifts are difficult to see or predict; yet other times it is not so difficult. The shift from propeller-driven aircraft to jets was easy to foretell. The shift from kerosene and gas lanterns to electric lights was easy to predict. Now we see a shift in design from open loop systems to closed loop controlled systems occurring as a result of emerging microelectronics, sensors, and actuators. Some trends are natural and obvious, others need to be forced. The natural and obvious ones are more likely to succeed.

One of the hardest things to determine is the level at which to support product research and development, both before product development and, later, after the product is on the market. In general, it is difficult to accurately assess the value of research and development. Usually, we tend to overestimate its value in the short term and underestimate it in the long term. One way to view research and development is as insurance against undesired events. Better information generally leads to better design. Thus, it is almost axiomatic that it is better to do the necessary research and development before system design begins, when its results can contribute to better selection of options. Later, after many design choices have been made, the potential value of better information is lessened by virtue of a restricted option space. Clearly the advice here is to err on the side of too much research and development prior to initiation of system design.

System Engineering and Design

One of the most important, yet not always well executed, elements of system engineering and design is a clear, concise, and full statement of the purpose of the system. Here we urge that at the very beginning of any system design effort, you develop a mathematical statement of the values against which system alternatives can be judged. Many products are chided for being nonresponsive to their market, that is, they fail to meet the customers' needs. Much of the reason for this lies in the fact that an adequate value space for the system was never developed or used.

Next, to the extent possible, it is important to avoid the use of constraints that are implemented merely to ease the engineering effort. Instead, it is recommended that you take the systems approach to design, wherein you develop system models in increasing levels of detail, and let the system design emerge from the models as they become more detailed. The modeling approach to engineering design is new to many, but it can be learned without too much difficulty, and its rewards are great. Indeed, the shift from constraint-based engineering design to model-based design may comprise the next major paradigm shift in the engineering design process itself. System models of the future will, no doubt, be computer based, and for each product or engineering system, the company with the best design software is likely to dominate the market. Future design software will integrate functions of computer aided design, finite element analysis of stresses, vibrations, and thermal conditions, product cost and market analysis, financial analysis, risk analysis, computer-aided manufacturing, and so on. The companies that start down this path the earliest will likely have the strongest position in the marketplace in the future.

The role of the systems engineer in the system engineering and design phase is that of synthesizer or integrator. The systems engineer should bring together specialists from the disparate disciplines needed for the overall system design and integrate their expertise productively. Again, the thesis presented here is that this can be accomplished through the modeling process. Through modeling, we can break the design process into its constituent elements, each of which is disciplinary in nature. By engaging disciplinary experts in the process of building models, we extract their expertise in a way that becomes relevant to the system design. Their expertise is contained in the model, it is used when we exercise the model. Again, we emphasize that all phases of product design, test and evaluation, manufacture, distribution and sales, use, maintenance and repair, and disposal are subject to modeling.

System Test and Evaluation

Most engineering system designs require a program of test and evaluation prior to acceptance, manufacturing, and marketing. One thing to keep in mind here is that the system design can influence the nature, duration, and cost of the test and evaluation program. These impacts should be modeled in the system design models. For example, use of standard parts, conventional materials and designs, or proven approaches generally lessen the extent of the test and evaluation process.

System Manufacturing

In the past, system design had been a process that was largely disconnected from the manufacturing process. A design group designed a product, and it was then handed to another group for manufacture. That is rapidly changing, and approaches such as concurrent engineering are emerging to take advantage of the very large gains possible by choosing system designs for their manufacturability as well as their functionality in use. The systems engineering approach to concurrent engineering is a modeling approach. Overall system models should include product manufacturing models so that the design engineer has immediately available the manufacturing implications, in terms of cost, marketability, or other variables, of various design choices.

It certainly seems true that the development of manufacturing models substantially lags the development of physical system models. Indeed, many aspects of the manufacturing process are nonphysical, such as the decision process for the selection of a manufacturing approach for a given product. Thus, the manufacturing process has not been the target of much modeling effort, as most modeling efforts have been spearheaded by physical scientists or discipline-oriented engineers. Modeling of the manufacturing process has been the domain of the industrial engineer. But to date, most of this effort has been undertaken based on the assumption that the system design is a given, and that the manufacturing approach—that is, all the steps in the manufacturing process—are preselected. Much research needs to be done in this area in order to realize effective systems engineering models, but even approximate and rudimentary models offer the possibility of considerable gain.

Distribution and Sales

The interaction between product or system design and its distribution and sales is not a topic that receives much attention in most engineering curricula. It can be both subtle and complex, and it can involve advanced technologies. Consider the modern soda can. The current can design, compared to earlier bottles, enables packaging with much higher density of product and with lower overall weight per unit of product. These attributes reduce shipping costs and improve display of the packaged product. The technology of the cans themselves is quite advanced. For one thing, they possess little inherent strength, and rely on the internal pressure under which the soda is packaged to provide the structural rigidity needed for shipping and display. Of the several technologies that have made the modern can possible, a key one is the ability to paint a label on the can after it has been formed and filled.

The engineer frequently faces tradeoffs between the structural integrity of a product and the need for protection of the product during shipping. Stronger products require less protection and thus less volume for storage and shipping. Consider the storage and shipping of a television set whose size is $0.65 \times 0.55 \times 0.4$ m. The volume of the set itself is less than 0.14 m^3. If the set is packaged with 5 cm of protective material on all sides, the resulting package occupies a volume of 0.243 cubic meters—1.70 times more volume than the volume of the set alone. Over 40 percent of the storage space required to maintain an inventory of such sets is occupied by the protective material around the television rather than the television itself. Still, it is not obvious what impact this might have. But suppose demand for the set is high, and production is limited by storage space for completed and packaged units. Now, the opportunity cost of the packaging could be considerable. It could keep the company from capturing a larger market share.

The advice to be offered here is to be sure to include storage, shipping, display, and other factors related to the distribution and sale of a product in your systems model, and take these factors into account in the design of the system or product.

System Operation, Maintenance, and Repair

System operation, maintenance, and repair all interact strongly with system design. We distinguish between routine maintenance and system repair due to malfunction. System design can make routine maintenance easy, or it can make it difficult. Some automobile designs make routine maintenance, such as sparkplug changing, very easy, while for others it is barely short of a major overhaul. When maintenance is relatively easy, you can expect that it will be done without problems. But in the case that it presents major difficulties, it may have to be performed by highly trained and skilled mechanics. Some automobile engines require removal of the intake manifold, and in one extreme case even lowering of the engine off its mounts, to change sparkplugs. This can result in improper replacement of the manifold and subsequent problems with several other subsystems of the vehicle. If you choose such designs, you should address the problem of providing adequately trained manpower for the necessary maintenance. If you fail to do this, the product reliability could suffer and the product could get a bad reputation.

Policies toward system operation and routine maintenance can have a very strong effect on system reliability. For example, proper system treatment and care can prolong life and reduce failure rates substantially. But sometimes customers need incentives to follow the appropriate procedures. When it comes to repair, system design dictates the nature of the repair process: replacement of parts or entire subsystems are two options. Depending upon such design choices and projected failure rates, you will need a replacement parts strategy.

The design of almost all systems should take into account operational reliability. This requires operational modeling of the system, with a focus on system reliability, failures, and repair procedures. Again, these models must take into account many nonphysical factors, such as failure recovery policies. Fortunately, however, operational and reliability modeling techniques are rather well developed and available for the purpose. Many of the techniques are presented in this book.

The major caution here is that operating and system management policies can drastically affect system economics. It is very important that careful consideration be given to such factors.

System Disposal

We have mentioned that environmental concerns are becoming an increasingly important element of system design. What do we do with the product after its use? What about waste products produced as a result of the use of a product? Some products, such as styrofoam cups, are relegated to the trash heap. But there are limits on how much of such waste the environment can handle, and on how much the public will tolerate. The need for system disposal can indeed limit the market for a product. Systems can produce waste in many ways. Disposal of a physical system can itself pose a problem. The system can produce waste products that can be difficult to dispose. This is the case of nuclear power plants, which produce radioactive spent fuel elements. CFCs are another example. System waste products can themselves produce undesirable products, such as in the photosynthesis of smog from auto emissions. Products that have already suffered from such problems include paints that use mercury to inhibit mold growth or lead pigment paints, leaded gasolines, automobiles, glass bottles and aluminum cans, and varieties of plastic products. You would do well to think about the disposal of the products you design and any waste products they create. In the future, products that are compatible with the environment will find larger and more lasting markets.

New Product Development

In these days of rapidly advancing technology, few products remain in production without change for more than 10 years. Many products, such as automobiles, computers, and consumer electronics, change much more rapidly. And as older products and designs phase out, the production of those products wanes. Thus, the company that wants to stay in business must always be thinking about its next product line. This is not an easy thing to do. A company may invest a considerable sum of money in the development, production, and marketing of a product, and the last thing it wants upon completion of this investment is to destroy the market for the product by introducing a more advanced product. Yet that is exactly what the market demands, and it is exactly what the company that wants to stay in business must do. Therefore, *part of every engineering system design effort should be a parallel effort to outdo the primary effort.*

Products can be bested in two ways. First, a marginally better product can be built. Second, however, a structurally different product can take its place. Take the case of airplane engines. As airplanes developed in the earlier part of the 1900s, piston engines got bigger, lighter, more powerful, and more reliable. By about 1955, 18-cylinder engines that produced 4,000 horsepower, with a weight of less than 2,000 kg, had emerged. They had achieved a lifetime of over 2,000 h between overhauls. Advancing technology was clearly leading to superior engines. But this advancing technology was not enough. The jet engine swept away the market for large piston engines, and totally replaced them in commercial aircraft.

There are two lessons here. First, a company must always be its own strongest competitor, always seeking to introduce better versions of its products, and always searching for product improvements on the margin. But, second, a company must also seek new products that will totally destroy the markets for its current products. These new products are even more dangerous to a company than are the marginally improved products. The latter merely nibble away at markets, allowing time to make adjustments. But new products can destroy a product's market overnight, leaving its producer high and dry without a salable product.

One more point needs to be made here. The company that does not always strive to achieve the best product at the lowest cost also leaves itself open for competition. Potential competitors searching for new markets are more likely to challenge an existing product if it can be easily improved upon or if it can be produced at significantly lower cost. Thus, the company that always treats all its markets as though they are highly competitive, even when there is no competition, protects its market position and discourages new entrants. On the other hand, the company that is sloppy regarding its products or that lolls around when profits are high opens the door for competition, and once competition is established the once high profits can become diminishingly small overnight. Some companies that fail to remain highly competitive during the time that they enjoy a monopoly position in the marketplace are overcome by competition when it arrives, and they are pushed entirely out of the market.

16.5 SYSTEM OPTIMIZATION AND IMPROVEMENT

The major problem that we face in the design of real engineering systems is that they are incredibly complex and the number of available design options may be virtually limitless. Indeed, the complexity of the design process, possibly more than any other factor, prevents us from reducing it from a combination of art and science to a purely scientific process. But, from a different point of view, it is precisely this complexity that makes engineering design so challenging and so interesting. Rare indeed is the situation in which an engineer can create an optimal design, and sit back with confidence in the knowledge that the design will never be bettered. Instead, engineering is a process of competition, where the design of one engineer is viewed by other engineers as a target to shoot at and better. In this competition, we need every tool we can muster, first to create a competitive design and second to improve it in a continuing process that keeps it competitive.

On the one hand, engineers might like to avoid the complexity of the mathematics of design as presented in this book. But, on the other hand, they must resort to it if they are to succeed at their jobs. Throughout this book, we present methods of modeling and analysis of engineering systems, and we use very special or simple problems for their illustration. Typical engineering system design is much more complex than these examples. Faced with this complexity, you can dismiss the methods as intellectual curiosities with little real application, or you seek to simplify and model to the extent possible. The trick to applying the methods of systems analysis, as in any case of engineering design, is to create workable models. Just as a model like $F = ma$ does not capture all the detail of the mechanical dynamics of a system, neither will most engineering systems design models. Nonetheless, both provide information that is useful in the design process.

We may also recall that some optimization procedures fail to give precise optima in the case of large problems. Nonetheless, they often give solutions that approach the optimum. This alone is useful, and given a good solution, the next issue is to improve on it. We do not always, or even often, need the best possible design. We need only a good design. But we also need to understand that the design is merely a good design, and that it is subject to further improvement. Then, we need to seek that improvement.

The continued improvement of product design is as important a process as obtaining a good design in the first place. And to improve a product, we do not have to define the entire option space, nor must we analyze all options that we do define. All we must do is find alternative designs and then identify them, one by one, as better or worse than the existing design. And, as we search for alternative designs, we should be aware that an engineering system consists of more than the design of the physical components of the system. It includes the nonphysical elements, such as the management algorithms by which the system is operated, as well.

Because of the complexity of most engineering systems, the software necessary to support good system design tends also to be very complex and expensive. Still, there are few things that have more value per unit mass than information.

16.6 LONG-TERM PLANNING

It is not uncommon for a company that holds a leadership position in a particular market for several years to lose that position to another company with a better idea. Often, the losing company is so secure in its position of dominance that it fails to see that it has been displaced even some time after its ex-customers are well aware of the fact. It is then not uncommon for this company to complain of unfair business practices by its competitor and to seek government protection. The question for the systems engineer is: could such problems be anticipated and perhaps even prevented by proper systems engineering, and can this be captured in our systems models?

To begin, let us identify two structurally different types of technology improvement that could pose problems to your competitive position in a marketplace: evolutionary and revolutionary. Evolutionary improvements are improvements on the margin—somewhat better materials, somewhat better manufacturing techniques, improved component designs, and the like. Often, we can analyze the impact of evolutionary improvements using our system models as they currently exist. Revolutionary improvements are different and much more difficult both to predict and to analyze. They tend to result in major changes in the system or even paradigm shifts, such as the changes noted above that accompanied the switch from vacuum tubes to transistors to integrated circuits. A model of a vacuum tube is not likely to provide insights on the shift to transistors. Analysis of this shift requires expertise in the new and emerging area. Thus, the job of the systems engineer in assuring that a product maintains its competitive position in a marketplace reduces to two parts: (1) seeking continuous evolutionary improvement of the product through marginal improvements, and (2) seeking revolutionary new approaches that could compete with the existing product.

Systems models can be used for more than system design. They can be used to play "what if" games. What if we could obtain a better material? What if we could reduce manufacturing tolerances? What if we could reduce friction by 5 percent, increase efficiency by 2 percent, improve the reliability of a part? Using our system model as a basis for analysis, we can pose and answer such questions. Through this process, we can gain insights on where our product is vulnerable to competition through technology improvement. Next, we can make an assessment of the likelihood of such improvements, and we can invest in areas that seem to have high potential payoff. In this way, we can guard our product against competition from evolutionary technological improvements.

Guarding against revolutionary improvements is not so easy for the reason that there is no simple way to predict where they will come from or what form they might take. You might think that one way to guard against revolutionary improvements is to employ expert advisors. But this might not work. In the case of the vacuum tube, if you employed an expert to tell you where the next major improvements might lie, the expert might miss the switch to transistors altogether because they are not vacuum tubes, and they require a different area of expertise to understand. Being an expert in vacuum tubes does not guarantee that a person even knows about the existence of transistors, let alone their future applications. It takes a very special type of person to foresee revolutionary technologies and their potential impacts on current products. That person is often a generalist, not an expert in one field, but well versed in many fields.

Perhaps the best insurance against revolutionary technology developments is to create, for each product or project team, a second but smaller team whose job it is to find new and better ways of accomplishing the same or a similar goal. You need to keep in mind that competition is a way of life. If your company does not compete against its own products, another company will. And sooner or later, someone will be successful. That someone will be a member of your company if your company promotes competition from within. If not, it could spell doom to your product line. We believe that the best policy is to be your own toughest competitor.

Finally, a word regarding short-term versus long-term strategies is in order. There is some tendency to think that the best long-term strategy is comprised of a set of short-term optimal strategies. But we see all too often that this is not necessarily the case. We need only look at the process of discounting to see that this is true. Emphasis is placed on short-term strategies by the use of very high discount rates, whereas long-term goals are emphasized by the use of low discount rates. But we know from the material of the preceding chapters that different discount rates lead to different optimal decisions. Therefore, what is optimal in the short term might not be optimal in the long term. It is very important to balance short-term and long-term objectives, and to take both into account in the design of an engineering system.

16.7 MODEL VALIDATION

The day of iconic models is all but gone! They are too expensive, too restrictive, too difficult to produce and use, worthless in many situations, and they fail to provide the information that an engineer needs. The systems engineering approach relies heavily on symbolic modeling. Indeed, symbolic modeling has many advantages. It enables us to test many design options very quickly and inexpensively and, in so doing, helps us to find more nearly optimal system designs. Second, it helps us understand the options that we are dealing with and their associated benefits and disbenefits. Modeling provides a basis for project management and communication. BUT: *It is very important to assure that the models you are using provide adequately accurate results. Therefore, you must always validate your models to the maximum possible extent.* It is virtually axiomatic that it costs less to validate a model than it would cost to correct design mistakes made by following the advice of an errant model.

While the need for model validation is clear, however, the methodology to accomplish the task is anything but clear. It is a task that will challenge you as much as anything you might do in engineering. To meet the challenge, you must use every trick you can muster. You must use every bit of data and every bit of knowledge or information available. The following are suggestions of actions to take in testing a model:

1. Make use of physical test data wherever such data are available.
2. Make use of existing alternative simulation models to test comparable model capabilities.
3. Use absurd cases as test cases. Consider what happens when particular variables go to zero or infinity. Concoct cases for which the results are intuitive, and see if the model results agree with your intuition.
4. Any time that the model results fail to agree precisely with your intuition, find an explanation. Do not let even minor differences lie.

16.8 THE USE OF MODEL RESULTS IN THE DESIGN PROCESS

As you gain proficiency in the techniques of systems engineering outlined in this book, you will find yourself turning more and more to models of engineering systems as a means of gaining information for system design decisions. Indeed, you will find that these models provide a profusion of information that was absent when you employed other engineering analysis methodologies. Your first reaction might be confusion over the plethora of options available, and the significant differences between them in terms of outcomes. And the information you will get by applying the systems engineering approach may frequently point you in design directions that are counterintuitive. But the model results may be powerful indicators of potentially desirable design decisions. You may feel strongly impelled to follow the advice of the models. That is why we provide a word of caution:

Never let a model do an engineering design for you. A model never provides an excuse to stop thinking. You, the engineer, make the design decisions, not the model. And you are responsible for the outcome of those decisions. Think about what you are doing, and be sure it makes sense.

A design should feel right. It should agree with your intuition. If a model produces a nonintuitive result, question the model. Nine times out of ten, the model will be wrong! Also keep in mind that models compare only those alternatives that they are asked and designed to compare. Another option, not examined, might be much better than all the options considered. *Do not be lulled into thinking that you have an optimal design when you have not analyzed all possible options.* Only when you have confidence that the model results are correct, and when you have altered your intuition accordingly, should you consider changing your mind to design according to a model's advice.

Sometimes, you will find that one system option stands out well above the rest. You might be inclined to accept that option as the best. But as often as not, it will not be the best, it will merely be pointing the way to a set of better designs in a region of the option space that you have failed to adequately examine. Keep in mind that the option space for virtually all engineering systems is far to complex to examine exhaustively, and better designs often lurk in hidden corners of the space. Always keep searching for these corners.

16.9 SUMMARY

In this chapter, we review the objectives of a firm and of consumers and then examine the phases in the life cycle of a product in terms of the design options they afford. We found that a firm can have many objectives besides or in addition to mere profit maximization, and we suggest that these too can be incorporated into the overall system utility function. We suggest a method of constructing a product demand function based on the utilities of consumers taken not as individuals, but rather as a set of classes, where each class is chosen such that it has a utility function. A demand function determined by this method could express demand for a product in terms of its attributes that are under the control of the design engineer.

Next, we took another look at the phases in the life cycle of a product, again with an eye out for options that are available for improvement of the overall product, its design, manufacture, distribution and sale, use, maintenance and repair, and disposal. We noted that part of the engineering system design effort for any product should be a competing effort to find designs and approaches that outdo the primary effort. We noted the difference between evolutionary and revolutionary product improvements, and we suggested approaches to predicting and coping with each. We observed once more that very few product designs are truly optimal, and that we can expect that they will be improved upon periodically. This process of continual product improvement gives rise to an environment of dynamic competition, where product utility to the consumers can be improved through better design and better technology, and the companies that provide the best product improvements tend to be the companies that survive in the marketplace.

Finally, we observed the powerful role that symbolic models can play in the overall product design and life cycle management process. But we gave two very strong words of caution: first, that models demand validation and, second, that engineers make design decisions, not models. In the end, it is the engineer who is responsible for the design of a product, not a model. And the engineer must be ready to accept responsibility for his or her decisions.

PROBLEMS

1. You are in charge of the design of a subsystem that is part of a larger system. The overall project is running over its projected cost, and your boss has told you that it is very important that your subsystem be brought in on budget. You face a design choice between options A and B. It is clear that both options will work, but option B could save some money and bring the project in under budget. But there is risk associated with both options. The costs of options A and B are estimated in the form of triangular distributions with the following parameters:

Parameter	Option A	Option B
Minimum	\$800,000	\$600,000
Most likely	\$950,000	\$700,000
Maximum	\$1,050,000	\$1,300,000

   Your budget for this project is \$1,000,000. Which option would you choose? Why?
2. Consider the conditions of problem 1. Assume that your boss gives you the information of problem 1 at 10 a.m. in the morning, but at 3 p.m. she calls an emergency meeting to tell you and the other subsystem managers that things are worse than thought. Everyone will have to cut costs, and you are given a new budget of only \$850,000. Now which choice do you take? Why?
3. You are designing toys for children aged 1 to 3 years. What are the phases in the life cycle of your products? What concerns should you have as regards each phase?
4. Pick a common product that you use frequently. It could be a food product such as soda or coffee, a ballpoint pen, or any other such product. List the life-cycle phases of this product. Then list options associated with each phase of the product's life cycle for improving the product.
5. You are a consultant called in to help a company improve their product, which is light bulbs. You believe that it is important that all engineers design according to a single utility function in order that their decision making be consistent. What is the form of the utility function that you recommend? What data will you need to estimate all the parameters of the utility function?

17

Game Theory

17.1 THE PURPOSE OF THIS CHAPTER

Thus far we have examined engineering systems design as a process of decision making. We have acknowledged that the keys to good decision making are good information and the knowledge to use it, and we have examined the use of symbolic models to obtain good information. We then presented some formal methods of decision analysis. To this point we have viewed engineering design as a "game against nature." In this game, the systems engineer strives to design and implement a system that maximizes value, worth, or utility. Competing against this goal, nature places limitations on designs and imposes uncertainties on outcomes. Sometimes this is an accurate model of the engineering design process. Public works projects or defense systems are possible examples. But other times engineering design takes place in an environment that includes more than nature, it includes competing designers. We do not have to look far to see engineering systems whose design clearly has been affected by the fact that they have been produced in a competitive environment. Automobiles compete in styling, road handling, and power. Televisions compete in screen size and resolution or clarity. Personal computers compete in terms of memory, speed, and price. So we would be remiss if we failed to include in this book some explanation of the formal methods of decision making under conditions of competition.

First, consider the process of engineering design against nature. The objective of our design often can be expressed in the form of a utility function that enables us to rank alternatives, and to distinguish between choices. Nature, on the other hand, does not have a utility function. Nature simply is. On the one hand, nature demands that we obey her laws, and she imposes uncertainties in many forms: storms, floods, earthquakes, solar flares, and so on, all of which can influence the operation, reliability, safety, and lifetime of engineering systems. But, on the other hand, she does not try to produce competing systems.

Competition with people is different. People have preferences, which we believe can be reduced to utilities. Thus, when people compete against each other, their choices are taken each according to their individual utility measures. Often, but not always, these utilities are in conflict. For example, if you play a game of chess with a friend, you want to win *and* your friend wants to win. But only one of you can win. In this game, each of you make decisions that consciously affect the other. Some games are cooperative. A man and woman choosing to get married are involved in a cooperative game, where both can be winners. When we design a product in competition with other firms, we consciously make design decisions that affect our competition. When two or more companies agree to establish an "industry standard," they make choices that affect their mutual well being. Both of these situations comprise games. Specifically, we define a *game* as *any activity that involves more than one individual, where the actions of each individual in the game affects others who are also in the game*. Examples of games frequently encountered in engineering practice are the following:

- Engineering system design in a competitive marketplace
- Contract and subcontract negotiation
- Setting a research and development policy in support of a product that is produced in a competitive marketplace
- Proposal writing and bidding
- Purchasing engineering goods and services
- Setting standards for a class of products or a service
- Establishing cooperative agreements for the design and manufacture of an engineering system

In each of the above cases, the significant characteristic of the activity that identifies it as a game is that it involves the activities of two or more people, each with his/her own unique objectives, and each acting in his/her own behalf. In one sense, then, we can view a game as a problem in optimization where there exists two or more objective or utility functions and, corresponding to each utility function, a separate set of control variables. We have already seen the analogies between optimization and design. Our objective in this chapter is to extend these analogies by presenting the fundamentals of game theory as they apply to engineering design. We present here a formal mathematical framework for the analysis of engineering system design decisions in a competitive environment. This mathematical framework is called *game theory*.

17.2 THE DESCRIPTION OF A GAME

A game is mathematically described in terms of the following: players, actions, information, strategies, payoffs, outcomes, and equilibria. Taken together, the players, actions, and outcomes are referred to as the *rules of the game*. We will now examine each of the elements of a game. In this description, we will largely adopt the notation of Rasmusen (1992), who provides a very readable description of game theory.

A ***player*** is an individual or goup acting collectively as an individual who is participating in a game. We will denote players by the subscript i.

An ***action*** is a move or choice made by a player. We denote by a_i a choice that player i can make, and we will refer to the set of choices available to the player as player i's *action set.*

$$A_i = \{a_i\}$$

For all n players in a game, we define their *action combination*, $\boldsymbol{a}$, as the set

$$\boldsymbol{a} = \{a_i\}\text{x}, \ (i = 1,2,\ldots,n)$$

To complete the specification of each player's available actions, we must also specify when they are available, that is, we must specify the *order of play*.

Information is defined in game theory in much the same way as we have defined it in the earlier chapters of this book. In the context of a game, however, we can formulate an even more precise definition. We will define player i's information set, ω_i, to be his or her state of information at the time he or she takes an action. We will elaborate on this definition shortly.

A ***strategy***, s_i, is a rule to tell player i which of the available actions to choose, subject to available information, each time he or she makes an action. The strategies available to player i are referred to as the player's *strategy set or strategy space.*

$$S_i = \{s_i\}$$

For all n players in a game, we define the *strategy combination*, $\boldsymbol{s}$, as the set

$$\boldsymbol{s} = (s_1,s_2,\ldots,s_n)$$

Also, we will use the notation $\boldsymbol{s}_{-i}$ to represent the strategies of the players other than player i.

$$\boldsymbol{s}_{-i} = (s_1,\ldots,s_{i-1},s_{i+1},\ldots,s_n)$$

The ***payoff*** of a game for each of its players, $u_i(s_1,\ldots,s_n) = u_i(\boldsymbol{s})$, can be defined in either of two ways:

a) as the utility received at the end of the game, or
b) as the expected utility obtained as a function of $\boldsymbol{s}$.

The ***outcome*** of a game is any set of elements, obtained upon completion of the game, that are of interest to the modeler. These elements could be chosen, for example, from the actions or payoffs of the players.

The ***equilibria*** of a game are those strategy combinations,

$$s^* = (s_1{}^*, s_2{}^*, \ldots, s_n{}^*)$$

consisting of a best strategy for each player. We note that many games do not have a unique equilibrium. We define here also an *equilibrium concept*,

$$F: \{S_1, \ldots, S_n, u_1, \ldots, u_n\} \rightarrow s^*$$

as a rule that defines an equilibrium based on the possible strategy space and payoffs.

17.3 TYPES OF GAMES

There are several different types of games that have become widely known by their names, even outside the context of game theory. We will discuss a few here.

A *zero-sum game* is a game in which a player wins or gains only at the expense of others in the game. Gambling games, such as poker, are often thought of as zero-sum games where the sum of winnings equals the sum of losses. Hillier and Lieberman (1990) define a two-person zero-sum game as a game in which, what one person wins, the other person loses. Rasmusen (1992) defines a zero-sum game as "a game in which the sum of the payoffs of all the players is zero whatever strategies they choose." We must be careful with this definition, however, because it refers to measures of physical quantities or their equivalents (money), not to utilities.

A *non-zero-sum game* is any game that is not a zero-sum game.

A *cooperative game* is any game in which players can make binding commitments.

A *noncooperative game* is any game that is not a cooperative game, that is, any game in which the players are not permitted to make binding commitments.

We want to clarify here that the distinction between a cooperative game and a noncooperative game does not depend on the presence or absence of conflict. Cooperative games can involve conflict, and noncooperative games can be devoid of conflict. We define *conflict* as the case that occurs when a player cannot choose the strategy that could result in the greatest possible payoff and obtain that payoff because the strategies that other players would choose in that case would result in a loss in the player's payoff.

We also define *side payments* as transfers made between players after a game is completed in order that the players can share in the benefits of a game.

17.4 EXAMPLE GAMES

We will now apply the mathematical concepts to a few illustrative games in systems engineering.

A Game of Dominant Strategies

The best known game of dominant strategies is the prisoner's dilemma. In this game, two criminals are apprehended by the police, who have strong evidence that they are guilty of one crime, but only weak evidence that they are guilty of a second. Without a confession by one criminal or the other, a conviction for the second crime cannot be obtained. Upon capture, the criminals are immediately separated for interrogation. They are both told that existing evidence will convict them of the first crime, and for that they will each serve 5 years in jail. But they are offered the opportunity to confess to the second crime. If one confesses and the other does not, the confessor will have his sentence reduced to 3 years, while the other will have his sentence lengthened to 10 years. But if both confess, both their sentences will be lengthened to 8 years. Their payoffs (utilities are assumed to be the negative of the time spent in jail) are shown in the payoff matrix of Fig. 17.1. The numbers in the boxes in this matrix give the payoffs to A and B, respectively.

		Prisoner B: Confess	Prisoner B: Withhold
Prisoner A	Confess	-8,-8	-3,-10
	Withhold	-10,-3	-5,-5

Figure 17.1 A payoff matrix for the prisoner's dilemma game.

If the prisoners could hold each other to binding agreements, both would be better off if neither confessed. But, acting alone, the better strategy for each is to confess. If A chooses to confess, B gets 10 years if he withholds, 8 if he confesses. If A chooses to withhold, B gets 5 years if he withholds and only 3 if he confesses. In short, no matter what A does, B is better off by confessing. The same applies to A. Thus, both prisoners confess, and both are worse off for their actions. Note further the strength of this solution. Even if B knows that A will withhold, B will confess and reduce his sentence. And the same for A. We refer to confess-confess as a strictly dominant strategy, because it is best for both players regardless of the other's action. A strictly dominant strategy satisfies the condition

$$u_i(s_i^*, s_{-i}) > u_i(s_i', s_{-i}) \ \forall s_{-i}, \forall s_i' \neq s_i^*$$

which is read, the dominant strategy is that strategy that yields strictly higher utility than any other, regardless of the strategies chosen by other players. We can define a *strictly dominant strategy equilibrium* as the strategy combination that consists of the dominant strategies for all players. The choices made by both prisoners comprise a strictly dominant strategy equilibrium.

A variation of the prisoner's dilemma that applies to systems engineering is the following: Two competing producers equally share the market for a product. Each earns a current profit from production and sale of the product with a net present value of \$10 million. Then one day a discovery is announced that could be incorporated into the product, improving the product in the eyes of the consumers. Both producers hear about the discovery. Unfortunately, the improvement costs more to incorporate into the product than the consumers are willing to pay. Thus, if a producer incorporates this improvement into the product, the company will earn less money for each unit of product sold. For each producer, if their markets remain the same, incorporation of the improvement into the product will result in a loss of total value, with a resulting net present value of production and sale of only \$7 million. However, if one producer incorporates the improvement and the other does not, that producer will quickly capture a larger market share, with a resulting gain in total net present value, bringing present value of earnings up to \$12 million. But this gain will come at the expense of the other producer, whose present value of earnings will drop to \$5 million. What actions do the producers take?

The payoff matrix for this game is shown in Fig. 17.2. Just as in the case of the prisoner's dilemma game, both producers choose to implement the product improvement, even though doing so lowers the net present value of their production by \$3 million. They choose this option not because they obtain an economic gain, but because by so doing they prevent an even larger loss. This simple example illustrates the power of game theory to explain as perfectly rational the seemingly irrational behavior of two players in a competitive game. And it shows the shortcomings of a conventional economic analysis that

		Producer B	
		Improve	Withhold
Producer A	Improve	7,7	12,5
	Withhold	5,12	10,10

Figure 17.2 A payoff matrix for the producer's dilemma game.

does not account for the impact of technology on the competitive position of the company in the marketplace.

A Game of Iterated Dominance

Next, we will consider a variation of the preceding game. The basic conditions of the game remain the same, but producers A and B are no longer equally capable. Producer A has the same capabilities as in the previous case, but producer B is at a disadvantage in implementing the improvement. This could be because producer A relies more on flexible manufacturing, for example, and can more easily adapt to the new design. Because of this disadvantage, B's production costs for the improved product are higher, and the resulting payoff matrix is as shown in Fig. 17.3. This matrix shows that A's dominant strategy is to implement the improvement, but B would be indifferent to the improvement if A does not improve its product.

We can still find an equilibrium solution to this game. Note that producer A's optimal strategy is to implement the improvement, regardless of B's decision. Therefore, A will choose to improve. Since this will be A's choice, we can delete the producer A "withhold" row from the matrix, leaving B the choice to improve or withhold given that A will improve. In this case, B receives greatest utility by improving. So, once more, the improve-improve strategy is an equilibrium. And we see that neither producer would find it beneficial to unilaterally move off of this strategy.

We say that the equilibrium strategy for this game is a *weakly dominant strategy* because, rather than satisfying the condition of a strictly dominant strategy, it satisfies the condition

$$u_i(s_i^*, s_{-i}) \geq u_i(s_i', s_{-i}) \; \forall s_{-i}, \forall s_i'$$

		Producer B: Improve	Producer B: Withhold
Producer A	Improve	7,5	12,4
	Withhold	5,10	10,10

Figure 17.3 A payoff matrix for the producer's dilemma game with A at an advantage.

and

$$u_i(s_i^*, s_{-i}) > u_i(s_i', s_{-i})\ \forall s_i', \qquad \text{for some} \quad s_{-i}$$

which is read, the weakly dominant strategy is that strategy that provides at least as much utility as any other strategy regardless of the strategies of the other players, and more utility than any other strategy for at least one strategy of the other players. We can find weakly dominant strategies by deleting from the strategy set of one player a strategy that is at least weakly dominated by other strategies. We then reexamine the payoff matrix to find other weakly dominant strategies, and repeat this process until all weakly dominant strategies are removed or until only one strategy for each player remains.

Getting to the Marketplace

We sometimes notice that producers have a tendency to push a product into the marketplace before it is ready, while it still has defects, or with such haste that it is poorly engineered. Many of us have noticed that this seems to be the case in particular with new software. An explanation for this behavior can be found in game theory. Producers A and B will compete in the same marketplace with a new product. The producer who introduces his product first will have access to a larger potential customer base because none of the customers will have purchased a product as yet. This gives the first producer to the marketplace a *market advantage*. Ways to get to the marketplace sooner include, for example, reduced product testing, the design of a simpler product, and the use of an inferior design. We can begin to analyze this situation with the assumption that both producers have equal capabilities and facilities. Their choices depend upon their payoffs. One possible set of payoffs is shown in the payoff matrix of Fig. 17.4. We see immediately that this is the prisoner's dilemma game, and that both producers' best strategy is to get an inferior product into the marketplace quickly. You could try other payoffs and see how their choices depend upon their perceptions of the payoffs. Note that these are not necessarily zero-sum games, as both producers stand to lose in the long term as a result of producing an inferior product.

17.5 INFORMATION IN THE CONTEXT OF GAME THEORY

As we shall see, information in the context of game theory is closely related to our definition of information in the context of decision making. Indeed, we will use decisions and decision/event trees to define information in the context of game theory.

For the above games, we were able to conveniently represent the game in a pictorial form called the payoff matrix, which shows the payoffs from each possible strategy combination. But these were simple, two-person, one-move games. For more compli-

		Producer B	
		Good but late	Poor but early
Producer A	Good but late	10,10	4,13
	Poor but early	13,4	6,6

Figure 17.4 A payoff matrix for competing producers seeking to get a product to market.

cated games, the payoff matrix is not a convenient representation. We will resort instead to a decision/event tree representation that we call the *extensive form*. The extensive form representation of a game looks much like the decision/event tree that we use for decision analysis, except that now we include decisions made by more than a single decision maker. As before, the decision/event tree consists of nodes and branches, and we will take the following definitions:

- A *node* is either
  A *decision point* at which a player makes a move or takes a choice, or
  an *event* at which nature takes an action producing a random outcome.
- A *branch* is an action taken from a node.
- A *path* is a sequence of nodes connected by branches.
- A *successor node* to node *n* is a node that follows node *n* in a path.
- A *predecessor node* to node *n* is a node that precedes node *n* in a path.
- A *starting node* is a node that has no predecessor nodes.
- An *end node* or *end point* is a node or point in the game tree that has no successor nodes.

We will add that we generally define a path as consisting of a sequence of nodes and branches taken from a starting node and reaching to an end node.

We now examine a game with sequential moves. Let player A begin by choosing between red, blue, green, and yellow. Following A's move, B chooses between right, left, and straight. Figure 17.5 shows the extensive form or decision/event tree for this game. Now, instead of denoting decisions by solid squares, we use boxes inside of which we indicate who is the decision maker. The payoffs, given at the end nodes, denote the payoffs to each player for reaching that node, in the form (u_A, u_B). A typical path is shown in this figure by the heavily drawn branches.

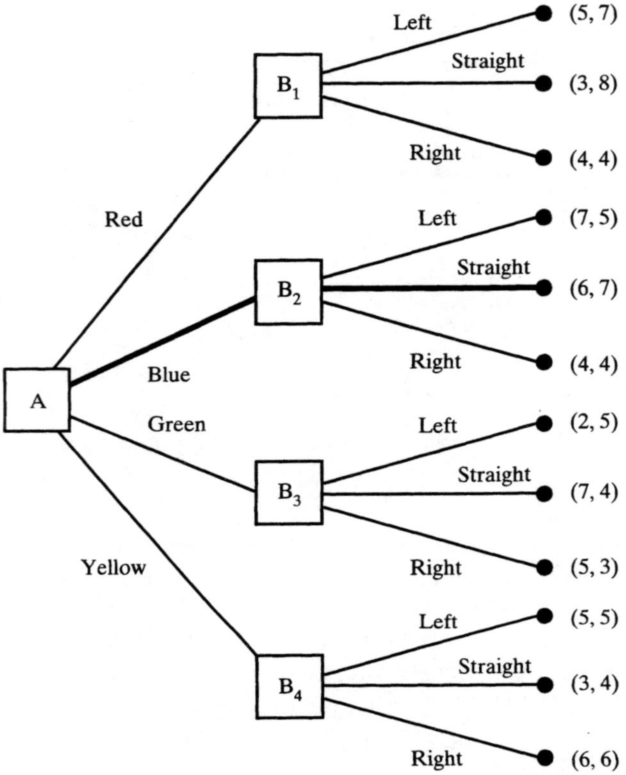

Figure 17.5 A two-person game with sequential decisions.

Information Sets

In the context of the extensive form, we can define the informational structure of a game. We begin by defining the information set, ω_i,

- An *information set* for player *i*, at a particular point in a game, is the set of decision nodes at which the player knows she is, but between which she cannot distinguish without additional information.

We can understand this definition through the above example. Figure 17.6 shows Fig. 17.5 redrawn to indicate that player B can distinguish between player A's choice of red, blue-green, or yellow. That is to say, if player B is at a point in the game within the dashed oval, she cannot determine whether A chose blue or green. Therefore, she cannot determine whether she is at node B_2 or node B_3. The set $\{B_2, B_3\}$ is her information set at this point in the game.

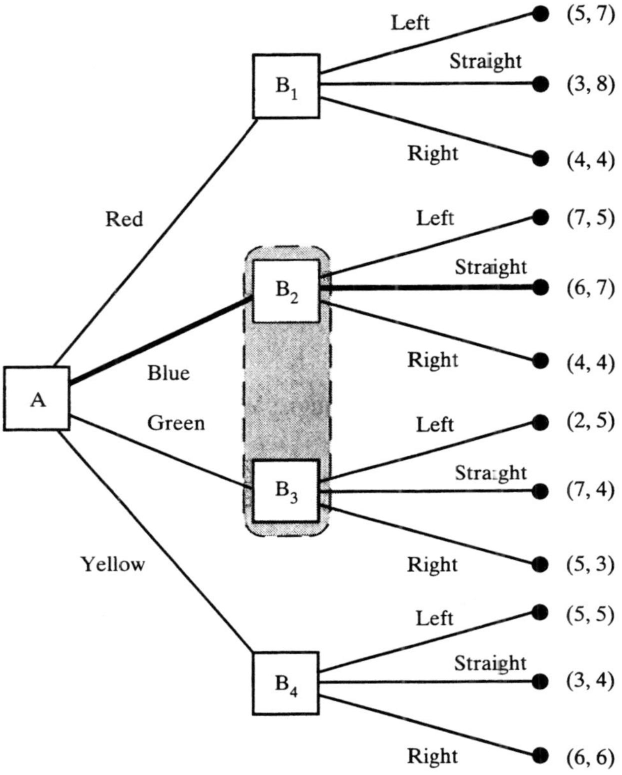

Figure 17.6 A two-person game with sequential decisions showing an information set of player B.

Next, we define an *information partition* to be the collection of information sets at a particular stage in a game. These represent the positions between which a player knows she can distinguish at a particular stage. Referring to Fig. 17.6, node A_1 comprises the first stage of the game. It is an information partition by itself. Nodes B_1, B_2, B_3, and B_4 comprise the second stage of the game. Player A's information partition for this stage is $(\{B_1\},\{B_2\},\{B_3\},\{B_4\})$. But player B's information partition for the same stage is $(\{B_1\},\{B_2,B_3\},\{B_4\})$. We can see easily that A's information is better than B's because A's information set at this stage has more elements than B's, and because B has in her information partition an information set with more elements than the corresponding information sets of A. We say that A's information set is *finer* than B's, or that B's information set is *coarser* than A's. As elements are added to an information set in an information partition, we say that the information set is *coarsened*, and as elements are removed from an information set, we say that it is *refined*. The ultimate degree of refinement is to obtain only one node in an information set. This we refer to as a *singleton*.

Having identified potential differences in information between players, we can introduce the notion of common information (sometimes referred to as common knowledge, which we will not use to prevent confusion with knowledge as we have defined it).

- *Common information* is information that is held by all players in a game, such that each player knows that all other players also hold the information, and that each player knows that all other players know that all players hold the information, and so on ad infinitum.

Categories of Information

We can classify games according to their use of information. We identify four classes:

- A game of *perfect* information is a game in which every information set is a singleton. Thus, every player in the game knows at exactly which node they lie at every point in the game.
- A game of *certain* information is a game in which nature makes no moves after any player moves. Thus, no uncertainty is introduced after any player has moved.
- A game of *symmetric* information is a game in which no player, when he or she moves or at the end nodes, has information that other players do not have.
- A game of *complete* information is a game in which nature does not move first or, if she does move first, her move is observed by every player.

Let us look at some of the games we have examined, and determine their informational classes. The games of dominant strategies are two-player, single-move games in which nature makes no moves. Since nature makes no moves, these games are both certain and complete. Also, each player's information set is a singleton, therefore, they are games of perfect information, and they are symmetric. The example provided in Fig. 17.6 is a game of certain and complete information—nature does not make any moves—but it is neither perfect—one of player B's information sets is not a singleton—nor symmetric.

Bayesian Games

Consider a game in which player A selects a coin from a bucket that contains some fair coins, heads on one side and tails on the other, and some unfair coins, coins that have heads on both sides. We assume that player A deliberately chooses one type coin or the other, and knows his choice. Player B, however, does not know which type coin is chosen. Player A then flips the coin, and player B is permitted to see the outcome of the flip. Player B then places a bet on the outcome of another flip of the coin, and player A then flips the coin a second time. The object of the game is for player B to guess the correct outcome of the second flip. If B guesses heads and is correct, B's payoff is 1. If B guesses tails and is correct, B's payoff is 6. Otherwise, B receives 0. If B guesses wrong, A receives 1. Otherwise, A receives 0. The strategies available to the players are as follows:

For player A
- Choose the fair coin.
- Choose the unfair coin.

For player B
- If the first flip results in tails, choose tails, or if the first flip results in heads, choose heads.
- If the first flip results in tails, choose heads, or if the first flip results in heads, choose tails.
- Choose heads regardless of the result of the first flip.
- Choose tails regardless of the result of the first flip.

This game is illustrated in Fig. 17.7. Nature moves after player A, and again after player B. Thus, this is a game of uncertain and incomplete information. In addition, player B's information sets at the third stage of the game are not singletons. Therefore, it is a game of imperfect information. And, at the third stage of the game, player A has information that player B does not have, making it a game of asymmetric information. We can analyze this game using the technique of backward induction introduced in Sec. 12.3. However, to do this, it is convenient to simplify the extensive form of Fig. 17.7. We notice that A's initial action is, essentially, to choose the probability of heads on a given flip of the coin. By choosing the fair coin, A chooses P(heads) = 0.5, and by choosing the unfair coin, A chooses P(heads) = 1.0. B then sees the game as beginning with the first flip, whose outcome can be heads or tails.

The game can be examined from B's point of view using the decision/event tree shown in Fig. 17.8. First, however, we need to obtain probabilities for all of the branches that follow actions of nature. It is obvious that, if A selects the fair coin, the probability of heads on the second flip will be 0.5, whereas it will be 1.0 if A selects the unfair coin. The first flip provides information on which coin is selected. Clearly, if the first flip is tails, the fair coin has been selected. Thus, our estimate of the probabilities on the outcome of the second flip must depend on B's observation of the first flip. In short, these probabilities are *conditioned* upon the information provided by the first flip. They are obtained using Bayes formula as derived in Sec. 3.6. Using the notation F and U to denote fair and unfair coins, and H and T to denote heads and tails, we first determine the probability that a fair coin was selected, conditioned upon either a heads or tails outcome of the first flip. The relevant equations are

$$P(F|H) = \frac{P(H|F)P(F)}{P(H|F)P(F) + P(H|U)P(U)}$$

$$P(F|T) = \frac{P(T|F)P(F)}{P(T|F)P(F) + P(T|U)P(U)}$$

where, for example, $P(F|T)$ is the probability that the coin is fair given that the outcome of the first flip is tails, and $P(T|F)$ is the probability of a tail occurring upon the flip of a

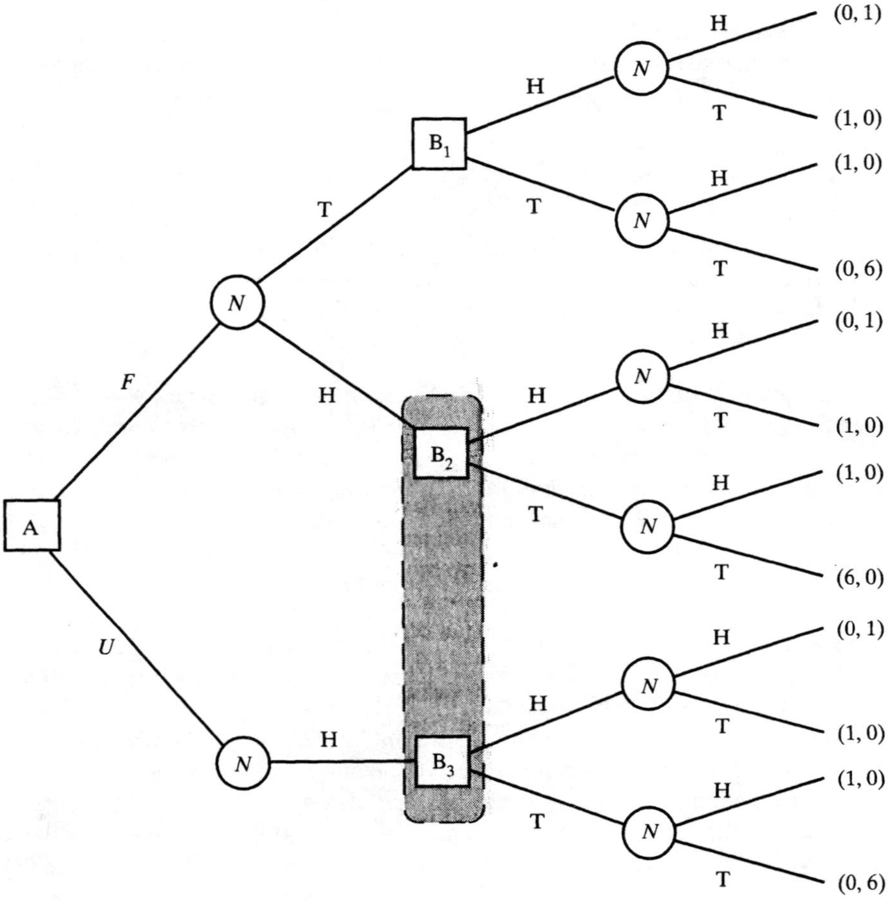

Figure 17.7 The extensive form of a coin-flip game.

coin given that it is a fair coin, namely 0.5. $P(F)$ is the prior probability, or the belief before the first flip, that the coin is a fair coin, and $P(U) = 1 - P(F)$ is B's prior probability that the coin is unfair. So we see that the probabilities we seek depend on both the outcome of the first flip *and* B's prior belief about which coin A will select. If we assume that B thinks it equally likely that A will select either coin, then $P(U) = P(F) = 0.5$, and we obtain

$$P(F|H) = \frac{0.5 \times 0.5}{0.5 \times 0.5 + 1 \times 0.5} = \frac{1}{3}$$

$$P(F|T) = \frac{0.5 \times 0.5}{0.5 \times 0.5 + 0 \times 0.5} = 1$$

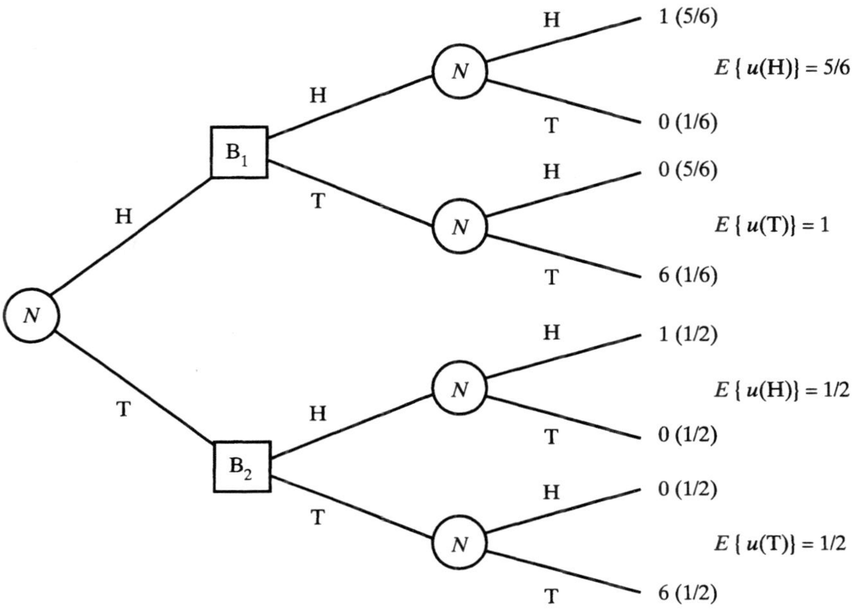

Figure 17.8 A decision/event tree for the coin-flip game from the perspective of player B.

Now we can compute the probabilities of the outcome of the second flip conditional upon the outcome of the first flip. The probability of heads on the second flip, given heads on the first flip is

$$P(H|H) = P(H|F)P(F) + P(H|U)P(U)$$

$$= \frac{1}{2} \times \frac{1}{3} + 1 \times \frac{2}{3} = \frac{5}{6}$$

and the probability of heads on the second flip given tails on the first flip is obviously 0.5 (it is a fair coin). We see from Fig. 17.8 that, subject to B's prior assessment of the probability that A will select a fair coin, 0.5, B will always choose to bet tails. Given this decision on the part of B, A now faces the decision/event tree shown in Fig. 17.9, where the event of nature is the second coin flip. It is obvious that A would choose the unfair coin. But this is also obvious to B. Therefore, B would choose instead to bet heads, as this would certainly win. But this also would be obvious to A, so A would not choose to select the unfair coin. But this is obvious to B, and so B would choose to bet tails, and so on. There simply is no equilibrium solution to this game as posed. We can see this from the payoff matrix for this game, Fig. 17.10. When a player must choose purely one strategy or another, we refer to the game as a game of *pure strategy*.

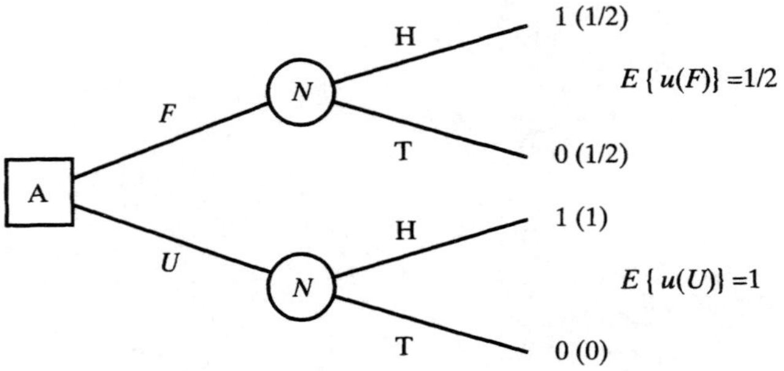

Figure 17.9 A decision/event tree for the coin-flip game from the perspective of player A.

		Player B	
		Heads	Tails
Player A	Fair	0.5,0.5	0.5,3
	Unfair	0,1	1,0

Figure 17.10 An expected payoff matrix for the coin-flip game.

17.6 MIXED STRATEGY GAMES

The above coin-flip game causes us to think more about optimal game strategies. The aspect of this game that prevents finding an equilibrium strategy is its predictability. Consider, for a moment, the game of professional football. Some attempt has already been made to use statistical approaches to improving the performance of teams. The performance of teams in various situations has been statistically analyzed in considerable detail. So, for example, if a team is on its opponent's 35-yard line with a third-and-seven situation (third down and seven yards to go), for this team the play most likely to obtain a first down might be a screen pass. You might think, then, that this is the play to call in this situation—every time. The problem with calling the play every time in this situation is that then the other team knows what you plan to do and can defend against it. Hence, the play loses its effectiveness. It is to the advantage of the offensive team to introduce uncertainty into the play call, even though the play that may be called has less likelihood of success than the optimal play. And the same thing is true of the above game. Perhaps we can find an equilibrium solution if we allow player A to introduce randomness in the selection of a coin.

Suppose we subtly modify the game so that, instead of choosing a particular coin, player A chooses the *probability* that a fair coin is selected. Suppose player A chooses probability 0.7, and this number is then disclosed to B. Next a random number, not known to player B, is chosen. If this number is less than 0.7, a fair coin is selected. Otherwise, an unfair coin is chosen. We will also simplify the game at this point by dropping the first coin flip and structuring the game such that, after nature selects the coin, B places a bet, and that is followed by a flip of the coin to determine the winner. Remember, B knows the probability chosen by A. This game, shown in Fig. 17.11, is referred to as a game of *mixed strategies* because player A chooses a *mixture* of the available strategies rather than one particular strategy. The mixture is characterized by a probability:

Choose a fair coin with probability $P(F)$, and an unfair coin with probability $1 - P(F)$.

Given $P(F)$, we can calculate the probability of heads and tails on a given toss of the coin.

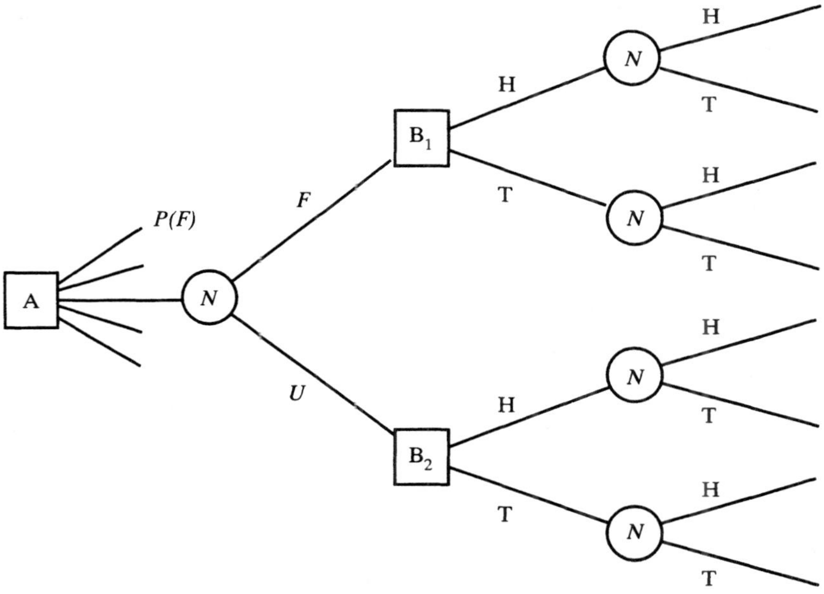

Figure 17.11 The coin-flip game reformulated as a game of mixed strategies.

$$P(H) = 1 - \frac{1}{2}P(F)$$

$$P(T) = \frac{1}{2}P(F)$$

Thus, we can reduce the game as seen by player B to the decision/event tree of Fig. 17.12. From this tree, we can derive the expected utilities of heads or tails decisions by player B.

$$E\{u_B(H)\} = 1 - \frac{1}{2}P(F)$$

$$E\{u_B(T)\} = 3\,P(F)$$

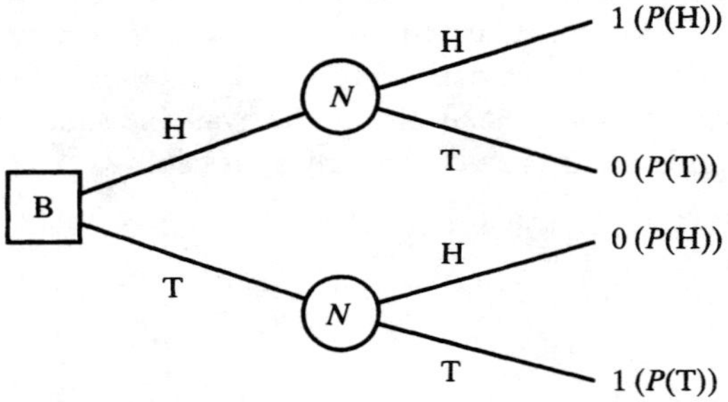

Figure 17.12 The mixed-strategy coin-flip game from the perspective of player A.

We know that player B will take the choice that maximizes expected utility. Setting $u_B(H) = u_B(T)$, we find that B's decision rule is:

If $P(F) < 2/7$, choose heads,
If $P(F) > 2/7$, choose tails.

Now, knowing how B will choose, we can look at the game from player A's point of view.

$$E\{u_A[P(F)]\} = 1 - \frac{1}{2}P(F) \qquad \text{for} \quad P(F) > \frac{2}{7}$$

$$E\{u_A[P(F)]\} = \frac{1}{2}P(F) \qquad \text{for} \quad P(F) < \frac{2}{7}$$

The solution for $P(F) = 2/7$ is indeterminant because player B is indifferent to the choice of heads or tails. We can plot the utilities of both players as shown in Fig. 17.13, and we see that the best choice for player A is a value of $P(F)$ that is just slightly larger than 2/7, resulting in B's choice of tails. Interestingly, B knows that this is A's best choice of $P(F)$, even if it is not disclosed to B. It is the value of $P(F)$ that maximizes the expected utility of the game to A and minimizes the expected utility of the game to B. And its selection by A provides a stable equilibrium.

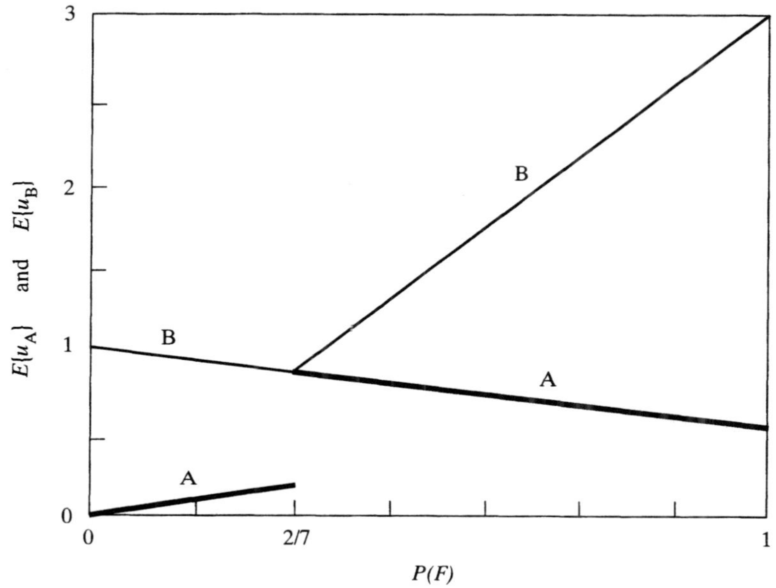

Figure 17.13 Expected utility of the mixed-strategy coin-flip game for each player as a function of $P(F)$.

It is relatively intuitive that a mixed strategy solution is needed if the game is played repeatedly. It is not so obvious that this is also the case if the game is played only once. But, unless B believes that A's choice is random, the game possesses no equilibrium. In fact, the game will have no equilibrium if there is any way for B to *deduce* A's strategy. The only way to guarantee that B cannot deduce A's strategy is if it is indeed random.

Many times in a competitive situation we see our competitor choosing an action that seems irrational or illogical. Such actions can confuse us and thwart our plans. Now we see, through the application of game theory, that such actions can be quite rational and, in fact, even beneficial.

A game of mixed strategy that relates to engineering involves the inspection of parts. Workers or suppliers will tend to shirk if they think that their work will not be checked. They will be diligent if they are sure they will be checked. But there will be no need to check them if they are diligent. The optimal amount of checking can be determined by treating inspection as a game, and the equilibrium strategy will be found as a mixed strategy.

17.7 THE NASH MODEL

A very powerful technique of game theory is the Nash model. We will briefly review this model here.

Nash Equilibrium

We define a *Nash equilibrium* as a strategy combination s_i^* for which no player has an incentive to deviate unilaterally from his or her strategy. Mathematically,

$$\forall i, u_i(s_i^*, s^*_{-i}) \geq u_i(s_i', s^*_{-i}), \ \forall s_i'$$

Generally, to find a Nash equilibrium, we propose a strategy and test to see if it is optimal against the options available to other players. For example, return to the prisoner's dilemma shown in Fig. 17.1. The strategy combination confess-confess is a Nash equilibrium because neither player has an incentive to unilaterally change his strategy. That is, neither player gains by changing his strategy given that the other player does not change his strategy. Note that this does not exclude the possibility that both players could gain by moving together. Here, however, we assume that they do not move together.

The Nash equilibrium strategy in the case of the prisoner's dilemma is a strictly dominant strategy. Every strictly dominant strategy is a Nash equilibrium. But not every Nash equilibrium is a strictly dominant strategy. We see an example of the latter statement in the example of Fig. 17.3.

Evolutionary Stable Strategies

A Nash equilibrium exists when no player has an incentive to change his or her strategy so long as no other player changes his or her strategy. This is to say that the equilibrium strategy combination (ESC) is in equilibrium against a move by a single player (a unilateral move) as shown in Fig. 17.14. As we see from the prisoner's dilemma game, this can prevent the players from reaching a more desired outcome. Indeed, if the players in this game are allowed to cooperate, that is, to form a binding agreement, they could reach a much more desirable strategy combination. In this case, both players must move from the

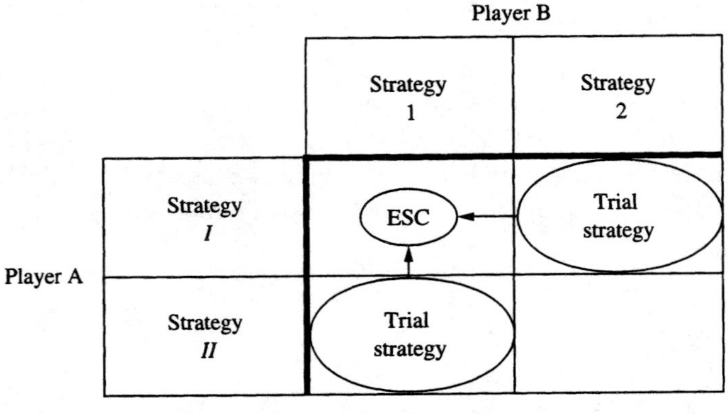

Figure 17.14 Test for a Nash equilibrium.

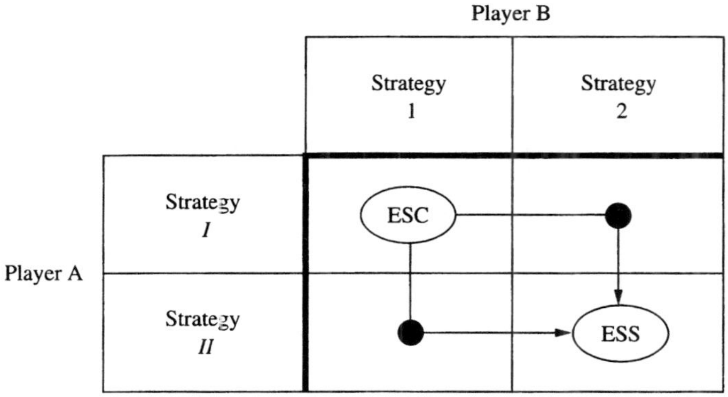

Figure 17.15 Determination of an evolutionary stable strategy.

equilibrium strategy combination in coordination as shown in Fig. 17.15, and this requires cooperation between the players. A solution thus obtained is called an *evolutionary stable strategy* (ESS). The mathematical condition for an ESS is either

$$u_i(s_i^*, s_{-i}) > u_i(s_i', s_{-i}) \ \forall s'_{-i},\ \forall s_i \neq s_i^*$$

which is the condition for a strictly dominant strategy equilibrium, or if

$$u_i(s_i^*, s_{-i}) = u_i(s_i', s_{-i}) \qquad \text{for any} \quad s'_{-i}$$

then

$$u_i(s_i^*, s'_{-i}) > u_i(s_i', s'_{-i}) \qquad \text{for at least one} \quad s'_{-i}$$

Satisfaction of the condition for a strictly dominant strategy yields a strong Nash equilibrium.

We now consider a game of *pure coordination* that makes use of an ESS to find the most desired strategy combination. Consider two firms that manufacture competitive products. Both firms have decided that their mutual business would benefit if they could agree on standardization for their products. Two alternative standardizations are possible. We shall refer to them here as large and small. A payoff matrix is shown in Fig. 17.16. Standardization is achieved when both firms agree to a small-small or a large-large strategy. Both small-small and large-large are Nash equilibria. Without coordination, however, if an agreement is obtained to implement the small-small strategy combination, then the large-large strategy cannot be achieved. Coordination provided by the ESS enables the firms to move from the small-small strategy to the large-large strategy.

Nash Existence Theorem

The Nash existence theorem, published in 1950, assures the existence of a Nash equilibrium strategy in a broad class of games.

		Firm B	
		Large	Small
Firm A	Large	10,10	5,3
	Small	3,5	7,7

Figure 17.16 A payoff matrix for product standardization.

Every game that has a finite number of pure strategies has at least one Nash equilibrium, possibly using a mixed strategy.

The coin-flip game presented above is an example of a game that has no equilibrium solution that is a pure strategy. We saw, however, that we could obtain an equilibrium solution as a mixed strategy. A simpler game that illustrates this same principle is the game of odds-evens. In this two-player game, one player is odd, and the other is even. Upon the count of three, the players each show either one finger or two. If the total number of fingers showing is odd (one-two or two-one), player odd wins. If the total number of fingers showing is even (one-one or two-two), player even wins. A payoff matrix for the game is shown in Fig. 17.17. We easily see that this game has no pure strategy equilibrium. But if we allow mixed strategies for both players, taking P_O and P_E to be the probabilities of odd and even selecting one finger respectively, we can write the expected utilities of the game for each player.

$$E\{u_O\} = P_O(1 - 2P_E) + (1 - P_O)(2P_E - 1)$$

$$E\{u_E\} = P_E(1 - 2P_O) + (1 - P_E)(2P_O - 1)$$

Each player can maximize his expected utility by proper selection of his probability of showing one finger. We take partial derivatives:

$$\frac{\partial E\{u_O\}}{\partial P_O} = 1 - 2P_E - 2P_E + 1 = 2 - 4P_E$$

$$\frac{\partial E\{u_E\}}{\partial P_E} = 1 - 2P_O - 2P_O + 1 = 2 - 4P_O$$

		Even: 1 Finger	Even: 2 Fingers
Odd	1 Finger	-1,-1	-1,-1
	2 Fingers	-1,-1	-1,-1

Figure 17.17 A payoff matrix for the game of odds-evens.

Notice that these equations, if set to zero, do not provide a condition for a maximum. This is because the maximum occurs on a boundary, $P_O = 0$ or $P_O = 1$ for player odd, and $P_E = 0$ or $P_E = 1$ for player even. But these solutions do not provide an equilibrium. We note, however, that if player odd selects $P_O = 0.5$, then player even is indifferent in his choice of one or two fingers. Likewise, if player even selects $P_E = 0.5$, player odd is indifferent in his choice of one or two fingers. Thus, the mixed strategy equilibrium solution is $P_O = P_E = 0.5$. This problem can also be solved graphically, the same as we solved the coin-flip game in Fig. 17.13.

Nash Bargaining Solution

Suppose that two individuals are in a position to barter with each other. They could be trading goods, or one could be providing a service to the other for payment. Both will benefit if they can agree on a trade. Neither will benefit if they cannot come to an agreement. The Nash bargaining solution models such situations.

We will refer to the individuals doing the trading as A and B, and the commodities being traded as X and Y. Initially, A and B will possess quantities of X and Y given by (x_A^*, y_A^*) and (x_B^*, y_B^*) respectively. We refer to this position as their status quo. After the trade, they will achieve quantities (x_A', y_A') and (x_B', y_B'). We will also refer to their utilities for these bundles of commodities before and after the trade as u^*, v^* and u', v'. The quantities

$$x_A^* + x_B^* \geq x_A' + x_B'$$

$$y_A^* + y_B^* \geq y_A' + y_B'$$

denote the limits of possible trades. A and B may obtain any combination of x and y subject to the above constraints. In utility space, these constraints define a convex bargaining region, R. The Nash bargaining solution now proceeds as follows:

1. Define for A and B new utility functions u_0 and v_0 such that

$$u_0(x_A{}^*, y_A{}^*) = 0$$

$$v_0(x_B{}^*, y_B{}^*) = 0$$

Given the utilities $u(x_A, y_A)$ and $v(x_B, y_B)$, u_0 and v_0 can be found by means of an affine transformation. For example,

$$u_0(x_A, y_A) = u(x_A, y_A) - u(x_A{}^*, y_A{}^*)$$

$$v_0(x_A, y_A) = v(x_A, y_A) - v(x_A{}^*, y_A{}^*)$$

2. Determine the quantities $x_A{}'$, $x_B{}'$, $y_A{}'$, and $y_B{}'$ that maximize the product $u_0 v_0$.

This is conceptually a simple enough procedure. A solution is shown in Fig. 17.18. The product space, $uv = c$, is a set of hyperbolas. The solution is the point in the bargaining region R that is just tangent to the curve of maximum attainable c in the product space. The question is, why, with all possible rules for division of X and Y, should we choose this one? The answer is because it satisfies four rather intuitive assumptions, and it is the only rule that satisfies these assumptions.

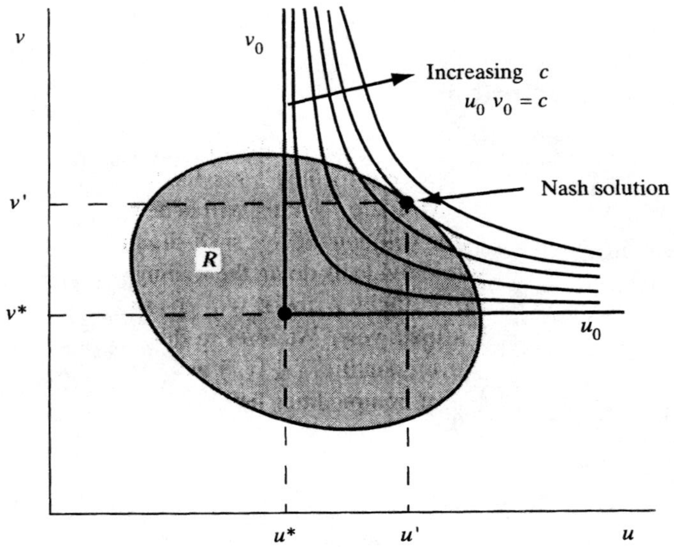

Figure 17.18 Illustration of the Nash bargaining solution.

Assumption 1: The solution stated in terms of the division of X and Y should remain the same if the utility functions, $u(x_A,y_A)$ and $v(x_B,y_B)$ are subjected to affine transformations.

Assumption 2: The solution should have the properties:

(i) The utilities after the trade should satisfy

$$u' \geq u_0$$

$$v' \geq v_0$$

(ii) The point u',v' should be contained in R

(iii) The solution point should not be bettered by any other solution. Namely there is no point u, v in R such that both

$$u \geq u'$$

and

$$v \geq v'$$

This condition assures that the solution is Pareto optimal.

Assumption 3: If the bargaining space R is changed in such a way as to enlarge the set of feasible trades, but such that it does not change the players' status quo, then either the solution remains the same or it is in the new region of R. This assumption says that presence or absence of feasible trades that are not desired as solutions should not affect the result.

Assumption 4: The solution should not depend on the labeling of the players. That is, if we relabel player A to be player B and player B to be player A, the solution stated in terms of quantities of X and Y obtained by each player should not change. That is, the outcome of the bargaining process should be independent of how that player is labeled.

These assumptions appear quite reasonable. We reiterate that the Nash bargaining solution satisfies these assumptions, and it is the only solution that satisfies these assumptions.

17.8 THE MINIMAX THEOREM

We will now restrict our discussion to the case of two-person, zero-sum games. We will refer to the players as player 1 and player 2. Player 1 has available a set of m pure strategies

$$A = \{\alpha_1,\alpha_2,\ldots,\alpha_m\}$$

and player 2 has available a set of n pure strategies

$$B = \{\beta_1,\beta_2,\ldots,\beta_n\}$$

Associated with each strategy combination is a payoff obtained by player 1, $M(\alpha_i,\beta_j)$. As this is a zero-sum game, for the same strategy combination, player 2 receives payoff $-M(\alpha_i,\beta_j)$. We shall abbreviate $M(\alpha_i,\beta_j)$ and write it simply as a_{ij}. We have seen from the above examples that not all games of pure strategy possess an equilibrium strategy. They do, however, possess an equilibrium strategy in mixed strategies. We will define a mixed strategy for player 1 as $\boldsymbol{x}$,

$$\boldsymbol{x} = \{x_1\alpha_1, x_2\alpha_2, \ldots, x_m\alpha_m\}$$

where

$$x_i \geq 0 \qquad \text{for} \quad i = 1,2,\ldots,m$$

$$\sum_{i=1}^{m} x_i = 1$$

Likewise, for player 2,

$$\boldsymbol{y} = \{y_1\beta_1, y_2\beta_2, \ldots, y_n\beta_n\}$$

where

$$y_j \geq 0, \qquad \text{for} \quad j = 1,2,\ldots,n$$

$$\sum_{j=1}^{n} y_j = 1$$

We can now write the mixed strategy expected payoff for player 1 as the sum

$$M(\boldsymbol{x},\boldsymbol{y}) = \sum_{i=1}^{m}\sum_{j=1}^{n} a_{ij}x_iy_j$$

And, as this is a zero-sum game, the payoff for player 2 is simply $-M(\boldsymbol{x}, \boldsymbol{y})$. The rules of the game require each player to select a strategy in complete ignorance of the other player's selection. Thus, player 1's objective is to maximize $M(\boldsymbol{x}, \boldsymbol{y})$, by selecting $\boldsymbol{x}$, for any strategy that might be selected by player 2. And, since player 1's gain is player 2's loss, player 2 seeks to minimize $M(\boldsymbol{x}, \boldsymbol{y})$, by selecting $\boldsymbol{y}$, for any strategy that might be selected by player 1.

Let X_m and Y_n represent the strategy sets of $\boldsymbol{x}$ and $\boldsymbol{y}$ respectively. Then we can define player 1's security level as

$$v_1(\boldsymbol{x}) = \min_{\mathrm{y}} M(\boldsymbol{x}, \boldsymbol{y})$$

And we observe that

$$M(\boldsymbol{x},\boldsymbol{y}) = \sum_{j=1}^{n} y_j\left(\sum_{i=1}^{m} a_{ij}x_i\right) = \sum_{j=1}^{n} y_jM(\boldsymbol{x},\beta_j)$$

is a weighted average of the payoffs $M(\boldsymbol{x},\beta_j)$. It is minimized by assigning all of the weight to the least of the $M(\boldsymbol{x},\beta_j)$:

$$v_i(\boldsymbol{x}) = \min[M(\boldsymbol{x},\beta_1),(\boldsymbol{x},\beta_2),\ldots,(\boldsymbol{x},\beta_n)]$$

An interpretation of this expression is that it is the expected payoff to player 1 if he disclosed to player 2 that $\boldsymbol{x}$ is his choice, and if player 2 is allowed to choose his best response to this choice. Thus, the strategy that maximizes player 1's security level is the strategy that maximizes $v_1(\boldsymbol{x})$. Mathematically,

$$v_1(\boldsymbol{x}^*) \geq v_1(\boldsymbol{x}), \qquad \forall \boldsymbol{x} \in X_m$$

We will denote $v_1(\boldsymbol{x}^*) = v_1$. Then we see that

$$v_1 = v_1(\boldsymbol{x}^*) = \max_x v_1(\boldsymbol{x}) = \max_x \min_y M(\boldsymbol{x}, \boldsymbol{y})$$

It follows that the strategy $\boldsymbol{x}^*$ guarantees to player 1 an expected payoff of at least v_1.

$$M(\boldsymbol{x}^*, \boldsymbol{y}) \geq v_1$$

This strategy is called a maximin strategy. Maximin strategies always exist, but they are not necessarily unique.

Player 1's gain is player 2's loss. By symmetry with the above, if player 2 uses strategy $\boldsymbol{y}$, player 1's maximum payoff cannot exceed

$$v_2(\boldsymbol{y}) = \max_x M(\boldsymbol{x}, \boldsymbol{y})$$

and we find

$$v_2 = v_2(\boldsymbol{y}^*) = \min_y v_2(\boldsymbol{y}) = \min_y \max_x M(\boldsymbol{x}, \boldsymbol{y})$$

This strategy, for player 2, guarantees that player 2 can limit player 1's gain to v_2. This strategy is called a minimax strategy. For this strategy, we see that

$$M(\boldsymbol{x}, \boldsymbol{y}^*) \leq v_2$$

If player 1 uses strategy $\boldsymbol{x}^*$, he guarantees himself of an expected payoff of at least v_1. And if player 2 uses strategy $\boldsymbol{y}^*$, he guarantees that player 1 cannot achieve an expected payoff greater than v_2. Thus, $v_1 \leq v_2$.

We say that a strategy combination $(\boldsymbol{x}', \boldsymbol{y}')$ is in equilibrium if neither player has an incentive to unilaterally change his strategy, namely if

$$M(\boldsymbol{x}, \boldsymbol{y}') \leq M(\boldsymbol{x}', \boldsymbol{y}') \leq M(\boldsymbol{x}', \boldsymbol{y})$$

Equivalently, we can write

$$\max_x M(\boldsymbol{x}, \boldsymbol{y}') = M(\boldsymbol{x}', \boldsymbol{y}') = \min_y M(\boldsymbol{x}', \boldsymbol{y})$$

Thus

$$v_1 = v_2$$

We can now state the *minimax theorem*:

In a two-person, zero-sum game, each of the following three conditions implies the other two:

1. *An equilibrium pair exists.*
2. $v_1 = v_2$.

3. *There exists a real number v such that player 1 can guarantee himself an expected payoff of at least v, and player 2 can guarantee that player 1 gets at most an expected payoff of v.*

The minimax theorem was first stated by von Neumann in 1928.

17.9 SOLUTION OF A TWO-PLAYER, ZERO-SUM GAME BY LINEAR PROGRAMMING

The expression of a two-player, zero-sum game in the form of a linear programming problem is a direct consequence of the minimax theorem. We note that the expected payoff to player 1 is given as

$$M(\mathbf{x},\mathbf{y}) = \sum_{i=1}^{m}\sum_{j=1}^{n} a_{ij}x_iy_j$$

subject to the constraints

$$x_i \geq 0$$

$$\sum_{i=1}^{m} x_i = 1$$

and

$$y_j \geq 0$$

$$\sum_{j=1}^{n} y_j = 1$$

Furthermore, player 1's strategy is optimal if

$$\sum_{i=1}^{m}\sum_{j=1}^{n} a_{ij}x_iy_j \geq v_1 = v$$

for all strategies $\mathbf{y}$ available to player 2. We can rewrite this in the form

$$\sum_{j=1}^{n} y_j\left(\sum_{i=1}^{m} a_{ij}x_i\right) \geq \sum_{j=1}^{n} y_jv = v$$

Because this must apply for all $\mathbf{y} \in Y_n$, it must apply for the case where

$$\mathbf{y} = (0,\ldots,0,\alpha_j,0,\ldots,0)$$

Thus, we can interpret the above inequality as n inequalities. Combined with the constraints on x_i and y_j, these form a legitimate linear programming problem, and we have reduced the problem of finding an optimal mixed strategy for player 1 to the problem of finding a feasible solution to the above linear programming problem.

Two problems remain. We do not have an objective function, and we do not know the value of v. We can solve both at once by selecting $x_{m+1} = v$, and maximizing x_{m+1}. Thus, the linear programming problem takes the form

$$\min(-x_{m+1})$$

subject to the constraints

$$a_{11}x_1 + a_{21}x_2 + \ldots + a_{m1}x_m - x_{m+1} \geq 0$$

$$a_{12}x_1 + a_{22}x_2 + \ldots + a_{m2}x_m - x_{m+1} \geq 0$$

$$\cdot$$

$$\cdot$$

$$a_{1n}x_1 + a_{2n}x_2 + \ldots + a_{mn}x_m - x_{m+1} \geq 0$$

$$-(x_1 + x_2 + \ldots + x_m) = -1$$

and

$$x_i \geq 0 \qquad \text{for} \quad i = 1,2,\ldots,m$$

The objective function for player 2 derives in an analogous manner. We obtain

$$\max(-y_{n+1})$$

subject to the constraints

$$a_{11}y_1 + a_{12}y_2 + \ldots + a_{1n}y_n - y_{n+1} \geq 0$$

$$a_{21}y_1 + a_{22}y_2 + \ldots + a_{2n}y_n - y_{n+1} \geq 0$$

$$\cdot$$

$$\cdot$$

$$a_{m1}y_1 + a_{m2}y_2 + \ldots + a_{mn}y_n - x_{n+1} \geq 0$$

$$-(y_1 + y_2 - \ldots + y_n) = -1$$

and

$$y_j \geq 0 \qquad \text{for} \quad j = 1,2,\ldots,n$$

We see here that the two linear programming problems, one for player 1 and the other for player 2 are *dual* to each other. Thus, if we solve one by means of the simplex method, we obtain the other solution at the same time. We also see the relationship between duality in optimization and two-person, zero-sum games.

7.10 SUMMARY

In this chapter, we introduce some of the basic concepts of game theory. We began by defining a game as any activity that involves more than one individual, where the actions of each individual in the game affect other individuals who are also in the game. We then established a language and notation for the discussion and mathematical analysis of games. The simplest games, we found, are games of dominant strategies, where all players agree strongly upon an equilibrium strategy combination. We found it possible to extend our concept of an equilibrium strategy combination in the case of a dominant strategy game to a game of iterated dominance, where the players preferences for one strategy over another might be stated only weakly.

Next, we considered the effect of information on games. We saw that the definition of information in the context of game theory is virtually identical to the definition of information as we have applied it to decision making in engineering design. Then we classified games in terms of their informational properties, as games of perfect or imperfect information, certain or uncertain information, symmetric or asymmetric information, and complete or incomplete information. We examined the extensive form, which is much like the decision/event tree that we introduced for purposes of decision analysis, for the analysis of games involving information. And we saw that the method of backward induction applies to the analysis of games expressed in their extensive form.

To this point we considered only pure strategies, that is, strategies that are made up solely of the selection of one available strategy or another. But then we observed that some games do not possess equilibrium strategy combinations in terms of pure strategies. We showed that these games have instead equilibrium strategy combinations that are comprised of mixed strategies. Mixed strategies are strategies that are random combinations of pure strategies.

Next, we reviewed the Nash model, beginning by defining a Nash equilibrium. We saw that the definition of the Nash equilibrium guarantees that all games that have a finite number of strategy combinations have an equilibrium strategy combination, sometimes in the form of a mixed strategy. Then we examined the Nash bargaining solution, which is the only bargaining solution that satisfies four intuitively satisfying assumptions.

Finally, we turned to an examination of two-person, zero-sum games. This included first a review of the minimax theorem and then the use of the minimax theorem to express a two-person, zero-sum game as a problem in linear programming. In the end, we found a very close relationship between two-person, zero-sum games and duality theory in optimization.

REFERENCES

1. Fudenberg, D. and J. Tirole, *Game Theory*, The MIT Press, Cambridge, MA, 1992.
2. Hillier, F. S. and G. J. Lieberman, *Introduction to Operations Research*, McGraw-Hill Publishing Company, New York, 1990.
3. Luce, R. D. and H. Raiffa, *Games and Decisions*, John Wiley, New York, 1957.
4. Nash, J. F., "The Bargaining Problem," *Econometrica*, vol. 18, pp. 155–162, 1950.
5. Rasmusen, E., *Games and Information, An Introduction to Game Theory*, Blackwell Publishers, Oxford, UK, 1992.
6. Von Neumann, J., "Zur Theorie der Gesellschaftsspiele," *Mathematische Annalen*, vol. 100, pp. 295–320, 1928.

PROBLEMS

1. Which of the following activities comprise games:
   a. Setting price in a competitive market.
   b. Bidding for a contract in competition with other firms.
   c. Designing an automobile.
   d. Setting a standard for a class of products such as high definition television
   e. Setting engineering standards such as building codes.
   f. Negotiating to purchase raw materials for your company.
2. Suppose two competing companies supply a good to a single market. Each company's goal is to maximize its net profit. Company A's marginal cost of supply is given by

$$S_A = a_0 + a_1 q_A$$

and company B's marginal cost of supply is given by

$$S_B = b_0 + b_1 q_B$$

Demand for the product is given by

$$D = d_0 + d_1 P$$

where P is the price. Assume that all consumers are fully informed and make rational decisions. What is the equilibrium price?

3. Two companies decide to collaborate on a bid for a job that is too big for either company to perform alone. However, they disagree on how to split the job, that is, which company gets what share of the work should they win the contract. The companies are of equal size and have similar utility functions, that is, the utility function of one can be written as an affine transformation of the utility function of the other. What is the equilibrium solution for splitting the job?
4. Your company produces a part that requires careful machining. But your machinists tend to be sloppy and do a poor job unless they are carefully checked. You could order that every part be carefully inspected, and then the machinists would be careful and produce only good parts. On the other hand, it costs a good deal of money to inspect each part. Suppose that the cost of producing each part is \$1,000, and the cost of inspecting each part is \$100. What is your optimal strategy? Make whatever assumptions you must but state them.
5. You are in charge of the development of a major new computer service. It has been estimated that the average present value to the company of each new customer, referenced to the date that the customer first purchases the service, is \$1,000, and that there are approximately 2 million potential customers. A competing company is reported to be developing a similar service, and you expect that customers who purchase their service from your competitor will not purchase it from you. You expect that early sales will be particularly important in determining market share, with first-month sales projected at 100,000, and diminishing by 5 percent per month as the product penetrates the market. You feel that your competitor is working on a schedule similar to yours, and that his costs and profits would be about the same as yours. You have two choices; to complete development of the product thoroughly before release, or to release the product prematurely, while it still has some problems. The latter option will get your product into the marketplace about six months earlier than the former option. You feel that your competitor also faces the same options. If you elect to enter the market early, you estimate that the problems in the software will dampen sales, such that after the first month, sales will diminish

by 8 percent per month. The same will apply to your competitor. But, if your competitor waits to introduce a better product, his sales will diminish at only five percent per month, and initial month sales will be reduced in proportion to the size of the remaining market. Further obscuring the issue, you expect that problems in the software will reduce the present value of each new customer to $800, instead of the full $1,000. What is your best strategy?

6. You have gotten a job as a consultant to a professional football team. You have analyzed the success of the team's plays and advise the coach on the best play for each situation. Your team has reached the playoffs, but it is losing by 5 points with 30 seconds to go. It is fourth down and 5 yards to go for a first down, and 15 yards to go for a touchdown. Your options are pass or run. The defensive team's options are to defend against the pass or defend against the run. If your team passes, and the opposing team defends against a run, your team will have a 0.7 probabil-ity of gaining the needed yardage. But if the opposing team defends against a pass, your team will have only a 0.2 chance of success. On the other hand, if your team runs and the other team guards against a pass, your team will have a 0.5 probability of success. And if the other team guards against the run, your team will have a 0.35 probability of success. What advice do you give to the coach? If the other team is equally sophisticated, what do you think they will guard against?
7. Determine the categories of information of the games described in problems 5 and 6.
8. Draw the extensive form for the game of problem 6.
9. You are bidding to produce a high-tech defense system. In terms of information, classify the game you are in.
10. Why should a company do product improvement R&D? Suppose that the company is a monopoly supplier. How does this affect your response?
11. Does manufacturing R&D, the goal of which is to lower production cost, always benefit the company? If not, why do it? If it does, why not do more?

18

Management of Engineering Systems Design and Operation

18.1 THE PURPOSE OF THIS CHAPTER

Today's technological world has become highly competitive, and it is ever more important that the systems engineer squeeze every possible bit of value out of a system. This means that the systems engineer must be able to find and use potential value wherever it may be. Potential for increased value lies in every phase of the life cycle of a product, but particularly in the design and operation of a system. Therefore, the systems engineer must be aware of the potential for system improvement and capable of evaluating options for possible use. Suggestions for improvements can come from many sources, but a very substantial and often untapped source is the people who are responsible for the design, manufacture, and maintenance of the system. Too often, companies fail to use the expertise that exists directly under their noses. Sometimes they fail to use this expertise because management does not want to recognize that it exists, and sometimes management simply does not know how to use it. But, if the systems approach to engineering emphasizes one thing above all else, it is that consideration of as many options as possible is very important. And a key source of options lies in the employees who design, build, maintain, and use the system.

One problem that prevents the use of expertise of the people who design, manufacture, maintain and use a system is that their expertise often spans many disciplines, whereas the project manager's disciplinary expertise is generally much narrower. Making broad use of disciplinary expertise can be a very significant challenge. Several problems prevent this. First, there is the problem of identifying who the experts really are. While many people tout themselves as experts, few actually are. Second, disciplinary experts tend to be very narrowly focused. That is how they got to be experts. But, because they tend to be so narrowly focused, they have a language of their own, and they have trouble talking to people of different disciplines. Thus, communications across disciplinary boundaries tends to be very difficult. Third, the manager, in order to enlist the aid of disciplinarians, must divide the project along established disciplinary boundaries. This requires that the systems engineer know and understand the requisite disciplinary boundaries. Fourth, results, suggestions, or advice provided by disciplinary experts tend to be highly specialized and narrowly focused. Integrating them into information that is useful in the context of overall project management poses major problems. And, fifth, the evaluation of results, suggestions, and advice from experts can also be a very difficult exercise.

The purpose of this chapter is to emphasize the problems inherent in the management of engineering systems projects both large and small, and to suggest approaches for improved project management.

18.2 THE MANAGEMENT OF ENGINEERING SYSTEMS DESIGN

Good management of engineering systems design is in part good management of human, monetary, and material resources. We can also think of it as good decision making, by which we mean the identification of good options and effective selection among the options. These are the capabilities that the systems engineering approach is keyed to provide. The systems engineer should know how to obtain and make use of human resources, and how to integrate the inputs from disparate experts to make rational programmatic decisions in the best interest of accomplishing the project goals. There are many tools provided in the earlier chapters of this book that will help you.

Throughout this book, we have emphasized that engineering systems design should not take place in isolation from the full set of phases in the life cycle of the system: research, development, design, test and evaluation, manufacture, distribution and sale, operation or use, maintenance and repair, and disposal, for example. From the very beginning of a project, you should seek expertise in areas of all life cycle phases of the system. The first problem you are likely to run up against, after identifying the life cycle phases, will be obtaining the necessary expertise.

One view, popularized in the business school mentality, is that it is possible to manage something that you do not understand provided only that you understand good principles of management. This view is based on the notion that expertise can always be bought and put to use by the skilled manager. The view is at least partly correct. Indeed, expertise can be bought. Lots of people sell themselves as experts or consultants. All the manager has to do is find one. But the problem is that for every true expert there may be as many as 10 people who claim to be experts. Furthermore, the would-be experts often sound better than the true experts because the true experts know about the problems they are likely to encounter in trying to meet system goals, and they tend to be rather conservative in promising solutions to these problems. The would-be experts, on the other hand, might not know of the existence of these problems at all, or they are willing to ignore the fact that they could pose real threats to project success. Thus, the problem with buying expertise is knowing it when you see it. The availability of expertise is simply not an asset if it is randomly mixed with non-experts and disguised so that it cannot be identified, and the chance of hiring a non-expert is far greater than the chance of hiring an expert.

Many successful companies deal with advanced technologies by maintaining an in-house expertise, and they draw their managers from among the technical experts in the company. Some companies have taken the strategy of maintaining high-class research laboratories not for the product of the research produced in the labs, but rather as a mechanism to attract the very best scientists and engineers, who contribute to the day-to-day activities of the company. Indeed, if you recognize that you cannot simply buy expertise on a moment's notice when it is needed, then you also might come to the conclusion that the best way to obtain expertise is to grow it internally in the company. In the end analysis, expertise can be bought, but it is kind of like borrowing money; it is easy to borrow money if you do not need it. Likewise, it is easy to identify an expert in a given discipline if you are one—*it takes one to know one*. Therefore, good project management begins long before the project begins by assuring that your company has the expertise that the project demands. This does not mean that you cannot add to that expertise by hiring additional employees or by retaining consultants. But it does mean that you need to have an internal expertise so that you can make intelligent selections on *adding* to that expertise.

Once you have assembled a project team, your next important job is to provide an environment that is conducive to getting the best possible performance from the team. This means that you must provide a comfortable and secure or nonthreatening work environment; you must promote the flow of information among team members; you must instill enthusiasm for the project in the team members; you must insist on openness and honesty; you must provide an appropriate delegation of responsibilities properly matched with authority; and you should create a system of incentives and rewards that lead team members to the conclusion that what is best for the project is best for them. In seeking to meet these conditions, you will have to work within the framework of your company and the resources and restrictions that this imposes. Try as you may, you probably will not be able to meet the conditions as well as you would like. Still, it is important that you try.

Let us start with the work environment. Good communication between team members is aided by locating the team leaders close to each other. Quality and location of office space is important. Informal communications are often more productive than formal ones, such as meetings. Consider how this can be achieved: over lunch, group sports activities, bull or rap sessions, by providing a small meeting room where people can congregate to discuss problems or opportunities informally. Physical comfort is also important. Proper lighting, ventilation, heat and air conditioning, the need for a smoke-free environment for some, and standards of dress that are appropriate to the job all affect job performance. For example, some people are allergic to wool, and they will produce far less output dressed neatly in a wool suit than they would dressed more comfortably in cotton. It is important that you be aware of such factors and sensitive to them and the options that you have available, and be reasonable in assessing what is important to you.

Few things can promote the success of a project as well as enthusiasm. People are far more likely to give their best effort to a project if they believe that the project is worthy of it than if not. Enthusiasm comes from several sources: a feeling that the project is the right thing at the right time, a feeling that the project is well managed, a feeling that the inputs from each individual are important to the success of the project, and a feeling that the project is important to the company and its upper management. But, in the end, it is also the case that enthusiasm is contagious. So if you can get a few enthusiastic people on your project team, their enthusiasm can often instill enthusiasm in others.

The need for openness and honesty should be self-evident. But often it is not achieved. There can be several reasons why this is so: management punishes failure rather than rewarding honesty, even when the failure is not the fault of an individual; people seek self-aggrandizement through exaggeration of their accomplishments; too much weight is placed on what people say they will do rather than on what they actually do; and because a person is not up to the job and is trying to cover up. One of the worst things that can happen is when you are faced with one or two team members who cannot adequately perform their jobs and know it. Such people sometimes try to cover up their shortcomings through secrecy, lack of openness, or outright dishonesty. These people can poison the enthusiasm of other team members, and they must be identified and taken off the team or given jobs that are more compatible with their capabilities.

Lastly, appropriate incentive and reward structures are extremely important factors in promoting the goals of a project. People tend to act in their own best interest, and the project will benefit if the best interests of the team members can be aligned with the best interests of the project. Since the objective here is to align the interests of the project and the team members, the first job is to carefully determine and express the goals and objectives of the project. Now we see that utility theory can be a very important tool in project management. You may begin by setting project goals in the form of written statements. Next, translate these into a utility function, which may be a multiattribute utility function. Then test this function carefully to see if it truly does express utility over the broadest possible range of system attributes. Remember that the utility function now becomes your statement of what is sought out of the project. It is, in essence, a list of desires. And any time that you list your desires you must be very careful because you just might get them. When you are satisfied that the utility function truly represents your goals for the project, then set goals for the members of the project team. From these goals, you can establish a structure of incentives and rewards. Finally, test the incentive and reward structure to be sure that it is in full accord with the utility function set for the overall project.

Two things have come out of this last exercise. First, you should have an incentive and reward structure that aligns the interests of the team members with the goals of the project. Second, you will have a utility function that expresses with mathematical precision the goals and objectives of the project. Use of this utility function by all team members should help them in their design decision making. It instills a rationality to the project decision-making process. Of course, nothing is unchangeable, and it may occur that you encounter a decision that goes against the utility function, yet that you are confident is correct. In this case, revise the utility function. Just as a design is subject to constant revision throughout the design process, so too is the project utility function.

The design process of many companies is a top-down process. An overall system is divided by the imposition of constraints into subsystems, components, and devices, and when it is finely enough divided engineers set about the design of its respective elements. To the extent that such constraints are imposed only in a downward direction, that is, from the larger system to smaller and smaller segments of the system, management disables and suppresses the contribution that could be made by lower level workers. But if a person is worthy of working on a project, that person has a contribution to make to the project. Part of your job as systems engineer is to solicit that contribution. In other words, the flow of information during the design process should be in two directions, both downward and upward. Failure to promote the upward flow of information denies management of the expertise present in the company, and in the process alienates lower level people and stifles enthusiasm. But allowing bidirectional flow of information has its dangers. It enables lower level workers to demonstrate their abilities, which may sometimes exceed those of their management. In some organizations, this is taken as a threat to the security of the management. Therefore, it is necessary to assure the security of the managers in order to promote an upward flow of information. This is not always easy to do, but it is substantially easier if the managers are technical experts themselves and confident of their own abilities. It might also help to keep in mind the following facetious definition: *a manager is a person who knows how to take credit for other people's work.* Give the manager credit for the work of the people who he or she manages. But this does not require that you take the credit from the people themselves.

It is human nature that people do a better job if they feel that someone is looking over their shoulder and grading them on their work. It is true that overdoing this can have negative effects, but done properly, uniformly, and fairly, oversight is highly beneficial to overall project success. A very effective way to accomplish oversight is by means of independent peer review. We recommend that all aspects of a project be subject to thorough and periodic peer review. And, although the project manager or systems engineer can obtain peer review on his or her own initiative, it is better done at a higher level in the organization. It is best that the peer review team be selected by and report to a manager at a higher level than the project itself, preferably at the level of the president or even the board of directors. An independent assessment of a project can assure that the highest standards are promoted and maintained in all aspects of the project, and it can anticipate problems that might otherwise degrade the project.

Taken together, the recommendations given above on project management can prove somewhat expensive. But expense is relative. Manufacturing and other costs nearly always dominate total project costs. Engineering is usually a rather small component of overall costs. So, even if these costs are doubled, the total project costs might not be severely affected. Further, there is always significant opportunity for cost savings or improved sales as a result of good design. It is just plain good business to do a good job of design, and skimping on the engineering design budget generally is not good management practice.

The systems engineering approach outlined in this book enlightens us on many aspects of good systems engineering management. We see that good design requires that we have good options to choose from and that we make good choices from among these options. Thus, we first want to assure that we elicit as broad a set of design options as possible. Everyone on the design team can contribute to the process of suggesting design options. Sometimes options emerge when different members of the design team get together to discuss problems. Sometimes they come from a single person either from pure inspiration or in response to a problem. The key, however, is that you provide an environment that encourages people to come forth with options. Team members should feel that it is part of their job to suggest design options. Customers can also be an important resource for design options, but we will discuss them later.

Given design options, the next job for the systems engineer is to evaluate them for selection. The techniques of design evaluation are the major subject of this book. Here, we will limit our discussion to emphasizing once again the need for a systems model. Without a systems model, it can be difficult or impossible to accurately compare options.

We have already noted the importance of having a two-way flow of information, both down and up through the organization, during the design process. Here we will place the emphasis on the word *information*. When a system is designed by constraints, management discussions tend to focus on the constraints. Initially, the discussions focus on establishing constraints, and later they focus on meeting the constraints. To be sure, the constraints contain information on the design itself: the design is an artifact of the constraints. But discussions relating to the constraints do not contain *information* relative to the selection of options, that is, they do not provide expectations on the outcomes of various choices. Remember that we have defined information as relating to specific decisions, and a numerical expectation provides information only when it is accompanied by a probability. Reorienting the design process to the systems engineering approach based on modeling promotes the exchange of information between members of the design team. Furthermore, the models can be developed along disciplinary lines, enabling the systems engineer to make effective use of the expertise of disciplinary experts. The modeling approach also gives the opportunity to workers below the management level to make inputs to the system design in a way that affects the entire system, not merely the subsystem, component, or device that they are responsible for.

Good managers accomplish many things, but key among them is that they make good decisions regarding the selection of system design options. Without good information, the selection of system design options is an art. With good information, it is easy, and the best choices are obvious. Good information is obtained largely by means of the modeling process. Again, we see the need for good system models that evaluate design options against the system goals, that is, against a system utility function, and that span the full life cycle of the system.

18.3 THE MANAGEMENT OF ENGINEERING SYSTEMS OPERATION

A phase in the life cycle of an engineering system that is frequently neglected by engineers is its use phase or its operation. Although this seems absurd because the main purpose of any system is its use, engineers are perhaps more removed physically from the use phase of a system than any other phase. Indeed, they are intimately involved with research, development, design, testing, and manufacture. They get feedback on distribution and sales. And they can be involved in maintenance, repair, and disposal. But normal use tends to be in the hands of the customer, and feedback on this phase of a system's life cycle can require deliberate action. If a system works well, why should the customers bother to report it? This is what they expect. If there are problems, how do they report them? Most companies do not make it easily possible to report problems with their products. If a customer has a suggestion for product improvement, what benefit do they get from it? There are two distinctly different types of systems that we need to consider here, systems that are turned over to the customer to use as they wish, such as an automobile, a toaster, or a personal computer, and systems that are operated only by highly trained persons in accord with accepted engineering standards. These systems include nuclear power plants, large commercial aircraft, spacecraft, and other typically very expensive systems. We will discuss these types of systems separately.

First, we will discuss systems that are turned over to customers for use as they see fit. These systems span a wide range of product cost and complexity, from something as simple as a pencil or coffee cup through appliances such as a hair dryer, washing machine, or dishwasher, up to systems that cost a major fraction of a worker's annual income, such as an automobile, a boat, or even a private airplane. Virtually all of these systems have one thing in common: they could be improved upon. This means that they could be more reliable; they could have longer life; they could be less expensive; they could be more convenient; and so on. Sometimes, products have significant design defects, and some such defects can be easy to fix. But the design engineers will not know to fix them unless they know about the problems or opportunities. Thus, one of the most important things that a systems engineer needs to be concerned with regarding the use phase of a product is obtaining information on the product: What is good about the product? What is bad about it? What problems and complaints do customers have? There are two sources of such information, customers at large and employees of the company that manufactures the product. The former source is larger, and it will have a broader purview on the product. But the latter source is more informed and has a greater stake in seeing the product improved. It is good practice to take advantage of both sources of information, but pay particular attention to the latter source. Employees can do much more than merely design and manufacture a product. They also can be its toughest critics.

To get full value from employees, they need incentives to buy and use the company's products and to report problems and possible solutions. Incentives to buy can include discounts and other special considerations. Incentives to report problems and potential solutions can include monetary rewards for good or novel ideas. Most important, however, is a sense of belonging to the company—a sense of pride in the company—and the knowledge that reporting a problem is a good thing, not a bad thing. All too often, management makes it clear to the workers below them that they do not want to hear about problems with their products, and that people who raise problems will be punished. This attitude quickly stifles all upward flow of information and deprives the company of one of its best sources of information.

We turn now to systems that are very expensive and that require close engineering supervision. We might think that, by assigning engineers to oversee the use phase of these systems, we will have done all that we can to assure their effective operation. But this is not the case. Effective operation of a system begins with its design. Indeed, the design and operation of a system are closely related, and their interaction is subject to analysis through modeling. The process of modeling the use phase or operation of a system is called *operational analysis*. The tools of operational analysis are discussed above, most importantly including simulation and reliability analysis. Cartoons and scenarios are useful graphical aids in developing a simulation for operational analysis. It is important to realize that the performance of a system can vary widely as a function of its use, and that such variations in performance can have substantial impact on system design. To put it differently, there generally exist many options in the use of a system, and these options interact strongly with options in the physical design of a system.

Good design demands that we investigate operational options as well as physical design options. To do this, we must first identify operational options. This is not difficult, but it must be done as a part of the overall process of option identification. Frequently, it is neglected. Then, operational options need to be evaluated and selected just as do physical design options. Again, the key to good decision making regarding operational options is good information. This can be obtained by modeling.

18.4 CONTINUING PRODUCT IMPROVEMENT

In our modern technology-oriented world, most products enjoy a rather short period during which they are competitive in the marketplace, and before they are superseded by more advanced products. For some products, such as computers, this period might be only one or two years. For other products, such as beer or wine, it could be longer. While there is not much you can do to stem the tide of technology advancement, there are a few things that you can do to take advantage of a favorable market position, even if it is only temporary.

First, you should recognize that, no matter how good your product is, sooner or later there will be a better one, and it will be manufactured either by your company or by another company. Your company will manufacture the advanced product only if it has the technology to do so. Therefore, the ability of your company to stay in its line of business depends upon maintaining a superior technology. In the long run, a company maintains a superior technology only if it is its own fiercest competitor.

When a company competes with itself, its management and researchers must recognize that technology development takes two forms, evolutionary and revolutionary. Evolutionary technology development is the result of incremental improvements on the existing technology. A revolutionary technology development leaves the previous technology behind and steps beyond its bounds. Many of the physical limitations that applied to the previous technology do not apply to the new technology, and it often provides an avenue to new levels of performance that could not be achieved by the previous technology. For example, early vacuum tubes enjoyed improved performance through more effective designs, better materials, better manufacturing techniques, and so on. Over time, tubes became smaller, were more reliable, had longer life, lower cost, and they had better operating characteristics. Earlier tubes could not compete, and still there was much room for improvement. But then the transistor appeared, and it evolved into the integrated circuit. Today, almost no vacuum tubes are produced, and in the United States only one vacuum tube manufacturer, Motorola, successfully made the transition from vacuum tubes to integrated circuits.

There are many stories about companies that were highly successful at competition within their product line. But they focused their attention so much on their existing product line that they failed to see that revolutionary developments would change that line altogether. The manufacturers of oil lamps all failed to recognize the emergence of the electric light bulb. The operators of telegraph systems failed to recognize the emergence of the telephone. Only one carriage manufacturer, Fischer, successfully made the transition from horse drawn carriages to the automobile. The manufacturers of bias ply tires failed to recognize the emergence of radial ply tires. It is interesting to examine this process of evolutionary versus revolutionary technology development.

Upon first emergence, we can think of all technologies as revolutionary. And when a technology first emerges, it tends to be in a very primitive state. The Wright brothers first airplane could fly only about 200 m at speeds of only about 40 km per hour. The first commercial integrated circuits contained only about 1,000 transistors. The first atomic bombs had explosive yields more than 1,000 times less than current bombs. Early automobiles were expensive, unreliable, and inconvenient. They had no self starter, no automatic transmission, poor handling characteristics, they were loud and dirty, and so on.

Generally, when a technology first emerges, it emerges on the margin of feasibility. There are many, many possible approaches to improving the technology. Some will be easy to obtain, others will be difficult. The promise of the technology is our incentive to pursue such improvements. It is not likely that the Wright brothers had in mind that their airplane would lead to the Boeing 747-400, capable of transporting 350 people nonstop one-third of the way around the world, or to the Space Shuttle, capable of circling the Earth once every 90 minutes at an altitude of over 250 km. But they could see the possibility of staying aloft for 10s of minutes, reaching altitudes of 1,000 m or more, and flying distances of more than 25 km. This was good enough.

The availability of a wide range of possible improvements to a technology tends to lead to rapid improvement during the period shortly after the emergence of the technology, particularly if there is a sufficient incentive. Again, if we look at the airplane, at the beginning of World War I some 10 years after the Wright brothers first flight, the airplane was a curiosity used mainly for observation. It was not armed, it had little payload or range, and it was terribly unreliable. But it soon was recognized as a potentially powerful weapon, and substantial efforts were made to improve the technology. Four years later, by the end of the War, airplanes were capable of carrying a 450 kg bomb load a distance of 1,600 km, and achieving altitudes approaching 7,000 m. The War ended in November 1918, and as soon as weather permitted, in May 1919, the Atlantic Ocean was successfully crossed by an airplane designed during the War.

The modern computer is another success story that goes beyond all reasonable expectation. Today, you can purchase a personal computer that has greater capability than the best of the multimillion dollar supercomputers of the early 1960s at a cost that is less than the cost of a mechanical calculator of the early 1950s. As a result, we now use computers for tasks such as word processing, which we would not have conceived of when the cost of computational capability was five orders of magnitude greater.

Eventually, we exhaust the supply of easy approaches to improving a technology. Technologies tend to be subject to physical limits. But these limits are rarely reached. Instead, as they are approached, each additional step toward them becomes more difficult and more expensive. This has clearly been the case with microelectronics. Each 4 years, for the past 20 years, has seen a doubling of the number of transistors that can be placed on a 1 cm^2 chip. But this doubling of capability has come only at great expense. Today, a state-of-the-art microelectronics fabrication facility can cost in excess of $2 billion. And still, the technology seems to be some distance from the ultimate physical limits.

If we plot for a typical technology, the technological advance obtained as a function of investment in the technology, we get a curve that closely resembles the Gompertz curve as shown in Fig. 18.1. Indeed, the Gompertz curve can be a good predictor of technological advancement versus investment if we can properly identify the physical limits of the technology. The major characteristic of a revolutionary technology is that its physical limits are significantly different than the current technology. For example, the explosive power that can be delivered in an atomic bomb is many orders of magnitude greater than that of conventional explosives; the underwater range of a nuclear submarine is virtually unlimited compared to only 10s of kilometers for conventional submarines; the top speed of jet airplanes ranges up to perhaps eight times the speed of sound versus subsonic maximum speeds for propeller-driven airplanes; and the bandwidth of fiberoptic cables is orders of magnitude greater than copper wires. This shift in physical limitations provides the opportunity for much greater system performance using the new technology, and so we often see new, revolutionary technologies replacing the conventional or existing tech-

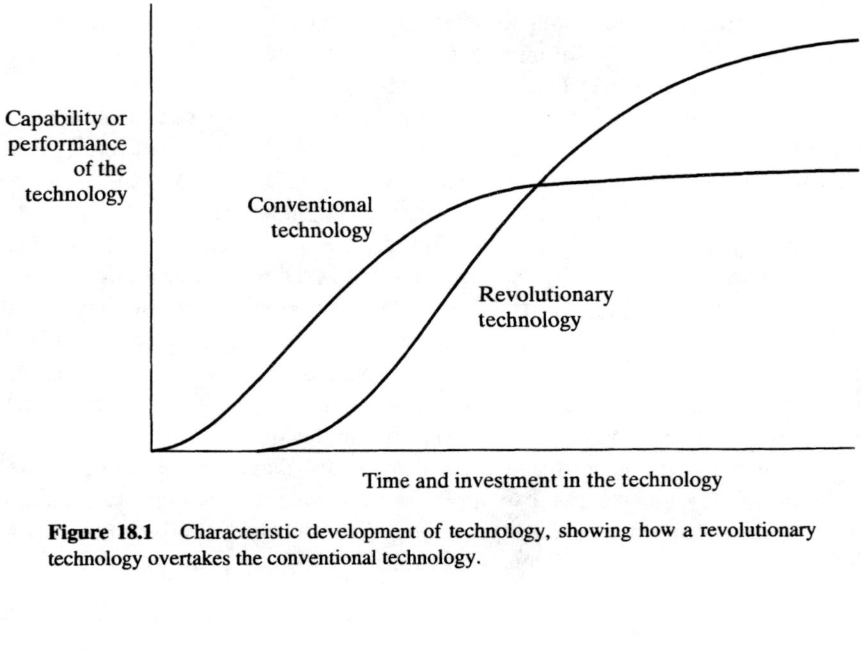

Figure 18.1 Characteristic development of technology, showing how a revolutionary technology overtakes the conventional technology.

nology until it too is superseded by yet another revolutionary technology. This phenomena is also shown in Fig. 17.1.

The lessons here for the systems engineer are the following: (1) When a technology first emerges, there is a chance, but not a certainty, that it will experience rapid improvement. (2) You can get considerable insight into the potential for the new technology by examining the shift in physical limitations made possible by the new technology. (3) You can expect that, with some probability but not certainty, the new technology will result in changes that you least expect. (4) If you make use of the new technology, you can expect that it will experience rapid improvement, requiring that you invest in its development in order to remain competitive. (5) You can expect that, at some future time, even this new technology will be superseded.

Technological improvements are to be found in all aspects of a technology, not merely in design. For example, in the area of microelectronics, U.S. companies focused on the design and fabrication of ever smaller devices such as transistors, leading to lower costs per device because smaller devices could be packaged more densely on the chip. This led, however, to yield rates on chip fabrication sometimes lower than 50 percent. Japanese companies focused on the problems of yield rate, and raised yield rates to over 90 percent. This resulted in halving the cost of a functional chip, and also lowered the perdevice cost. If you have a good systems model, you will be able to examine all such aspects of a technology and make informed decisions on investments in technology development.

With respect to existing systems, the systems engineer needs to keep the following in mind:

> *Product design is not done when the product enters production. It must continue throughout the lifetime of the product, incorporating improvements as they become available.*

At the same time, however, the systems engineer must remain aware of the fact that revolutionary technologies are lurking everywhere. And while it is often bad policy to forego development efforts on evolutionary advances in a technology, it can be equally bad policy to neglect the possibility of revolutionary advances.

It can be very difficult to understand the impact of a revolutionary technology. Even experts in the conventional technology might not understand the true impact of a revolutionary technology. The examples given above illustrate this over and over. It takes an expert in a technology to understand the technology. Therefore, if you want to maintain an understanding of the potential for revolutionary technologies and how they might compete against your product line, there is no substitute for maintaining an in-house expertise. This almost demands that the long-term preservation of a product line requires research in the areas of potential competing technologies. Again, in order to survive, a company must be its own fiercest competitor.

Today, competition takes place on a global scale. You cannot expect to produce a product in isolation from the rest of the world and remain competitive. Still, many high-level managers of major corporations think they can. Economic competition is much like running a race, and there are two ways to win a race: (1) simply run faster than anyone else in the race, and (2) shoot all the competitors so that none of them cross the finish line.

Many economic policies seek to accomplish the latter approach. These include imposing high tariffs or other trade barriers or limitations on imports. Many managers desire shoot-the-competition policies. But are they good policies, and do they really protect or foster healthy domestic industries?

A person who is running a race at the fastest possible pace has no time for anything but running. Any other activity, even talking, slows the pace. Certainly, a person who tries to shoot the other runners cannot do so while running at maximum speed. Thus, people who seek to win races by shooting the competition are very unlikely to win by running fast. To win, they must rely on shooting everyone else in the race, or at least the better competitors. To do this, at the very least, it is necessary to know who is the competition. In an economic race, this can be very difficult. Vacuum tube manufacturers thought the competition was other tube manufacturers. The manufacturers of bias ply tires thought their competition was the other manufacturers of bias ply tires. The manufacturers of oil lamps thought their competition was other oil lamp manufacturers. In the 1970s, many companies did not think of the Japanese as economic competitors. All too often, we are mistaken in our identification of the competition. And all too often, we fail to recognize that revolutionary technologies can pop up at any time, vastly broadening the field of competition. Thus, in this book, we believe that the most competitive policy is the run-fast policy.

What does it take to run fast in high-technology economic competition? First, it takes a thorough understanding of your own product line. You cannot know the competition if you don't know yourself. Again, system models help. They provide a basis of understanding that is very difficult to obtain from other sources. Second, you need to know what is going on in the world around you. To accomplish this, you need a global orientation. Corporate secrecy is necessary at many stages in the life cycle of a product. But openness is also important as a means of obtaining information from other sources. It is necessary to balance these opposing objectives. Third, you need to travel and to be an active member of your profession. You need friends, and you need to communicate with them. You need relationships with other companies and other nations. Like the runner who uses the person in front to reduce drag and save strength for the home stretch, economic competitors can share information to their mutual benefit.

Finally, there is no substitute for excellence. Consumers want quality, and companies that provide quality products tend to sustain longer in the marketplace. The pursuit of quality demands an in-depth understanding of the technology, and it is this understanding that helps the company understand its competitive position. One bad product can give a company a bad reputation that can destroy markets for other good products. It is important that a company act in a responsible and respectful way toward its customers. If you have been responsible and honest in the design of a product, when problems develop it is much easier to deal with them in an honest manner. Most complex systems have some type of problem over their life cycle. Automobiles sometimes need recall, airplanes develop problems that require modification, appliances may be found unsafe. But experience shows that the company that treats its customers with respect and honesty can sustain such problems. Companies that try to hide problems from their customers generally lose in the end.

18.5 SUMMARY

Whether you are a manager or not, as a systems engineer, part of your job is project management. Your decisions and your designs represent an allocation of resources; material, human, and capital. You are in a key position as regards the success or failure of your company's product. And you are more likely to do a good job if you elicit a broad range of options and provide good information for their selection or rejection.

It is not possible to physically try most options. Therefore, the best way to understand them is through systems models. In general, the best models are symbolic models, and most modern symbolic models are implemented on computers. Such models can be extremely broad reaching, and they can provide entirely new levels of understanding for decision making. As a systems engineer, your major role is to assure that the project is supported by the best possible models. It is difficult to overemphasize the importance of good models.

PROBLEMS

1. You are in charge of the design of a system. Ten people work for you, and your budget is $2,000,000. If you can complete the design under your budget, you will receive a bonus of 10 percent of the cost savings, and you will be promoted and given a 20 percent salary increase. If the design cost exceeds your budget, you will be demoted and receive no increase in salary. If, however, the system doesn't work, you will be fired. What is your utility function for completion of the design? In a stroke of total empowerment, your boss gives you complete authority over the compensation of the people who work under you on your project. What incentive and reward structure for these people will make it in their best interest to work toward your best interest?
2. You are in charge of the design of a kitchen appliance that is already in production. Design an information system, including incentives and rewards, to promote an upward flow of information to you from both employees of your company and customers buying the appliance that will enable you to continually improve your product.
3. Your company manufactures automobiles. A problem has arisen in which a high failure rate has occurred for a particular bearing in the suspension system. It has been determined that the principal cause of this failure is the failure of owners to properly maintain the automobile and provide routine lubrication of the bearing. Nonetheless, the failures are giving your company a bad name. What options might be available to solve the problem?
4. Your company produces industrial diesel engines. What areas of technological advance should you worry about from the point of view of future competition?
5. Pick a product. Discuss what this product might be like in 25 years. What technological advances would enable the changes you predict? Are they likely? What other products might be introduced to displace from the marketplace the product you have chosen?
6. Consider a product that is designed by a team of engineers in a company that does not provide a utility function for design decision making. What can you say about the resulting product? Is the product "rational," that is, is it optimal?

7. In general, when engineers do design, whose objective function are they seeking to maximize: the company's, the customers', government regulators, their own, their supervisor's, or someone else's?
8. What are reasonable objectives of engineering management? What are the costs of failing to provide management to accomplish all these objectives?
9. Your company has asked you to chair a committee to develop a long-range strategic plan. For a 10-year long-range plan, what would be important issues to address? Suppose instead of a 10-year plan, you seek to do a 25-year plan. How does your list of issues change? What about a 200-year long-range plan? Which of these makes the most sense?
10. How can the design of a product to be produced during the next year affect a company's competitive position 10 years in the future?
11. Two engineers are asked separately but at the same time to design similar products. One engineer is given the customers' needs and requirements in the form of constraints that the product should satisfy. The other is given extensive market survey data, providing information on demand for the product as a function of its attributes and price. How might the engineers use the data they are given? Which has more information? Which is likely to obtain a better design?
12. What do customers want? Can an engineer design to provide it?
13. Global Technologies, Inc. (GloTech) develops a new product with a potentially excellent market. What factors should GloTech consider in setting up a production facility? Specifically, consider the risks of too much versus too little production capacity.

19

Case Studies

19.1 THE PURPOSE OF THIS CHAPTER

This chapter provides examples of the systems engineering approach.

19.2 SOLUTION OF THE BEAN JAR PROBLEM

The bean jar problem is described in Chap. 1. A jar full of beans is presented to a group of people who can place bets on how many beans are in the jar. The person who guesses the number of beans that is closest to but not greater than the actual number of beans in the jar wins. The problem-solving approach would be to compute the number of beans in the jar and use this number as your guess. But the precise number of beans in the jar is a random variable, and it cannot be determined with certainty. As a result, the use of the computed number of beans does not give the best chance of winning. A decision-making approach is needed to estimate the best guess.

To begin, we will model the number of beans in the jar. As given in Chap. 1, a model for the number of beans in the jar, n, is

$$n = \frac{V_j}{V_b} f \qquad V_b = \frac{4}{3} \pi a^2 b$$

V_j is the volume of the jar, V_b is the average volume of a bean, and f is a "packing" factor or the fraction of the jar that is filled with beans as opposed to air. This model assumes that beans are oblate spheroids with semimajor axes a and semiminor axis b. This is a descriptive model. Thus, if these variables were known precisely, we could determine n precisely. But we have no way of knowing them precisely, and thus we cannot determine n with precision.

The number of beans actually in the jar is, of course, of interest to us. But what we are even more interested in is knowing what guess or range of guesses provides our best chance of winning the game. To analyze the choice of a guess, we need to simulate the bean guessing game. A flowchart for a Monte Carlo model to do this, and at the same time estimate the number of beans in the jar, is shown in Fig. 19.1. We will begin by simulating the game K times. Next, we will compute an estimate of the number of beans in the jar, n. We will, for now, assume that this is the actual number of beans in the jar, N. Next, assuming that the game is being played against M other people, we simulate their guesses by drawing guesses from our estimate of a distribution on how they will guess. Let each guess of a competing player be given by g. Let g^* represent the winning guess, assuming that we have not made a guess as yet. For each simulation of the game, we save N and g^*. If our guess, i, lies between or equal to g^* and N, we will win the game. Otherwise, we will lose. Thus, after generating a table of sampled values of N and g^*, we examine our probability of winning as a function of our guess, shown in Fig. 19.1 as i. The count $j(i)/K$ is our probability of winning as a function of our guess, i. To end the simulation, we plot $j(i)/K$ versus i, and the probability of N, which is equivalent to the probability of n.

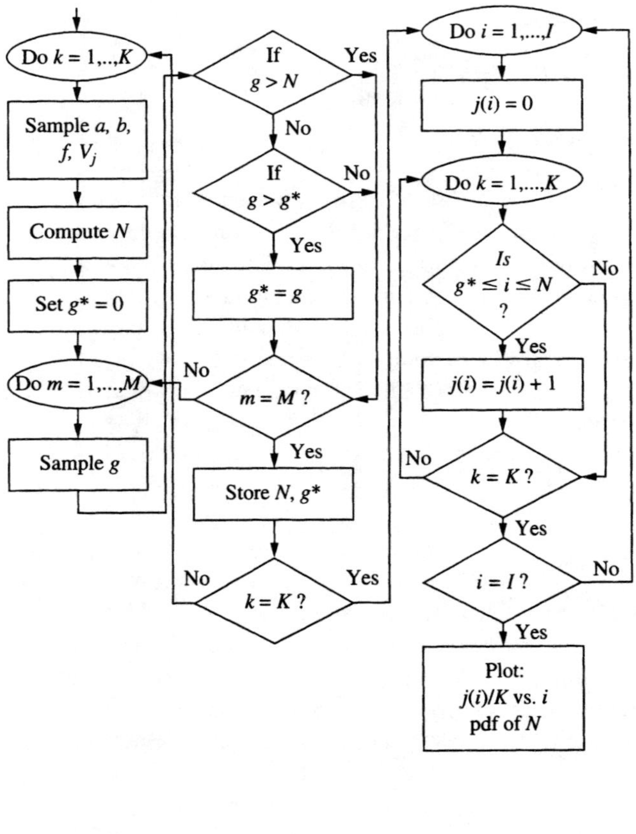

Figure 19.1 Flowchart of a Monte Carlo simulation to determine the optimal guess in the bean-jar game.

The Fortran code to execute this model is given in Table 19-1, and the results are shown in Figs. 19.2 and 19.3. Figure 19.2 shows the probability of the number of beans in the jar. The best estimate of about 470 beans for the example case has a probability of about 0.007 that it is correct. This means that there is probability 0.993 that this number is wrong! Furthermore, the probability distribution is skewed to the high side of the most likely value. Therefore, the mean of the distribution is higher than the most likely value. We see that this distribution alone does not provide the information we need to choose a guess.

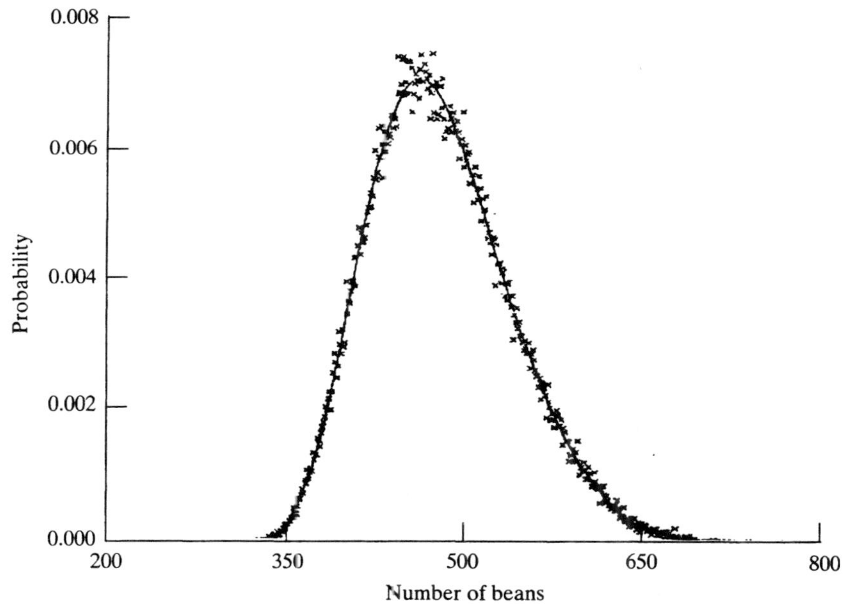

Figure 19.2 A probability density function on the number of beans in the jar.

The results of a simulation for a game involving 40 people are shown in Fig. 19.3. Here we can see that the probability of winning depends rather strongly on our guess, with the best guess having a probability of winning of slightly over 0.05. If everyone guessed randomly, we would have a probability of winning of 1/40 or 0.025. Thus, the simulation predicts that we can roughly double our chance of winning by making an intelligent guess based on the simulation results.

The example results shown here were provided by a student who played this game with the other members of his class. His was the best simulation, and he updated his guess based on his simulation results and won.

TABLE 19-1 FORTAN CODE FOR SIMULATION OF THE BEAN GUESSING GAME

```
C
C   MONTE CARLO MODELING BY YOONHO KIM    MAR., 30,
    1993
C
C     INPUT : A, B, MU, VJ AND OTHER_N'S
C           MINIMUM, PEAK AND MAXIMUM VALUES
C
C     OUTPUT : PDF OF NO. OF BEANS BY MONTE CARLO
      MODEL-
C           ING AND PDF OF CHANCE OF WINNING
C
C
C
      REAL*8 SEEED
      REAL PA(100),PB(100),PMU(100),PVJ(100),POTH(100),
     *     CA(100),CB(100),CMU(100),CVJ(100),COTH(100),
     *     AMIN,APEAK,AMAX,BMIN,BPEAK,BMAX,FMUMIN,
     *     FMUPEAK,FMUMAX,
     *     VJMIN,VJPEAK,VJMAX,OTHMIN,OTHPEAK,OTHMAX
      INTEGER  NMY(3000),NWIN(3000),NBIN
C
C    NITER : NO. OF ITERATION
C    NBIN  : NO. OF BINS
C
      SEEED=12345./2147483647.
      NITER=100000
      NBIN=100
C
C
      PI=3.141592
      WRITE(*,10)
      READ(*,*) AMIN,APEAK,AMAX
      WRITE(*,*) AMIN,APEAK,AMAX
      WRITE(*,20)
      READ(*,*) BMIN,BPEAK,BMAX
      WRITE(*,*) BMIN,BPEAK,BMAX
      WRITE(*,30)
      READ(*,*) FMUMIN,FMUPEAK,FMUMAX
      WRITE(*,*) FMUMIN,FMUPEAK,FMUMAX
      WRITE(*,40)
      READ(*,*) VJMIN,VJPEAK,VJMAX
      WRITE(*,*) VJMIN,VJPEAK,VJMAX
      WRITE(*,50)
      READ(*,*) OTHMIN,OTHPEAK,OTHMAX
      WRITE(*,*) OTHMIN,OTHPEAK,OTHMAX
```

```
C
  10 FORMAT(' INSERT A"S MIN, PEAK, MAX')
  20 FORMAT(' INSERT B"S MIN, PEAK, MAX')
  30 FORMAT(' INSERT MU"S MIN, PEAK, MAX')
  40 FORMAT(' INSERT VJ"S MIN, PEAK, MAX')
  50 FORMAT(' INSERT OTHERS" MIN, PEAK, MAX')
C
C
C   GET THE PDF AND CDF OF A, B, MU, VJ AND OTHERS
C
    CALL CDF(PA,CA,AMIN,APEAK,AMAX)
    CALL CDF(PB,CB,BMIN,BPEAK,BMAX)
    CALL CDF(PMU,CMU,FMUMIN,FMUPEAK,FMUMAX)
    CALL CDF(PVJ,CVJ,VJMIN,VJPEAK,VJMAX)
    CALL CDF(POTH,COTH,OTHMIN,OTHPEAK,OTHMAX)
C
C
    DELTAA=(AMAX-AMIN)/100.
    DELTAB=(BMAX-BMIN)/100.
    DELTAM=(FMUMAX-FMUMIN)/100.
    DELTAV=(VJMAX-VJMIN)/100.
    DELTAO=(OTHMAX-OTHMIN)/100.
C
C   INITIALIZE NMY(3000) AND NWIN(3000)
C
    DO 60 II=1,3000
    NMY(II)=0
    NWIN(II)=0
  60 CONTINUE
C
C
    DO 100 I=1,NITER
C
C   FIND A,B,MU AND VJ TO GET NREAL
C
    CALL INVERSE(CA,AMIN,DELTAA,XA,SEEED)
    CALL INVERSE(CB,BMIN,DELTAB,XB,SEEED)
    CALL INVERSE(CMU,FMUMIN,DELTAM,XM,SEEED)
    CALL INVERSE(CVJ,VJMIN,DELTAV,XV,SEEED)
C
C   MODELING
C
    NREAL=NINT(3.*XV*XM/(4.*PI*XA*XA*XB))
    NMY(NREAL)=NMY(NREAL)+1

    NALPHA=0
    DO 120 J=1,39
    CALL LEWIS(SEEED)
```

```
      DO 130 K=1,NBIN
      IF (SEEED .LE. COTH(K)) THEN
      NOTHER=INT(REAL(K)*DELTAO+OTHMIN)
      GOTO 140
      ENDIF
  130 CONTINUE
  140 CONTINUE
      IF ((NOTHER .LE. NREAL) .AND. (NOTHER .GT. NALPHA))
     *  THEN
      NALPHA=NOTHER
      ENDIF
  120 CONTINUE
      DO 150 JJ=NALPHA+1,NREAL
        NWIN(JJ)=NWIN(JJ)=1
  150 CONTINUE
  100 CONTINUE
      DO 200 I=1,3000
      IF (NMY(I) .NE. 0) WRITE(*,*)
     *    I,REAL(NMY(I))/REAL(NITER)
  200 CONTINUE
      DO 210 J=1,3000
      IF (NWIN(J) .NE. 0) WRITE(*,*)
     *    J,REAL(NWIN(J))/REAL(NITER)
  210 CONTINUE
      STOP
      END
C
C
C   RANDON NUMBER GENERATOR
C
      SUBROUTINE LEWIS(SEED)
      REAL*8 SEED
C
      SEED=SEED*2147483647.
      SEED=DMOD(SEED*16807.,2147483647.)
      SEED=SEED/2147483647.
      RETURN
      END
C
C   GET THE VALUE OF THE PDF AND CDF
C
      SUBROUTINE CDF(X,CX,XMIN,XPEAK,XMAX)
      REAL X(100),CX(100),XMIN,XPEAK,XMAX,HIGH,DELTAX,A
      INTEGER NPEAK
C
      HIGH=2./(XMAX-XMIN)
      DELTAX=(XMAX-XMIN)/100.
      NPEAK=NINT((XPEAK-XMIN)/DELTAX)
C
```

```
C   GET THE VALUE FOR PDF
C
      DO 1000 I=1,NPEAK
      A=HIGH/(XPEAK-XMIN)
      X(I)=REAL(I)*DELTAX*A
 1000 CONTINUE
      DO 1100 J=NPEAK+1,100
      B=-HIGH/(XMAX-XPEAK)
      X(J)=REAL(J-NPEAK)*DELTAX*B+X(NPEAK)
 1100 CONTINUE
C
C   INITIAL VALUE FOR THE CDF
C
      CX(1)=X(1)
      DO 1200 K=1,99
      CX(K+1)=CX(K)+X(K+1)
 1200 CONTINUE
      CXCONST=CX(100)
      DO 1300 L=1,100
      CX(L)=CX(L)/CXCONST
 1300 CONTINUE
      RETURN
      END

C
C   GET THE CDF VALUE FOR THE RANDOM NUMBER
C
      SUBROUTINE INVERSE(X,XMIN,DELTAX,XN,SEEED)
      REAL*8 SEEED
      REAL X(100),XMIN,XN
C
      CALL LEWIS(SEEED)
      DO 1400 I=1,100
      IF (SEEED .LE. DBLE(X(I))) THEN
        XN=REAL(I)*DELTAX+XMIN
        GOTO 1500
      ENDIF
 1400 CONTINUE
 1500 CONTINUE
      RETURN
      END
```

19.3 AN UNDERSEA CABLE DESIGN

This example deals with the design of an undersea fiberoptic cable. Statement of the problem is as follows. The company you are working for plans to install a fiberoptic cable for communication from the mainland of Korea to Cheju Island, over a total distance of

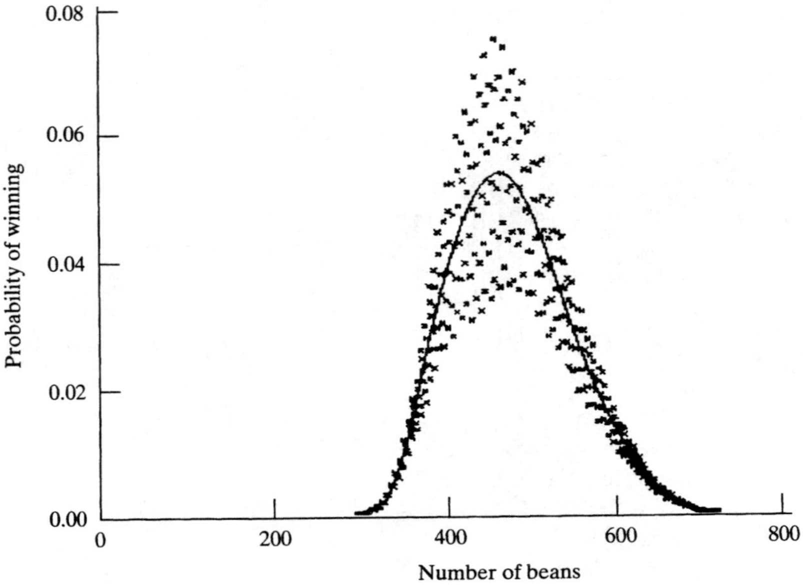

Figure 19.3 Simulation results for the bean guessing game showing the probability of winning as a function of our guess.

200 km. You are in charge of the systems engineering for the cable. A state-of-the-art fiber cable has been selected for its ability to withstand service in the sea. It can transmit an optical signal for a maximum distance of 10 km, but then a repeater is needed to boost the signal strength. A reliable repeater has been designed, but still failures occur. Failures also occur in the cable itself, sometimes the result of damage by fishing boats and sometimes the result of damage caused by fish. When a repeater or cable section fails, the cable must be brought to the surface and the repeater or an entire section of cable replaced. The repair operations can take up to 10 days for a cable failure, and five days for a repeater failure.

The cable will be used for telephone and television service, and it is expected to have a revenue generation capability of 10,000,000 won/h, plus or minus 20 percent. However, expected advances in signal processing and data compression should increase the revenue generation capability by about 5 percent per year. In addition, capacity of the system could be increased by 10 percent for each 1 km reduction in distance between repeaters. Initially, however, it is expected that demand will be less than the capacity of the cable, and that actual revenue generation will be only 50 percent of capability. The demand is expected to grow at a rate of 15 percent per year, plus or minus about 5 percent. At no time, however, will the cable be fully utilized. Even when demand exceeds capacity, the utilization of the cable will be less than 100 percent. We can assume that the utilization will not exceed 90 percent. The projected life of the cable is 25 years.

The cost of the cable is 200,000 won/m. The cost of a single repeater is 250,000,000 won, and the initial installation cost for the cable is 150,000,000,000 won. If redundant repeaters are used, each repeater station will require a sensor and a switch. The sensor cost is 20,000,000 won, and the switch cost is 50,000,000 won plus 20,000,000 won for each additional repeater (that is, 70 million won for 2 repeaters, 90 million won for 3 repeaters, and so on). The repair ship costs 1,250,000 won/h, charged from the time it begins to sail to the cable break until the time it returns to port having fixed the cable.

The cable has an MTBF of 1,000 years/km. Each repeater has an MTBF of 100 years when it is in use and 500 years when it is not in use. The sensor has an MTBF of 1500 years, and the switch has a probability of 0.98 of successfully switching each time it is called upon to switch, for the case of two repeaters, 0.97 for the case of three repeaters, 0.96 for the case of four repeaters, and so on. Each time the cable is raised to repair a repeater, there is a 0.1 probability that the fiber cable will break and need repair.

What is the optimal system design?

We begin by stating the objective of our design: it is to maximize the utility of the cable to the decision maker in charge of providing the cable service. Let us assume that utility to the decision maker relates to profitability of the cable system. But it is important to recognize that other utility measures also might be appropriate.[1] We will assume a decision maker who is somewhat risk averse and take the following utility function

$$u = \ln(N + N_0)^\alpha$$

where N is the present value of net revenues generated by the project. Note that we could alternatively assume that utility is equivalent to the present value of net revenues, but this, in essence, assumes that the project is small compared to the overall wealth of the decision maker or, alternatively, that the decision maker is risk neutral. For simplicity, we will take $\alpha = 1$, and N_0 will be a number large enough to assure that $N + N_0$ is always positive. In this case, $N_0 = 0$ is adequate. Our task remaining is to model u as a function of the system design parameters.

The system that we are to design has the general configuration shown in Fig. 19.4. Overall, it consists of a sequence of cable sections separated by repeater boxes. In order for the system to work, each section of cable and each repeater box must work. Repeater boxes can be designed with redundant repeaters as shown in the figure. Thus, this figure also gives the reliability diagram for the system. The depth of redundancy is one of the design parameters that we wish to optimize. We also wish to optimize the number of repeater boxes or, alternatively, the distance between each pair of repeater boxes. It should not be difficult to convince yourself that the best design will have an equal distance between repeater boxes. This maximizes the capacity of the cable for a given number of repeater boxes, and hence also maximizes revenue generation. On the other hand, spacing repeater boxes differently has no impact on system reliability or cost.

[1]For example, the decision maker could have limited access to investment capital. In this case, a reasonable objective function might be to minimize the probability of cumulative negative cash flow exceeding available investment capital. More simply stated, minimize the probability of going bankrupt.

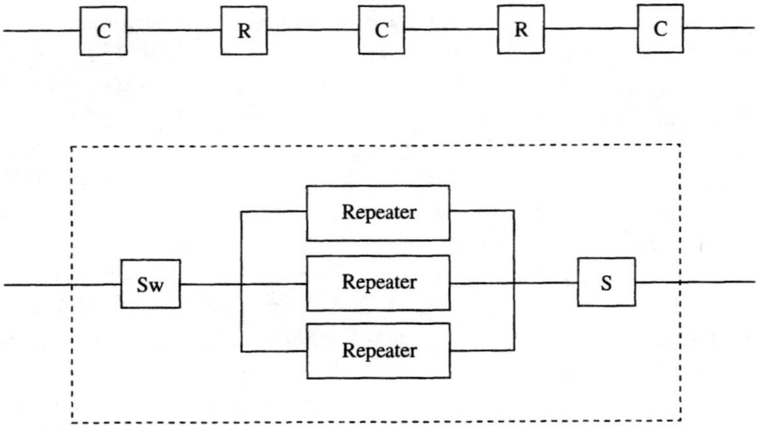

Figure 19.4 General configuration of the cable system showing detail of a repeater box.

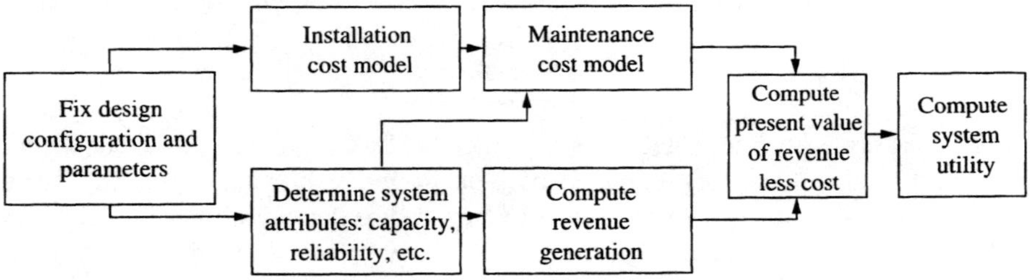

Figure 19.5 Overview of a systems engineering model for a cable design.

Figure 19.5 provides a very simplistic overview of a model for u. We begin with a specified system design. This design specification determines both the attributes of the system, such things as capacity and reliability, and it determines the installation cost of the The system capacity and reliability determine, for this problem, revenue generation. Also, system maintenance cost is a function of the system reliability. Revenues and costs, taken as functions of time, can be combined to obtain the present value of net revenues (revenues less costs). And present value of net revenues can be converted to a system utility by the formula given above.

Uncertainty enters our analysis in two ways. First, there are a number of exogenous variables, variables over which we have no control and which we cannot know precisely. These include demand growth, capacity growth, system utilization, and perhaps several cost elements. It also includes the failure rates of the various components of the system. We see that failures can occur that will cause the cable to become inoperative and to require repair. The possibility of failures infers that both revenue and costs are uncertain, and hence the resulting utility includes risk. It follows that we must actually model expected utility, $E\{u\}$, as a function of system design, and use this as our measure of system worth against which we will make system choices.

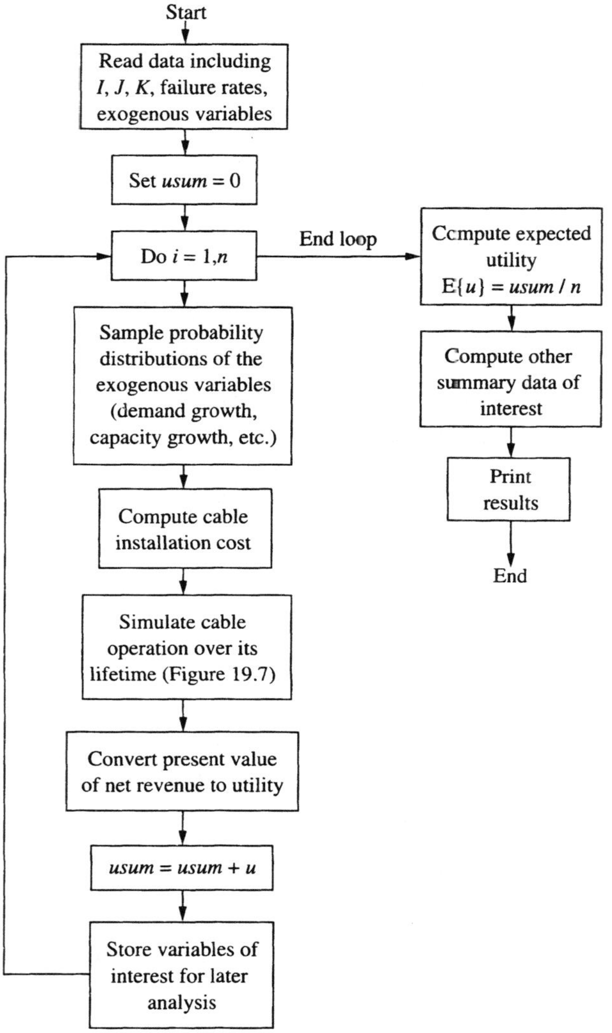

Figure 19.6 Flowchart for a Monte Carlo simulation of cable installation and operation.

First, we will develop a detailed flowchart of a system model. This is done in Figs. 19.6, 19.7, and 19.8. Figure 19.6 shows a Monte Carlo simulation for the installation and operation of the cable over its projected lifetime. The simulation begins by reading, as data, the variables that describe the system design, in this case I and J, which give the number of repeater boxes and the redundancy within each repeater box, and exogenous variables or variables over which the designer has no control. These include component failure rates, component cost data and other cost data, and failure rates. Note that we may specify each of these data as a probability distribution rather than a deterministic value. It is convenient to use triangular distributions, for which we need specify only a minimum,

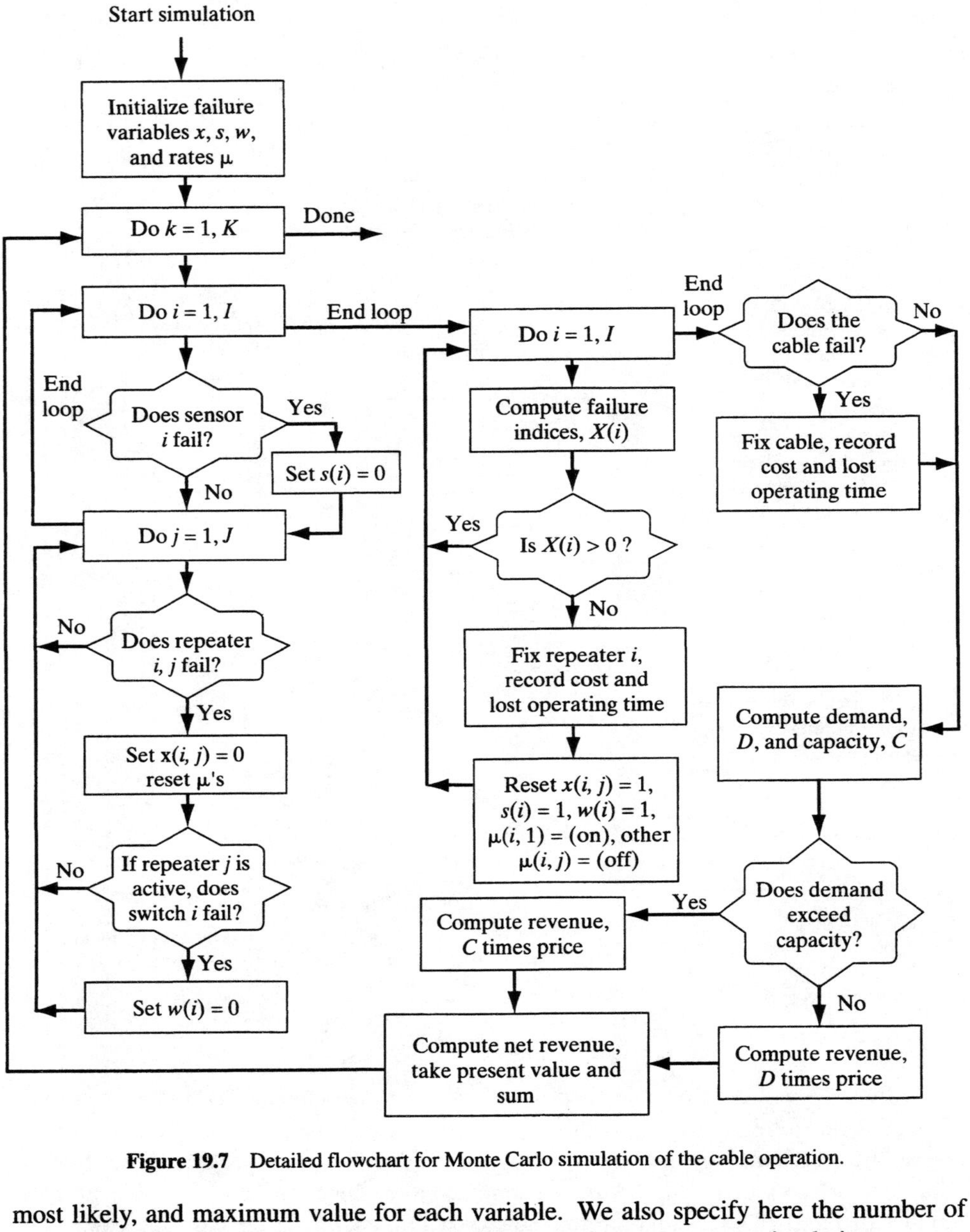

Figure 19.7 Detailed flowchart for Monte Carlo simulation of the cable operation.

most likely, and maximum value for each variable. We also specify here the number of Monte Carlo simulations that we want to perform, n, for each system simulation.

Next, we will set a summation variable, *usum*, which we shall use to determine expected utility. Note that the probability of each outcome of a Monte Carlo simulation is simply $1/n$. Thus, we can compute the expected utility by summing utilities obtained for each simulation, and dividing the sum by n. This is the purpose of the variable *usum*.

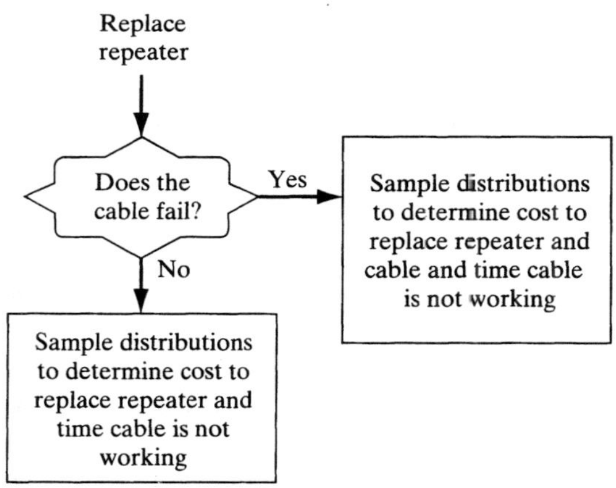

Figure 19.8 Detailed flowchart for Monte Carlo simulation of repeater box replacement.

At this point, we have obtained all necessary data, and we are ready to begin the Monte Carlo simulation. We will repeat the simulation n times. Each individual simulation consists of two basic parts: the simulation of the cable installation and the simulation of the cable operation. First, however, we begin by sampling distributions of demand and capacity growth so that we can later compute the revenue generation. Second, we compute the cable cost by sampling the distributions of the cost components and summing these. Then we perform a simulation of the operation of the cable. The flowchart for this simulation is shown in Fig. 19.7. Upon completion of this simulation, we have a sampled value of net revenue generation. This we convert to a utility using a utility function such as that provided above. Finally, we increment *usum* and store any other variables that we might like later to analyze.

When n simulations have been completed, we compute expected utility and any other output variables of interest. Although expected utility is the objective function according to which we will make system decisions, it is both reassuring and insightful to have access to other system data as well. We are interested basically in two questions: Is the model output believable? And, have we overlooked important options that we should consider?

Now we turn to a more detailed flowchart for the simulation of the operation of the cable, Fig. 19.7. We will define variables $x(i,j)$ to refer to the jth repeater in the ith repeater box and $s(i)$ and $w(i)$ to refer to the sensor and switch in the ith repeater box. In each case, the variable is given the value one if its associated component is functional and zero if it is nonfunctional or broken. Then we note that the ith repeater box is functional if

$$X(i) = s(i)\, w(i) \sum_{j=i}^{J} x(i,j) > 0$$

Thus, in the first part of our simulation, we will determine the variables $x(i,j)$, $s(i)$, and $w(i)$. In doing this, we must be careful to account for the fact that failures can occur while components are idle. Failure rates for components (in this example all failure distributions are memoryless) are given by the variables μ. Accordingly, we will set the values of the μ's depending upon whether the component is working or idle. To implement this procedure, we begin by setting all $x(i,j)$, $s(i)$, and $w(i)$ to one. As components fail, we will reset these variables to zero. And, when we replace a repeater box, we reset the corresponding variables back to one.

We then initiate three do loops. The first loops on time, performing a time-driven simulation for K time intervals. The second and third loops enable all components in the repeater boxes to be checked. First, the sensor is checked, and then the repeaters. If an active repeater fails, the switch must be activated to switch to a spare repeater, if one exists. Any failures that occur are flagged by setting the corresponding variable to zero. After checking all repeater box components, we then determine if repeater boxes or cable need replacement. If they do, we compute the cost of replacement and the lost operating time. From this, we compute the net revenue. There is one slight trick to the computation of the cost of repeater replacement. We observe that sometimes, when the cable is lifted to replace a repeater box, the cable itself breaks and must be replaced. The simulation logic for this is shown in Fig. 19.8.

The Fortran code to implement the above flowcharts is given in Table 19-2, and typical results from a simulation are plotted in Figs. 19.9 through 19.15. Our objective is to examine the consequences, including uncertainty, of different design choices. The major design choices, and the ones we examine here, are the number of repeater boxes to be used over the length of the cable and the level of redundancy within each repeater box. We want to know how these choices affect the utility of the system to the decision maker. But we are also interested in understanding the system as best we can. Through a better understanding of the system, we may be able to identify beneficial design options that we may have overlooked.

Figures 19.9, 19.10, and 19.11 give data on the present value of net revenues generated by the cable over its projected 25-year lifetime. These numbers were generated using only 200 simulations for each case. A case is defined here by a number of repeater boxes and the number of repeaters per repeater box, that is, the level of redundancy. For this example, we have examined cases including from 19 to 40 repeater boxes and from one to three repeaters per box. In the case of one repeater per box, there is no redundancy. The plots given in these figures convey not only information on the expected value of the present value of net revenues, but on the uncertainty in this quantity as well. The shaded region depicts a one-standard-deviation band of uncertainty around the expected value, and the upper and lower bound lines depict the extreme values encountered in the simulation. The jaggedness evident in these curves is a result of the Monte Carlo simulation procedure. We can expect that the upper and lower bound curves will remain jagged regardless of the number of simulations performed. But the expected value curve and the one-standard-deviation band will smooth as we perform more and more simulations per case. However, the smoothness of this curve is quite acceptable even with only 200 simulations per case.

Comparing these curves, we can see a few things. First, there is a clear advantage to increasing the number of repeater boxes. This advantage continues to build even up to

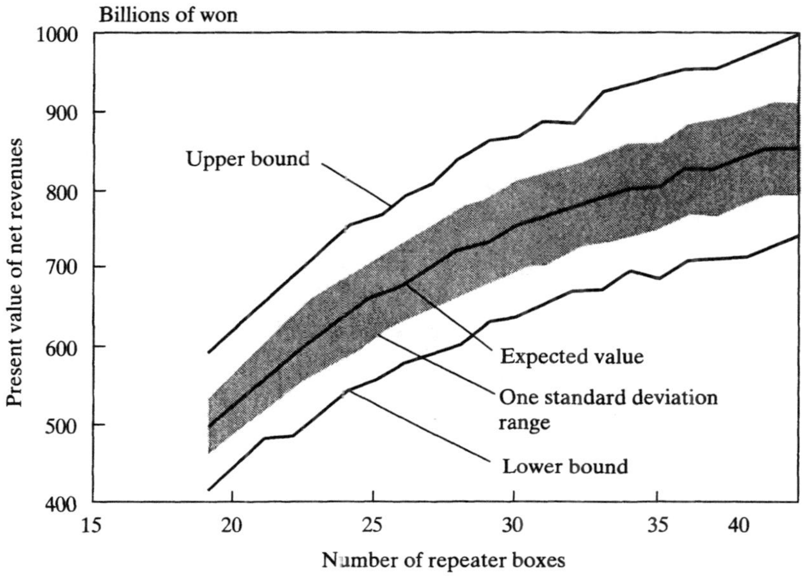

Figure 19.9 Present value of net revenues for system designs.

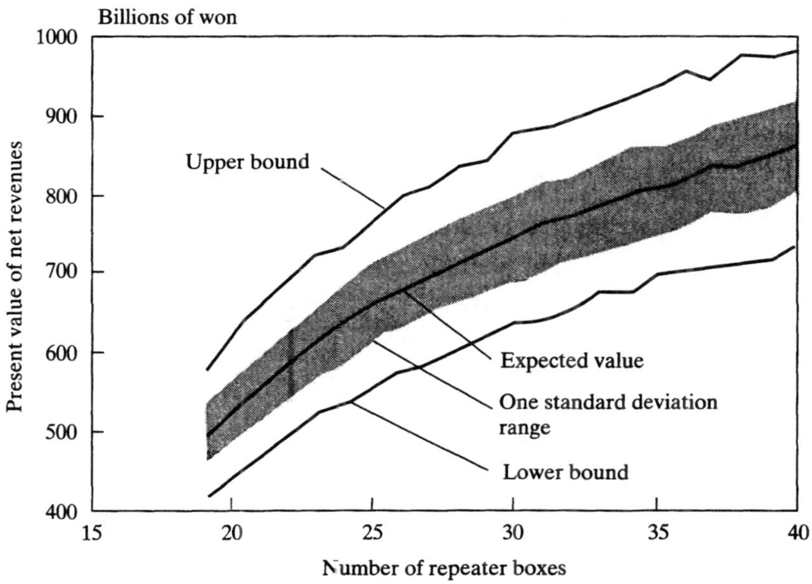

Figure 19.10 Present value of net revenues for the case of single redundancy, that is, two repeaters per repeater box.

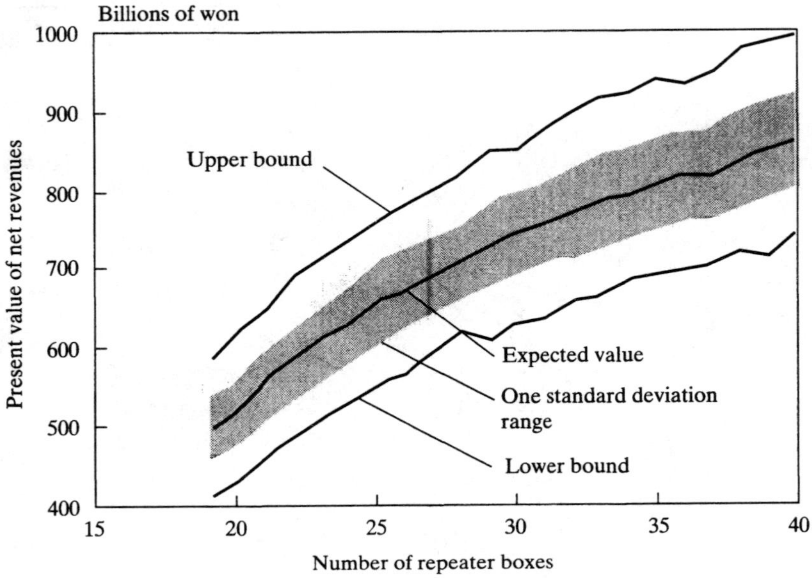

Figure 19.11 Present value of net revenues in the case of double redundancy, that is, three repeaters per repeater box.

40 repeater boxes, and we could be inclined to believe that even more than 40 repeater boxes would lead to further increases in present value of net revenues. This is a curious result, as we might expect that the addition of repeater boxes would reduce the reliability of the overall system. We will return to this point after examining more output data.

Second, we can see little difference between the three levels of redundancy examined in this analysis. Indeed, the three plots look quite similar. We will need to examine other variables to determine the differences between the different levels of redundancy. We do this in the later figures.

Third, we observe some, but not much uncertainty. Quite frequently, uncertainty will be very significant, and there will be a substantial probability that the present value of net revenues will be negative. The small amount of uncertainty present in these figures is partly a result of our input data, which do not include very much uncertainty. In a real example, we could expect much more uncertainty in the output data. A comment here on our use of the word "uncertainty" is appropriate. Normally, we use the word uncertainty to refer to the variability of the input data. We say that we are *uncertain* as regards many input variables to a model. We use the word "risk" to denote variability of the output variables. In particular, we refer to the variability of utility as *risk*. Here, however, we have misused the word uncertainty intentionally because we feel that the use of the word "risk" would lead to confusion.

Given an understanding of the risk present in our modeling of the system, we now turn in Fig. 19.12 to a more direct comparison of the expected present value of net revenues. Again, this figure confirms our notion that there is little difference in the value of

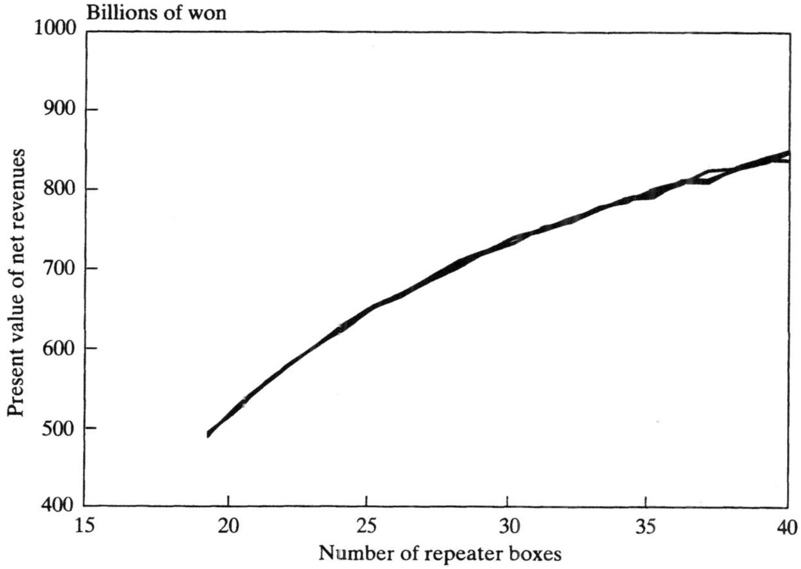

Figure 19.12 Comparison of the expected present values of net revenues for three levels of system redundancy.

the system as a function of the level of redundancy in each repeater box, but that there is a clear advantage to having a large number of repeater boxes. We can deduce, from our model, that the reason that a large number of repeater boxes is advantageous is that this system design enables the cable to produce greater revenues in the period after the demand grows to exceed the capacity of the cable. Increasing the number of repeater boxes increases the capacity of the cable. This effect seems to overshadow the impact of reduced reliability obtained by increasing the number of repeater boxes.

To examine system reliability, we turn to Figs. 19.13 and 19.14. These figures show the number of outages resulting from cable failures and repeater box failures, and the number of days that the cable is unavailable because of these failures. It is clear that there is some advantage in terms of reliability to including a spare repeater in each repeater box. There is no apparent advantage, however, to having two spare repeaters in each box. Notice also that repeater redundancy does not affect the number of outages resulting from cable failures.

Finally, Fig. 19.15 sums up our findings by plotting the expected utility for each system option examined. In this case, the conclusions that we draw from an examination of expected utility are pretty much the same as those we drew from the earlier figures. In the case that there is more risk, however, our conclusions might be different. We see a clear advantage to a design with more repeater boxes, and little to be gained or lost from added redundancy. Probably the best conclusion that we could obtain from these data are that we have failed to examine all reasonable options. In short, we should broaden our option space to include examination of options that allow for future installation of additional cables as demand grows to exceed capacity. We might expect that this would lead to a

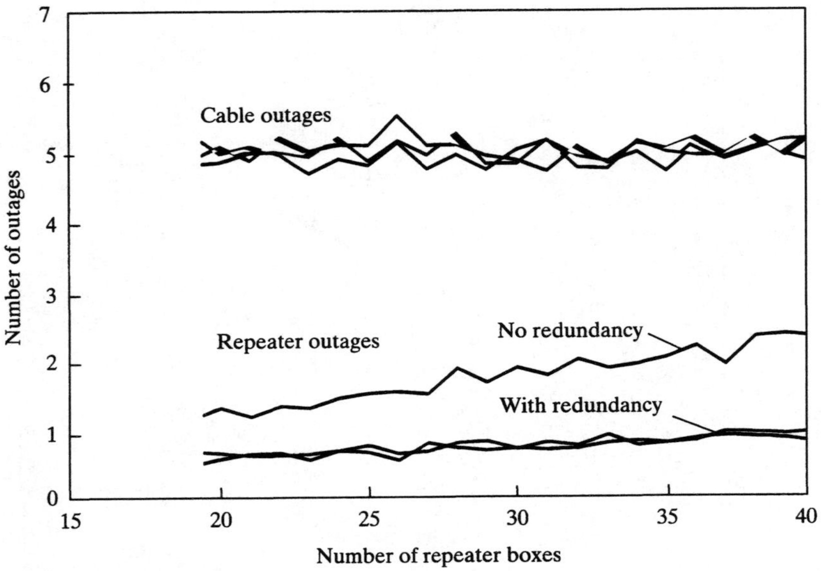

Figure 19.13 Expected numbers of outages for different system designs.

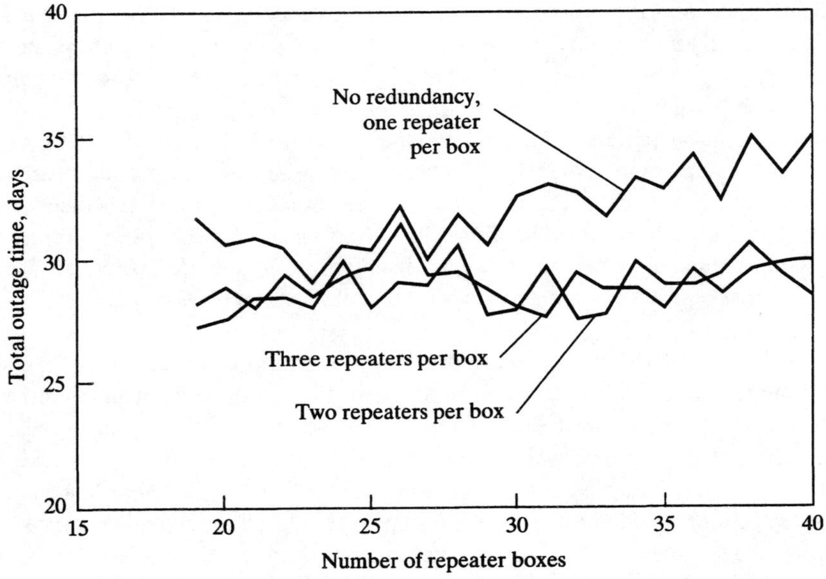

Figure 19.14 Total expected duration of outages resulting from cable and repeater box failures.

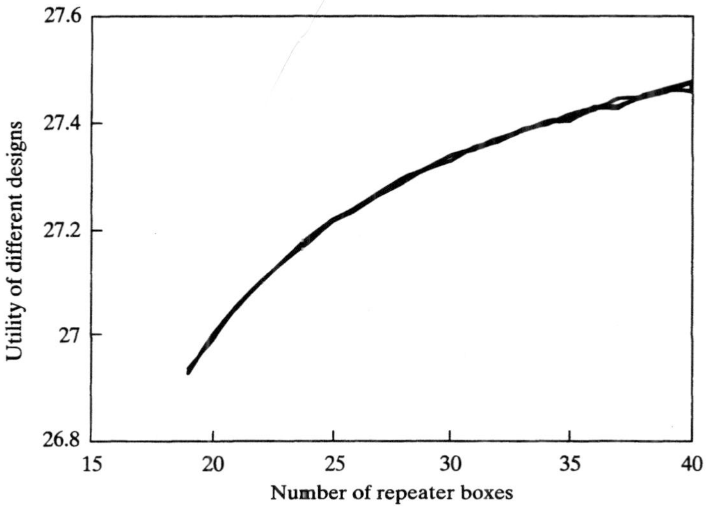

Figure 19.15 Expected utilities for different system designs.

design that uses fewer repeater boxes. It is hard to say how this option would affect our conclusions regarding redundancy. At this point, then, we would conclude that we need to perform more analyses before fixing on a design.

And so we see that the main impact of the above analysis was to discover that we had not considered a sufficient breadth of options. But it is clear that different designs can have a very large impact on the present value of net revenues. Suppose our gut feeling at the beginning of this design was that we needed to go with the minimum number of repeater boxes and doubly or triply redundant repeaters in each box. We could have designed this system and calculated its reliability to discover that the outage rate would be acceptable. We might then reasonably estimate its net present value at about 500 billion won. By means of a systems analysis, however, we see that a well designed system might be quite different, and it might have a value that is closer to $1,000 billion won—about twice as much! If investment in this system depends on a high net present value, the difference between a good system design and an acceptable system design could be the difference between choosing to build the cable and choosing not to build it.

TABLE 19-2 FORTAN CODE FOR SIMULATION OF AN UNDERSEA CABLE

```
C
C  MONTE CARLO MODEL FOR AN UNDERSEA CABLE DESIGN
C         DECEMBER 21, 1993
C
C  INPUT VARIABLES :
C
C  DESIGN VARIABLES
C
```

```
C     I — NUMBER OF REPEATER BOXES
C     J — NUMBER OF REDUNDANT REPEATERS IN EACH BOX
C     CLENGTH — LENGTH OF CABLE, METERS
C
C   SIMULATION CONTROL VARIABLES
C
C     K — NUMBER OF TIME STEPS
C     N — NUMBER OF MONTE CARLO SIMULATIONS
C     T — TIME DURATION OF CABLE OPERATION, YRS
C     K — NUMBER OF TIME STEPS IN THE SIMULATION
C       DELTAT = T/K, YRS
C
C   EXOGENOUS VARIABLES
C
C     GAMMA — CAPACITY GROWTH, 1/YRS
C     BETA — DEMAND GROWTH, 1/YRS
C     DEM0 — INITIAL DEMAND, FRACTION OF CAPACITY
C
C     MURON — MTBF OF A REPEATER WHEN ON, YRS
C     MUROF — MTBF OF A REPEATER WHILE OFF, YRS
C     MUSEN — MTBF OF A SENSOR, YRS
C     RELSW — RELIABILITY OF A SWITCH
C     MUCAB — MTBF OF CABLE PER KILOMETER, YRS
C     CFAIL — PROBABILITY OF A CABLE FAILURE WHILE
C         FIXING A REPEATER BOX
C
C     TREPR — TIME TO REPLACE A REPEATER BOX, DAYS
C     TCABR — TIME TO REPLACE A SECTION OF CABLE, DAYS
C     TFRAC — FRACTION OF REPAIR TIME THAT CABLE IS
C         INOPERATIVE
C     TCBBR — ADDED TIME TO FIX CABLE IF CABLE BREAKS
C         WHEN REPEATER IS REPLACED, DAYS
C
C     DRATE — DISCOUNT RATE, 1/YR
C
C   COST DATA
C
C     CSTIN — INITIAL SYSTEM INSTALLATION COST
C     CSTRB — REPEATER BOX COST
C     CSTRP — ADDITIONAL REPEATER COST
C     CSTSN — SENSOR COST
C     CSTSW — SWITCH COST
C     CSTCB — CABLE COST PER METER
C     CSTSP — REPAIR SHIP COST, PER HOUR
C
C   REVENUE GENERATION
C
C     REVGN — MAXIMUM REVENUE GENERATION PER HOUR
C     CAPFT — CAPACITY FACTOR (FRACTION OF CAPACITY
C         ACTUALLY USED)
C
C   UNCERTAINTY
C     EACH OF THE ABOVE EXOGENOUS VARIABLES IS SPECIFIED
```

```
C     AS A PROBABILITY DISTRIBUTION TAKEN HERE TO BE A
C     TRIANGULAR DISTRIBUTION SPECIFIED
C     IN TERMS OF A MINIMUM VALUE, A MOST LIKELY VALUE,
C     AND A MAXIMUM VALUE.  THESE INPUTS ARE DESIGNATED
C     BY THE SUFFIXES MN, ML, AND MX RESPECTIVELY.  FOR
C     EXAMPLE,
C       GAMMAMN = MINIMUM POSSIBLE CAPACITY GROWTH
C       GAMMAML = MOST LIKELY CAPACITY GROWTH
C       GAMMAMX = MAXIMUM POSSIBLE CAPACITY GROWTH
C
      REAL*8 SEED
      REAL GAMMA,BETA,MURON,MUROF,MUSEN,RELSW,MUCAB,CFAIL,
     * CSTN,CSTRB,CSTRP,CSTSN,CSTSW,CSTCB,CSTSP,REVGN,CAPFT,
     * DRATE,CLENGTH,DEM0
      REAL GAMMAMN,GAMMAML,GAMMAMX,BETAMN,
     * BETAML,BETAMX,MURONMN,
     * MURONML,MURONMX,MUROFMN,MUROFML,MUROFMX,MUSENMN,
     * MUSENML,MUSENMX,RELSWMN,RELSWML,RELSWMX,MUCABMN,
     * MUCABML,MUCABMX,CFAILMN,CFAILML,CFAILMX,CSTINMN,
     * CSTINML,CSTINMX,CSTRBMN,CSTRBML,CSTRBMX,CSTRPMN,
     * CSTRPML,CSTRPMX,CSTSNMN,CSTSNML,CSTSNMX,CSTSWMN,
     * CSTSWML,CSTSWMX,CSTCBMN,CSTCBML,CSTCBMX,CSTSPMN,
     * CSTSPML,CSTSPMX,REVGNMN,REVGNML,REVGNMX,CAPFTMN,
     * CAPFTML,CAPFTMX
      REAL DEM0MN,DEM0ML,DEM0MX,TREPRMN,TREPRML,TREPRMX,
     * TCABRMN,TCABRML,TCABRMX,TFRACMN,TFRACML,TFRACMX,
     * TCBBRMN,TCBBRML,TCBBRMX
      COMMON/SEEED/SEED
      COMMON/VARS/GAMMA,BETA,MURON,MUROF,
     * MUSEN,RELSW,MUCAB,CFAIL,
     * CSTN,CSTRB,CSTRP,CSTSN,CSTSW,CSTCB,CSTSP,REVGN,CAPFT,
     * DRATE,CLENGTH,DEM0,DELTAT
      COMMON/REPAIR/TREPRMN,TREPRML,TREPRMX,TCABRMN,TCABRML,
     * TCABRMX,TFRACMN,TFRACML,TFRACMX,TCBBRMN,TCBBRML,
     * TCBBRMX
      COMMON/OUTS/TOUT,KREP,KCAB
C
C   SET SEED
C
      SEED = 12345.
    1 CONTINUE
C
C   READ DATA
C
      WRITE (*,201)
  201 FORMAT (' INPUT I, J')
      READ (*,*) I,J
C
C   ****************** DATA *****************
C
      CLENGTH = 200000.
      K = 300
      T = 25.
```

```
N = 200
GAMMAMN = 0.04
GAMMAML = 0.05
GAMMAMX = 0.06
BETAMN = 0.10
BETAML = 0.15
BETAMX = 0.20
DEM0MN = 0.5
DEM0ML = 0.5
DEM0MX = 0.5
MURONMN = 100.
MURONML = 100.
MURONMX = 100.
MUROFML = 500.
MUROFML = 500.
MUROFMX = 500.
MUSENMN = 1500.
MUSENML = 1500.
MUSENMX = 1500.
XI = I
RELSWMN = 0.99-0.01*XI
RELSWML = 0.99-0.01*XI
RELSWMX = 0.99-0.01*XI
 MUCABMN = 1000.
MUCABML = 1000.
MUCABMX = 1000.
CFAILMN = 0.1
CFAILML = 0.1
CFAILMX = 0.1
TREPRMN = 2.0
TREPRML = 4.0
 TREPRMX = 5.0
TCABRMN = 5.0
TCABRML = 7.5
TCABRMX = 10.0
TFRACMN = 0.7
TFRACML = 0.7
TFRACMX = 0.8
TCBBRMN = 5.0
TCBBRML = 5.0
TCBBRMX = 5.0
DRATE = 0.08
CSTINMN = 140.E09
CSTINML = 150.E09
CSTINMX = 200.E09
CSTRBMN = 240.E06
CSTRBML = 250.E06
CSTRBMX = 300.E06
CSTRPMN = 18.E06
CSTRPML = 20.E06
CSTRPMX = 25.E06
CSTSNMN = 18.E06
CSTSNML = 20.E06
```

```
      CSTSNMX = 25.E06
      CSTSWMN = 40.E06
      CSTSWML = 50.E06
      CSTSWMX = 70.E06
      CSTCBMN = 175000.
      CSTCBML = 200000.
      CSTCBMX = 250000.
      CSTSPMN = 1250000.
      CSTSPML = 1250000.
      CSTSPMX = 1250000.
      REVGNMN = 8.E06*(1.+.1*(10.-CLENGTH/1000./(XI+1.)))
      REVGNML = 10.E06*(1.+.1*(10.-CLENGTH/1000./(XI+1.)))
      REVGNMX = 12.E06*(1.+.1*(10.-CLENGTH/1000./(XI+1.)))
      CAPFTMN = 0.9
      CAPFTML = 0.9
      CAPFTMX = 0.9
C
C     ************** END DATA **************
C
C     INITIALIZE MONTE CARLO SIMULATION
      USUM = 0.
      VSUM = 0.
      VSUM2 = 0.
      XK = K
      XN = N
      DELTAT = T/XK
      XTOUT = 0.
      XKREP = 0.
      XKCAB = 0.
C
C     BEGIN MONTE CARLO SIMULATION
C
      DO 100 NSIM=1,N
C
C     SAMPLE EXOGENOUS VARIABLES
C
      GAMMA = (SAMP(GAMMAMN,GAMMAML,GAMMAMX)+1.)**DELTAT
      BETA  = (SAMP(BETAMN,BETAML,BETAMX)+1.)**DELTAT
      MURON = SAMP(MURONMN,MURONML,MURONMX)/DELTAT
      MUROF = SAMP(MUROFMN,MUROFML,MUROFMX)/DELTAT
      MUSEN = SAMP(MUSENMN,MUSENML,MUSENMX)/DELTAT
      RELSW = SAMP(RELSWMN,RELSWML,RELSWMX)
      MUCAB = SAMP(MUCABMN,MUCABML,MUCABMX)/DELTAT
      CFAIL = SAMP(CFAILMN,CFAILML,CFAILMX)
      CSTIN = SAMP(CSTINMN,CSTINML,CSTINMX)
      CSTRB = SAMP(CSTRBMN,CSTRBML,CSTRBMX)
      CSTRP = SAMP(CSTRPMN,CSTRPML,CSTRPMX)
      CSTSN = SAMP(CSTSNMN,CSTSNML,CSTSNMX)
      CSTSW = SAMP(CSTSWMN,CSTSWML,CSTSWMX)
      CSTCB = SAMP(CSTCBMN,CSTCBML,CSTCBMX)
      CSTSP = SAMP(CSTSPMN,CSTSPML,CSTSPMX)
      REVGN = SAMP(REVGNMN,REVGNML,REVGNMX)
      CAPFT = SAMP(CAPFTMN,CAPFTML,CAPFTMX)
```

```
      DEM0 = SAMP(DEM0MN,DEM0ML,DEM0MX)
C
C     COMPUTE INSTALLATION COST
C
      CINST = CSTIN+CLENGTH*CSTCB+I*CSTRB
      IF (J.GT.1) CINST = CINST+(J-1)*CSTRP+CSTSN+CSTSW
C
C     SIMULATE CABLE OPERATION
C
      CALL SIMCAB(I,J,K,VALUE)
C
C     CONVERT NPV OF VALUE TO UTILTY
C
      VALUE = VALUE-CINST
      UTILITY = LOG(VALUE)
      USUM = USUM+UTILITY
      VSUM = VSUM+VALUE
      VSUM2 = VSUM2+VALUE**2
      XTOUT = XTOUT+TOUT
      XK = KREP
      XKREP = XKREP+XK
      XK = KCAB
      XKCAB = XKCAB+XK
       IF (NSIM.EQ.1) THEN
      VMIN = VALUE
      VMAX = VALUE
      END IF
      IF (VALUE.GT.VMAX) VMAX = VALUE
      IF (VALUE.LT.VMIN) VMIN = VALUE
C
C     END MONTE CARLO SIMULATION
C
  100 CONTINUE
C
C     COMPUTE EXPECTED UTILITY
C
      EXPUTIL = USUM/XN
      EXPVAL = VSUM/XN
      SIGVAL = SQRT(VSUM2/XN-EXPVAL**2)
      XTOUT = XTOUT/XN
      XKREP = XKREP/XN
      XKCAB = XKCAB/XN
C
C     WRITE RESULTS
C
      WRITE (*,*) I,J,EXPUTIL
      WRITE (*,*) EXPVAL,SIGVAL,VMIN,VMAX
      WRITE (*,*) XTOUT,XKREP,XKCAB
C
      WRITE (*,501)
  501 FORMAT ('   ANOTHER CASE?')
      READ (*,*) NYORN
      IF (NYORN.EQ.1) GO TO 1
```

```
      STOP
      END
C
C
C   RANDOM NUMBER GENERATOR
C
      FUNCTION RLEWIS(X)
      REAL*8 SEED
      COMMON/SEEED/SEED
      SEED = DMOD(SEED*16807.,2147483647.)
      RLEWIS = SEED/2147483647.
      RETURN
      X = 1.
      END
C
C
C   SAMPLE A DISTRIBUTION
C
      FUNCTION SAMP(XMIN,XML,XMAX)
      IF (XMAX.LE.XMIN) GO TO 2
      R = RLEWIS(1.)
      IF (R.LE.((XML-XMIN)/(XMAX-XMIN))) GO TO 1
      SAMP = XMIN+SQRT(R*(XML-XMIN)*(XMAX-XMIN))
      RETURN
    1 SAMP = XMAX-SQRT((1.-R)*(XMAX-XML)*(XMAX-XMIN))
      RETURN
    2 SAMP = XML
      RETURN
      END
C
C
C   SIMULATION OF CABLE OPERATION
C
      SUBROUTINE SIMCAB(II,JJ,KK,VALUE)
C
C   DEFINE VARIABLES
C
      REAL*8 SEED
      REAL GAMMA,BETA,MURON,MUROF,MUSEN,RELSW,MUCAB,CFAIL,
     * CSTN,CSTRB,CSTRP,CSTSN,CSTSW,CSTCB,CSTSP,REVGN,CAPFT,
     * DRATE,CLENGTH
      REAL XMU(200,10)
      INTEGER*4 X(200,10),S(200),W(200)
      COMMON/SEEED/SEED
      COMMON/VARS/GAMMA,BETA,MURON,MUROF,
     * MUSEN,RELSW,MUCAB,CFAIL,
     * CSTN,CSTRB,CSTRP,CSTSN,CSTSW,CSTCB,CSTSP,REVGN,CAPFT,
     * DRATE,CLENGTH,DEM0,DELTAT
      COMMON/REPAIR/TREPRMN,TREPRML,TREPRMX,TCABRMN,TCABRML,
     * TCABRMX,TFRACMN,TFRACML,TFRACMX,TCBBRMN,TCBBRML,
     * TCBBRMX
      COMMON/OUTS/TOUT,KREP,KCAB
C
```

```
C   INITIALIZE FAILURE VARIABLES
C
      DO 10 I=1,II
      S(I) = 1
      W(I) = 1
      DO 11 J=1,JJ
      X(I,J) = 1
      IF (I.EQ.1) XMU(I,J) = MURON
      IF (I.GT.1) XMU(I,J) = MUROF
   11 CONTINUE
   10 CONTINUE
      KREP = 0
      KCAB = 0
      TOUT = 0.
      VALUE = 0.
C
C   LOOP ON TIME
C
      DO 100 K=1,KK
C
C   LOOP ON REPEATER BOXES
C
      DO 200 I=1,II
C
C   TEST FOR FAILURE OF THE SENSOR IN THE ITH BOX
C
      IF (RLEWIS(1.).LE.(1./MUSEN).AND.JJ.GT.1) S(I) = 0
C
C   LOOP ON REDUNDANCY IN REPEATER BOXES
C
      DO 201 J = 1,JJ
C
C   TEST FOR FAILURE OF REPEATERS
C
      IF (RLEWIS(1.).GT.(1./XMU(I,J)).OR.X(I,J).EQ.0)
     *  GO TO 201
      X(I,J) = 0
      IF (XMU(I,J).NE.MURON.OR.J.GE.JJ) GO TO 201
      IF (X(I,J+1).EQ.0) GO TO 203
      XMU(I,J+1) = MURON
C
C   CHECK SWITCH
C
      IF (RLEWIS(1.).LE.RELSW) W(I) = 0
      GO TO 201
C
C   REPEATER (I,J+1) IS BROKEN
C
  203 J2 = J+2
  202 IF (J2.GT.JJ) GO TO 201
      IF (X(I,J).NE.0) THEN
      XMU(I,J2) = MURON
      GO TO 201
```

```
      END IF
      J2 = J2+1
      GO TO 202
  201 CONTINUE
  200 CONTINUE
C
C     SIMULATE REPLACEMENT OF REPEATER BOXES
C     TLOST — TIME CABLE IS INOPERATIVE WHILE REPLACING
C        CABLE OR REPEATER BOXES
C
      TLOST = 0.
      CSTREP = 0.
      DO 300 I = 1,II
C
C     COMPUTE FAILURE INDICES
C
      ISUM = 0
      DO 301 J =1,JJ
  301 ISUM = ISUM+X(I,J)
      XI = S(I)*W(I)*ISUM
C
      IF (XI.GT.0) GO TO 300
C
C     REPLACE REPEATER BOX
C
      TREPR = SAMP(TREPRMN,TREPRML,TREPRMX)
      TFRAC = SAMP(TFRACMN,TFRACML,TFRACMX)
C
C     WE ASSUME THAT NO EXTRA TIME IS REQUIRED IF THE
C     CABLE BREAKS DURING REPEATER BOX REPLACEMENT
C
      TLOST = TLOST+TREPR*TFRAC
      CSTREP = CSTREP+CSTRB+TREPR*24.*CSTSP
      XJ1 = J-1
      IF (J.GT.1) CSTREP = CSTREP+XJ1*CSTRP+CSTSN+CSTSW
      IF (RLEWIS(1.).LE.CFAIL) THEN
      XI = II+1
      TCBBR = SAMP(TCBBRMN,TCBBRML,TCBBRMX)
      CSTREP = CSTREP+CSTCB*CLENGTH/XI+TCBBR*24.*CSTSP
      TLOST = TLOST+TCBBR
      END IF
C
C     RESET FAILURE VARIABLES
C
      S(I) = 1
      W(I) = 1
      DO 302 J = 1,JJ
      XMU(I,J) = MUROF
  302 X(I,J) = 1
      XMU(I,1) = MURON
      KREP = KREP+1
C
C     END OF REPEATER FAILURE CHECK AND REPLACEMENT
```

```
C
 300 CONTINUE
C
C   CHECK FOR CABLE FAILURES
C   ASSUME THAT ONLY ONE SECTION OF CABLE CAN FAIL
C   DURING A TIME INTERVAL
C
    IF (RLEWIS(1.).GT.0.001*CLENGTH/MUCAB) GO TO 350
    TFRAC = SAMP(TFRACMN,TFRACML,TFRACMX)
    TCABR = SAMP(TCABRMN,TCABRML,TCABRMX)
    XI = II+1
    CSTREP = CSTREP+CSTCB*CLENGTH/XI+TCABR*24.*CSTSP
    TLOST = TLOST+TCABR*TFRAC
    KCAB = KCAB+1
 350 CONTINUE
    IF (TLOST.GT.DELTAT*365.) TLOST = DELTAT*365.
C
C   COMPUTE DEMAND AND CAPACITY
C
    DEM = DEM0*GAMMA**K
    CAPAC = BETA**K
    IF (DEM.GT.CAPAC*CAPFT) REVNET = (DELTAT-TLOST/365.)*
    *     REVGN*8760.*CAPFT-CSTREP
    IF (DEM.LE.CAPAC*CAPFT) REVNET = (DELTAT-TLOST/365.)*
    *     DEM*REVGN*8760.*CAPFT-CSTREP
    XK = K
    VALUE = VALUE+REVNET*(1.+DRATE)**(-XK*DELTAT)
    TOUT = TOUT+TLOST
 100 CONTINUE
    RETURN
    END
```

19.4 SUMMARY OF THE SYSTEMS ENGINEERING APPROACH

The preceding examples point to many features of the systems engineering approach. First, they outline an approach to system design and operation decision making that is all inclusive. That is, the system engineering approach includes consideration of all life-cycle phases of a system. And it is an approach that can account for uncertainty and risk both in the modeling process itself and in the information that we have regarding the design and its implementation. The systems engineering approach takes into consideration manufacturability, installation, operation, maintenance and repair, and disposal. Through demand and expected utility analysis, it also takes into account customers' preferences regarding the system. All these things are done in a rigorous mathematical framework. Thus, the systems engineering approach encompasses approaches such as just-in-time, concurrent engineering, and total quality management. And, whereas these other approaches to engineering decision making are neither complete nor mathematically rigorous, the systems engineering approach is both.

The specific benefits of the systems engineering approach, evident in the previous examples, include the following:

The *systems engineering approach* provides a framework for organization of our thoughts regarding a project.

The *systems engineering approach* forces us to define the goals and/or objectives of the project.

The *systems engineering approach* focuses our attention of the goals and objectives of the project as the object of our design decisions.

The *systems engineering approach* provides a framework for making use of the expertise of experts from a disparate set of disciplines. It does this by enabling models to provide the interface between the disciplines.

The *systems engineering approach* provides a list of the input data needed to conduct a system design. Frequently, engineers choose to use other design methods because it can be difficult to obtain the data demanded by the systems engineering approach. In this case, however, the engineer is simply conducting the design without sufficient information to properly evaluate design choices.

The *systems engineering approach* provides a framework for the use of data on the many disparate variables that impact design choices. These include variables such as the corporate discount rate and corporate risk preferences. It also includes data regarding customer or consumer preferences.

The *systems engineering approach* provides an index of goodness for comparing system alternatives.

The *systems engineering approach* details areas of uncertainty and their impact on risk. It points to areas where better information is needed to effect better system design.

The *systems engineering approach* provides a consistent and comprehensive decision-making approach to engineering design.

The *systems engineering approach* solicits information on design options and enables its use in engineering design decision making. Thus, it is an approach to information-based engineering design.

20

Concluding Remarks

In this book I put forth a theory of engineering design. This theory is based on the notion that engineering design is a decision-making process, coupled with an analytic approach that relies heavily on mathematics and computer modeling. It is my thesis that all aspects of an engineering system are subject to modeling and analysis, not merely the physical elements of the system. But, while I believe that all aspects of a system can be modeled and simulated, I also believe that such models and simulations do not, cannot, and never will give precise results. Nonetheless, by modeling an entire system, including its design, manufacture, use, maintenance and repair, disposal, and other related factors, I believe that better design decisions can be made. Specifically, we seek to express the total value or worth of a system in the form of a system utility, which is ultimately a function of the system design choices and options. Then, optimal design is simply a matter of choosing those options or parameters that maximize utility. I use the word "simply" rather loosely here, as this process, although conceptually simple, is often far from simple in practice. Indeed, realistic treatment and analysis of engineering design options is a very challenging task. Still, despite complexities in application, this approach goes a long way toward the reduction of engineering design from an art to a science, and it provides a rigorous framework for the teaching or learning of engineering design.

Our view of the systems engineering process can be summed up, albeit rather simplistically, by Fig. 20.1. We recognize *explicitly* that engineering design is a decision-making process that seeks to result in a system that maximizes utility, to the decision maker, through optimal choice of design parameters, $\boldsymbol{x}$. The parameters, $\boldsymbol{x}$, fully describe the system and its use or operation. Through analysis of the physical system, we can estimate the system attributes, $\boldsymbol{a}$, and its costs. We recognize that our determination of $\boldsymbol{a}$ is generally corrupted by exogenous variables $\boldsymbol{y}$, such as the weather, over which we have no control. Particularly, the estimation of system attributes is the core of current engi-

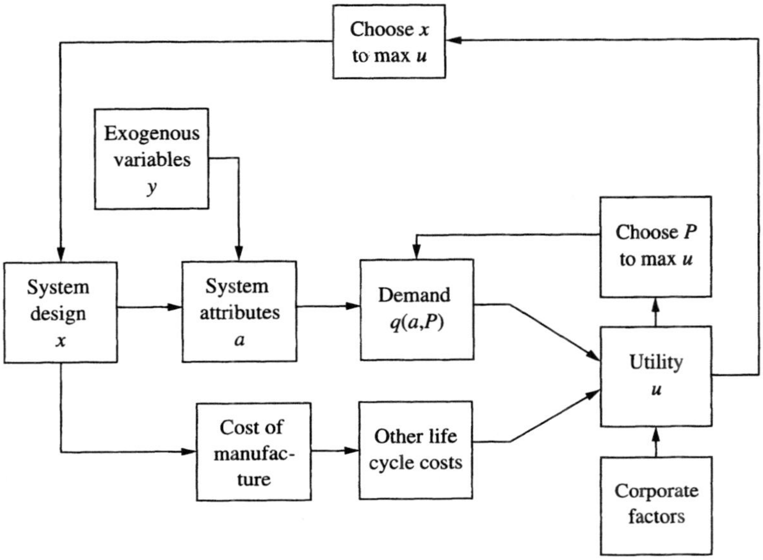

Figure 20.1 A simplistic overview of the systems engineering process.

neering education and the focus of nearly all engineering modeling. Yet, we see that it is only a small part of a systems engineering model, although an important part. System attributes, combined with price and exogenous factors that describe the marketplace, determine the demand for the system. But price is under the control of the system provider, and can be chosen to maximize utility to the decision maker. Demand and price combined determine the revenue stream that the system generates, and utility is determined by combining revenues and costs, subject to the decision maker's discount rate and risk preferences. Finally, utility is taken as the measure of system goodness, based upon which the systems engineer makes design choices.

It is my belief that rigorous procedures such as those described above take much of the guesswork out of the design process, enabling the designer to approach more closely an optimal design. Sometimes, this approach will enable improvements on the margin, providing a small but needed competitive edge. Other times the designer might enjoy very substantial benefits from this approach. Benefits can derive from a newfound ability to fine tune the design, manufacture, use, and disposal of a system, or they might derive from entirely new insights gained from the system modeling process itself. The point, however, is that *all* available information relating to a system and its design *can* be gathered together into a system model that explicitly depicts overall system utility as a function of system design and decision variables. With this model, the designer can measure and monitor the impact of design choices and system alternatives directly. Indeed, we could envision ultimately having available in a computer-aided design (CAD) system, directly on screen, a measure of overall system utility, so that the design engineer could monitor this descriptor of the worth of a design in real time throughout the design process.

We can liken systems engineering to the process of information gathering while driving a car. Determination of the attributes of a car might give us the information that the car has four doors, front-wheel drive, power steering, and automatic transmission. But these are not the data that we need to guide the car safely down a street. We need instead to know where the car is, which way it is headed relative to the street, how fast the car is going, and whether there are obstructions ahead. The modeling of physical system attributes provide information equivalent to the description of the car. For design, however, we need the information equivalent to that for driving the car. This is the information that we gain from systems modeling.

Over the past few decades, the complexity of engineering systems has grown immensely. No longer can a single engineer understand the full complexity of a design, and a wide variety of disciplinary skills are needed to effect the design of most modern systems. We show in Chap. 7 that, any time a group of individuals works together, it is likely that there will be inconsistencies or intransitivities in their decision-making process. One job of the systems engineer is to transcend these intransitivities, or we might even say *irrationalities* of the design process, and evolve a more rational design. Through the use of the modeling approaches developed in this book, the systems engineer can integrate the knowledge held by a diverse group of individuals, and apply it to a specified end. This ability places the systems engineer in a new position: that of integrator, synthesizer, enabler, and manager, with the task of developing systems that better serve the company, the nation, and society.

This book is at least in one sense a strange collection of thoughts and theories. We have drawn the theory from a wide range of disciplines, really only scratching the surface of each. In no case did we stretch the frontiers of knowledge in any of the disciplines from which we have drawn this material. Rather the uniqueness of this book derives from the breadth of this collection and its direct application to the process of engineering design. One immediate benefit of this is that you could quite easily extend the theory presented here merely by delving more deeply into its constituent subjects.

To be sure, this book is not the first attempt to bring together a rigorous theory of engineering design. But I believe that it is the first book to be based solidly on the notion that engineering design is no more and no less than decision making, and that to do it well requires good information. Nor will the reader find this book either complete or flawless. As with any emerging theory, it has its limitations. Many areas of this theory could be improved, and many of the explanations provided in this book could be simplified or clarified, or possibly even corrected. Further development of this theory will require research, discussion, and debate. In this regard, I welcome contributions to this theory and invite personal communications.

The test of any theory lies not in what it describes but in what it predicts. If this theory and the computational approaches presented here lead to better design decisions, then it will prove worthwhile. I personally am convinced that better design information can be obtained by application of this theory, and I believe that its use will only improve the results of the design process. Yet, I cannot know what specific benefits this theory will bring you. This remains for you to discover. I wish you luck and good fortune.

Appendix A

Vectors and Matrices

A.1 DEFINITIONS

When we work with engineering systems models, we frequently encounter numerous variables that can be conveniently ordered and identified by subscripts. For example, suppose you wish to record the dimensions of various rooms. Each room has a length, width, and height. For each room, we can designate these dimensions by the indexed variables x_1, x_2, and x_3. It is often convenient to write these three variables in condensed form as the single variable $\boldsymbol{x}$. The use of bold print reminds us that this variable refers not to a single value, but to an ordered set of values. We refer to an ordered set of values as a ***vector***, and we recognize two types of vectors, a column vector written

$$\boldsymbol{x} = \begin{bmatrix} x_1 \\ x_2 \\ x_3 \end{bmatrix}$$

and a row vector written

$$\boldsymbol{x}^T = [x_1, x_2, x_3]$$

The superscript T on $\boldsymbol{x}$ refers to the transpose operation, which converts a column vector to a row vector or vice versa. We will discuss this operation more later. Each value in a vector, x_1, x_2, ... , x_i, ... and so on, is referred to as an element of the vector. The *dimensionality* of a vector is the number of elements that it contains, and this is not limited to three. For example,

$$\boldsymbol{x}^T = [x_1, x_2, \ldots, x_m]$$

and we would say that x has dimensionality m.

*A **vector** is an ordered set of numbers.*

We also define the vector $\mathbf{0}$ to be the origin of the vector space.

$$\mathbf{0}^T = [0,0,...,0]$$

Suppose, in our example, that we had recorded the dimensions of several rooms. We could designate the dimensions of each room by the variable $\boldsymbol{x}$, and to distinguish between the rooms, we could subscript $\boldsymbol{x}$, $\boldsymbol{x}_j$. As j takes on the numbers 1, 2, 3, ... ,n, it denotes to which room we are referring. Now we can define a new variable X as follows

$$X = [\boldsymbol{x}_1, \boldsymbol{x}_2, ..., \boldsymbol{x}_n]$$

Thus X is an ordered set of vectors, shown here as a "row vector" of column vectors. Each of the n vectors, indexed by j, in X contains m elements, indexed by i. We refer to X as a ***matrix***. The elements of X are

$$X = \begin{bmatrix} x_{1,1} & x_{1,2} & \cdots & x_{1,n} \\ x_{2,1} & x_{2,2} & \cdots & x_{2,n} \\ & \cdot & & \\ & \cdot & & \\ x_{m,1} & x_{m,2} & \cdots & x_{m,n} \end{bmatrix} = \|x_{i,j}\| = \|x_{ij}\|$$

We say that X is an $m \times n$ matrix. The sets of elements of this matrix, each for a specified value of i, are referred to as the rows of the matrix. The sets of elements, each for a specified value of j, are referred to as the columns of the matrix. The above matrix has m rows and n columns. In the case that $m = n$, we say that X is a *square matrix*. We frequently use capital letters to designate matrices.

*A **matrix** is an ordered rectangular array of numbers.*

A special case exists when a vector has dimensionality 1 or in the case of a 1×1 matrix. In these cases, the vector or matrix has only one element. We then refer to this matrix as a ***scalar***.

A.2 VECTOR OPERATIONS

In this section, we will consider arithmetic operations that we can perform on vectors. These include addition, subtraction, and multiplication. Division is not defined for vectors.

Vector Addition

Consider three vectors, $\boldsymbol{x}$, $\boldsymbol{y}$, and $\boldsymbol{z}$. The vector sum is defined as follows:

$$\boldsymbol{x} + \boldsymbol{y} = \begin{bmatrix} x_1 \\ x_2 \\ \cdot \\ \cdot \\ x_m \end{bmatrix} + \begin{bmatrix} y_1 \\ y_2 \\ \cdot \\ \cdot \\ y_m \end{bmatrix} = \begin{bmatrix} x_1 + y_1 \\ x_2 + y_2 \\ \cdot \\ \cdot \\ x_m + y_m \end{bmatrix} = \begin{bmatrix} z_1 \\ z_2 \\ \cdot \\ \cdot \\ z_m \end{bmatrix} = \boldsymbol{z}$$

Thus we see that the vector sum is defined only for vectors of equal dimensionality, and that each element of the sum is comprised of the sums of the corresponding elements of the added vectors.

Vector addition has the property

$$\boldsymbol{x} + \boldsymbol{0} = \boldsymbol{x}$$

It is also commutative

$$\boldsymbol{x} + \boldsymbol{y} = \boldsymbol{y} + \boldsymbol{x}$$

and associative

$$(\boldsymbol{x} + \boldsymbol{y}) + \boldsymbol{z} = \boldsymbol{x} + (\boldsymbol{y} + \boldsymbol{z}) = \boldsymbol{x} + \boldsymbol{y} + \boldsymbol{z}$$

Vector addition has a graphical interpretation, which we can illustrate in two dimensions as shown in Fig. A.1. A vector can be drawn as a line with an arrow on the tip, denoting both magnitude and direction. Vector addition is a process of arranging lines denoting vectors, beginning at the origin, and extending tip-to-tail, vector-to-vector. The vector sum is then the line drawn from the origin to the tip of the last vector in the sequence.

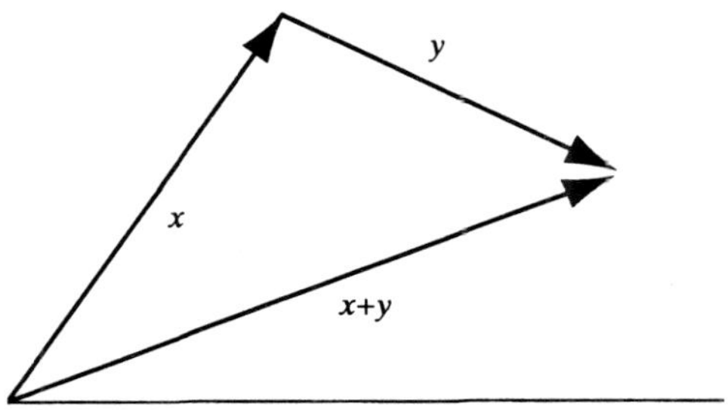

Figure A.1 Illustration of vector addition.

Vector Subtraction

Vector subtraction is defined as vector addition replacing $\boldsymbol{y}$ with $(-\boldsymbol{y})$. That is:

$$\boldsymbol{x} - \boldsymbol{y} = \begin{bmatrix} x_1 \\ x_2 \\ \cdot \\ \cdot \\ x_m \end{bmatrix} - \begin{bmatrix} y_1 \\ y_2 \\ \cdot \\ \cdot \\ y_m \end{bmatrix} = \begin{bmatrix} x_1 - y_1 \\ x_2 - y_2 \\ \cdot \\ \cdot \\ x_m - y_m \end{bmatrix} = \begin{bmatrix} z_1 \\ z_2 \\ \cdot \\ \cdot \\ z_m \end{bmatrix} = \boldsymbol{z}$$

Vector subtraction also has an analogous graphical representation, except that in vector subtraction, we first reverse the direction of the vector that we subtract—we place the arrow on the opposite end of the vector (its negative end) and then proceed as in the case of addition.

Vector Multiplication

There are several different multiplication operations involving vectors. We will review three that are important in systems engineering. The first is the case of a vector multiplied by a scalar. Let $\boldsymbol{x}$ be a vector and k denote a scalar. Then

$$k\boldsymbol{x} = \begin{bmatrix} kx_1 \\ kx_2 \\ \cdot \\ \cdot \\ kx_m \end{bmatrix}$$

That is to say, the scalar multiplies each and every element of the vector. The same applies in the case of a row vector. Also,

$$k(\boldsymbol{x} + \boldsymbol{y}) = k\boldsymbol{x} + k\boldsymbol{y}$$

Second, we will examine the *scalar product* or *dot product* of two vectors. This product is written in the following form:

$$\boldsymbol{x} \cdot \boldsymbol{y} = \boldsymbol{x}^T\boldsymbol{y} = [x_1, x_2, \ldots, x_m] \begin{bmatrix} y_1 \\ y_2 \\ \cdot \\ \cdot \\ y_m \end{bmatrix} = x_1y_1 + x_2y_2 + \ldots + x_my_m$$

The scalar product is commutative

$$x \cdot y = y \cdot x$$

Graphically, the scalar product is the magnitude of the projection of the first vector on the second, times the magnitude of the second. Thus, if the two vectors are perpendicular to each other, that is, if they are *orthogonal*, their dot product is zero.

Third, we have the matrix product of two vectors:

$$\boldsymbol{x}\boldsymbol{y}^T = \begin{bmatrix} x_1 \\ x_2 \\ \cdot \\ \cdot \\ x_m \end{bmatrix} [y_1, y_2, \ldots, y_m] = \begin{bmatrix} x_1y_1 & x_1y_2 & \cdot\cdot & x_1y_m \\ x_2y_1 & x_2y_2 & \cdot\cdot & x_2y_m \\ & & \cdot & \\ & & \cdot & \\ x_my_1 & x_my_2 & \cdot\cdot & x_my_m \end{bmatrix}$$

This product is not commutative except in the special case when $\boldsymbol{x}_i\boldsymbol{y}_j = \boldsymbol{x}_j\boldsymbol{y}_i$ for all i and j.

Note that the dot product and matrix product of vectors is defined only for vectors of equal dimension.

A.3 MATRIX OPERATIONS

We will now define similar algebraic operations on matrices. In addition to addition, subtraction, and multiplication, for square matrices we can also perform an operation referred to as the inverse.

Matrix Addition

Taking the elements of the matrix A to be a_{ij}, and the elements of the matrix B to be b_{ij}, then their sum is given as follows:

$$A + B = C$$

where the elements of C are given by

$$c_{ij} = a_{ij} + b_{ij}$$

Matrix addition is both commutative and associative:

$$A + B = B + A$$

and

$$(A + B) + C = A + (B + C) = A + B + C$$

Matrix Subtraction

As with vector subtraction, matrix subtraction is the same as matrix addition, replacing B with $(-B)$.

$$A - B = C$$

where the elements of C are given by

$$c_{ij} = a_{ij} - b_{ij}$$

Matrix Multiplication

We write the product of two matrices in the form:

$$AB = C$$

where the elements of the matrix C are defined as follows:

$$c_{ij} = \sum_{k=1}^{n} a_{ik} b_{kj}$$

Thus, we see that, if A is an $m \times n$ matrix, then B must be an $n \times p$ matrix, and the result, C, is an $m \times p$ matrix. Otherwise, the multiplication operation is not defined. We see that matrix multiplication is not commutative, but it is associative.

$$(AB)C = A(BC) = ABC$$

A special case arises when $p = 1$. In this case, both B and C reduce to vectors. We write

$$A\boldsymbol{b} = \boldsymbol{c}$$

The multiplication of a matrix by a scalar is also defined.

$$\alpha B = C$$

where the elements of C are given by

$$c_{ij} = \alpha \boldsymbol{b}_{ij}$$

The Matrix Inverse

We next define a special matrix called the *identity matrix*:

$$I = \begin{bmatrix} 1 & 0 & \ldots & 0 \\ 0 & 1 & \ldots & 0 \\ & & \cdot & \\ & & \cdot & \\ 0 & 0 & \ldots & 1 \end{bmatrix}$$

where I must be a square matrix. I is defined such that

$$AI = A$$

or that

$$IA = A$$

recognizing that the I's in the above two equations may have different dimensionality. Given the identity matrix, and for a square matrix A, we can define an inverse matrix, A^{-1}, which we refer to as the *inverse of* A, such that

$$A^{-1}A = AA^{-1} = I$$

We note that the inverse matrix is analogous to the notion that, for any non-zero x,

$$x^{-1}x = \frac{x}{x} = 1$$

And, just as with the scalar operation above, the inverse matrix does not always exist. We refer to the scalar case when $x = 0$ as a singular case. Likewise, when A^{-1} does not exist, we say that the matrix A is singular. We will now examine the conditions necessary for the existence of A^{-1}, and we will then determine the elements of the A^{-1} matrix.

First we must define *linear independence* of vectors. We say that a set of vectors $\boldsymbol{x}_i$, $i = 1,2,...,m$, is linearly dependent if there exist m numbers c_m, some of which are not zero, such that

$$c_1\boldsymbol{x}_1 + c_2\boldsymbol{x}_2 + ... + c_m\boldsymbol{x}_m = \mathbf{0}$$

Otherwise, we say that the vectors $\boldsymbol{x}_i$ are linearly independent. In the case that the vectors $\boldsymbol{x}_i$ are not linearly independent, one or more of the vectors can be expressed as a linear combination of the other vectors. Next we define the *rank* of a matrix as follows:

> The ***row rank*** of a matrix is the number of linearly independent rows in the matrix. The ***column rank*** of a matrix is the number of linearly independent columns in the matrix. The *row rank* and the *column rank* of a matrix are equal.

We may now make the following statement:

> A matrix is *nonsingular* if its rank equals the number of rows and the number of columns in the matrix. Otherwise, the matrix is *singular*.

Given a matrix that is non-singular, the elements of its inverse matrix are given as the matrix of cofactors divided by the determinant of the matrix. First, the determinant of a square matrix A,

$$A = \begin{bmatrix} a_{1,1} & a_{1,2} & \cdots & a_{1,m} \\ a_{2,1} & a_{2,2} & \cdots & a_{2,m} \\ & & \cdot & \\ & & \cdot & \\ a_{m,1} & a_{m,2} & \cdots & a_{m,m} \end{bmatrix} = \|a_{i,j}\|$$

namely det(A), is given by

$$\det(A) = \begin{vmatrix} a_{1,1} & a_{1,2} & \cdots & a_{1,m} \\ a_{2,1} & a_{2,2} & \cdots & a_{2,m} \\ & & \cdot & \\ & & \cdot & \\ a_{m,1} & a_{m,2} & \cdots & a_{m,m} \end{vmatrix}$$

$$= \sum_{j_1,\ldots,j_m=1}^{m} \varepsilon_{j_1 j_2 \ldots j_n} a_{j_1 1} a_{j_2 2} \ldots a_{j_m m}$$

where

$$\varepsilon_{j_1,j_2,\ldots,j_m} = \text{sgn}\ \phi\ (j_1,j_2,\ldots,j_m)$$

is an alternating function of $j_1, j_2,\ldots, j_m$, which assumes only the values $-1, 0, +1$. Taking the values of $j_1, j_2,\ldots, j_m$ to be restricted to the numbers $1,2,\ldots,m$, we have

$$\varepsilon_{j_1,j_2,\ldots,j_m} = 0$$

unless the numbers $j_1, j_2,\ldots, j_m$ are distinct, that is, unless they form a permutation of the numbers $1,2,\ldots,m$. We define a distinct set of numbers $j_1, j_2,\ldots, j_m$, from 1 to m, as an *even permutation* of the numbers $1,2,\ldots,m$ if the set $j_1, j_2,\ldots, j_m$ can be rearranged into sequential order by an even number of interchanges of two elements. In the case of an even permutation,

$$\varepsilon_{j_1,j_2,\ldots,j_m} = +1$$

We refer to the distinct set of numbers $j_1, j_2,\ldots, j_m$ as an *odd permutation* if they require an odd number of interchanges of two elements to rearrange into sequential order. In the case of an odd permutation,

$$\varepsilon_{j_1,j_2,\ldots,j_m} = -1$$

Having computed the determinant of a matrix, we can proceed to obtain its inverse. The inverse is given by

$$A^{-1} = \frac{1}{|A|}\ (\text{cof}\ A)^T$$

where cof A is the matrix of cofactors of A. The ij element of the matrix of cofactors is determined by deleting from the matrix A the row i and the column j and taking the determinant of the remaining matrix, with appropriate sign.

$$(\text{cof } A)_{2,3} = \text{cof}\begin{bmatrix} a_{1,1} & a_{1,2} & a_{1,3} & a_{1,4} & \cdots \\ a_{2,1} & a_{2,2} & a_{2,3} & a_{2,4} & \cdots \\ a_{3,1} & a_{3,2} & a_{3,3} & a_{3,4} & \cdots \\ a_{4,1} & a_{4,2} & a_{4,3} & a_{4,4} & \cdots \\ \cdot & \cdot & \cdot & \cdot & \\ \cdot & \cdot & \cdot & \cdot & \\ \cdot & \cdot & \cdot & \cdot & \end{bmatrix}$$

$$= (-1)^{i+j}\begin{vmatrix} a_{1,1} & a_{1,2} & a_{1,4} & \cdots \\ a_{3,1} & a_{3,2} & a_{3,4} & \cdots \\ a_{4,1} & a_{4,2} & a_{4,4} & \cdots \\ \cdot & \cdot & \cdot & \\ \cdot & \cdot & \cdot & \\ \cdot & \cdot & \cdot & \end{vmatrix}$$

This process appears very complicated. Indeed, for large matrices, it can involve a vast number of computations. However, many fine computer programs are available for obtaining matrix inverses, even for very large matrices.

The Transpose Operator

The transpose operator, denoted by a superscript T, has already been defined for the case of vectors. In the case of a matrix,

$$A = \begin{bmatrix} a_{1,1} & a_{1,2} & \cdots & a_{1,n} \\ a_{2,1} & a_{2,2} & \cdots & a_{2,n} \\ & & \cdot & \\ & & \cdot & \\ a_{m,1} & a_{m,2} & \cdots & a_{m,n} \end{bmatrix} = \|a_{i,j}\|$$

The transpose is given by

$$A^T = \begin{bmatrix} a_{1,1} & a_{2,1} & \cdots & a_{m,1} \\ a_{1,2} & a_{2,2} & \cdots & a_{m,2} \\ & & \cdot & \\ & & \cdot & \\ a_{n,1} & a_{n,2} & \cdots & a_{n,m} \end{bmatrix} = \|a_{j,i}\|$$

That is, the transpose of a matrix is obtained by interchanging its rows and columns. The following rules apply to the use of the transpose:

$$(AB)^T = B^T A^T$$

and

$$\det A^T = \det A$$

A.4 LINEAR ALGEBRA

The power of matrix notation and matrix operations becomes evident when we begin to deal with the solution of linear equations. For example, the set of linear equations

$$\begin{aligned} a_{1,1}x_1 + a_{1,2}x_2 + \ldots + a_{1,m}x_m &= b_1 \\ a_{2,1}x_1 + a_{2,2}x_2 + \ldots + a_{2,m}x_m &= b_2 \\ \cdot \quad \cdot \quad \cdot \quad \cdot \quad \cdot \quad \cdot \quad & \cdot \quad \cdot \quad \cdot \\ a_{m,1}x_1 + a_{m,2}x_2 + \ldots + a_{m,m}x_m &= b_m \end{aligned}$$

can be written in matrix notation as

$$A\boldsymbol{x} = \boldsymbol{b}$$

And this equation has the solution

$$A^{-1}A\boldsymbol{x} = I\boldsymbol{x} = \boldsymbol{x} = A^{-1}\boldsymbol{b}$$

Thus we have reduced an enormous amount of algebra to one simple statement. Of course, the matrix inverse in this equation is still a complex quantity to compute. But you will rarely see this complexity in practice. And, written in matrix form, we can often gain considerable insight into the solution of equations that appear very complex when written in scalar notation. Also, this is only a very minimal introduction to linear algebra.

REFERENCES

1. Courant, R. and F. John, *Introduction to Calculus and Analysis,* vol. II, John Wiley, New York, 1974.
2. Duderstadt, J. J. and L. J. Hamilton, *Nuclear Reactor Analysis,* John Wiley, New York, 1976.
3. Hillier, R. S. and G. J. Lieberman, *Introduction to Operations Research,* McGraw-Hill, New York, 1990.
4. Powell, J. L. and B. Crasemann, *Quantum Mechanics,* Addison-Wesley, Reading, MA, 1961.

Appendix B

A Test for the Utility Independence of Attributes

Let $X = \{X_1, X_2, ..., X_i, ..., X_r\}$ be a set of attributes that describe an engineering system. Let x_{ij} represent a measure of attribute i obtained by system design j, and $\mathbf{x}_j$ refer to the vector of attribute measures for system design j. If the attributes X_i are *utility independent*, we can express the system utility in the form

$$U_j = \sum_{i=1}^{r} a_i u_i (x_{ij})$$

where the a_i are weighting factors, and the u_i are utility functions, each dependent only on x_{ij}. That is, for example, $u_i(x_{i1})$ depends only on the value of x_{i1}, and not on the values of x_{i2}, x_{i3}, and so on. This condition does not imply that $u_i(x_{ij})$ is a linear function of x_{ij}. Thus, if the condition of utility independence is met, we can determine the functions $u_i(x_{ij})$ with reference only to the ith attribute.

We will now provide a test for the utility independence of system attributes. The test is based on the work of Fishburn (1965), and makes use of lotteries. Consider a set G of pairs of gambles of the form (ω_1, ω_2), where

$$\omega_1 = (p_1 x_1, p_2 x_2, .., p_m x_m)$$

$$\omega_2 = (q_1 y_1, q_2 y_2, .., q_n y_n)$$

and where

$$\sum_{k=1}^{m} p_k = 1 \qquad \text{for} \quad k = 1,2,..,m$$

$$\sum_{l=1}^{n} q_l = 1 \qquad \text{for} \quad l = 1,2,..,n$$

Then, the pair of gambles (ω_1,ω_2) is an element of G if and only if $\omega_1 \neq \omega_2$, that is, if not every element of ω_1 equals a corresponding element of ω_2, and both ω_1 and ω_2 have the same total probability for any $\boldsymbol{x}_i \in X$ that appears in either ω_1 or ω_2. To illustrate what we mean by total probability, consider the following example for $r = 3$.

$$\omega_1 = \left[\frac{2}{3}(x_{1,1}, x_{2,1}, x_{3,1}) + \frac{1}{3}(x_{1,2}, x_{2,2}, x_{3,2})\right]$$

$$\omega_2 = \left[\frac{1}{3}(x_{1,2}, x_{2,1}, x_{3,1}) + \frac{1}{3}(x_{1,1}, x_{2,2}, x_{3,1}) + \frac{1}{3}(x_{1,1}, x_{2,1}, x_{3,2})\right]$$

Notice that ω_1 has probability 2/3 for $x_{1,1}$, $x_{2,1}$, and $x_{3,1}$, and probability 1/3 for $x_{1,2}$, $x_{2,2}$, and $x_{3,2}$. Also, ω_2 has probability 2/3 for $x_{1,1}$, $x_{2,1}$, and $x_{3,1}$, and probability 1/3 for $x_{1,2}$, $x_{2,2}$, and $x_{3,2}$. Hence, since $\omega_1 \neq \omega_2$, $(\omega_1,\omega_2) \in G$. Given the total attribute set X, the r $\boldsymbol{x}_i$ are mutually independent in the utility sense—they are *utility independent*—if we are indifferent between the lotteries ω_1 and ω_2, that is, if

$$\omega_1 \sim \omega_2$$

for every $(\omega_1,\omega_2) \in G$.

In summary then, the concept of utility independence is that the utility associated with a particular attribute is independent of the values of other attributes.

REFERENCE

1. Fishburn, P. C., "Independence in Utility Theory with Whole Product Sets," *Operations Research*, vol. 13, no. 2, March–April, 1965, pp. 28–45.

Appendix C

Determination of the Weighting Factors in a Linearly Additive Utility Function

The purpose of this appendix is to provide a methodology for determining the weighting factors, a_m, in a linearly additive utility function of the form

$$u = \sum_{m=1}^{M} a_m u_m(x_m)$$

where $u_m(x_m)$ is the utility associated with the value of the mth attribute.

To distinguish between choices made under conditions of risk, that is, conditions under which the outcome of a decision cannot be known with certainty in advance of the decision and conditions of certainty or riskless choice, we will define the riskless preference function

$$v = \sum_{m=1}^{M} \lambda_m b_m(x_m)$$

In this case, we use the symbol b_m to denote the deterministic benefit associated with the value of the mth attribute, x_m, and we will restrict b_m to lie within the interval $0 \le b_m(x_m) \le 1$ for values of x_m on the range $x_{m*} \le x_m \le x_m{}^*$, where

$$x_{m*} \le x_m \le x_m{}^*$$

x_{m*} and $x_m{}^*$ are, respectively, the worst and best values of the mth attribute.

We define further the negative riskless preference function, $\bar{v}$

$$\bar{v} = -v = -\sum_{m=1}^{M} \lambda_m b_m(x_m)$$

and note that, under conditions of no risk, the same preferences are obtained by minimizing the negative riskless preference function as by maximizing the preference function, v. The negative preference function and the regret function,

$$R = \sum_{m=1}^{M} w_m \frac{B_m^* - B_m(x_m)}{B_m^*}$$

can be shown to be equivalent, where $B_m(x_m)$ is the benefit, on an arbitrary scale, associated with attribute benefit x_m, and $B_m{}^*$ is the maximum attainable benefit of B_m. This equivalence requires a proper relationship between λ_m and w_m. We also define B_{m*} as the minimum attainable value of B_m. Note that $-\infty \leq B_{m*} \leq B_m \leq B_m{}^* \leq +\infty$ while

$$x_{m*} \preceq x_m \preceq x_m^*$$

Then

$$b_m = \frac{B_m - B_{m*}}{B_m^* - B_{m*}}$$

and

$$w_m = \alpha\lambda_m \frac{B_m^*}{B_m^* - B_{m*}}$$

where α is an arbitrary, positive, non-zero constant. This assures that maximize (v), minimize ($\overline{v}$) and minimize (R) all lead to the same preference orderings.

It is essential that we keep in mind that we are dealing only with linearly additive utility or value functions. To examine this restriction, we need to define the following notation. Let

$$\underline{x}_* = (x_{1*}, x_{2*}, \ldots, x_{I*})$$

$$\underline{x}^* = (x_1^*, x_2^*, \ldots, x_I^*)$$

That is $\underline{x}_*$ and $\underline{x}^*$ are respectively the worst possible and the best possible outcomes. Also define

$$\underline{x}_*(-m) = (x_{1*}, \ldots, x_{m-1*}, x_{m+1*}, \ldots, x_{M*})$$

$$\underline{x}^*(-m) = (x_1^*, \ldots, x_{m-1}^*, x_{m+1}^*, \ldots, x_M^*)$$

so that

$$\underline{x}_* = \underline{x}_*(-m), x_{m*}$$

$$\underline{x}^* = \underline{x}^*(-m), x_m^*$$

And, finally,

$$\underline{x}_{m*} = (x_{1*}, \ldots x_{m-1*}, x_m^*, x_{m+1*}, \ldots, x_{M*})$$

$$\underline{x}_m^* = (x_1^*, \ldots, x_{m-1}^*, x_{m*}, x_{m+1}^*, \ldots, x_M^*)$$

Now we can state that, in order for a preference function to be linearly additive, the condition of preference independence must apply. That is, preferences between any two attributes, x_i and x_j, must be independent of the values of the other attributes, $\underline{x}(-i, j)$. Thus, if

$$\lambda_i b_i + \lambda_j b_j \geq \lambda_i b_i' + \lambda_j b_j'$$

it is inferred that

$$x_i, x_j, \underline{x}'(-i,j) \succeq x_i', x_j', \underline{x}(-i,j)$$

independent of $\underline{x}(-i,j)$. More specifically, the above must apply for any $\underline{x}(-i,j)$,

$$\underline{x}_*(-i,j) \preceq \underline{x}(-i,j) \preceq \underline{x}^*(-i,j)$$

Note, however, that it is absolutely *not* required that $B_m(x_m)$ be a linear function of x_m.

Frequently, one deals with linearizations of nonlinear preference or utility surfaces. The regret function is often used in this manner. In so doing, it is implicitly assumed that, for all $i, j, i \neq j$, contained in the set of attributes, M, the permissible range of $\underline{x}(-i,j)$ is sufficiently small, and consequently the range of $\underline{B}(-i,j)$ is also sufficiently small, that preference orderings between pairs (x_i,x_j), (x_i',x_j'), (x_i'',x_j''), etc. are independent of $\underline{x}(-i,j)$.

In the case of utility functions, if the above preference order appplies for any $\underline{x}(-i,j)$ between $\underline{x}_*(-i,j)$ and $\underline{x}^*(-i,j)$, then we can say that the preferences satisfy the condition of *weak conditional utility independence*. If, furthermore, l, l', l'', etc. are lotteries on consequences of the form $[x_i, x_j, \underline{x}(-i,j)]$, $[x_i', x_j', \underline{x}(-i,j)]$, $[x_i'', x_j'', \underline{x}(-i,j)]$, and the preference orderings for these lotteries are independent of the utility of $\underline{x}(-i,j)$ between $\underline{x}_*(-i,j)$ and $\underline{x}^*(-i,j)$, then we can say that the preferences satisfy the condition of *strong conditional utility independence*.

The remainder of this appendix deals with methods for evaluating the a_m or λ_m in the above utility or preference functions. The issue of determining $u_m(x_m)$ or $B_m(x_m)$, while an important issue, is not dealt with in this appendix.

C.1 DETERMINATION OF UTILITY FUNCTION WEIGHTING FACTORS IN THE CASE OF RISKLESS CHOICE

We will distinguish here between choices made under conditions of certainty and those made under conditions of risk, that is, under the condition that the outcome of a choice cannot be known with certainty before the choice is taken. We will use the term *value function* to denote a preference ordering function that is valid only in the case of choice under conditions of certainty.

We turn now to an interpretation of preference statements that will lead to mathematical relationships between the λ's of linearly additive value functions. Since, due to the conditions of linearity, we need deal only with preferences between two attributes simultaneously, we seek first to identify preferences between pairs of outcomes. Such pairs may be, for example, $\underline{x}_{i*}$ versus $\underline{x}_{j*}$ for $i \neq j$. Suppose $x_{i*} \succeq x_{j*}$, then, by the above preference inequality, we obtain the condition

$$\lambda_i \cdot 1 + \lambda_j \cdot 0 \geq \lambda_i \cdot 0 + \lambda_j \cdot 1$$

or simply

$$\lambda_i \geq \lambda_j$$

Note that b_i and b_j, in this comparison take on values only of zero or one. Recognize that we are not saying that $B_i^* - B_{i*}$ is greater than $B_j^* - B_{j*}$, but only that we prefer to raise B_{i*} to B_i^* at least as much as raising B_{j*} to B_j^*. In summary then,

$$\underline{x}_{i*} \succeq \underline{x}_{j*} \qquad \text{implies } \lambda_i \geq \lambda_j$$

Furthermore, note that the above is a convenience of the construct of the linearly additive value function, that is, in terms of b_m,

$$0 \geq b_m \geq 1$$

Thus, it is possible to obtain inequalities on all pairs of λ's by repeated application of preference declarations such as that above. In addition, because of the homogeneity of the riskless preference function in the λ_m, an arbitrary condition such as the following is possible.

$$\sum_{m=1}^{M} \lambda_m = 1$$

Or, alternatively, for example,

$$\lambda_1 = 1$$

In examining the inequalities created by the above preference orderings, potential intransitive preference orderings might be uncovered. These will take on a form such as

$$\lambda_i \geq \lambda_j \geq \lambda_k \geq \lambda_i$$

Now either of two conditions might apply. On the one hand, it is quite possible that the preferences cannot be resolved better than these inequalities would indicate. In this case, you should admit to indifference between the corresponding outcomes. This yields

$$\lambda_i = \lambda_j = \lambda_k$$

On the other hand, however, you might insist that preferences, not indifferences, exist. In this case, at least one of the preferences is irrational, and it must be resolved.

Assuming that we can resolve all intransitivities, we can uniquely rank attributes according to the importance associated with raising B_{m*} to B^*_m. For simplicity, say $\lambda_1 \geq \lambda_2 \geq \lambda_3 \geq \lambda_4 \geq \ldots$ Then, raising B_{1*} to B^*_1 is at least as important as raising B_{2*} to B^*_2, etc. The question now is: by how much is $\lambda_1 > \lambda_2$, $\lambda_2 > \lambda_3$, and so on? There are two approaches to answering this question, an equality approach and an inequality approach. Taking the equality approach, we note that $\lambda_1 \geq \lambda_2$ implies $\underline{x}_{1*} \succeq \underline{x}_{2*}$. Since this is true, it implies that there is a value of x_1, $x_{1*} \preceq x_1 \preceq x^*_1$, such that

$$x_1, \underline{x}_* (-1) \sim \underline{x}_{2*}$$

We must now find the value of x_1 that satisfies this indifference. Then, from our original inequality on the λ's, we obtain

$$\lambda_2 = \frac{B_1 - B_{1*}}{B_1^* - B_{1*}} \lambda_1$$

The process may be repeated until all λ's are determined.

The equality procedure may, unfortunately, be somewhat difficult to apply, since it may be difficult to decide on the proper values of the x_i. Thus, we will look at the continued development of inequalities. Let

$$\underline{x}_{i,j*} = (x_{1*},\ldots,x_{i-1*},x_i^*,x_{i+1*},\ldots,x_{j-1*},x_j^*,x_{j+1*},\ldots,x_{M*})$$

and so on. Now make comparisons between $\underline{x}_{m*}$ and $\underline{x}_{i,j*}$ or $\underline{x}_{i,j,k*}$, etc. Suppose we know $\lambda_1 \geq \lambda_2 \geq \lambda_3 \geq \lambda_4 \geq \ldots$ Then compare x_{1*} and $\underline{x}_{2,3*}$. If $\underline{x}_{2,3*} \succeq x_{1*}$,

$$\lambda_2 + \lambda_3 \geq \lambda_1$$

Next compare $\underline{x}_{1*}$ and $\underline{x}_{2,4*}$, etc. until we obtain, say, $\underline{x}_{1*} \succeq \underline{x}_{2,6*}$. Then

$$\lambda_1 \geq \lambda_2 + \lambda_6$$

By continued comparisons of this sort, we obtain a set of inequalities that substantially bound the λ_m. Further determination may not be necessary; indeed, it might not be possible.

It is interesting to note that the above procedure could be equally well applied to comparisons between $\underline{x}_i^*$ and $\underline{x}_j^*$ instead. This may be helpful if $\underline{x}^*$ is easy to establish and $\underline{x}_*$ more difficult, and if you can think more easily of foregoing benefits rather than receiving benefits.

C.2 DETERMINATION OF UTILITY FUNCTION WEIGHTING FACTORS

The basic difference between a utility function and a value function is that utility functions apply in cases of choice with risk whereas value functions apply only in the case of riskless choice. In the limit, however, as risk becomes diminishingly small, utility function preference orderings (assuming a continuous utility function) should become equivalent to value function preference orderings. Thus, we note that the utility function at zero risk is a value function. A value function is not necessarily ever a utility function, however. So long as the utility function becomes a value function in the zero-risk limit and is linearly additive, all the results of the previous section apply. That is, we can make statements of the form that if $\underline{x}_{i*} \succeq \underline{x}_{j*}$ it is implied that $a_i \geq a_j$ given that we have normalized the functions, $u_m(x_m)$, so that for all m, $m = 1,2,\ldots,M$, $u_m(x_{m*}) = 0$ and $u_m(x_m^*) = 1$.

In addition to the above results, however, utility functions also apply under conditions of risky choice; and it is therefore permissible to use utilities, in fact expected utilities, to obtain preferences in the case of lotteries. Thus, suppose we know that $\underline{x}_{i*} \succeq x_{j*}$. This infers the existence of a probability p, $0 \leq p \leq 1$, such that the decision maker is indifferent between receiving $\underline{x}_{j*}$ directly or obtaining a lottery yielding $\underline{x}_{i*}$ with probability p and $\underline{x}_*$ with probability $1 - p$.

$$\underline{u}_{j*} = p\underline{u}_{i*} + (1 - p)\underline{u}_*$$

Then, combining this equation with the first equation of this appendix and recalling that $u_m(x_{m*}) = 0$ and $u_m(x_m{}^*) = 1$,

$$a_j = pa_i$$

Thus, we see that, in the case of utilities, the lottery approach yields an equality relationship between the a_m.

The lottery approach can, as with the value function approach, also be applied in the reverse sense. Compare $\underline{x}_{i*}$ to $\underline{x}_{j*}$. Suppose $\underline{x}_i{}^* \succeq \underline{x}_{j*}$. Then we can form the lottery

$$\underline{u}_i^* = p\underline{u}_j^* + (1 - p)\underline{u}^*$$

This yields the condition

$$a_i = pa_j$$

It is also possible to develop compound lotteries or lotteries involving more than one attribute. For example, suppose $\underline{x}_{i,j*} \succeq \underline{x}_{k*}$. We then form the lottery

$$\underline{u}_{k*} = p\underline{u}_{i,j*} + (1 - p)\underline{u}_*$$

which yields

$$a_k = p(a_i + a_j)$$

Lotteries of this form can be used to check the accuracy of the a_m after they have been determined by lotteries of the earlier form. Of course, the comparison between $\underline{x}_{i,j}{}^*$ and $\underline{x}_k{}^*$ could also be made.

When is a preference function a utility function, as opposed to a value function, and, thus, when is it permissible to apply the lottery technique? The answer lies in the determination of the functions $u_m(x_{m*})$. If these functions obey conditions required of utility functions in the case of a single attribute, and furthermore if the strong condition of utility independence applies, then we can be assured that the functions, $u_m(x_{m*})$, are valid utility functions.

What are the conditions required of a utility function, $u_m(x_{m*})$, of a single attribute? The answer lies basically in the fact that $u_m(x_{m*})$ must describe your preferences for x_m under conditions of risk. To do this, first, your preferences must be transitive, otherwise $u_m(x_{m*})$ does not exist. Thus, if $x_m \succeq x_m{}'$ and $x_m{}' \succeq x_m{}''$, it must also be true that $x_m \succeq x_m{}''$ for all sets x_m, $x_m{}'$, and $x_m{}''$. The utility function is then such that

$$x_m \succeq x_m{}' \quad \text{iff } u_m(x_m) \geq u_m(x_m{}')$$

Furthermore, preferences must be described by expected utilities. Thus, if $x_m \succeq x_m{}'$ and $x_m{}' \succeq x_m{}''$ a probability p exists, $0 \leq p \leq 1$, such that

$$u_m(x_m') = pu_m(x_m) + (1 - p)\, u_m(x_m'')$$

The utility function must be such that you are indifferent to receiving x_m' directly or a lottery wherein you receive x_m with probability p and x_m'' with probability $p - 1$. If the above conditions are met, $u_m(x_{m^*})$ can be interpreted as a utility and the linearly additive utility form applies.

Note in closing that the method outlined above works only if all the $u_m(x_{m^*})$ can be expressed adequately on the interval $0 \leq u_m(x_{m^*}) \leq 1$. This is not always possible. In the case that it is not possible, we may need to restrict the range of x_m over which the utility function is valid.

REFERENCES

1. Fishburn, P. C., "Independence in Utility Theory with Whole Product Sets," Presented at the 1964 CORS/ORSA Joint Conference, Montreal, Canada, May 27, 1964.
2. Hazelrigg, G. A., "Determination of Weighting Factors in Linearly Additive Utility Functions," ER Memo No. 6, The Aerospace Systems Laboratory, Princeton University, Princeton, NJ, December 30, 1975.
3. Raiffa, H., "Preferences for Multi-Attributed Alternatives," The RAND Corporation, Memorandum RM-5868-DOT RC, April 1969.

Name Index

A

Abramowitz 49, 60
Ackoff 286
Anderson 286
Ansoff 286
Arrow 164
Atherton 173

B

Baker 286
Bazaraa 121
Beaver 60, 200
Bernoulli 167
Blanchard 14, 18, 30
Bolza 111
Brown 8
Bush 14

C

Carlson 287
Chipman 174
Çinlar 60
Clemen 312
Cooper 121
Courant 121, 449
Crasemann 449
Cyert 312

D

DeGroot 312
Devons 286
Draper 200
Duderstadt 449
Dyer 312

F

Fabrycky 14, 18, 30
Ferguson 15
Firstenberg 15
Fishburn 14, 175, 312
Fisher 287
Fletcher 89, 90, 91
Ford 103
Fromm 286
Fudenberg 312, 390
Fulkerson 103

G

Gee 286
Gero 15
Gillett 121
Gompertz 191, 192
Grayson 15
Greenberg 175, 256, 257, 287

H

Hamilton 449
Hammond 312
Hausner 312
Hazelrigg 15, 153, 175, 257, 287, 303, 325, 339

Heal 175
Hertz 287
Hildebrand 121
Hillier 121, 153, 364, 390
Hirsch 287
Hitchcock 287
Holt 312
Hotelling 148, 153
Howard 175
Huband 339
Hurwicz 175

J

John 121, 449

K

Karmarkar 95
Karl Compton 7
Keeny 312
Koen 15

L

Lagrange 110
Layard 287
Lewis 15
Lieberman 121, 153, 364, 390, 449
Lietzke 153
Lindgren 312
Lomask 15
Louis XV 7
Luce 167, 175, 312, 390

M

Magee 312
Matheson 175, 312
Mayer 111
Mendenhall 60, 200
Miles 312
Miller 312
Mishan 287

N

Napoleon 7
Nash 390

P

Perronet 7
Pontryagin 114, 118
Pound 286
Powell 89, 90, 91, 449

R

Raiffa 169, 175, 312, 390
Rasmusen 364, 390
Richter 175
Ridker 287
Rivett 203, 208, 226
Roederer 15
Ross 60
Roths 312
Rozonoér 118, 121
Rubinstein 15, 287

S

Samuel 15
Samuelson 125, 153
Savage 307, 312
Settle 286
Shetty 121
Silk 287
Smith 200
Sonnenschein 175
Starr 175
Stegun 49, 60
Steinberg 121, 153

T

Taguchi 15
Takayama 121, 153

Thesen 66, 77
Tirole 312, 390
Travis 77
Tribus 15

V

von Neumann 143, 388, 390

W

Weierstrass 114
Wonnacott 60, 200

Y

Yoonho Kim 409

Topic Index

A

Abstraction error 214
Accuracy 202
Action 363
Action combination 363
Action set 363
Added use benefits 274
Adjoined function 93
Adjoint 93
Adjoint variable 93
Algorithmic error 214
Alternative approach 326
Alternative hypothesis 58
Analog models 205
Annuity 144
Applied research 4, 7, 8
Arrow's impossibility theorem 164, 300
Assignment problem 101
Augmented 93

B

Backward induction 293
Basic research 4
Bayes' actions 311
Bayes' formula 55
Bayes' law 334
Bayesian decision making 310
Bayesian games 372
Benefit-cost ratio 283, 288
Benefit to society 348
Benefits 273
Bernoulli's paradox 167, 302
Beta distribution 269
Binomial distribution 48
Branch 369
Break-even time 285, 290

C

Calculus of variations 105
Capital 143, 261
Cardinal utilities 159, 346
Cartoons 25, 241, 400
Cash flow 346
Central limit theorem 45
Central moment 41
CER 264–266
Certainty equivalent value 295
Coefficient of correlation 186
Column vector 439
Common information 372
Compound lottery 169
Computational error 214
Concurrent engineering 351
Conditional probabilities 38
Confidence level decision making 306
Conflict 364
Conjugate gradient methods 89
Constant-year dollars 146
Constrained optimization 91
Constraints 79, 217–219
Consumer price index 146
Consumer surplus 134
Continuous random variable 39
Continuous uniform distribution 51
Cooperative game 364
Corporate objective function 224
Cost 260

Cost estimating 263, 266, 270
Cost estimating relationship 264
Cost learning 266, 267
Cost savings benefit 274
Cost spreading 268
Credibility 348
Cumulative cash flow 348
Cumulative distribution function 39, 48
Customer loyalty 348

D

Data error 214
Decision 3, 292, 293
Decision analysis 292
Decision/event tree 247, 254, 292
Decision/event trees 369
Decision making 2, 3, 52, 125, 165, 210, 340, 345, 361, 369, 394
Decision point 369
Demand curves 133
Demand elasticity 133
Descriptive model 208, 209
Design synthesis 10, 18
Design validation 220
Discount rate 142
Discounting 141
Discrete random variable 39
Discrete uniform distribution 46
Disjoint events 33
Disposal 28, 223, 354
Distribution 222
Division of two random variables 43
Dominant strategy 365
Dot product 442

E

Economics 124
Economic competitiveness 20
Elements of an R&D program 327
End node 369
End point 369
Engel curve 126
Engineering 3
Enthusiasm 395
Environmental concerns 353
Equal budget benefits 279
Equal capability benefits 279
Equilibria of a game 364
Equilibrium concept 364
Equilibrium price 136
Equivalent subprojects 331
Euler-Lagrange condition 119
Euler-Lagrange equation 110, 150
Event 35, 369
Event-driven Monte Carlo 252, 253
Evolutionary stable strategy 380
Evolutionary technology advance 356, 401
Excluded subprojects 332
Expectations 346
Experiment 34
Exponential distribution 54, 230
Extensive form 369
Externalities 279

F

Feasibility analysis 29
First order iteration 86
Fletcher-Powell 89, 90
Flood Control Act of 1936 282
Forecasting 178, 271
Forecasts 146
Forward and backward differences 87
Forward differences 87
Functional 105
Future event clock 252

G

Game 362
Game of certain information 372
Game of complete information 372
Game of mixed strategy 376
Game of perfect information 372
Game of pure coordination 381
Game of pure strategy 375
Game of symmetric information 372
Gamma distribution 49
Gamma function 53, 269
Gaussian distribution 52

Generic problem area 326
Geometric distribution 49
Global maxima 80
Global search 84
Gomory's algorithm 100
Gompertz curve 190, 283, 319, 402

H

Hamiltonian 113, 119, 150
Handty-Whitman index 453
Hessian matrix 90
Histogram 66, 74
Hyami-Peterson model 147
Hypothesis testing 57

I

Iconic models 206
Identification of non-inituitive cases 202
Identity matrix 444
Incentive and reward structures 396
Incidence matrix 327
Independent 38
Independent attributes 162
Indifference curve 129
Indifference curves 127
Information 4, 169, 170, 340, 345
Informational entropy 343, 363, 369, 398
Information partition 371
Information set 370
Integer programming 99
Interest 143
Interior-point algorithm 95
Intersection of two sets 33
Intransitive 156
Irrational 156
Iterated dominance 368
Iterative solution procedures 84

K

Kertosis 41
Knowledge 4

L

Labeling process 103
Labor 143, 260
Lack of fit 192
Lagrangian method 93
Lagrangian multiplier 101
Law of diminishing marginal utility 158
Law of equal marginal utilities 158
Law of large numbers 180
Leapfrogging 330
Lewis algorithm 68, 93
Lexicographic preference ordering 162
Linear algebra 448
Linear congruential generators 66
Linear equations 448
Linear independence 182
Linear independence of vectors 445
Linear programming 94
Linearly additive utility 162
Linearly additive utility function 161
Local maxima 80
Log-linear utility 164
Logarithmic utility 303
Logistic curves 190
Long-term elasticities 134
Look-see research 5
Lottery 443

M

Maintenance 3, 222
Manufacturing 221
Marginal value 127
Matrix 440
Matrix addition 443
Matrix inverse 444
Matrix multiplication 444
Matrix subtraction 444
Maximum 79
Maximum principle 113
Mean 41, 73
Mean time between failure 229
Mean value theorem 117
Memoryless system 230
Microeconomics 124
Minimax 307

Minimax theorem 385
Minimum 85
Morrill Act 7
Model validation 358
Model 205
Moments of a distribution 41
Monopoly rent 137, 141
Monte Carlo 63, 71, 199, 217, 236, 245, 247, 248, 267, 268, 274, 280, 301, 302, 318, 322, 413
Most likely 42
MTBF 79, 229, 230
Multi-attribute utility functions 161
Multi-variate regression 188
Multiplication 43
Multiplicative utility 163
Mutually exclusive 33

N

Nash bargaining solution 383
Nash equilibrium 380
Nash existence theorem 381
Natural resources 143
Nature 3
Necessary condition 82
Needs and wants 125
Negative definite 83
Net benefit 283, 284
Network flow optimization 101
New capability benefits 278
Newton's method 88
Node 369
Non-cooperative game 364
Non-recurring costs 261
Non-zero-sum game 110
Normal distribution 52, 230
Null hypothesis 58
Null set 33

O

Objective functions 346
Objective probability 37
OPEC 148
Operational analysis 400
Opportunity costs 270
Optimization 235
Option space 9, 18, 261, 282, 345
Ordinal utilities 158
Outcome 293, 363

P

Packaging complexity 23
Parameterization 10, 18
Pareto optimality 166, 274
Path 369
Payoff 363
Peer review 397
Player 362
Poisson distribution 49
Pontryagin maximum principle 114, 118
PRA 252–255
Predecessor node 369
Predictive model 209, 210
Preference 156
Preferences 129–131
Prerequisite subprojects 332
Present value 141, 142
Price 261
Prisoner's dilemma 365
Probability 34, 37
Probability density function 40
Probablistic forecast 182
Probablistic risk assessment 252
Problem 2
Problem of Bolza 111
Problem of Lagrange 111
Problem of Mayer 111
Problem solving 180
Producer surplus 137
Product availability 348
Product liability 26, 29
Product life cycle 349
Profit 347
Pure error 192

R

R&D 324
RADSIM 325

Random number 66
Random variable 39
Randu 67
Rank of a matrix 445
Rationality 155
Real interest 143
Recurring costs 261
Redundancy 234
Regression analysis 181
Regret 307, 308, 423
Regulated monopolies 140
Relative error 215
Reliability 228
Reliability diagram 232, 235
Rensselaer Polytechnic Institute 7
Rent 142
Repair 20
Research 2
Research and development 139, 140, 324
Resolution 202, 216
Resources 261
Resource economics 148
Return on investment 284, 285, 348
Revolutionary technology advance 357, 400
ROI 285, 347
Row vector 439
Rules of a game 362, 363

S

Sales 222
Sample space 35
Sampling a distribution 69
Scalar 440
Scalar product 442
Scale indifference of utility preferences 159
Scenario 241, 247
Scenarios 400
Science 1
Second order methods 89
Sequential decisions 295, 296
Set 32
Shadow prices 151
Short-term elasticities 134
Simple event 35
Simple lottery 171
Simplex 95
Singleton 373
Social discount rate 131
Sources of modeling error 211
Space Shuttle 241
Square matrix 440
Standard deviation 41, 74
Standard normal distribution 48, 52
Starting node 369
State transition matrix model 315
STM 315
Strategic plan 21
Strategy 363
Strategy combination 363
Strategy space 363
Strictly dominant strategy equilibrium 366
Strong conditional utility independence 425
Strong law of large numbers 45
Strong preference 156
Strong variation 114
Subjective probability 38
Suboptimal designs 20
Subset 33
Successor node 369
Sufficient condition 82
Sum of two random variables 42
Sunk costs 262, 271
Supply curves 135
Symbolic model 218
Symbolic models 205
System 204
System life cycle 270
System objective function 224, 225
System objectives 20
System operation 25
System reliability 231
Systems engineering 6

T

Taxes 140
Taylor series 81, 82, 86, 90
Technological breakthroughs 330
Technology 37
Technology advancement 330
Technology advancement factor 330, 331
Time-driven Monte Carlo 250, 251
Transcendental 83

Transitive 158
Transportation problem 104
Transpose 388, 397
Transversality condition 110
Trend line 182
Triangular distribution 71, 79
Type I error 58
Type II error 58

U

Uncertainty 36
Union of two sets 34
Utility 127, 156, 225, 231, 237, 274, 299, 346, 362, 397
Utility independence 421

V

Value 127
Value function 426
Value space 129, 131
Values 153, 346, 364
Variance 41
Variation 107
Vector 440
Vector addition 440
Vector multiplication 442
Vector subtraction 442
Venn diagram 33
Von Neumann 144
Von Neumann-Morgenstern lotteries 167

W

Wages 143
Weak conditional utility independence 425
Weak preference 156
Weak variation 107
Weakly dominant strategy 368
Weibull distribution 55, 229
Weierstrass condition 114, 150
Weierstrass E function 117
Wholesale price index 146

Z

Zero-base benefit-cost ratio 285, 286
Zero-base net benefit 283, 285
Zero-sum game 364
Zeroth order methods 84